THE UNIVERSAL HISTORY OF NUMBERS

THE WORLD'S FIRST NUMBER-SYSTEMS

GEORGES IFRAH, now aged fifty, was the despair of his maths teachers at school – he lingered near the bottom of the class. Nevertheless he grew up to become a maths teacher himself and, in order to answer a pupil's question as to where numbers came from, he devoted some ten years to travelling the world in search of the answers, earning his keep as a night clerk, waiter, taxi-driver. Today he is a maths encyclopaedia on two legs, and his book has been translated into fourteen languages.

DAVID BELLOS is Professor of French at Princeton University and author of *Georges Perec: A Life in Words* and *Jacques Tati*. E. F. HARDING has taught at Aberdeen, Edinburgh and Cambridge and is a Director of the Statistical Advisory Unit at Manchester Institute of Science and Technology.

THE UNIVERSAL HISTORY OF
NUMBERS

THE WORLD'S FIRST
NUMBER-SYSTEMS

GEORGES IFRAH

Translated from the French by
David Bellos and E. F. Harding

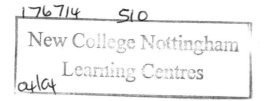

THE HARVILL PRESS
LONDON

For you my wife,
the admirably patient witness of the joys and agonies that this hard labour has
procured me, or to which you have been subjected, over so many years.
For your tenderness and for the intelligence of your criticisms.
For you, Hanna, to whom this book and its author owe so much.

And for you, Gabrielle and Emmanuelle,
my daughters, my passion.

* *

First published in France with the title *Histoire universelle des chiffres*
by Editions Robert Laffont, Paris, in 1994

Part One of the original edition
First published in Great Britain in 1998
and reissued as a two-volume set in 2000 by
The Harvill Press,
2 Aztec Row, Berners Road
London N1 0PW

www.harvill.com

This translation has been published with the financial support
of the European Commission and of the
French Ministry of Culture and Communications

1 3 5 7 9 8 6 4 2

A CIP catalogue record for this book
is available from the British Library

ISBN 1 86046 790 3

Designed and typeset in Stone Print at
Libanus Press, Marlborough, Wiltshire

Printed and bound in Italy

SUMMARY TABLE OF CONTENTS

Foreword   vii
Introduction   ix
Where "Numbers" Come From

All illustrations, with the exception of Fig. 1.30–36 and 2.10 by Lizzie Napoli,
have been drawn, or recopied, by the author.

FOREWORD

The main aim of this work is to provide in simple and accessible terms the full and complete answer to all and any questions that anyone might want to ask about the history of numbers and of counting, from prehistory to the age of computers.

More than ten years ago, an American translation of the predecessor of *The Universal History of Numbers* appeared under the title *From One to Zero*, translated by Lowell Bair (Viking, 1985). The present book – translated afresh – is many times larger, and seeks not only to provide a historical narrative, but also, and most importantly, to serve as a comprehensive, thematic encyclopaedia of numbers and counting. It can be read as a whole, of course; but it can also be consulted as a source-book on general topics (for example, the Maya, the numbers of Ancient Egypt, Arabic counting, or Greek acrophonics) and on quite specific problems (the proper names of the nine mediaeval *apices*, the role of Gerbert of Aurillac, how to do a long division on a dust-abacus, and so on).

The second volume, which accompanies this present one and forms an integral part of the book, includes the bibliography and index of names and subjects covering the two volumes.

The bibliography has been divided into two sections: sources available in English; and other sources. In the text, references to works listed in the bibliography give just the author name and the date of publication, to avoid unnecessary repetition. Abbreviations used in the text, in the captions to the many illustrations, and in the bibliography of this volume are explained in the second volume, immediately before the bibliography.

INTRODUCTION

Where "Numbers" Come From

TEACHER LEARNS A LESSON

This book was sparked off when I was a schoolteacher by questions asked by children. Like any decent teacher, I tried not to leave any question unanswered, however odd or naive it might seem. After all, a curious mind often is an intelligent one.

One morning, I was giving a class about the way we write down numbers. I had done my own homework and was well-prepared to explain the ins and outs of the splendid system that we have for representing numbers in Arabic numerals, and to use the story to show the theoretical possibility of shifting from base 10 to any other base without altering the properties of the numbers or the nature of the operations that we can carry out on them. In other words, a perfectly ordinary maths lesson, the sort of lesson you might have once sat through yourself – a lesson taught, year in, year out, since the very foundation of secondary schooling.

But it did not turn out to be an ordinary class. Fate, or Innocence, made that day quite special for me.

Some pupils – the sort you would not like to come across too often, for they can change your whole life! – asked me point-blank all the questions that children have been storing up for centuries. They were such simple questions that they left me speechless for a moment:

"Sir, where do numbers come from? Who invented zero?"

Well, where do numbers come from, in fact? These familiar symbols seem so utterly obvious to us that we have the quite mistaken impression that they sprang forth fully formed, as gods or heroes are supposed to. The question was disconcerting. I confess I had never previously wondered what the answer might be.

"They come ... er ... they come from the remotest past," I fumbled, barely masking my ignorance.

But I only had to think of Latin numbering (those Roman numerals which we still use to indicate particular kinds of numbers, like sequences of kings or millionaires of the same name) to be quite sure that numbers have not always been written in the same way as they are now.

"Sir!" said another boy, "Can you tell us how the Romans did their sums? I've been trying to do a multiplication with Roman numerals for days, and I'm getting nowhere with it!"

"You can't do sums with those numerals," another boy butted in. "My dad told me the Romans did their sums like the Chinese do today, with an abacus."

That was almost the right answer, but one which I didn't even possess.

"Anyway," said the boy to the rest of the class, "if you just go into a Chinese restaurant you'll see that those people don't need numbers or calculators to do their sums as fast as we do. With their abacuses, they can even go thousands of times faster than the biggest computer in the world."

That was a slight exaggeration, though it is certainly true that skilled abacists can make calculations faster than they can be done on paper or on mechanical calculating machines. But modern electronic computers and calculators obviously leave the abacus standing.

I was fortunate and privileged to have a class of boys from very varied backgrounds. I learned a lot from them.

"My father's an ethnologist," said one. "He told me that in Africa and Australia there are still primitive people so stupid that they can't even count further than two! They're still cavemen!"

What extraordinary injustice in the mouth of a child! Unfortunately, there used to be plenty of so-called experts who believed, as he did, that "primitive" peoples had remained at the first stages of human evolution. However, when you look more closely, it becomes apparent that "savages" aren't so stupid after all, that they are far from being devoid of intelligence, and that they have extraordinarily clever ways of coping without numbers. They have the same potential as we all do, but their cultures are just very different from those of "civilised" societies.

But I did not know any of that at the time. I tried to grope my way back through the centuries. Before Arabic numerals, there were Roman ones. But does "before" actually mean anything? And even if it did, what was there before those numerals? Was it going to be possible to use an archaeology of numerals and computation to track back to that mind-boggling moment when someone first came up with the idea of counting?

Several other allegedly naive questions arose as a result of my pupils' curious minds. Some concerned "counting animals" that you some-times see at circuses and fairs; they are supposed to be able to count (which is why some people claim that mathematicians are just circus artistes!) Other pupils put forward the puzzle of "number 13", alternately considered an omen of good luck and an omen of bad luck. Others wondered what was in the minds of mathematical prodigies, those phenomenal beings who can perform very complex operations in their heads at high speed – calculating the cube root of a fifteen-digit number, or reeling off all the prime numbers between seven million and ten million, and so on.

In a word, a whole host of horrendous but fascinating questions exploded in the face of a teacher who, on the verge of humiliation, took the full measure of his ignorance and began to see just how inadequate the teaching of mathematics is if it makes no reference to the history of the subject. The only answers I could give were improvised ones, incomplete and certainly incorrect.

I had an excuse, all the same. The arithmetic books and the school manuals which were my working tools did not even allude to the history of numbers. History textbooks talk of Hammurabi, Caesar, King Arthur, and Charlemagne, just as they mention the travels of Marco Polo and Christopher Columbus; they deal with topics as varied as the history of paper, printing, steam power, coinage, economics, and the calendar, as well as the history of human languages and the origins of writing and of the alphabet. But I searched them in vain for the slightest mention of the history of numbers. It was almost as if a conspiracy of obviousness aimed to make a secret, or, even worse, just to make us ignorant of one of the most fantastic and fertile of human discoveries. Counting is what allowed people to take the measure of their world, to understand it better, and to put some of its innumerable secrets to good use.

These questions had a profound impact on me, beginning with this lesson in modesty: my pupils, who were manifestly more inquisitive than I had been, taught me a lesson by spurring me on to study the history of a great invention. It turned out to be a history that I quickly discovered to be both universal and discontinuous.

THE QUEST FOR THE MATHEMATICAL GRAIL

I could not now ever let go of these questions, and they soon drew me into the most fascinating period of learning and the most enthralling adventure of my life.

My desire to find the answers and to have time to think about them persuaded me, not without regrets, to give up my teaching job. Though I had only slender means, I devoted myself full-time to a research project that must have seemed as mad, in the eyes of many people, as the mediaeval quest for the Holy Grail, the magical vessel in which the blood of Christ on the cross was supposed to have been collected.

Lancelot, Perceval, and Gawain, amongst many other valiant knights of Christendom, set off in search of the grail without ever completing their quest, because they were not pure enough or lacked sufficient faith or chastity to approach the Truth of God.

I couldn't claim to have chastity or purity either. But faith and calling led me to cross the five continents, materially or intellectually, and to glimpse horizons far wider than those that the cloistered world of mathematics usually allows. But the more my eyes opened onto the wider world, the more I realised the depth of my ignorance.

Where, when and how did the amazing adventure of the human intellect begin? In Asia? In Europe? Or somewhere in Africa? Did it take place at the time of Cro-Magnon man, about thirty thousand years ago, or in the Neanderthal period, more than fifty thousand years ago? Or could it have been half a million years ago? Or even – why not? – a million years ago?

What motives did prehistoric peoples have to begin the great adventure of counting? Were their concerns purely astronomical (to do with the phases of the moon, the eternal return of day and night, the cycle of the seasons, and so on)? Or did the requirements of communal living give the first impulse towards counting? In what way and after what period of time did people discover that the fingers of one hand and the toes of one foot represent the same concept? How did the need for calculation impose itself on their minds? Was there a chronological sequence in the discovery of the cardinal and ordinal aspects of the integers? In which period did the first attempts at oral numbering occur? Did an abstract conception of number precede articulated language? Did people count by gesture and material tokens before doing so through speech? Or was it the other way round? Does the idea of number come from experience of the world? Or did the idea of number act as a catalyst and make explicit what must have been present already as a latent idea in the minds of our most distant ancestors? And finally, is the concept of number the product of intense human thought, or is it the result of a long and slow evolution starting from a very concrete understanding of things?

These are all perfectly normal questions to ask, but most of the answers cannot be researched in a constructive way since there is no longer any trace of the thought-processes of early humans. The event, or, more probably, the sequence of events, has been lost in the depths of pre-historic time, and there are no archaeological remains to give us a clue.

However, archaeology was not necessarily the only approach to the problem. What other discipline might there be that would allow at least a stab at an answer? For instance, might psychology and ethnology not have some power to reconstitute the origins of number? The Quest for Number? Or a quest for a wraith? That was the question. It was not easy to know which it was, but I had set out on it and was soon to conquer the whole world, from America to Egypt, from India to Mexico, from Peru to China, in my search for more and yet more numbers. But as I had no financial backer, I decided to be my own sponsor, doing odd jobs (delivery boy, chauffeur, waiter, night watchman) to keep body and soul together.

As an intellectual tourist I was able to visit the greatest museums in the world, in Cairo, Baghdad, Beijing, Mexico City, and London (the British Museum and the Science Museum); the Smithsonian in Washington, the Vatican Library in Rome, the libraries of major American universities (Yale, Columbia, Philadelphia), and of course the many Paris collections at the Musée Guimet, the Conservatoire des arts et métiers, the Louvre, and the Bibliothèque nationale. I also visited the ruins of Pompeii and Masada. And took a trip to the Upper Nile Valley to see Thebes, Luxor, Abu Simbel, Gizeh. Had a look at the Acropolis in Athens and the Forum in Rome. Pondered on time's stately march from the top of the Mayan pyramids at Quiriguá and Chichén Itzá. And from here and from there I gleaned precious information about past and present customs connected with the history of counting.

When I got back from these fascinating ethno-numerical and archaeo-arithmetical expeditions I buried myself in popularising and encyclopaedic articles, plunged into learned journals and works of erudition, and fired off thousands of questions to academic specialists in scores of different fields.

At the start, I did not get many replies. My would-be correspondents were dumbfounded by the banality of the topic.

There are of course vast numbers of oddballs forever pestering specialists with questions. But I had to persuade them that I was serious. It was essential for me to obtain their co-operation, since I needed to be kept up to date about new and recent discoveries in their fields, however apparently insignificant, and as an amateur I needed their help in avoiding misinterpretations. And since I was dealing with many specialists who were far outside the field of mathematics, I had not only to persuade them that I was an honest toiler in a respectable field, but also to get them to accept that "numbers" and "mathematics" are not quite the same thing. As we shall see ...

All this work led me to two basic facts. First, a vast treasure-house of documentation on the history of numbers does actually exist. I owe a great deal to the work of previous scholars and mention it frequently throughout this book. Secondly, however, the articles and monographs in this store of knowledge each deal with only one specialism, are addressed to other experts in the same field, and are far from being complete or comprehensive accounts. There were also a few general works, to be sure, which I came across later, and which also gave me some help. But as they describe the state of knowledge at the time they were written, they had been long overtaken by later discoveries in archaeology, psychology, and ethnography.

No single work on numbers existed which covered the whole of the available field, from the history of civilisations and religions to the history of science, from prehistoric archaeology to linguistics and philology, from mythical and mathematical interpretation to ethnography, ranging over the five continents.

Indeed, how can one successfully sum up such heterogeneous material without losing important distinctions or falling into the trap of simplification? The history of numbers includes topics as widely divergent as the perception of number in mammals and birds, the arithmetical use of prehistoric notched bones, Indo-European and Semitic numbering systems, and number-techniques among so-called primitive populations in Australia, the Americas, and Africa. How can you catch in one single net things as different as finger-counting and digital computing? Counting with beads and Amerindian or Polynesian knotted string? Pharaonic epigraphy and Babylonian baked clay tablets? How can you talk in the same way about Greek and Chinese arithmetic, astronomy and Mayan inscriptions, Indian poetry and mathematics, Arabic algebra and the mediaeval quadrivium? And all of that so as to obtain a coherent overall vision of the development through time and space of the defining invention of modern humanity, which is our present numbering system? And where do animals fit into what is already an enormously complex field? Not to mention human infants ...

What I had set out to do was manifestly mad. The topic sat at the junction of all fields of knowledge and constituted an immense universe of human intellectual evolution. It covered a field so rich and huge that no single person could hope to grasp it alone.

Such a quest is by its nature unending. This book will occupy a modest place in a long line of outstanding treatises. It will not be the last of them, to be sure, for so many more things remain undiscovered or not yet understood. All the same, I think I have brought together practically everything of significance from what the number-based sciences, of the logical and historical kinds, have to teach us at the moment. Consequently, this is also probably the only book ever written that gives a more or less universal and comprehensive history of numbers and numerical calculation, set out in a logical and chronological way, and made accessible in plain language to the ordinary reader with no prior knowledge of mathematics.

And since research never stands still, I have been able to bring new solutions to some problems and to open up other, long-neglected areas of the universe of numbers. For example, in one of the chapters you will find a solution to the thorny problem of the decipherment of Elamite numbering, used nearly five thousand years ago in what is now Iran. I have also shown that Roman numbering, long thought to have been derived from the Greek system, was in fact a "prehistoric fossil", developed from the very ancient practice of notching. There are also some new contributions on Mesopotamian numbering and arithmetic, as well as a quite new way of looking at the fascinating and sensitive topic of how "our" numbers evolved from the unlikely conjunction of several great ideas. Similarly, the history of mechanical calculation culminating in the invention of the computer is entirely new.

A VERY LONG STORY

If you wanted to schematise the history of numbering systems, you could say that it fills the space between One and Zero, the two concepts which have become the symbols of modern technological society.

Nowadays we step with careless ease from Zero to One, so confident are we, thanks to computer scientists and our mathematical masters, that the Void always comes before the Unit. We never stop to think for a moment that in terms of time it is a huge step from the invention of the number "one", the first of all numbers even in the chronological sense, to the invention of the number "zero", the last major invention in the story of numbers. For in fact the whole history of humanity is spread out backwards between the time when it was realised that the void was "nothing" and the time when the sense of "oneness" first arose, as humans became aware of their individual solitude in the face of life and death, of the specificity of their species as distinct from other living beings, of the singularity of their selves as distinct from others, or of the difference of their sex as distinct from that of their partners.

But the story is neither abstract nor linear, as the history of mathematics is sometimes (and erroneously) imagined to be. Far from being an impeccable sequence of concepts each leading logically to the next, the history of numbers is the story of the needs and concerns of enormously diverse social groupings trying to count the days in the year, to make deals and bargains, to list their members, their marriages, their bereavements, their goods and flocks, their soldiers, their losses, and even their prisoners, trying also to record the date of the foundation of their cities or of one of their victories.

Goatherds and shepherds needed to know when they brought their flocks back from grazing that none had been lost; people who kept stocks of tools or arms or stood guard over food supplies for a community needed to know whether the complement of tools, arms or supplies had remained the same as when they last checked. Or again, communities with hostile neighbours must have been concerned to know whether, after each military foray, they still had the same number of soldiers, and, if not, how many they had lost in the fight. Communities that engaged in trading needed to be able to "reckon" so as to be able to buy or barter goods. For harvesting, and also in order to prepare in time for religious ceremonies, people needed to be able to count and to measure time, or at the very least to develop some practical means of managing in such circumstances.

In a word, the history of numbers is the story of humanity being led by the very nature of the things it learned to do to conceive of needs that could only be satisfied by "number reckoning". And to do that, everything and anything was put in service. The tools were approximate, concrete, and empirical ones before becoming abstract and sophisticated, originally imbued with strange mystical and mythological properties, becoming disembodied and generalisable only in the later stages.

Some communities were utilitarian and limited the aims of their counting systems to practical applications. Others saw themselves in the infinite and eternal elements, and used numbers to quantify the heavens and the earth, to express the lengths of the days, months and years since the creation of the universe, or at least from some date of origin whose meaning had subsequently been lost. And because they found that they needed to represent very large numbers, these kinds of communities did not just invent more symbols, but went down a path that led not only towards the fundamental rule of position, but also onto the track of a very abstract concept that we call "zero", whence comes the whole of mathematics.

THE FIRST STEPS

No one knows where or when the story began, but it was certainly a very long time ago. That was when people were unable to conceive of numbers as such, and therefore could not count. They were capable, at most, of the concepts of *one*, *two*, and *many*.

As a result of studies carried out on a wide range of beings, from crows to humans as diverse as infants, Pygmies, and the Amerindian inhabitants of Tierra del Fuego, psychologists and ethnologists have been able to establish the absolute zero of human number-perception. Like some of the higher animals, the human adult with no training at all (for example, learning to recognise the 5 or the 6 at cards by sight, through sheer practice) has direct and immediate perception of the numbers 1 to 4 only. Beyond that level, people have to learn to count. To do that they need to develop, firstly, advanced number-manipulating skills, then, for the purposes of memorisation and of communication, they need to develop a linguistic instrument (the names of the numbers), and, finally, and much later on, they need to devise a scheme for writing numbers down.

However, you do not have to "count" the way we do if what you want to do is to find the date of a ceremony, or to make sure that the sheep and the goats that set off to graze have all come back to the byre. Even in the complete absence of the requisite words, of sufficient memory, and of the abstract concepts of number, there are all sorts of effective substitute devices for these kinds of operation. Various present-day populations in Oceania, America, Asia, and Africa whose languages contain only the words for *one*, *two*, and *many*, but who nonetheless understand one-for-one parities perfectly well, use notches on bones or wooden sticks to keep a tally. Other populations use piles or lines of pebbles, shells, knucklebones, or sticks. Still others tick things off by the parts of their body (fingers, toes, elbows and knees, eyes, nose, mouth, ears, breasts, and chest).

THE EARLIEST COUNTING MACHINES

Early humanity used more or less whatever came to hand to manage in a quantitative as well as a qualitative universe. Nature itself offered every cardinal model possible: birds with two wings, the three parts of a clover-leaf, four-legged animals, and five-fingered hands ... But as everyone began counting by using their ten fingers, most of the numbering systems that were invented used base 10. All the same, some groups chose base 12. The Mayans, Aztecs, Celts, and Basques, looked down at their feet and realised that their toes could be counted like fingers, so they chose base 20. The Sumerians and Babylonians, however, chose to count on base 60, for reasons that remain mysterious. That is where our present division of the hour into 60 minutes of 60 seconds comes from, as does the division of a circle into 360 degrees, each of 60 minutes divided into 60 seconds.

The very oldest counting tools that archaeologists have yet dug up are the numerous animal bones found in western Europe and marked with one or more sets of notches. These tally sticks are between twenty thousand and thirty-five thousand years old.

The people using these bones were probably fearsome hunters, and, for each kill, they would score another mark onto the tally stick. Separate counting bones might have been used for different animals – one tally for bears, another for bison, another for wolves, and so on.

They had also invented the first elements of accounting, since what they were actually doing was writing numbers in the simplest notation known.

The method may seem primitive, but it turned out to be remarkably robust, and is probably the oldest human invention (apart from fire) still in use today. Various tallies found on cave walls next to animal paintings leave us in little real doubt that we are dealing with an animal-counting device. Modern practice is no different. Since time immemorial, Alpine shepherds in Austria and Hungary, just like Celtic, Tuscan, and Dalmatian herdsmen, have checked off their animals by scoring vertical bars, Vs and Xs on a piece of wood, and that is still how they do it today. In the eighteenth century, the same "five-barred gate" was used for the shelf marks of parliamentary papers at the British House of Commons Library; it was used in Tsarist Russia and in Scandinavia and the German-speaking countries for recording loans and for calendrical accounts; whereas in rural France at that time, notched sticks did all that present-day account books and contracts do, and in the open markets of French towns they served as credit "slates". Barely twenty years ago a village baker in Burgundy made notches in pieces of wood when he needed to tot up the numbers of loaves each of his customers had taken on credit. And in nineteenth-century Indo-China, tally sticks were used as credit instruments, but also as signs of exclusion and to prevent contact with cholera victims. Finally, in Switzerland, we find notched sticks used, as elsewhere, for credit reckonings, but also for contracts, for milk deliveries, and for recording the amounts of water allocated to different grazing meadows.

The long-lasting and continuing currency of the tally system is all the more surprising for being itself the source of the Roman numbering system, which we also still use alongside or in place of Arabic numerals.

The second concrete counting tool, the hand, is of course even older. Every population on earth has used it at one stage or another. In various places in Auvergne (France), in parts of China, India, Turkey, and the former Soviet Union, people still do multiplication sums with their fingers, as the numbers are called out, and without any other tool or device. Using joints and knuckles increases the possible range, and it allowed the Ancient Egyptians, the Romans, the Arabs and the Persians (not forgetting Western Christians in the Middle Ages) to represent concretely all the numbers from 1 to 9,999. An even more ingenious variety of finger-reckoning allowed the Chinese to count to 100,000 on one hand, and to one million using both hands!

But the story of numbers can be told in other ways too. In places as far apart as Peru, Bolivia, West Africa, Hawaii, the Caroline Islands, and Ryû-Kyû, off the Japanese coast, you can find knotted string used to represent numbers. It was with such a device that the Incas sorted the archives of their very effective administration.

A third system has a far from negligible role in the history of arithmetic – the use of pebbles, which really underlies the beginning of calculation. The pebble-method is also the direct ancestor of the abacus, a device still in wide use in China, Japan and Eastern Europe. But it is the very word *calculation* that sends us back most firmly to the pebble-method: for in Latin the word for pebble is *calculus*.

THE FIRST NUMBERS IN HISTORY

The pebble-method actually formed the basis for the first written numbering system in recorded history. One day, in the fourth millennium BCE, in Elam, located in present-day Iran towards the Persian Gulf, accountants had the idea of using moulded, unbaked clay tokens in the place of ordinary or natural pebbles. The tokens of various shapes and sizes were given conventional values, each different type representing a unit of one order of magnitude within a numbering system: a stick shape for 1, a pellet for 10, a ball for 100, and so on. The idea must have been in the air for a long time, for at about the same period a similarly clay-based civilisation in Sumer, in lower Mesopotamia, invented an identical system. But since the Sumerians counted to base 60 (sexagesimal reckoning), their system was slightly different: a small clay cone stood for 1, a pellet stood for 10, a large cone for 60, a large perforated cone stood for 600, a ball meant 3,600, and so on.

These civilisations were in a phase of rapid expansion but remained exclusively oral, that is to say without writing. They relied on the rather limited potential of human memory. But the accounting system that was developed from the principles just explained turned out to be very serviceable. In the first development, the idea arose of enclosing the tokens in a spherical clay case. This allowed the system not only to serve for actual arithmetical operations, but also for keeping a record of inventories and transactions of all kinds. If a check on past dealings was needed, the clay cases could be broken open. But the second development was even more pregnant. The idea was to symbolise on the outside of the clay case the objects that were enclosed within it: one notch on the case signified that there was one small cone inside, a pellet was symbolised by a small circular perforation, a large cone by a thick notch, a ball by a circle, and so on. Which is how the oldest numbers in history, the Sumerian numerals, came into being, around 3200 BCE.

This story is obviously related to the origins of writing, but it must not be confused with it entirely. Writing serves not only to give a visual representation to thought and a physical form to memory (a need felt by all advanced societies), but above all to record articulated speech.

THE COMMON STRUCTURE OF THE HUMAN MIND

It is extraordinary to see how peoples very distant from each other in time and space used similar methods to reach identical results.

All societies learned to number their own bodies and to count on their fingers; and the use of pebbles, shells and sticks is absolutely universal. So the fact that the use of knotted string occurs in China, in Pacific island communities, in West Africa, and in Amerindian civilisations does not require us to speculate about migrations or long-distance travellers in prehistory. The making of notches to represent number is just as widespread in historical and geographical terms. Since the marking of bone and wood has the same physical requirements and limitations wherever it is done, it is no surprise that the same kinds of lines, Vs and Xs are to be seen on armbones and pieces of wood found in places as far apart as Europe, Asia, Africa, Oceania and the Americas. That is also why these marks crop up in virtually identical form in civilisations as varied as those of the Romans, the Chinese, the Khâs Boloven of Indo-China, the Zuñi Indians of New Mexico, and amongst contemporary Dalmatian and Celtic herdsmen. It is therefore not at all surprising that some numbers have almost always been represented by the same figure: 1, for instance, is represented almost universally by a single vertical line; 5 is also very frequently, though slightly less universally, figured by a kind of V in one orientation or another, and 10 by a kind of X or by a horizontal bar.

Similarly, the Ancient Egyptians, the Hittites, the Greeks, and the Aztecs worked out written numbering systems that were structurally identical, even if their respective base numbers and figurations varied considerably. Likewise the common system of Sumerian, Roman, Attic, and South Arabian numbering. Several family groupings of the same kind can be found in other sets of unrelated cultures. There is no need to hypothesise actual contact between the cultures in order to explain the similarities between their numbering systems.

So it would seem that human beings possess, in all places and at all times, a permanent capacity to repeat an invention or discovery already made elsewhere, provided only that the society or individual involved encounters cultural, social, and psychological conditions similar to those that prevailed when the invention was first made.

This is what explains why in modern science, the same discovery is sometimes made at almost the same time by two different scientists working in complete isolation from each other. Famous examples of such coincidences of invention include the simultaneous development of analytical geometry by Descartes and Fermat, of differential calculus by Newton and Leibnitz, of the physical laws of gasses by Boyle and Mariotte, and of the principles of thermodynamics by Joule, Mayer, and Sadi Carnot.

NUMBERS AND LETTERS

Ever since the invention of alphabetic writing by the Phoenicians (or at least, by a northwestern Semitic people) in the second millennium BCE, letters have been used for numbers. The simplicity and ingenuity of the alphabetic system led to its becoming the most widespread form of writing, and the Phoenician scheme is at the root of nearly every alphabet in the world today, from Hebrew to Arabic, from Berber to Hindu, and of course Greek, which is the basis of our present (Latin) lettering.

Given their alphabets, the Greeks, the Jews, the Arabs and many other peoples thought of writing numbers by using letters. The system consists of attributing numerical values from 1 to 9, then in tens from 10 to 90, then in hundreds, etc., to the letters in their original Phoenician order (an order which has remained remarkably stable over the millennia).

Number-expressions constructed in this way worked as simple accumulations of the numerical values of the individual letters. The mathematicians of Ancient Greece rationalised their use of letter-numbers within a decimal system, and, by adding diacritic signs to the base numbers, became able to express numbers to several powers of 10.

In poetry and literature, however, and especially in the domains of magic, mysticism, and divination, it was the sum of the number-values of the letters in a word that mattered.

In these circumstances, every word acquired a number-value, and conversely, every number was "loaded" with the symbolic value of one or more words that it spelled. That is why the number 26 is a divine number in Jewish lore, since it is the sum of the number-values of the letter that spell YAHWEH, the name of God:

$$\text{ה ו ה י} \quad = 5 + 6 + 5 + 10$$

The Jews, Greeks, Romans, Arabs (and as a result, Persians and Muslim Turks) pursued these kinds of speculation, which have very ancient origins: Babylonian writings of the second millennium BCE attribute a numerical value to each of the main gods: 60 was associated with Anu, god of the sky; 50 with Enlíl, god of the earth; 40 with Ea, god of water, and so forth.

The device also allowed poets like Leonidas of Alexandria to compose quite special kinds of work. It is also the basis for the art of the chronogram (verses that express a date simultaneously in words and in numbers) that can be found amongst the poets and stone-carvers of North Africa, Turkey, and Iran.

From ancient times to the present, the device has given a rich field to cabbalists, Gnostics, magicians, soothsayers, and mystics of every hue, and innumerable speculations, interpretations, calculations and predictions have been built on letter-number equivalences. The Gnostics, for example, thought they could work out the "formula" and thus the true name of God, which would enable them to penetrate all the secrets of the divine. Several religious sects are based on beliefs of this kind (such as the Hurufi or "Lettrists" of Islam) and they still have many followers, some of them in Europe.

The Greeks and Jews who first established a number-coded alphabet certainly could not have imagined that fifteen hundred or two thousand years later a Catholic theologian called Petrus Bungus would churn out a seven-hundred page numerological treatise "proving" (subject to a few spelling improvements!) that the name of Martin Luther added up to 666. It was a proof that the "isopsephic" initiates knew how to read, since according to St John the Apostle, 666 was the number of the "Beast of the Apocalypse", that is to say the Antichrist. Bungus was neither the first nor the last to make use of these methods. In the late Roman Empire, Christians tried to make Nero's name come to 666; during World War II, would-be numerological prophets managed to "prove" that Hitler was the real "Beast of the Apocalypse". A discovery that many had already made without the help of numbers.

THE HISTORY OF A GREAT INVENTION

Logic was not the guiding light of the history of number-systems. They were invented and developed in response to the concerns of accountants, first of all, but also of priests, astronomers, and astrologers, and only in the last instance in response to the needs of mathematicians. The social categories dominant in this story are notoriously conservative, and they probably acted as a brake on the development and above all on the accessibility of numbering systems. After all, knowledge (however rudimentary it may now appear) gives its holders power and privilege; it must have seemed dangerous, if not irreligious, to share it with others.

There were also other reasons for the slow and fragmentary development of numbers. Whereas fundamental scientific research is pursued in terms of scientists' own criteria, inventions and discoveries only get developed and adopted if they correspond to a perceived social need in a civilisation. Many scientific advances are ignored if there is, as people say, no "call" for them.

The stages of mathematical thought make a fascinating story. Most peoples throughout history failed to discover the rule of position, which was discovered in fact only four times in the history of the world. (The rule of position is the principle of a numbering system in which a 9, let's say, has a different magnitude depending on whether it comes in first, second, third ... position in a numerical expression.) The first discovery of this essential tool of mathematics was made in Babylon in the second millennium BCE. It was then rediscovered by Chinese arithmeticians at around the start of the Common Era. In the third to fifth centuries CE, Mayan astronomers reinvented it, and in the fifth century CE it was rediscovered for the last time, in India.

Obviously, no civilisation outside of these four ever felt the need to invent zero; but as soon as the rule of position became the basis for a numbering system, a zero was needed. All the same, only three of the four (the Babylonians, the Mayans and the Indians) managed to develop this final abstraction of number: the Chinese only acquired it through Indian influences. However, the Babylonian and Mayan zeros were not conceived of as numbers, and only the Indian zero had roughly the same potential as the one we use nowadays. That is because it is indeed the Indian zero, transmitted to us through the Arabs together with the number-symbols that we call Arabic numerals and which are in reality Indian numerals, with their appearance altered somewhat by time, use and travel.

Our knowledge of the history of numbers is of course only fragmentary, but all the pieces converge inexorably towards the system that we now use and which in recent times has conquered the whole planet.

COMPUTATION, FIGURES, AND NUMBERS

Arithmetic has a history that is by no means limited to the history of the figures we use to represent numbers. In this history of computation, figures arose quite late on; and they constitute only one of many possible ways of representing number-concepts. The history of numbers ran parallel to the history of computation, became part of it only when modern written arithmetic was invented, and then separated out again with the development of modern calculating machines.

Numbers have become so integrated into our way of thinking that they often seem to be a basic, innate characteristic of human beings, like walking or speaking. But that is not so. Numbers belong to human culture, not nature, and therefore have their own long history. For Plato, numbers were "the highest degree of knowledge" and constituted the essence of outer and inner harmony. The same idea was taken up in the Middle Ages by Nicholas Cusanus, for whom "numbers are the best means of approaching divine truths". These views all go back to Pythagoras, for whom "numbers alone allow us to grasp the true nature of the universe".

In truth, though, it is not numbers that govern the universe. Rather, there are physical properties in the world which can be expressed in abstract terms through numbers. Numbers do not come from things themselves, but from the mind that studies things. Which is why the history of numbers is a profoundly human part of human history.

IN CONCLUSION

Once a person's curiosity, on any subject, is aroused it is surprising just how far it may lead him in pursuit of its object, how readily it overcomes every obstacle. In my own case my curiosity about, or rather my absolute fascination with, numbers has been well served by a number of assets with which I set out: a Moroccan by birth, a Jew by cultural heritage, I have been afforded a more immediate access to the study of the work of Arab and Hebrew mathematicians than I might have obtained as a born European. I could harmonise within myself the mind-set of Eastern metaphysics with the Cartesian logic of the West. And I was able to identify the basic rules of a highly complex system. Moreover I possessed a sufficient aptitude for drawing to enable me to make simple illustrations to help clarify my text. I hope that the reader will recognise in this History that numbers, far from being tedious and dry, are charged with poetry, are the very vehicle for traditional myths and legends – and the finest witness to the cultural unity of the human race.

THE UNIVERSAL HISTORY OF NUMBERS

CHAPTER 1

EXPLAINING THE ORIGINS

*Ethnological and Psychological Approaches
to the Sources of Numbers*

WHEN THE SLATE WAS CLEAN

There must have been a time when nobody knew how to count. All we can surmise is that the concept of number must then have been indissociable from actual objects – nothing very much more than a direct apperception of the plurality of things. In this picture of early humanity, no one would have been able to conceive of a number as such, that is to say as an abstraction, nor to grasp the fact that sets such as "day-and-night", a brace of hares, the wings of a bird, or the eyes, ears, arms and legs of a human being had a common property, that of "being two".

Mathematics has made such rapid and spectacular progress in what are still relatively recent periods that we may find it hard to credit the existence of a time without number. However, research into behaviour in early infancy and ethnographic studies of contemporary so-called primitive populations support such a hypothesis.

CAN ANIMALS COUNT?

Some animal species possess some kind of notion of number. At a rudimentary level, they can distinguish concrete quantities (an ability that must be differentiated from the ability to count numbers in abstract). For want of a better term we will call animals' basic number-recognition the *sense of number*. It is a sense which human infants do not possess at birth.

Humans do not constitute the only species endowed with intelligence: the higher animals also have considerable problem-solving abilities. For example, hungry foxes have been seen to "play dead" so as to attract the crows they intend to eat. In Kenya, lions that previously hunted alone learned to hunt in a pack so as to chase prey towards a prepared ambush. Monkeys and other primates, of course, are not only able to make tools but also to learn how to manipulate non-verbal symbols. A much-quoted example of the first ability is that of the monkey who constructed a long bamboo tube so as to pick bananas that were out of reach. Chimpanzees have been taught to use tokens of different shapes to obtain bananas, grapes, water, and so on, and some even ended up hoarding the tokens against future needs. However, we must be careful not to be taken in by the kind of "animal intelligence" that you can see at the circus and the fairground. Dogs that can "count" are examples of effective training or (more likely) of clever trickery, not of the intellectual properties of canine minds. However, there are some very interesting cases of number-sense in the animal world.

Domesticated animals (for instance, dogs, cats, monkeys, elephants) notice straight away if one item is missing from a small set of familiar objects. In some species, mothers show by their behaviour that they know if they are missing one or more than one of their litter. A sense of number is marginally present in such reactions. The animal possesses a natural disposition to recognise that a small set seen for a second time has undergone a numerical change.

Some birds have shown that they can be trained to recognise more precise quantities. Goldfinches, when trained to choose between two different piles of seed, usually manage to distinguish successfully between three and one, three and two, four and two, four and three, and six and three.

Even more striking is the untutored ability of nightingales, magpies and crows to distinguish between concrete sets ranging from one to three or four. The story goes that a squire wanted to destroy a crow that had made its nest in his castle's watchtower. Each time he got near the nest, the crow flew off and waited on a nearby branch for the squire to give up and go down. One day the squire thought of a trick. He got two of his men to go into the tower. After a few minutes, one went down, but the other stayed behind. But the crow wasn't fooled, and waited for the second man to go down too before coming back to his nest. Then they tried the trick with three men in the tower, two of them going down: but the third man could wait as long as he liked, the crow knew that he was there. The ploy only worked when five or six men went up, showing that the crow could not discriminate between numbers greater than three or four.

These instances show that some animals have a potential which is more fully developed in humans. What we see in domesticated animals is a rudimentary perception of equivalence and non-equivalence between sets, but only in respect of numerically small sets. In goldfinches, there is something more than just a perception of equivalence – there seems to be a sense of "more than" and "less than". Once trained, these birds seem to have a perception of intensity, halfway between a perception of quantity (which requires an ability to numerate beyond a certain point) and a perception of quality. However, it only works for goldfinches when the "moreness" or "lessness" is quite large; the bird will almost always confuse five and four, seven and five, eight and six, ten and six. In other words, goldfinches can recognise differences of intensity if they are large enough, but not otherwise.

Crows have rather greater abilities: they can recognise equivalence and non-equivalence, they have considerable powers of memory, and they can perceive the relative magnitudes of two sets of the same kind separated in time and space. Obviously, crows do not count in the sense that we do, since in the absence of any generalising or abstracting capacity they cannot conceive of any "absolute quantity". But they do manage to distinguish concrete quantities. They do therefore seem to have a basic number-sense.

NUMBERS AND SMALL CHILDREN

Human infants have few innate abilities, but they do possess something that animals never have: a potential to assimilate and to recreate stage by stage the conquests of civilisation. This inherited potential is only brought out by the training and education that the child receives from the adults and other children in his or her environment. In the absence of permanent contact with a social milieu, this human potential remains undeveloped – as is shown by the numerous cases of *enfants sauvages*. (These are children brought up by or with animals in the wild, as in François Truffaut's film, *The Wild Child*. Of those recaptured, none ever learned to speak and most died in adolescence.)

We should not imagine a child as a miniature adult, lacking only judgement and knowledge. On the contrary, as child psychology has shown, children live in their own worlds, with distinct mentalities obeying their own specific laws. Adults cannot actually enter this world, cannot go back to their own beginnings. Our own childhood memories are illusions, reconstructions of the past based on adult ways of thinking.

But infancy is nonetheless the necessary prerequisite for the eventual transformation of the child into an adult. It is a long-drawn-out phase of preparation, in which the various stages in the development of human intelligence are re-enacted and reconstitute the successive steps through which our ancestors must have gone since the dawn of time.

According to N. Sillamy (1967), three main periods are distinguished: *infancy* (up to three years of age), *middle childhood* (from three to six or seven); and *late childhood*, which ends at puberty. However, a child's intellectual and emotional growth does not follow a steady and linear pattern. Piaget (1936) distinguishes five well-defined phases:

1. a *sensory-motor period* (up to two years of age) during which the child forms concepts of "object" out of fragmentary perceptions and the concept of "self" as distinct from others;

2. a *pre-operative* stage (from two to four years of age), characterised by egocentric and anthropomorphic ways of thinking ("look, mummy, the moon is following me!");

3. an *intuitive period* (from four to six), characterised by intellectual perceptions unaccompanied by reasoning; the child performs acts which he or she would be incapable of deducing, for example, pouring a liquid from one container into another of a different shape, whilst believing that the volume also changes;

4. a stage of *concrete operations* (from eight to twelve) in which, despite acquiring some operational concepts (such as class, series, number, causality), the child's thought-processes remain firmly bound to the concrete;

5. a period (around puberty) characterised by the emergence of *formal operations*, when the child becomes able to make hypotheses and test them, and to operate with abstract concepts.

Even more precisely: the new-born infant in the cradle perceives the world solely as variations of light and sound. Senses of touch, hearing and sight slowly grow more acute. From six to twelve months, the infant acquires some overall grasp of the space occupied by the things and people in its immediate environment. Little by little the child begins to make associations and to perceive differences and similarities. In this way the child forms representations of relatively simple groupings of beings and objects which are familiar both by nature and in number. At this age, therefore, the child is able to reassemble into one group a set of objects which have previously been moved apart. If one thing is missing from a familiar set of objects, the child immediately notices. But the abstraction of number – which the child simply feels, as if it were a feature of the objects themselves – is beyond the child's grasp. At this age babies do not use their fingers to indicate a number.

Between twelve and eighteen months, the infant progressively learns to distinguish between one, two and several objects, and to tell at a glance the relative sizes of two small collections of things. However, the infant's numerical capabilities still remain limited, to the extent that no clear distinction is made between the numbers and the collections that they represent. In other words, until the child has grasped the generic principle of the natural numbers ($2 = 1 + 1$; $3 = 2 + 1$; $4 = 3 + 1$, etc.), numbers remain nothing more than "number-groupings", not separable from the concrete nature of the items present, and they can only be recognised by the principle of *pairing* (for instance, on seeing two sets of objects lined up next to each other).

Oddly enough, when a child has acquired the use of speech and learned to name the first few numbers, he or she often has great difficulty in symbolising the number three. Children often count from one to two and then miss three, jumping straight to four. Although the child can recognise, visually and intuitively, the concrete quantities from one to four, at this stage of development he or she is still at the very doorstep of abstract numbering, which corresponds to *one, two, many*.

However, once this stage is passed (at between three and four years of age, according to Piaget), the child quickly becomes able to count properly. From then on, progress is made by virtue of the fact that the abstract concept of number progressively takes over from the purely perceptual aspect of a collection of objects. The road lies open which leads on to the acquisition of a true grasp of abstract calculation. For this reason, teachers call this phase the "pre-arithmetical stage" of intellectual development. The child will first learn to count up to ten, relying heavily on the use of fingers; then the number series is progressively extended as the capacity for abstraction increases.

ARITHMETIC AND THE BODY

The importance of the hand, and more generally of the body in children's acquisition of arithmetic can hardly be exaggerated. Inadequate access to or use of this "counting instrument" can cause serious learning difficulties.

In earliest infancy, the child plays with his or her fingers. It constitutes the first notion of the child's own body. Then the child touches everything in order to make acquaintance with the world, and this also is done primarily with the hands. One day, a well-intentioned teacher who wanted arithmetic to be "mental", forbade finger-counting in his class. Without realising it, the teacher had denied the children the use of their bodies, and forbidden the association of mathematics with their bodies. I've seen many children profoundly relieved to be able to use their hands again: their bodies were at last accepted [...] Spatio-temporal disabilities can likewise make learning mathematics very difficult. Inadequate grasp of the notions of "higher than" and "lower than" affect the concepts of number, and all operations and relations between them. The unit digits are written to the right, and the hundred digits are written to the left, so a child who cannot tell left from right cannot write numbers properly or begin an operation at all easily. Number skills and the whole set of logical operations of arithmetic can thus be seriously undermined by failure to accept the body. [L. Weyl-Kailey (1985)]

NUMBERS AND THE PRIMITIVE MIND

A good number of so-called primitive people in the world today seem similarly unable to grasp number as an abstract concept. Amongst these populations, number is "felt" and "registered", but it is perceived as a *quality*, rather as we perceive smell, colour, noise, or the presence of a person or thing outside of ourselves. In other words, "primitive" peoples are affected only by changes in their visual field, in a direct subject-object relationship. Their grasp of number is thus limited to what their predispositions allow them to see in a single visual glance.

However, that does not mean that they have no perception of quantity. It is just that the plurality of beings and things is measured by them not in a quantitative but in a qualitative way, without differentiating individual items. Cardinal reckoning of this sort is never fixed in the abstract, but always related to concrete sets, varying naturally according to the type of set considered.

A well-defined and appropriately limited set of things or beings, provided it is of interest to the primitive observer, will be memorised with all its characteristics. In the primitive's mental representation of it, the exact number of the things or beings involved is implicit: it resembles a quality by which this set is different from another group consisting of one or several more or fewer members. Consequently, when he sets eyes on the set for a second time, the primitive knows if it is complete or if it is larger or smaller than it was previously. [L. Lévy-Bruhl (1928)]

ONE, TWO ... MANY

In the first years of the twentieth century, there were several "primitive" peoples still at this basic stage of numbering: Bushmen (South Africa), Zulus (South and Central Africa), Pygmies (Central Africa), Botocudos (Brazil), Fuegians (South America), the Kamilarai and Aranda peoples in Australia, the natives of the Murray Islands, off Cape York (Australia), the Vedda (Sri Lanka), and many other "traditional" communities.

According to E. B. Tylor (1871), the Botocudos had only two real terms for numbers: one for "one", and the other for "a pair". With these lexical items they could manage to express three and four by saying something like "one and two" and "two and two". But these people had as much difficulty conceptualising a number above four as it is for us to imagine quantities of a trillion billions. For larger numbers, some of the Botocudos just pointed to their hair, as if to say "there are as many as there are hairs on my head".

A. Sommerfelt (1938) similarly reports that the Aranda had only two number-terms, *ninta* (one), and *tara* (two). Three and four were expressed as *tara-mi-ninta* (one and two) and *tara-ma-tara* ("two and two"), and the number series of the Aranda stopped there. For larger quantities, imprecise terms resembling "a lot", "several" and so on were used.

Likewise G. Hunt (1899) records the Murray islanders' use of the terms
netat and *neis* for "one" and "two", and the expressions *neis-netat* (two +
one) for "three", and *neis-neis* (two + two) for "four". Higher numbers were
expressed by words like "a crowd of . . . "

Our final example is that of the Torres Straits islanders for whom *urapun*
meant "one", *okosa* "two", *okosa-urapun* (two-one) "three", and *okosa-okosa*
(two-two) "four". According to A. C. Haddon (1890) these were the only
terms used for absolute quantities; other numbers were expressed by the
word *ras*, meaning "a lot".

Attempts to teach such communities to count and to do arithmetic in the
Western manner have frequently failed. There are numerous accounts of
natives' lack of memory, concentration and seriousness when confronted
with numbers and sums [see, for example, M. Dobrizhoffer (1902)]. It
generally turned out much easier to teach primitive peoples the arts of
music, painting, and sculpture than to get them to accept the interest and
importance of arithmetic. This was perhaps not just because primitive
peoples felt no need of counting, but also because numbers are amongst the
most abstract concepts that humanity has yet devised. Children take longer
to learn to do sums than to speak or to write. In the history of humanity,
too, numbers have proved to be the hardest of these three skills.

PARITY BEFORE NUMBER

These primitive peoples nonetheless possessed a fundamental arithmetical
rule which if systematically applied would have allowed them to manipulate
numbers far in excess of four. The rule is what we call the *principle
of base 2* (or binary principle). In this kind of numbering, five is "two-
two-one", six is "two-two-two", seven is "two-two-two-one", and so on.
But primitive societies did not develop binary numbering because, as
L. Gerschel (1960) reminds us, they possessed only the most basic degree
of numeracy, that which distinguishes between the singular and the dual.

A. C. Haddon (1890), observing the western Torres Straits islanders,
noted that they had a pronounced tendency to count things in groups of two
or in couples. M. Codrington, in *Melanesian Languages*, noticed the same
thing in many Oceanic populations: "The natives of Duke of York's Island
count in couples, and give the pairings different names depending how
many of them there are; whereas in Polynesia, numbers are used although it
is understood that they refer to so many pairs of things, not
to so many things." Curr, as quoted by T. Dantzig (1930), confirms that
Australian aborigines also counted in this way, to the extent that "if two pins
are removed from a set of seven the aborigines rarely notice it, but they see
straight away if only one is removed".

These primitive peoples obviously had a stronger sense of parity than of number. To express the numbers three and four, numbers they did not grasp as abstracts but which common sense allowed them to see in a single glance, they had recourse only to concepts of *one* and *pair*. And so for them groups like "two-one" or "two-two" were themselves pairs, not (as for us) the abstract integers (or "whole numbers") "three" and "four". So it is easy to see why they never developed the binary system to get as far as five and six, since these would have required three digits, one more than the pair which was their concept of the highest abstract number.

THE LIMITS OF PERCEPTION

The limited arithmetic of "primitive" societies does not mean that their members were unintelligent, nor that their innate abilities were or are lesser than ours. It would be a grave error to think that we could do better than a Torres Straits islander at recognising number if all we had to use were our natural faculties of perception.

In practice, when we want to distinguish a quantity we have recourse to our memories and/or to acquired techniques such as comparison, splitting, mental grouping, or, best of all, actual counting. For that reason it is rather difficult to get to our natural sense of number. There is an exercise that we can try, all the same. Looking at Fig. 1.1, which contains sets of objects *in line*, try to estimate the quantity of each set of objects in a single visual glance (that is to say, *without* counting). What is the best that we can do?

FIG. 1.1.

Everyone can see the sets of one, of two, and of three objects in the figure, and most people can see the set of four. But that's about the limit of our natural ability to numerate. Beyond four, quantities are vague, and our eyes alone cannot tell us how many things there are. Are there fifteen or twenty plates in that pile? Thirteen or fourteen cars parked along the street? Eleven or twelve bushes in that garden, ten or fifteen steps on this staircase, nine, eight or six windows in the façade of that house? The correct answers cannot be just seen. *We have to count to find out!*

The eye is simply not a sufficiently precise measuring tool: its natural number-ability virtually never exceeds four.

There are many traces of the "limit of four" in different languages and cultures. There are several Oceanic languages, for example, which distinguish between nouns in the singular, the dual, the triple, the quadruple, and the plural (as if in English we were to say *one bird, two birdo, three birdi, four birdu, many birds*).

In Latin, the names of the first four numbers (*unus, duos, tres, quatuor*) decline at least in part like other nouns and adjectives, but from five (*quinque*), Latin numerical terms are invariable. Similarly, Romans gave "ordinary" names to the first four of their sons (names like Marcus, Servius, Appius, etc.), but the fifth and subsequent sons were named only by a numeral: Quintus (the fifth), Sixtus (the sixth), Septimus (the seventh), and so on. In the original Roman calendar (the so-called "calendar of Romulus"), only the first four months had names (Martius, Aprilis, Maius, Junius), the fifth to tenth being referred to by their order-number: Quintilis, Sextilis, September, October, November, December.*

Perhaps the most obvious confirmation of the basic psychological rule of the "limit of four" can be found in the almost universal counting-device called (in England) the "five-barred gate". It is used by innkeepers keeping a tally or "slate" of drinks ordered, by card-players totting up scores, by prisoners keeping count of their days in jail, even by examiners working out the mark-distribution of a cohort of students:

1	I	6	ЖĦ I	11	ЖĦ ЖĦ I
2	II	7	ЖĦ II	12	ЖĦ ЖĦ II
3	III	8	ЖĦ III	13	ЖĦ ЖĦ III
4	IIII	9	ЖĦ IIII	14	ЖĦ ЖĦ IIII
5	ЖĦ	10	ЖĦ ЖĦ	15	ЖĦ ЖĦ ЖĦ

FIG. 1.2. *The five-barred gate*

---

* The original ten-month Roman calendar had 304 days and began with *Martius*. It was subsequently lengthened by the addition of two further months, *Januarius* and *Februarius* (our January and February). Julius Caesar further reformed the calendar, taking the start of the year back to 1 January and giving it 365 days in all. Later, the month of *Quintilis* was renamed *Julius* (our July) in honour of Caesar, and *Sextilis* became *Augustus* in honour of the emperor of that name.

Most human societies the world has known have used this kind of number-notation at some stage in their development and all have tried to find ways of coping with the unavoidable fact that beyond four (IIII) nobody can "read" intuitively a sequence of five strokes (IIIII) or more.

ARAMAIC (Egypt)
Elephantine script: 5th to 3rd centuries BCE

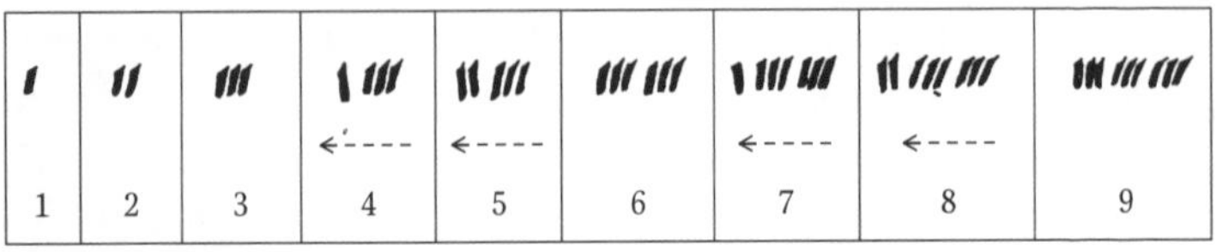

1	2	3	4	5	6	7	8	9

FIG. 1.3.

ARAMAIC (Mesopotamia)
Khatra script: First decades of CE

1	2	3	4	5	6	7	8	9

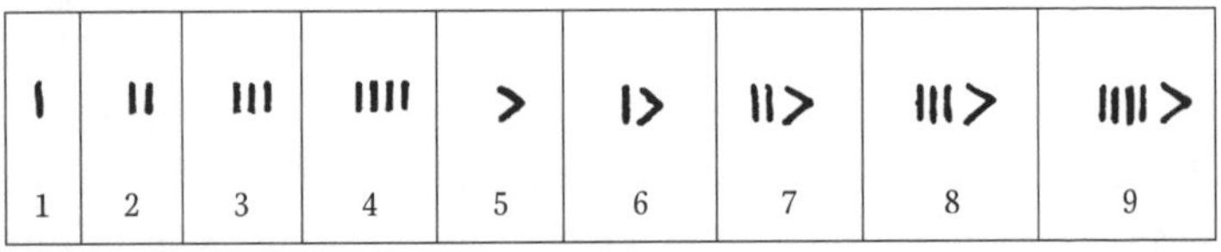

FIG. 1.4.

ARAMAIC (Syria)
Palmyrenean script: First decades of CE

1	2	3	4	5	6	7	8	9

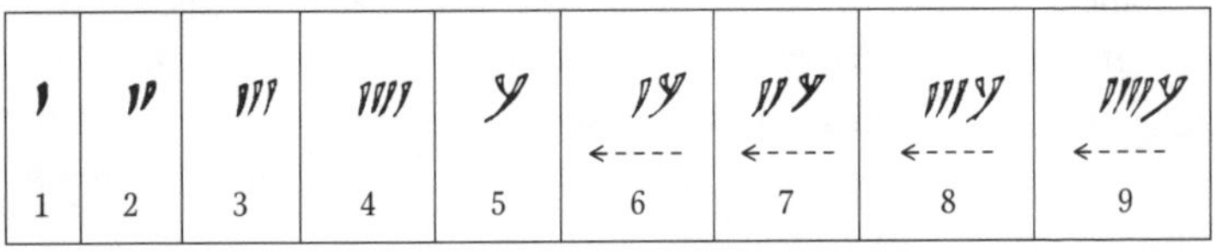

FIG. 1.5.

CRETAN CIVILISATION
Hieroglyphic script: first half of second millennium BCE

1	2	3	4	5	6	7	8	9

FIG. 1.6.

CRETAN CIVILISATION
Linear script: 1700–1200 BCE

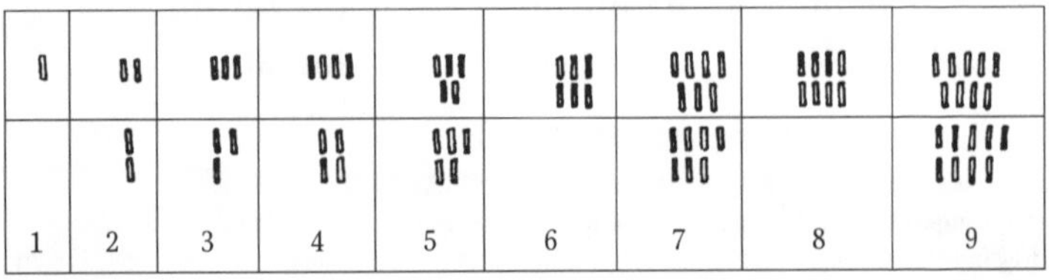

1	2	3	4	5	6	7	8	9

FIG. 1.7.

EGYPT
Hieroglyphic script: third to first millennium BCE

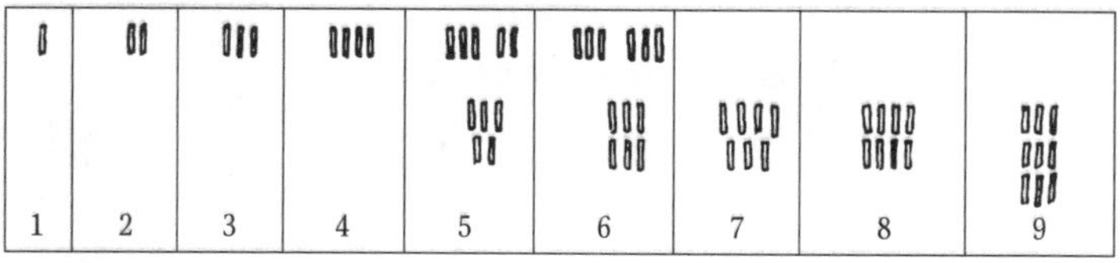

1	2	3	4	5	6	7	8	9

FIG. 1.8.

ELAM
"Proto-Elamite" script: Iran, first half of third millennium BCE

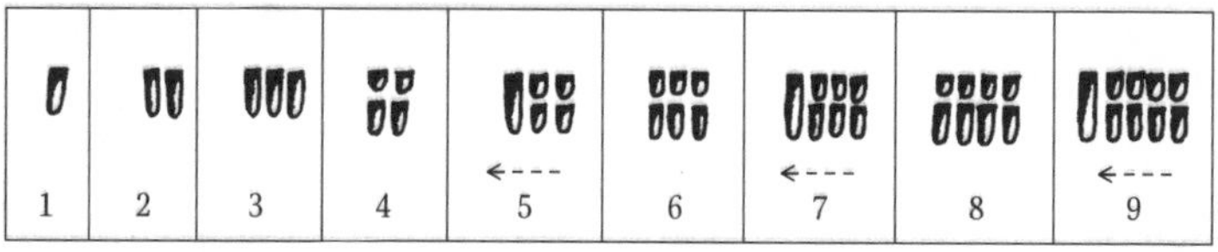

1	2	3	4	5	6	7	8	9

FIG. 1.9.

ETRUSCAN CIVILISATION
Italy, 6th to 4th centuries BCE

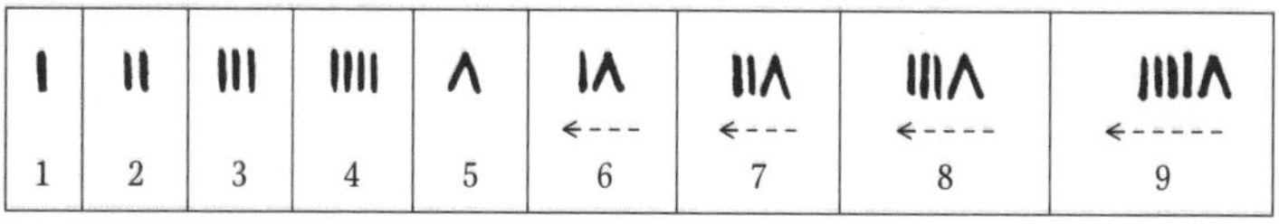

1	2	3	4	5	6	7	8	9

FIG. 1.10.

GREECE
Epidaurus and Argos, 5th to 2nd centuries BCE

1	2	3	4	5	6	7	8	9

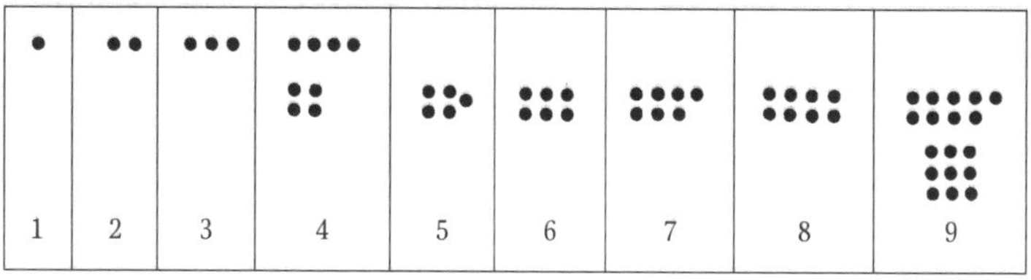

FIG. 1.11.

GREECE
Taurian Chersonesus, Chalcidy, Troezen, 5th to 2nd centuries BCE

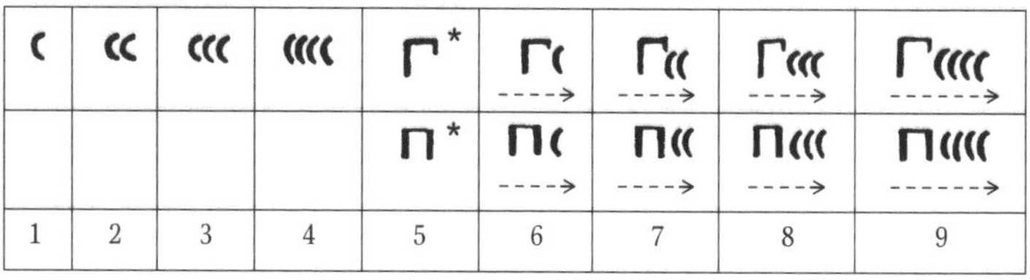

1	2	3	4	5	6	7	8	9

*π, initial of <u>pente</u>, five

FIG. 1.12.

GREECE
Thebes, Karistos, 5th to 1st centuries BCE

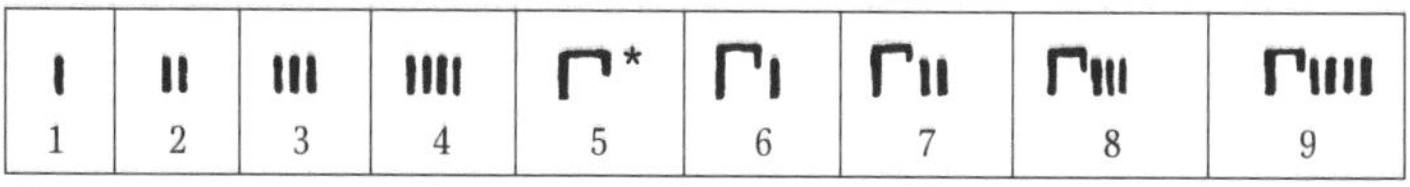

1	2	3	4	5	6	7	8	9

*π, initial of <u>pente</u>, five

FIG. 1.13.

INDUS CIVILISATION
2300–1750 BCE

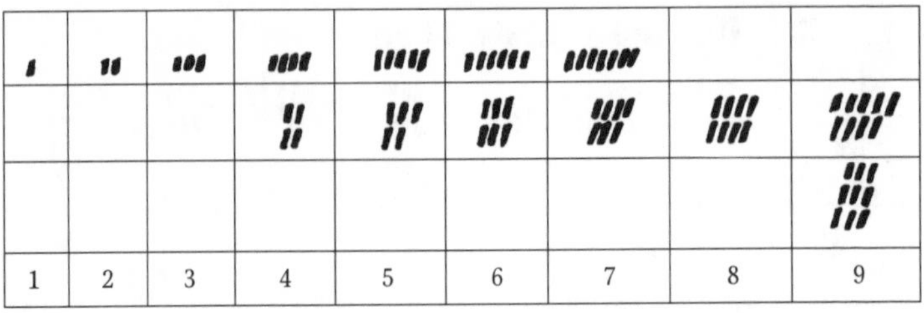

Fig. 1.14.

HITTITE CIVILISATION
Hieroglyphic: Anatolia, 1500–800 BCE

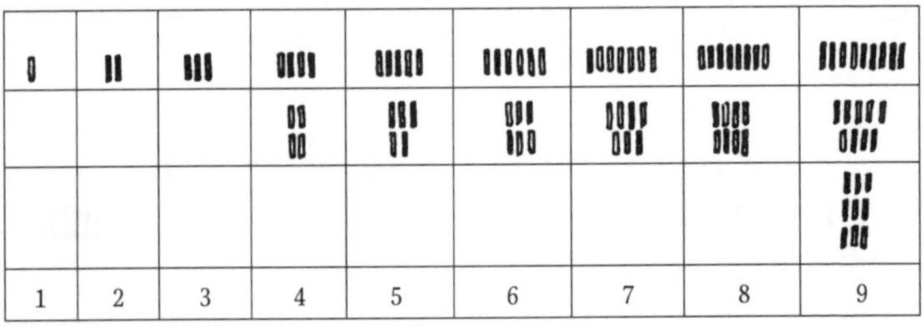

Fig. 1.15.

LYCIAN CIVILISATION
Asia Minor, first half of first millennium BCE

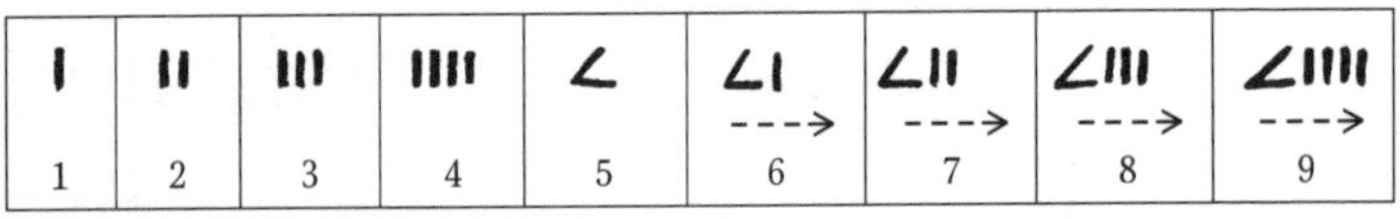

Fig. 1.16.

LYDIAN CIVILISATION
Asia Minor, 6th to 4th centuries BCE

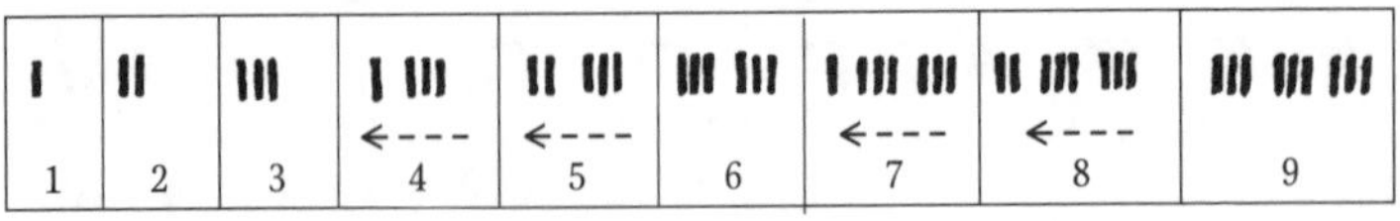

Fig. 1.17.

MAYAN CIVILISATION
Pre-Columbian Central America, 3rd to 14th centuries CE

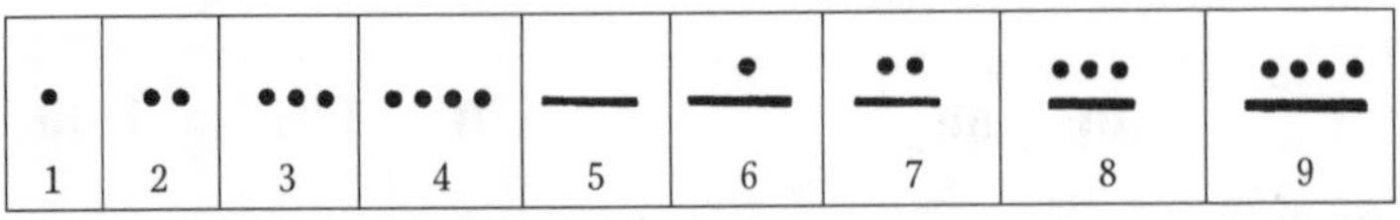

Fig. 1.18.

MESOPOTAMIA
Archaic Sumerian, beginning of third millennium BCE

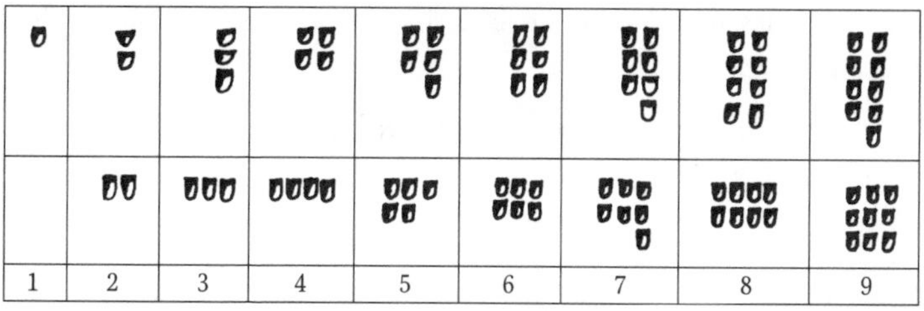

FIG. 1.19.

MESOPOTAMIA
Sumerian cuneiform, 2850–2000 BCE

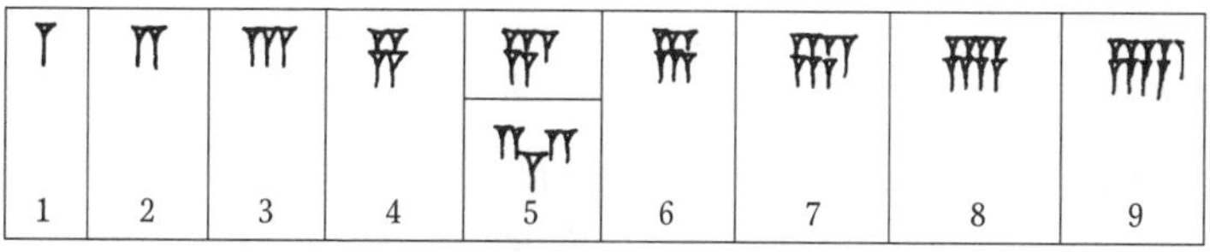

FIG. 1.20.

MESOPOTAMIA
Assyro-Babylonian cuneiform, second to first millennium BCE

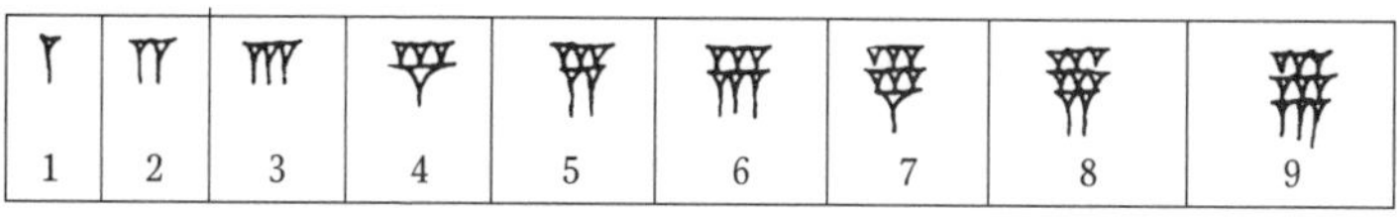

FIG. 1.21.

CIVILISATIONS OF MA'IN & SABA (SHEBA)
Southern Arabia, 5th to 1st centuries BCE

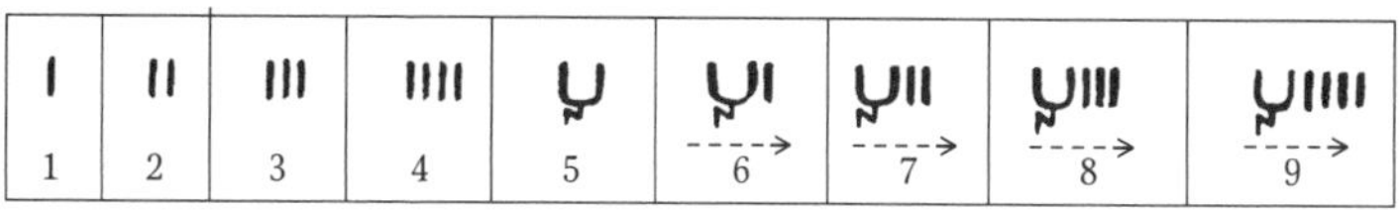

FIG. 1.22.

PHOENICIAN CIVILISATION
From 6th century BCE

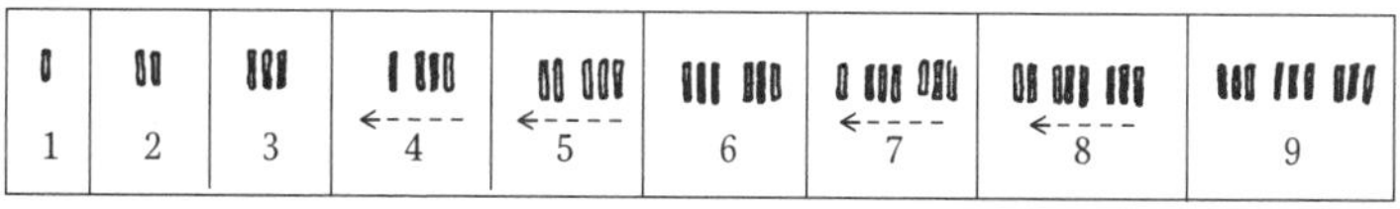

FIG. 1.23.

URARTU
Hieroglyphic script, Armenia, 13th to 9th centuries BCE

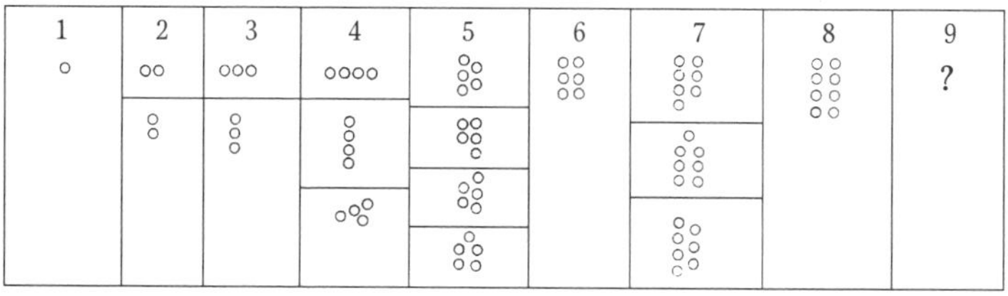

FIG. 1.24.

To recapitulate: at the start of this story, people began by counting the first nine numbers by placing in sequence the corresponding number of strokes, circles, dots or other similar signs representing "one", more or less as follows:

I	II	III	IIII	IIIII	IIIIII	IIIIIII	IIIIIIII	IIIIIIIII
1	2	3	4	5	6	7	8	9

Fig. 1.25.

But because series of identical signs are not easy to read quickly for numbers above four, the system was rapidly abandoned. Some civilisations (such as those found in Egypt, Sumer, Elam, Crete, Urartu, and Greece) got round the difficulty by grouping the signs for numbers from five to nine to 9 according to a principle that we might call *dyadic representation:*

I	II	III	IIII	III	III	IIII	IIII	IIIII
				II	III	III	IIII	IIII
1	2	3	4	5	6	7	8	9
				(3 + 2)	(3 + 3)	(4 + 3)	(4 + 4)	(5 + 4)

Fig. 1.26.

Other civilisations, such as the Assyro-Babylonian, the Phoenician, the Egyptian-Aramaean and the Lydian, solved the problem by recourse to a *rule of three*:

I	II	III	III	III	III	III	III	III
			I	II	III	III	III	III
						I	II	III
1	2	3	4	5	6	7	8	9
			(3 + 1)	(3 + 2)	(3 + 3)	(3 + 3 + 1)	(3 + 3 + 2)	(3 + 3 + 3)

Fig. 1.27.

And yet others, like the Greeks, the Manaeans and Sabaeans, the Lycians, Mayans, Etruscans and Romans, came up with an idea (probably based on finger-counting) for a special sign for the number five, proceeding thereafter on a *rule of five* or quinary system ($6 = 5 + 1$, $7 = 5 + 2$, and so on).

There really can be no debate about it now: *natural human ability to perceive number does not exceed four!*

So the basic root of arithmetic as we know it today is a very rudimentary numerical capacity indeed, a capacity barely greater than that of some animals. There's no doubt that the human mind could no more accede *by innate aptitude alone* to the abstraction of counting than could crows or goldfinches. But human societies have enlarged the potential of these very limited abilities by inventing a number of mental procedures of enormous fertility, procedures which opened up a pathway into the universe of numbers and mathematics ...

DEAD RECKONING

Since we can discriminate unreflectingly between concrete quantities only up to four, we cannot have recourse only to our natural sense of number to get to any quantity greater than four. We must perforce bring into play the device of abstract counting, the characteristic quality of "civilised" humanity.

But is it therefore the case that, in the absence of this mental device for counting (in the way we now understand the term), the human mind is so enfeebled that it cannot engage in any kind of numeration at all?

It is certainly true that without the abstractions that we call "one", "two", "three", and so on it is not easy to carry out mental operations. But it does not follow at all that a mind without numbers of our kind is incapable of devising specific tools for manipulating quantities in concrete sets. There are very good reasons for thinking that for many centuries people were able to reach several numbers without possessing anything like number-concepts.

There are many ethnographic records and reports from various parts of Africa, Oceania and the Americas showing that numerous contemporary "primitive" populations have numerical techniques that allow them to carry out some "operations", at least to some extent.

These techniques, which, in comparison to our own, could be called "concrete", enable such peoples to reach the same results as we would, by using *mediating objects* or *model collections* of many different kinds (pebbles, shells, bones, hard fruit, dried animal dung, sticks, the use of notched bones or sticks, etc.). The techniques are much less powerful and often more complicated than our own, but they are perfectly serviceable for establishing (for example) whether as many head of cattle have returned from grazing as went out of the cowshed. You do not need to be able to count by numbers to get the right answer for problems of that kind.

ELEMENTARY ARITHMETIC

It all started with the device known as "one-for-one correspondence". This allows even the simplest of minds to compare two collections of beings or things, of the same kind or not, without calling on an ability to count in numbers. It is a device which is both the prehistory of arithmetic, and the dominant mode of operation in all contemporary "hard" sciences.

Here is how it works: You get on a bus and you have before you (apart from the driver, who is in a privileged position) two sets: a set of seats and a set of passengers. In one glance you can tell whether the two sets have "the same number" of elements; and, if the two sets are not equal, you can tell just as quickly which is the larger of the two. This ready-reckoning of number without recourse to numeration is more easily explained by the device of one-for-one correspondence.

If there was no one standing in the bus and there were some empty seats, you would know that each passenger has a seat, but that each seat does not necessarily have a passenger: therefore, there are fewer passengers than seats. In the contrary case – if there are people standing and all the seats are taken – you know that there are more passengers than seats. The third possibility is that there is no one standing and all seats are taken: as each seat corresponds to one passenger, there are as many passengers as seats. The last situation can be described by saying that there is a mapping (or a *biunivocal correspondence*, or, in terms of modern mathematics, a *bijection*) between the number of seats and the number of passengers in the bus.

At about fifteen or sixteen months, infants go beyond the stage of simple observation of their environment and become capable of grasping the principle of one-for-one correspondence, and in particular the property of mapping. If we give a baby of this age equal numbers of dolls and little chairs, the infant will probably try to fit one doll on each seat. This kind of play is nothing other than mapping the elements of one set (dolls) onto the elements of a second set (chairs). But if we set out more dolls than chairs (or more chairs than dolls), after a time the baby will begin to fret: it will have realised that the mapping isn't working.

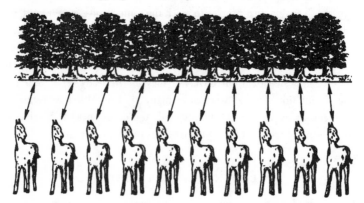

FIG. 1.28. *Two sets map if for each element of one set there is a corresponding single element of the other, and vice versa.*

This mental device does not only provide a means for comparing two groups, but it also allows its user to manipulate several numbers without knowing how to count or even to name the quantities involved.

If you work at a cinema box-office you usually have a seating plan of the auditorium in front of you. There is one "box" on the plan for each seat in the auditorium, and, each time you sell a ticket, you cross out one of the boxes on the plan. What you are doing is: mapping the seats in the cinema onto the boxes on the seating plan, then mapping the boxes on the plan onto the tickets sold, and finally, mapping the tickets sold onto the number of people allowed into the auditorium. So even if you are too lazy to add up the number of tickets you've sold, you'll not be in any doubt about knowing when the show has sold out.

To recite the attributes of Allah or the obligatory laudations after prayers, Muslims habitually use a string of prayer-beads, each bead corresponding to one divine attribute or to one laudation. The faithful "tell their beads" by slipping a bead at a time through their fingers as they proceed through the recitation of eulogies or of the attributes of Allah.

FIG. 1.29. *Muslim prayer-beads (*subha *or* sebha *in Arabic) used for reciting the 99 attributes of Allah or for supererogatory laudations. This indispensable piece of equipment for pilgrims and dervishes is made of wooden, mother-of-pearl or ivory beads that can be slipped through the fingers. It is often made up of three groups of beads, separated by two larger "marker" beads, with an even larger bead indicating the start. There are usually a hundred beads on a string (33 + 33 + 33 + 1), but the number varies.*

Buddhists have also used prayer-beads for a very long time, as have Catholics, for reciting *Pater noster, Ave Maria, Gloria Patri,* etc. As these litanies must be recited several times in a quite precise order and number, Christian rosaries usually consist of a necklace threaded with five times ten small beads, each group separated by a slightly larger bead, together with a chain bearing one large then three small beads, then one large bead and a cross. That is how the litanies can be recited without counting but without omission – each small bead on the ring corresponds to one *Ave Maria,* with a *Gloria Patri* added on the last bead of each set of ten, and a *Pater noster* is said for each large bead, and so on.

The device of one-for-one correspondence has thus allowed these religions to devise a system which ensures that the faithful do not lose count of their litanies despite the considerable amount of repetition required. The device can thus be of use to the most "civilised" of societies; and for the completely "uncivilised" it is even more valuable.

Let us take someone with no arithmetical knowledge at all and send him to the grocery store to get ten loaves of bread, five bottles of cooking oil, and four bags of potatoes. With no ability to count, how could this person be trusted to bring back the correct amount of change? But in fact such a person is perfectly capable of carrying out the errand provided the proper equipment is available. The appropriate kit is necessarily based on the principle of one-for-one correspondence. We could make ten purses out of white cloth, corresponding to the ten loaves, five yellow purses for the bottles of cooking oil, and four brown purses, for the bags of potatoes. In each purse we could put the exact price of the corresponding item of purchase, and all the uneducated shopper needs to know is that a white purse can be exchanged for a loaf, a yellow one for a bottle of oil and a brown one for a bag of potatoes.

This is probably how prehistoric humanity did arithmetic for many millennia, before the first glimmer of arithmetic or of number-concepts arose.

Imagine a shepherd in charge of a flock of sheep which is brought back to shelter every night in a cave. There are fifty-five sheep in this flock. But the shepherd doesn't know that he has fifty-five of them since he does not know the number "55": all he knows is that he has "many sheep". Even so, he wants to be sure that all his sheep are back in the cave each night. So he has an idea – the idea of a concrete device which prehistoric humanity used for many millennia. He sits at the mouth of his cave and lets the animals in one by one. He takes a flint and an old bone, and cuts a notch in the bone for every sheep that goes in. So, without realising the mathematical meaning of it, he has made exactly fifty-five incisions on the bone by the time the last animal is inside the cave. Henceforth the shepherd can check whether any sheep in his flock are missing. Every time he comes back from grazing, he lets the sheep into the cave one by one, and moves his finger over one indentation in the tally stick for each one. If there are any marks left on the bone after the last sheep is in the cave, that means he has lost some sheep. If not, all is in order. And if meanwhile a new lamb comes along, all he has to do is to make another notch in the tally bone.

So thanks to the principle of one-for-one correspondence it is possible to manage to count even in the absence of adequate words, memory or abstraction.

One-for-one mapping of the elements of one set onto the elements of a second set creates an abstract idea, entirely independent of the type or nature of the things or beings in the one or other set, which expresses a property common to the two sets. In other words, mapping abolishes the distinction that holds between two sets by virtue of the type or nature of the elements that constitute them. This abstract property is precisely why one-for-one mapping is a significant tool for tasks involving enumeration; but in practice, the methods that can be based on it are only suitable for relatively small sets.

This is why *model collections* can be very useful in this domain. Tally sticks with different numbers of marks on them constitute so to speak a range of *ready-made mappings* which can be referred to independently of the type or nature of the elements that they originally referred to. A stick of ivory or wood with twenty notches on it can be used to enumerate twenty men, twenty sheep or twenty goats just as easily as it can be used for twenty bison, twenty horses, twenty days, twenty pelts, twenty kayaks, or twenty measures of grain. The only number technique that can be built on this consists of choosing the most appropriate tally stick from the ready-mades so as to obtain a one-to-one mapping on the set that you next want to count.

However, notched sticks are not the only concrete *model collection*s available for this kind of matching-and-counting. The shepherd of our example could also have used pebbles for checking that the same number of sheep come into the cave every evening as went out each morning. All he needs to do to use this device would be to associate one pebble with each head of sheep, to put the resulting pile of pebbles in a safe place, and then to count them out in a reverse procedure on returning from the pasture. If the last animal in matches the last pebble in the pile, then the shepherd knows for sure that none of his flock has been lost, and if a lamb has been born meanwhile, all he needs to do is to add a pebble to the pile.

All over the globe people have used a variety of objects for this purpose: shells, pearls, hard fruit, knucklebones, sticks, elephant teeth, coconuts, clay pellets, cocoa beans, even dried dung, organised into heaps or lines corresponding in number to the tally of the things needing to be checked. Marks made in sand, and beads and shells, strung on necklaces or made into rosaries, have also been used for keeping tallies.

Even today, several "primitive" communities use parts of the body for this purpose. Fingers, toes, the articulations of the arms and legs (elbow, wrist, knee, ankle …), eyes, nose, mouth, ears, breasts, chest, sternum, hips and so on are used as the reference elements of one-for-one counting systems. Much of the evidence comes from the Cambridge Anthropological Expedition to Oceania at the end of the last century. According to Wyatt Gill, some Torres Straits islanders "counted visually" (see Fig. 1.30):

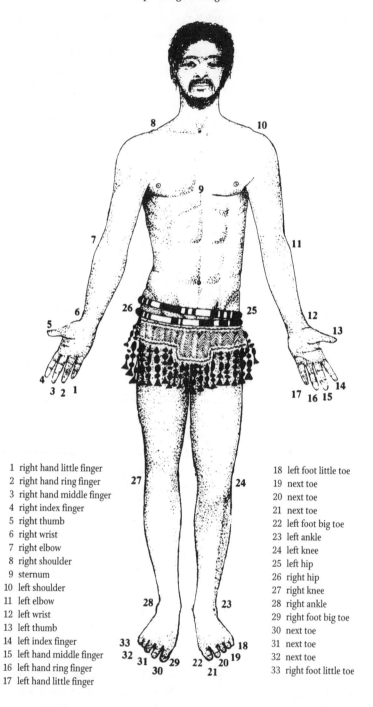

1 right hand little finger
2 right hand ring finger
3 right hand middle finger
4 right index finger
5 right thumb
6 right wrist
7 right elbow
8 right shoulder
9 sternum
10 left shoulder
11 left elbow
12 left wrist
13 left thumb
14 left index finger
15 left hand middle finger
16 left hand ring finger
17 left hand little finger

18 left foot little toe
19 next toe
20 next toe
21 next toe
22 left foot big toe
23 left ankle
24 left knee
25 left hip
26 right hip
27 right knee
28 right ankle
29 right foot big toe
30 next toe
31 next toe
32 next toe
33 right foot little toe

FIG. 1.30. *Body-counting system used by Torres Straits islanders*

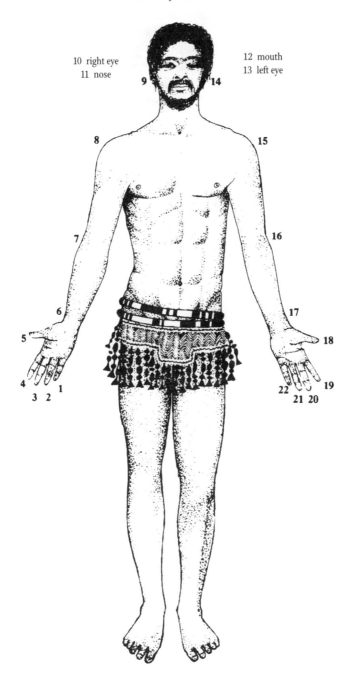

10 right eye
11 nose

12 mouth
13 left eye

Fɪɢ. 1.31. *System used by Papuans (New Guinea)*

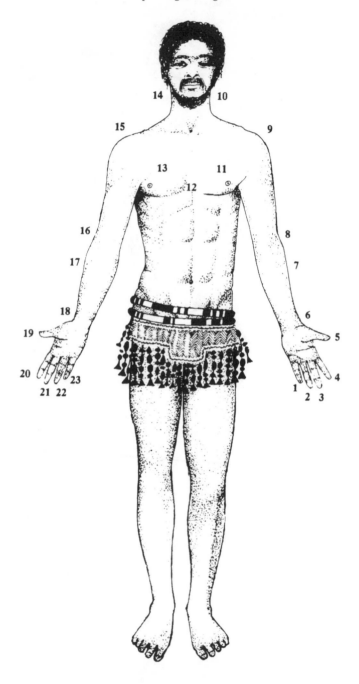

FIG. 1.32. *Body-counting system used by the Elema (New Guinea)*

They touch first the fingers of their right hand, one by one, then the right wrist, elbow and shoulder, go on to the sternum, then the left-side articulations, not forgetting the fingers. This brings them to the number seventeen. If the total needed is higher, they add the toes, ankle, knee and hip of the left then the right hand side. That gives 16 more, making 33 in all. For even higher numbers, the islanders have recourse to a bundle of small sticks. [As quoted in A. C. Haddon (1890)]

Murray islanders also used parts of the body in a conventional order, and were able to reach 29 in this manner. Other Torres Straits islanders used similar procedures which enabled them to "count visually" up to 19; the same customs are found amongst the Papuans and Elema of New Guinea.

NUMBERS, GESTURES, AND WORDS

The question arises: is the mere enumeration of parts of the body in regular order tantamount to a true arithmetical sequence? Let us try to find the answer in some of the ethnographic literature relating to Oceania.

The first example is from the Papuan language spoken in what was British New Guinea. According to the report of the Cambridge Expedition to the Torres Straits, Sir William MacGregor found that "body-counting" was prevalent in all the villages below the Musa river. "Starting with the little finger on the right hand, the series proceeds with the right-hand fingers, then the right wrist, elbow, shoulder, ear and eye, then on to the left eye, and so on, down to the little toe on the left foot." Each of the gestures to these parts of the body is accompanied, the report continues, by a specific term in Papuan, as follows:

NUMBER	NUMBER-GESTURE	GESTURE-WORD
1	right hand little finger	*anusi*
2	right hand ring finger	*doro*
3	right hand middle finger	*doro*
4	right hand index finger	*doro*
5	right thumb	*ubei*
6	right wrist	*tama*
7	right elbow	*unubo*
8	right shoulder	*visa*
9	right ear	*denoro*
10	right eye	*diti*
11	left eye	*diti*
12	nose	*medo*
13	mouth	*bee*
14	left ear	*denoro*

NUMBER	NUMBER-GESTURE	GESTURE-WORD
15	left shoulder	*visa*
16	left elbow	*unubo*
17	left wrist	*tama*
18	left thumb	*ubei*
19	left hand index finger	*doro*
20	left hand middle finger	*doro*
21	left hand ring finger	*doro*
22	left hand little finger	*anusi*

The words used are simply the names of the parts of the body, and strictly speaking they are not numerical terms at all. *Anusi*, for example, is associated with both 1 and 22, and is used to indicate the little fingers of both the right and the left hands. In these circumstances how can you know which number is meant? Similarly the term *doro* refers to the ring, middle and index fingers of both hands and "means" either 2 or 3 or 4 or 19 or 20 or 21. Without the accompanying gesture, how could you possibly tell which of these numbers was meant?

However, there is no ambiguity in the system. What is spoken is the name of the part of the body, which has its rank-order in a fixed, conventional sequence within which no confusion is possible. So there is no doubt that the mere enumeration of the parts of the body does not constitute a true arithmetical sequence unless it is associated with a corresponding sequence of gestures. Moreover, the mental counting process has no direct oral expression – you can get to the number required without uttering a word. A conventional set of "number-gestures" is all that is needed.

In those cases where it is possible to recover the original meanings of the names given to numbers, it often turns out that they retain traces of body-counting systems like those we have looked at. Here, for example, are the number-words used by the Bugilai (former British New Guinea) together with their etymological meanings:

1	*tarangesa*	left hand little finger
2	*meta kina*	next finger
3	*guigimeta kina*	middle finger
4	*topea*	index finger
5	*manda*	thumb
6	*gaben*	wrist
7	*trankgimbe*	elbow
8	*podei*	shoulder
9	*ngama*	left breast
10	*dala*	right breast

[Source: J. Chalmers (1898)]

E. C. Hawtrey (1902) also reports that the Lengua people of the Chaco (Paraguay) use a set of number-names broadly derived from specific number-gestures. Special words apparently unrelated to body-counting are used for 1 and 2, but for the other numbers they say something like:

3	"made of one and two"
4	"both sides same"
5	"one hand"
6	"reached other hand, one"
7	"reached other hand, two"
8	"reached other hand, made of one and two"
9	"reached other hand, both sides same"
10	"finished, both hands"
11	"reached foot, one"
12	"reached foot, two"
13	"reached foot, made of one and two"
14	"reached foot, both sides same"
15	"finished, foot"
16	"reached other foot, one"
17	"reached other foot, two"
18	"reached other foot, made of one and two"
19	"reached other foot, both sides same"
20	"finished, feet"

The Zuñi have names for numbers which F. H. Cushing (1892) calls "manual concepts":

1	*töpinte*	taken to begin
2	*kwilli*	raised with the previous
3	*kha'i*	the finger that divides equally
4	*awite*	all fingers raised bar one
5	*öpte*	the scored one
6	*topalik'ye*	another added to what is counted already
7	*kwillik'ya*	two brought together and raised with the others
8	*khailik'ya*	three brought together and raised with the others
9	*tenalik'ya*	all bar one raised with the others
10	*ästem'thila*	all the fingers
11	*ästem'thila topayä'thl' tona*	all the fingers and one more raised

and so on.

All this leads us to suppose that in the remotest past gestures came before any oral expression of numbers.

CARDINAL RECKONING DEVICES FOR CONCRETE QUANTITIES

Let us now imagine a group of "primitive" people lacking any conception of abstract numbers but in possession of perfectly adequate devices for "reckoning" relatively small sets of concrete objects. They use all sorts of model collections, but most often they "reckon by eye" in the following manner: they touch each other's right-hand fingers, starting with the little finger, then the right wrist, elbow, shoulder, ear, and eye. Then they touch each others' nose, mouth, then the left eye, ear, shoulder, elbow, and wrist, and on to the little finger of the left hand, getting to 22 so far. If the number needed is higher, they go on to the breasts, hips, and genitals, then the knees, ankles and toes on the right then the left sides. This extension allows 19 further integers, or a total of 41.

The group has recently skirmished with a rebellious neighbouring village and won. The group's leader decides to demand reparations, and entrusts one of his men with the task of collecting the ransom. "For each of the warriors we have lost", says the chief, "they shall give us as many pearl necklaces as there are from the little finger on my right hand to my right eye, as many pelts as there are from the little finger of my right hand to my mouth, and as many baskets of food as there are from the little finger of my right hand to my left wrist." What this means is that the reparation for each lost soldier is:

10 pearl necklaces
12 pelts
17 baskets of food

In this particular skirmish, the group lost sixteen men. Of course none amongst the group has a notion of the number "16", but they have an infallible method of determining numbers in these situations: on departing for the fight, each warrior places a pebble on a pile, and on his return each surviving warrior picks a pebble out of the pile. The number of unclaimed pebbles corresponds precisely to the number of warriors lost.

One of the leader's envoys then takes possession of the pile of remaining pebbles but has them replaced by a matching bundle of sticks, which is easier to carry. The chief checks the emissaries' equipment and their comprehension of the reparations required, and sends them off to parley with the enemy.

The envoys tell the losing side how much they owe, and proceed to enumerate the booty in the following manner: one steps forward and says:

"Bring me a pearl necklace each time I point to a part of my body," and he then touches in order the little finger, the ring finger, the middle finger, the index finger and the thumb of his right hand. So the vanquished bring him one necklace, then a second, then a third and so on up to the fifth. The envoy then repeats himself, but pointing to his right wrist, elbow, shoulder, ear and eye, which gets him five more necklaces. So without having any concept of the number "10" he obtains precisely ten necklaces.

Another envoy proceeds in identical fashion to obtain the twelve pelts, and a third takes possession of the seventeen baskets of food that are demanded.

That is when the fourth envoy comes into the equation, for he possesses the tally of warriors lost in the battle, in the form of a bundle of sixteen sticks. He sets one aside, and the three other envoys then repeat their operations, allowing him to set another stick aside, and so on, until there are no sticks left in the bundle. That is how they know that they have the full tally, and so collect up the booty and set off with it to return to their own village.

As can be seen, "primitives" of this kind are not using body-counting in exactly the same way as we might. Since we know how to count, a conventional order of the parts of the body would constitute a true arithmetical sequence; each "body-point" would be assimilated in our minds to a cardinal (rank-order) number, characteristic of a particular quantity of things or beings. For instance, to indicate the length of a week using this system, we would not need to remember that it contained as many days as mapped onto our bodies from the right little finger to the right elbow, since we could just attach to it the "rank-order number" called "right elbow", which would suffice to symbolise the numerical value of any set of seven elements.

That is because we are equipped with *generalising abstractions* and in particular with number-concepts. But "primitive" peoples are not so equipped: they cannot *abstract* from the "points" in the numbering sequence: their grasp of the sequence remains embedded in the specific nature of the "points" themselves. Their understanding is in effect restricted to one-for-one mapping; the only "operations" they make are to add or remove one or more of the elements in the basic series.

Such people do not of course have any abstract concept of the number "ten", for instance. But they do know that by touching in order their little finger, ring finger, middle finger, index finger and thumb on the right hand, then their right wrist, elbow, shoulder, ear, and eye, they can "tally out" as many men, animals or objects as there are body-points in the sequence. And having done so, they remember perfectly well which body-point any particular tally of things or people reached, and are able to repeat the operation in order to reach exactly the same tally whenever they want to.

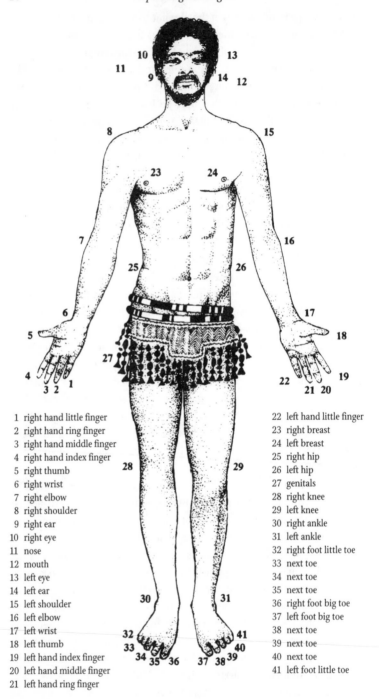

1 right hand little finger
2 right hand ring finger
3 right hand middle finger
4 right hand index finger
5 right thumb
6 right wrist
7 right elbow
8 right shoulder
9 right ear
10 right eye
11 nose
12 mouth
13 left eye
14 left ear
15 left shoulder
16 left elbow
17 left wrist
18 left thumb
19 left hand index finger
20 left hand middle finger
21 left hand ring finger

22 left hand little finger
23 right breast
24 left breast
25 right hip
26 left hip
27 genitals
28 right knee
29 left knee
30 right ankle
31 left ankle
32 right foot little toe
33 next toe
34 next toe
35 next toe
36 right foot big toe
37 left foot big toe
38 next toe
39 next toe
40 next toe
41 left foot little toe

Fig. 1.33.

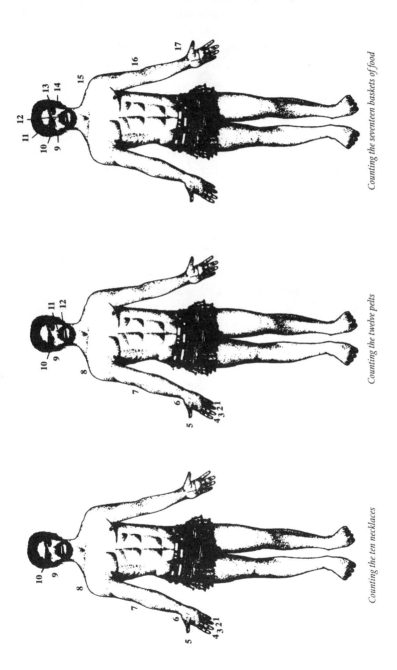

Counting the seventeen baskets of food

Counting the twelve pelts

Counting the ten necklaces

FIG. 1.34.

In other words, this procedure is a simple and convenient means of establishing ready-made mappings which can then be mapped one-to-one onto any sets for which a total is required. So when our imaginary tribe went to collect its ransom, they used only these notions, not any true number-concepts. They simply mapped three such ready-made sets onto a set of ten necklaces, a set of twelve pelts, and a set of seventeen baskets of food for each of the lost warriors.

These body-counting points are thus not thought of by their users as "numbers", but rather as the last elements of model sets arrived at after a regulated (conventional) sequence of body-gestures. This means that for such people the mere designation of any one of the points *is not sufficient to describe a given number of beings or things unless the term uttered is accompanied by the corresponding sequence of gestures*. So in discussions concerning such and such a number, no real "number-term" is uttered: instead, a given number of body-counting points will be enumerated, alongside the simultaneous sequence of gestures. This kind of enumeration therefore fails to constitute a genuine arithmetical series; participants in the discussion must also necessarily keep their eyes on the speaker!

All the same, our imaginary tribesmen have unknowingly reached quite large numbers, even with such limited tools, since they have collected:

$$16 \times 10 = 160 \text{ necklaces}$$
$$16 \times 12 = 192 \text{ pelts}$$
$$16 \times 17 = 272 \text{ baskets of food}$$

or *six hundred and twenty-four* items in all! (see Fig. 1.34)

There is a simple reason for this: they had thought of associating easily manipulated material objects with the parts of the body involved in their counting operations. It is true that they counted out the necklaces, pelts and food-baskets by their traditional body-counting method, but the determining element in calculating the ransom to be paid (the number of men lost in the battle) was "numerated" with the help of pebbles and a bundle of sticks.

Let us now imagine that the villagers are working out how to fix the date of an important forthcoming religious festival. The shaman who that morning proclaimed the arrival of the new moon also announced in the following way, accompanying his words with quite precise gestures of his hands, that the festival will fall on the *thirteenth day of the eighth moon thereafter*: "Many suns and many moons will rise and fall before the festival. The moon that has just risen must first wax and then wane completely. Then it must wax as many times again as there are from the little finger on my right hand to the elbow on the same side. Then the sun will rise and set as many times as there are from the little finger on my right hand to my mouth. That is when the sun will next rise on the day of our Great Festival."

This community obviously has a good grasp of the lunar cycle, which is only to be expected, since, after the rising and the setting of the sun, the moon's phases constitute the most obvious regular phenomenon in the natural environment. As in all *empirical calendars*, this one is based on the observation of the first quarter after the end of each cycle. With the help of model collections inherited from forebears, many generations of whom must have contributed to their slow development, the community can in fact "mark time" and compute the date thus expressed without error, as we shall see.

On hearing the shaman's pronouncement, the chief of the tribe paints a number of marks on his own body with some fairly durable kind of colouring material, and these marks enable him to record and to recognise the festival date unambiguously. He first records the series of reappearances that the moon must make from then until the festival by painting *small circles* on his right-hand little finger, ring finger, middle finger, index finger, thumb, wrist, and elbow. Then he records the number of days that must pass from the appearance of the last moon by painting a *thin line*, first of all on each finger of his right hand, then on his right wrist, elbow, shoulder, ear, and eye, then on his nose and mouth. To conclude, he puts a *thick line* over his left eye, thereby symbolising the dawn of the great day itself.

The following day at sunset, a member of the tribe chosen by the chief to "count the moons" takes one of the ready-made ivory tally sticks with thirty incised notches, the sort used whenever it is necessary to reckon the days of a given moon in their order of succession (see Fig. 1.35). He ties a piece of string around the first notch. The next evening, he ties a piece of string around the second notch, and so on every evening until the end of the moon. When he reaches the penultimate notch, he looks carefully at the night sky, in the region where the sun has just set, for he knows that the new moon is soon due to appear.

On that day, however, the first quarter of the new moon is not visible in the sky. So he looks again the next evening when he has tied the string around the last notch on the first tally stick; and though the sky is not clear enough to let him see the new moon, he decides nonetheless that a new month has begun. That is when he paints a little circle on his right little finger, indicating that one lunar cycle has passed.

At dusk the following day, our "moon-counter" takes another similar tally stick and ties a string around the first notch. The day after, he or she proceeds likewise with the second notch, and so on to the end of the second month. But at that month's end the tally man knows he will not need to scan the heavens to check on the rising of the new moon. For in this tribe, the knowledge that moon cycles end alternately on the penultimate and last notches of the tally sticks has been handed down for generations. And this knowledge is only very slightly inaccurate, since the average length of a lunar cycle is 29 days and 12 hours.

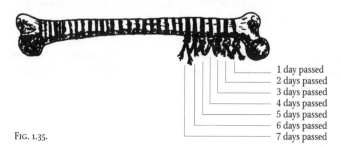

1 day passed
2 days passed
3 days passed
4 days passed
5 days passed
6 days passed
7 days passed

FIG. 1.35.

The moon-counter proceeds in this manner through alternating months of 29 and 30 days until the arrival of the last moon, when he paints a little circle on his right elbow. There are now as many circles on the counter's body as on the chief's: the counter's task is over: the "moon tally" has been reached.

The chief now takes over as the "day-counter", but for this task tally sticks are not used, as the body-counting points suffice. The community will celebrate its festival when the chief has crossed out all the *thin lines* from his little finger to his mouth and also the *thick line* over his left eye, that is to say on the thirteenth day of the eighth moon (Fig. 1.36)

This reconstitution of a non-numerate counting system conforms to many of the details observed in Australian aboriginal groups, who are able to reach relatively high numbers through the (unvocalised) numeration of parts of the body when the body-points have a fixed conventional order and are associated with manipulable model collections – knotted string, bundles of sticks, pebbles, notched bones, and so on.

Valuable evidence of this kind of system was reported by Brooke, observing the Dayaks of South Borneo. A messenger had the task of informing a number of defeated rebel villages of the sum of reparations they had to pay to the Dayaks.

> The messenger came along with some dried leaves, which he broke into pieces. Brooke exchanged them for pieces of paper, which were more convenient. The messenger laid the pieces on a table and used his fingers at the same time to count them, up to ten; then he put his foot on the table, and counted them out as he counted out the pieces of paper, each of which corresponded to a village, with the name of its chief, the number of warriors and the sum of the reparation. When he had used up all his toes, he came back to his hands. At the end of the list, there were forty-five pieces of paper laid out on the table.* Then

---

* Each finger is associated with one piece of paper and one village, in this particular system, and each toe with the set of ten fingers.

he asked me to repeat the message, which I did, whilst he ran through the pieces of paper, his fingers and his toes, as before.

"So there are our letters," he said. "You white folk don't read the way we do."

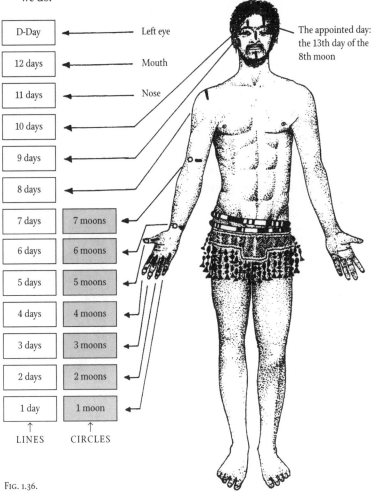

D-Day	← Left eye	The appointed day: the 13th day of the 8th moon
12 days	← Mouth	
11 days	← Nose	
10 days	←	
9 days	←	
8 days	←	
7 days	7 moons ←	
6 days	6 moons ←	
5 days	5 moons ←	
4 days	4 moons ←	
3 days	3 moons ←	
2 days	2 moons ←	
1 day	1 moon ←	
↑ LINES	↑ CIRCLES	

FIG. 1.36.

Later that evening he repeated the whole set correctly, and as he put his finger on each piece of paper in order, he said:

"So, if I remember it tomorrow morning, all will be well; leave the papers on the table."

Then he shuffled them together and made them into a heap. As soon as we got up the next morning, we sat at the table, and he re-sorted the pieces of paper into the order they were in the previous day, and repeated all the details of the message with complete accuracy. For almost a month, as he went from village to village, deep in the interior, he never forgot the different sums demanded. [Adapted from Brooke, *Ten Years in Sarawak*]

All this leads us to hypothesise the following evolution of counting systems:

First stage

Only the lowest numbers are within human grasp. Numerical ability remains restricted to what can be evaluated in a single glance. "Number" is indissociable from the concrete reality of the objects evaluated.*   In order to cope with quantities above four, a number of concrete procedures are developed. These include finger-counting and other body-counting systems, all based on one-for-one correspondence, and leading to the development of simple, widely-available ready-made mappings. What is articulated (lexicalised) in the language are these ready-made mappings, accompanied by the appropriate gestures.

Second stage

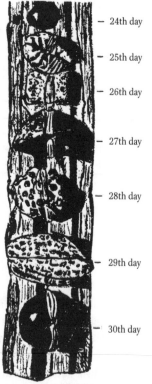

— 24th day

— 25th day

— 26th day

By force of repetition and habit, the list of the names of the body-parts in their numerative order imperceptibly acquire abstract connotations, especially the first five. They slowly lose their power to suggest the actual parts of the body, becoming progressively more attached to the corresponding number, and may now be applied to any set of objects. (L. Lévy-Bruhl)

— 27th day

Third stage

— 28th day

A fundamental tool emerges: numerical nomenclature, or the names of the numbers.

— 29th day

FIG. 1.37. *Detail from a "material model" of a lunar calendar formerly in use amongst tribal populations in former Dahomey (West Africa). It consists of a strip of cloth onto which thirty objects (seeds, kernels, shells, hard fruit, stones, etc.) have been sewn, each standing for one of the days of the month. (The fragment above represents the last seven days). From the Musée de l'Homme, Paris.*

— 30th day

---

* Thus as L. Lévy-Bruhl reports, Fijians and Solomon islanders have collective nouns for tens of arbitrarily selected items that express neither the number itself nor the objects collected into the set. In Fijian, *bola* means "a hundred dugouts", *koro* "a hundred coconuts", *salavo* "a thousand coconuts". Natives of Mota say *aka peperua* ("butterfly two dugout") for "a pair of dugouts" because of the appearance of the sails. See also Codrington, E. Stephen and L. L. Conant.

COUNTING: A HUMAN FACULTY

The human mind, evidently, can only grasp integers as abstractions if it has fully available to it the notion of distinct units as well as the ability to "synthesise" them. This intellectual faculty (which presupposes above all a complete mastery of the ability to analyse, to compare and to abstract from individual differences) rests on an idea which, alongside mapping and classification, constitutes the starting point of all scientific advance. This creation of the human mind is called "hierarchy relation" or "order relation": it is the principle by which things are ordered according to their "degree of generality", from *individual*, to *kind*, to *type*, to *species*, and so on.

Decisive progress towards the art of abstract calculation that we now use could only be made once it was clearly understood that the integers could be classified into a *hierarchised system of numerical units* whose terms were related as kinds within types, types within species, and so on. Such an organisation of numerical concepts in an invariable sequence is related to the generic principle of "recurrence" to which Aristotle referred (*Metaphysics* 1057, a) when he said that an integer was a "multiplicity measurable by the one". The idea is really that integers are "collections" of abstract units obtained successively by the adjunction of further units.

Any element in the regular sequence of the integers (other than 1) is obtained by adding 1 to the integer immediately preceding in the "natural" sequence that is so constituted (see Fig. 1.38). As the German philosopher Schopenhauer put it, any natural integer presupposes its preceding numbers as the cause of its existence: for our minds cannot conceive of a number as an abstraction unless it subsumes all preceding numbers in the sequence. This is what we called the ability to "synthesise" distinct units. Without that ability, number-concepts remain very cloudy notions indeed.

But once they have been put into a natural sequence, the set of integers permits another faculty to come into play: numeration. To numerate the items in a group is to assign to each a symbol (that is to say, a word, a gesture or a graphic mark) corresponding to a number taken from the natural sequence of integers, beginning with 1 and proceeding in order until the exhaustion of that set (Fig. 1.40). The symbol or name given to each of the elements within the set is the name of its order number within the collection of things, which becomes thereby a sequence or procession of things. The order number of the last element within the ordered group is precisely equivalent to the number of elements in the set. Obviously the number obtained is entirely independent of the order in which the elements are numerated – whichever of the elements you begin with, you always end up with the same total.

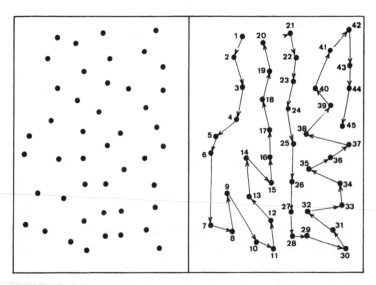

FIG. 1.38. *The generation of integers by the so-called procedure of recurrence*

FIG. 1.39. *Numeration of a "cloud" of dots*

For example, let us take a box containing "several" billiard balls. We take out one at random and give it the "number" 1 (for it is the first one to come out of the box). We take another, again completely at random, and give it the "number" 2. We continue in this manner until there are no billiard balls left in the box. When we take out the last of the balls, we give it a specific number from the natural sequence of the integers. If its number is 20, we say that there are "twenty" balls in the box. Numeration has allowed us to transform a vague notion (that there are "several" billiard balls) into exact knowledge.

In like manner, let us consider a set of "scattered" points, in other words dots in a "disordered set" (Fig. 1.39). To find out how many dots there are,

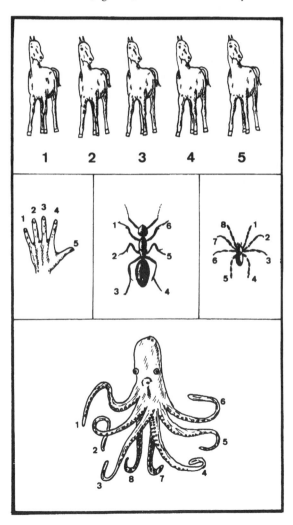

Fig. 1.40. *Numeration allowing us to advance from concrete plurality to abstract number*

all we have to do is to connect them by a "zigzag" line passing through each dot once and no dot twice. The points then constitute what is commonly called a chain. We then give each point in the chain an order-number, starting from one of the ends of the chain we have just made. The last number, given therefore to the last point in the chain, provides us with the total number of dots in the set.

So with the notions of succession and numeration we can advance from the muddled, vague and heterogeneous apperception of concrete plurality to the abstract and homogenous idea of "absolute quantity".

So the human mind can only "count" the elements in a set if it is in possession of all three of the following abilities:

1. the ability to assign a "rank-order" to each element in a procession;
2. the ability to insert into each unit of the procession the memory of all those that have gone past before;
3. the ability to convert a sequence into a "stationary" vision.

The concept of number, which at first sight seemed quite elementary, thus turns out to be much more complicated than that. To underline this point I should like to repeat one of P. Bourdin's anecdotes, as quoted in R. Balmès (1965):

> I once knew someone who heard the bells ring four as he was trying to go to sleep and who counted them out in his head, one, one, one, one. Struck by the absurdity of counting in this way, he sat up and shouted: "The clock has gone mad, it's struck one o'clock four times over!"

THE TWO SIDES OF THE INTEGERS

The concept of number has two complementary aspects: cardinal numbering, which relies only on the principle of mapping, and ordinal numeration, which requires both the technique of pairing and the idea of succession.

Here is a simple way of grasping the difference. January has 31 days. The number 31 represents the total number of days in the month, and is thus in this expression a cardinal number. However, in expressions such as "31 January 1996", the number 31 is not being used in its cardinal aspect (despite the terminology of grammar books) because here it means something like "the thirty-first day" of the month of January, specifying not a total, but a rank-order of a specific (in this case, the last) element in a set containing 31 elements. It is therefore unambiguously an ordinal number.

We have learned to pass with such facility from cardinal to ordinal number that the two aspects appear to us as one. To determine the plurality of a collection, i.e. its cardinal number, we do not bother any more to find a model collection with which we can match it – we *count* it. And to the fact that we have learned to identify the two aspects of number is due our progress in mathematics. For whereas in practice we are really interested in the cardinal number, this latter is incapable of creating an arithmetic. The operations of arithmetic are based on the tacit assumption that *we can always pass from any number to its successor*, and this is the essence of the ordinal concept.

And so matching by itself is incapable of creating an art of reckoning. Without our ability to arrange things in ordered succession little progress could have been made. Correspondence and succession, the two principles which permeate all mathematics – nay, all realms of exact thought – are woven into the very fabric of our number-system. [T. Dantzig (1930)]

TEN FINGERS TO COUNT BY

Humankind slowly acquired all the necessary intellectual equipment thanks to the ten fingers on its hands. It is surely no coincidence if children still learn to count with their fingers – and adults too often have recourse to them to clarify their meaning.

Traces of the anthropomorphic origin of counting systems can be found in many languages. In the Ali language (Central Africa), for example, "five" and "ten" are respectively *moro* and *mbouna*: *moro* is actually the word for "hand" and *mbouna* is a contraction of *moro* ("five") and *bouna*, meaning "two" (thus "ten" = "two hands").

CARDINAL ASPECT   ORDINAL ASPECT

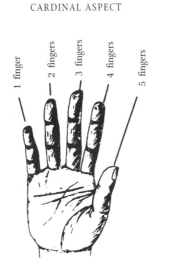 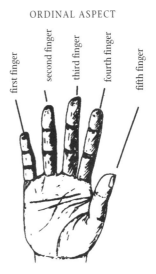

Fig. 1.41.

It is therefore very probable that the Indo-European, Semitic and Mongolian words for the first ten numbers derive from expressions related to finger-counting. But this is an unverifiable hypothesis, since the original meanings of the names of the numbers have been lost.

In any case, the human hand is an extremely serviceable tool and constitutes a kind of "natural instrument" well suited for acquiring the first ten numbers and for elementary arithmetic.

Because there are ten fingers and because each can be moved independently of the others, the hand provides the simplest "model collection" that people have always had – so to speak – to hand.

The asymmetric disposition of the fingers puts the hand in harmony with the normal limitation of the human ability to recognise number visually (a limit set at four). As the thumb is set at some distance from the index finger it is easy to treat it as being "in opposition" to the elementary set of four, and makes the first five numbers an entirely natural sequence. Five thus imposes itself as a basic unit of counting, alongside the other natural grouping, ten. And because each of the fingers is actually different from the others, the human hand can be seen as a true succession of abstract units, obtained by the progressive adjunction of one to the preceding units.

In brief, one can say that the hand makes the two complementary aspects of integers entirely intuitive. It serves as an instrument permitting natural movement between cardinal and ordinal numbering. If you need to show that a set contains three, four, seven or ten elements, you raise or bend *simultaneously* three, four, seven or ten fingers, using your hand as cardinal mapping. If you want to count out the same things, then you bend or raise three, four, seven or ten fingers *in succession*, using the hand as an ordinal counting tool (Fig. 1.41).

The human hand can thus be seen as the simplest and most natural counting machine. And that is why it has played such a significant role in the evolution of our numbering system.

CHAPTER 2

BASE NUMBERS
AND THE BIRTH OF NUMBER-SYSTEMS

NUMBERS AND THEIR SYMBOLS

Once they had grasped abstract numbers and learned the subtle distinction between cardinal and ordinal aspects, our ancestors came to have a different attitude towards traditional "numbering tools" such as pebbles, shells, sticks, strings of beads, or points of the body. Gradually these simple mapping devices became genuine numerical symbols, which are much better suited to the tasks of assimilating, remembering, distinguishing and combining numbers.

Another great step forward was the creation of names for the numbers. This allowed for much greater precision in speech and opened the path towards real familiarity with the universe of abstract numbers.

Prior to the emergence of number-names, all that could be referred to in speech were the "concrete maps" which had no obvious connection amongst themselves. Numbers were referred to by intuitive terms, often directly appealing to the natural environment. For instance, 1 might have been "sun", "moon", or "penis"; for 2, you might have found "eyes", "breasts", or "wings of a bird"; "clover" or "crowd" for 3; "legs of a beast" for 4; and so on. Subsequently some kind of structure emerged from body-counting. At the start, perhaps, you had something like this: "the one to start with" for 1; "raised with the preceding finger" for 2; "the finger in the middle" for 3; "all fingers bar one" for 4; "hand" for 5; and so on. Then a kind of anatomical mapping occurred, so that "little finger" = 1, "ring finger" = 2, "middle finger" = 3, "index finger" = 4, "thumb" = 5, and so on. However, the need to distinguish between the number-symbol and the name of the object or image being used to symbolise the number led people to make an ever greater distinction between the two names, so that eventually the connection between them was entirely lost. As people progressively learned to rely more and more on language, the sounds superseded the images for which they stood, and the originally concrete models took on the abstract form of number-words. The idea of a natural sequence of numbers thus became ever clearer; and the very varied set of initial counting maps or model collections turned into a real *system of number-names*. Habit and memory gave a concrete form to these abstract ideas, and, as T. Dantzig says (p. 8), that is how "mere words became measures of plurality".

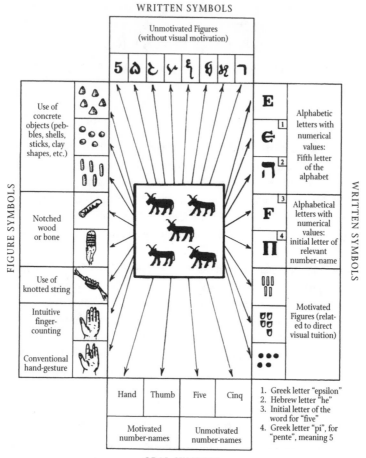

FIG. 2.1.

Of course, concrete symbols and spoken expressions were not the only devices that humanity possessed for mastering numbers. There was also writing, even if that did arise much later on. Writing involves *figures*, that is to say, graphic signs, of whatever kind (carved, drawn, painted, or scored on clay or stone; iconic signs, letters of the alphabet, conventional signs, and so on). We should note that figures are not numbers. "Unit", "pair of", "triad" are "numbers", whilst 1, 2, 3 are "figures", that is to say, conventional graphic signs that represent number-concepts. A figure is just one of the "dresses" that a number can have: you can change the way a number is written without changing the number-concept at all.

These were very important developments, for they allowed "operations" on things to be replaced by the corresponding operation on number-symbols. For numbers do not come from things, but from the laws of the human mind as it works on things. Even if numbers seem latent in the natural world, they certainly did not spring forth from it by themselves.

THE DISCOVERY OF BASE NUMBERS

There were two fundamental principles available for constructing number-symbols: one that we might call a *cardinal* system, in which you adopt a standard sign for the unit and repeat it as many times as there are units in the number; and another that we could call an *ordinal* system, in which each number has its own distinctive symbol.

In virtue of the first principle, the numbers 2 to 4 can be represented by repeating the name of the number 1 two, three or four times, or by laying out in a line, or on top of each other, the appropriate number of "unit signs" in pebbles, fingers, notches, lines, or dots (see Fig. 2.2).

The second principle gives rise to representations for the first four numbers (in words, objects, gestures or signs) that are each different from the others (see Fig. 2.3).

Either of these principles is an adequate basis for acquiring a grasp of ever larger sets – but the application of both principles quickly runs into difficulty. To represent larger numbers, you can't simply use more and more pebbles, sticks, notches, or knotted string; and the number of fingers and other counting points on the body is not infinitely extensible. Nor is it practicable to repeat the same word any number of times, or to create unique symbols for any number of numbers. (Just think how many different symbols you would need to say how many cents there are in a ten-dollar bill!)

To make any progress, people had first to solve a really tricky problem: What in practice is the smallest set of symbols in which the largest numbers can in theory be represented? The solution found is a remarkable example of human ingenuity.

The solution is to give one particular set (for example, the set of ten, the set of twelve, the set of twenty or the set of sixty) a special role and to classify the regular sequence of numbers in a hierarchical relationship to the chosen ("base") set. In other words, you agree to set up a *ladder* and to organise the numbers and their symbols on ascending *steps* of the ladder. On the first step you call them "first-order units", on the second step, "units of the second order", on the third step, "units of the third order", and so on. And that is all there was to the invention of a number-system that saves vast amounts of effort in terms of memorisation and writing-out. The system is called "the rule of position" (or "place-value system"), and its discovery marked the birth of numbering systems where the "base" is the number of units in the set that constitutes the unit of the

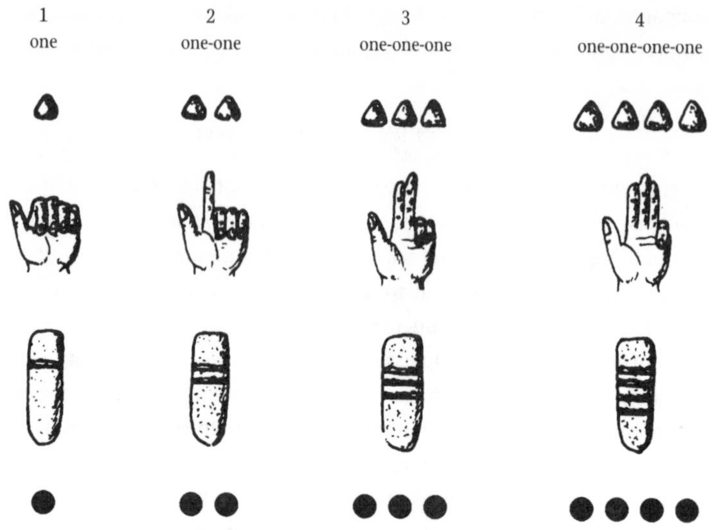

Base Numbers

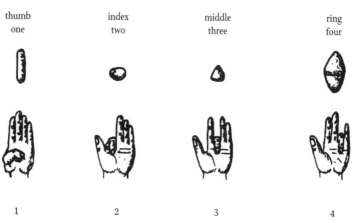

1	2	3	4
one	one-one	one-one-one	one-one-one-one

FIG. 2.2. *"Cardinal" representations of the first four numbers*

thumb	index	middle	ring
one	two	three	four

1	2	3	4

FIG. 2.3. *"Ordinal" representations of the first four numbers*

next order. The place-value system can be applied to material "relays", to words in a language, or to graphic marks – producing respectively concrete, oral, and written numbering systems.

WHY BASE 10?

Not so long ago shepherds in certain parts of West Africa had a very practical way of checking the number of sheep in their flocks. They would make the animals pass by one by one. As the first one went through the gate, the shepherd threaded a shell onto a white strap; as the second went through, he threaded another shell, and so on up to the ninth. When the tenth went through, he took the shells off the white strap and put one on a blue strap, which served for counting in tens. Then he began again threading shells onto the white strap until the twentieth sheep went through, when he put a second shell on the blue strap. When there were ten shells on the blue strap, meaning that one hundred sheep had now been counted, he undid the blue strap and threaded a shell onto a red strap, which was the "hundreds" counting device. And so he continued until the whole flock had been counted. If there were for example two hundred and fifty-eight head in the flock, the shepherd would have eight shells on the white strap, five on the blue strap and two on the red strap. There's nothing "primitive" about this method, which is in effect the one that we use now, though with different symbols for the numbers and orders of magnitude.

The basic idea of the system is the primacy of grouping (and of the rhythm of the symbols in their regular sequence) in "packets" of tens, hundreds (tens of tens), thousands (tens of tens of tens), and so on. In the shepherd's concrete technique, each shell on the white strap counts as a simple unit, each shell on the blue strap counts for ten, and each shell on the red strap counts for a hundred. This is what is called *the principle of base 10*. The shepherd's device is an example of a concrete decimal number-system.

Obviously, instead of using threaded shells and leather straps, we could apply the same system to words or to graphic signs, producing oral or written decimal numeration. Our current number system is just such, using the following graphic signs, often referred to as Arabic numerals:

$$1\ 2\ 3\ 4\ 5\ 6\ 7\ 8\ 9\ 0$$

The first nine symbols represent the simple units, or units of the first decimal order (or "first magnitude"). They are subject to the rule of position, or place-value, since their value depends on the place or position that they occupy in a written numerical expression (a 3, for instance, counts for three units, three tens, or three hundreds depending on its position in a three-digit numerical expression). The tenth symbol above represents what we call "zero", and it serves to indicate the absence of any unit of a particular decimal order, or order of magnitude. It also has the meaning of "nought" – for example, the number you obtain when you subtract a number from itself.

The base of ten, which is the first number that can be represented by two figures, is written as 10, a notation which means "one ten and no units".

The numbers from 11 to 99 are represented by combinations of two of the figures according to the rule of position:

11 "one ten, one unit"
12 "one ten, two units"

20   "two tens, no units"
21   "two tens, one unit"
30   "three tens, no units"
40   "four tens, no units"
50   "five tens, no units"

The *hundred*, equal to the square of the base, is written: 100, meaning "one hundred, no tens, no units", and is the smallest number that can be written with three figures.

Numbers from 101 to 999 are represented by combinations of three of the basic figures:

101   "one hundred, no tens, one unit"
358   "three hundreds, five tens, eight units"

There then comes the *thousand*, equal to the cube of the base, which is written : 1,000 ("one thousand, no hundreds, no tens, no units"), and is the smallest number that can be written with four figures. The following step on the ladder is the *ten thousand*, the base to the power of four, which is written 10,000 ("one ten thousand, no thousands, no hundreds, no tens, no units") and is the smallest number that can be written with five figures; and so on.

In oral (spoken) numeration constructed in the same way things proceed in very similar general manner, but with one difference that is inherent to the nature of language: all the numbers less than or equal to ten and also the several powers of ten (100, 1,000, 10,000, etc.) have individual names entirely unrelated to each other, whereas all other numbers are expressed by words made up of combinations of the various number-names.

In English, if we restrict ourselves for a moment to cardinal numbers, the system would proceed in theory as follows. For the first ten numbers we have individual names:

one	*two*	*three*	*four*	*five*	*six*	*seven*	*eight*	*nine*	*ten*
1	2	3	4	5	6	7	8	9	10

The first nine are "units of the first decimal order of magnitude" and the tenth constitutes the "base" of the system (and by definition is therefore the sign for the "unit of the second decimal order of magnitude"). To name the numbers from 11 to 19, the units are grouped in "packets" of ten and we proceed (in theory) by simple addition:

11   *one-ten*   (= 1 + 10)
12   *two-ten*   (= 2 + 10)
13   *three-ten*   (= 3 + 10)
14   *four-ten*   (= 4 + 10)

15	*five-ten*	(= 5 + 10)
16	*six-ten*	(= 6 + 10)
17	*seven-ten*	(= 7 + 10)
18	*eight-ten*	(= 8 + 10)
19	*nine-ten*	(= 9 + 10)

Multiples of the base, from 20 to 90, are the "tens", or units of the second decimal order, and they are expressed by multiplication:

20	*two-tens*	(= 2 × 10)
30	*three-tens*	(= 3 × 10)
40	*four-tens*	(= 4 × 10)
50	*five-tens*	(= 5 × 10)
60	*six-tens*	(= 6 × 10)
70	*seven-tens*	(= 7 × 10)
80	*eight-tens*	(= 8 × 10)
90	*nine-tens*	(= 9 × 10)

If the number of tens is itself equal to or higher than ten, then the tens are also grouped in packets of ten, constituting the "units of the third decimal order", as follows:

100	hundred	(= 10^2)
200	two hundreds	(= 2 × 100)
300	three hundreds	(= 3 × 100)
400	four hundreds	(= 4 × 100)
...	...	...

The hundreds are themselves then grouped into packets of ten, constituting "units of the fourth decimal order", or thousands:

1,000	one thousand	(= 10^3)
2,000	two thousands	(2 × 1,000)
3,000	three thousands	(3 × 1,000)
...	...	...

Then come the ten thousands, which used to be called *myriads,* corresponding to the "units of the fifth decimal order":

10,000	a myriad	(= 10^4)
20,000	two myriads	(= 2 × 10,000)
30,000	three myriads	(= 3 × 10,000)
...	...	...

Using only these words of the language, the names of all the other numbers are obtained by creating expressions that rely simultaneously on multiplication and addition in strict descending order of the powers of the base 10:

53,781

five-myriads	three-thousands	seven-hundreds	eight-tens	one
$(= 5 \times 10,000$	$+ 3 \times 1,000$	$+ 7 \times 100$	$+ 8 \times 10$	$+ 1)$

Such then are the general rules for the formation of the names of the cardinal numbers in the "base 10" system of the English language.

It must have taken a very long time for people to develop such an effective way of naming numbers, as it obviously presupposes great powers of abstraction. However, what we have laid out is evidently a purely theoretical naming system, which no language follows with absolute strictness and regularity. Particular oral traditions and the rules of individual languages produce a wide variety of irregularities: here are some characteristic examples from around the world.

Numbers in Tibetan

[For sources and further details, see M. Lalou (1950), S. C. Das (1915); H. Bruce Hannah (1912). Information kindly supplied by Florence and Hélène Béquignon]

Tibetan has an individual name for each of the first ten numbers:

gcig	gnyis	gsum	bzhi	lnga	drug	bdun	brgyad	dgu	bcu
1	2	3	4	5	6	7	8	9	10

For numbers from 11 to 19, Tibetan uses addition:

11	bcu-gcig	$(= 10 + 1)$
12	bcu-gnyis	$(= 10 + 2)$
13	bcu-gsum	$(= 10 + 3)$
14	bcu-bzhi	$(= 10 + 4)$
15	bcu-lnga	$(= 10 + 5)$
16	bcu-drug	$(= 10 + 6)$
17	bcu-bdun	$(= 10 + 7)$
18	bcu-brgyad	$(= 10 + 8)$
19	bcu-dgu	$(= 10 + 9)$

And for the tens, multiplication is applied:

20	gnyis-bcu	"two-tens"	$(= 2 \times 10)$
30	gsum-bcu	"three-tens"	$(= 3 \times 10)$
40	bzhi-bcu	"four-tens"	$(= 4 \times 10)$
50	lnga-bcu	"five-tens"	$(= 5 \times 10)$
60	drug-bcu	"six-tens"	$(= 6 \times 10)$
70	bdun-bcu	"seven-tens"	$(= 7 \times 10)$
80	brgyad-bcu	"eight-tens"	$(= 8 \times 10)$
90	dgu-bcu	"nine-tens"	$(= 9 \times 10)$

For a hundred (=10^2) there is the word *brgya*, and the corresponding multiples are obtained by the same principle of multiplication:

200	*gnyis-brgya*	"two-hundreds"	(= 2 × 100)
300	*gsum-brgya*	"three-hundreds"	(= 3 × 100)
400	*bzhi-brgya*	"four-hundreds"	(= 4 × 100)
500	*lnga-brgya*	"five-hundreds"	(= 5 × 100)
600	*drug-brgya*	"six-hundreds"	(= 6 × 100)
700	*bdun-brgya*	"seven-hundreds"	(= 7 × 100)
800	*brgyad-brgya*	"eight-hundreds"	(= 8 × 100)
900	*dgu-brgya*	"nine-hundreds"	(= 9 × 100)

There are similarly individual words for "thousand", "ten thousand" and so on, producing a very simple naming system for all intermediate numbers:

21:	*gnyis-bcu rtsa gcig*	"two-tens and one"
		(= 2 × 10 + 1)
560:	*lnga-brgya rtsa drug-bcu*	"five-hundreds and six-tens"
		(= 5 × 100 + 6 × 10)

Numbers in Mongolian

[Source: L. Hambis (1945)]

Numbering in Mongolian is similarly decimal, but with some variations on the regular system we have seen in Tibetan. It has the following names for the first ten numbers:

nigän	*qoyar*	*γurban*	*dörbän*	*tabun*	*ǰirγu'an*	*dolo'an*	*naiman*	*yisün*	*arban*
1	2	3	4	5	6	7	8	9	10

and proceeds in a perfectly normal way for the numbers from eleven to nineteen:

11	*arban nigän*	("ten-one")
12	*arban qoyar*	("ten-two")
. . .	. . .	. . .

However, the tens are formed rather differently. Instead of using analytic combinations of the "two-tens", "three-tens" type, Mongolian has specific words formed from the names of the corresponding units, subjected to a kind of "declension" or alteration of the ending of the word:

20	*qorin*	(from *qoyar* = 2)
30	*γučin*	(from *γurban* = 3)
40	*döčin*	(from *dörbän* = 4)
50	*tabin*	(from *tabun* = 5)

60	*jirin*	(from *jirɣu'an* = 6)
70	*dalan*	(from *dolo'an* = 7)
80	*nayan*	(from *naiman* = 8)
90	*jarin*	(from *yisün* = 9)

From one hundred, however, numbers are formed in a regular way based on multiplication and addition, as explained above:

100	*ja'un*	("hundred")
200	*qoyar ja'un*	("two-hundreds")
300	*ɣurban ja'un*	("three-hundreds")
400	*dörbän ja'un*	("four-hundreds")
. . .	. . .	. . .
1,000	*mingɣan*	("thousand")
2,000	*qoyar mingɣan*	("two-thousands")
3,000	*ɣurban mingɣan*	("three-thousands")
. . .	. . .	. . .
10,000	*tümän*	("myriad")
20,000	*qoyar tümän*	("two-myriads")
. . .	. . .	. . .

20,541	*qoyar tümän*	*tabun ja'un*	*döčin*	*nigän*
	"two myriads	five-hundreds	forty	one"
	(= 2 × 10,000	+ 5 × 100	+ 40	+ 1)

Ancient Turkish numbers

[Source: A. K. von Gabain (1950)]

This section describes the numerals in spoken Turkish of the eighth century CE as deduced from Turkish inscriptions found in Mongolia. The system has some remarkable features.

The first nine numbers are as follows:

bir	*iki*	*üč*	*tört*	*beš*	*alti*	*yĕti*	*säkiz*	*tokuz*
1	2	3	4	5	6	7	8	9

For the tens, the following set of names are used:

10	*on*
20	*yegirmi*
30	*otuz*
40	*kïrk*
50	*ällig*
60	*altmïš*
70	*yĕtmiš*

80   *säkiz on*
90   *tokuz on*

The tens from 20 to 50 do not seem to have any etymological relation with the corresponding units. However, *altmïš* (= 60) and *yetmiš*(= 70) derive respectively from *altï* (= 6) and *yeti* (= 7) by the addition of the ending *mïš* (or *miš*). The words for 80 and 90 also derive from the names of 8 and 9, but by analytical combination with the word for 10, so that they mean something like "eight tens" and "nine tens".

The word for 50, however, is very probably derived from the ancient method of finger-counting, since *ällig* is clearly related to *äl* (or *älig*), the Turkish word for "hand". (Turkish finger-counting is still done in the following way: using one thumb, you touch in order on the other hand the tip of the little finger, the ring finger, the middle finger and the index finger, which gets you to 4; for 5, you raise the thumb of the "counted" hand; then you bend back the thumb and raise in order the index finger, the middle finger, the ring finger, the little finger, and finally the thumb again, so that for 10 you have all the fingers of the "counted" hand stretched out. This technique represents the trace of an even older system in which the series was extended by raising one finger of the other hand for each ten counted, so that one hand with all fingers stretched out meant 10, and the other hand with all fingers stretched out meant 50.)

The system then gives the special name of *yüz* to the number 100, and proceeds by multiplication for the names of the corresponding multiples of a hundred:

100	*yüz*		
200	*iki yüz*	"two-hundreds"	(= 2 × 100)
300	*üč yüz*	"three-hundreds"	(= 3 × 100)
400	*tört yüz*	"four-hundreds"	(= 4 × 100)
500	*beš yüz*	"five-hundreds"	(= 5 × 100)
600	*altï yüz*	"six-hundreds"	(= 6 × 100)
700	*yeti yüz*	"seven-hundreds"	(= 7 × 100)
800	*säkiz yüz*	"eight-hundreds"	(= 8 × 100)
900	*tokuz yüz*	"nine-hundreds"	(= 9 × 100)

The word for a thousand is *bïng* (which in some Turkic dialects also means "a very large amount"), and the multiples of a thousand are similarly expressed by analytical combinations of the same type:

1,000	*bïng*		
2,000	*iki bïng*	"two-thousands"	(= 2 × 1,000)
3,000	*üč bïng*	"three-thousands"	(= 3 × 1,000)

4,000	*tört bïng*	"four-thousands"	(= 4 × 1,000)
5,000	*beš bïng*	"five-thousands"	(= 5 × 1,000)
6,000	*altï bïng*	"six-thousands"	(= 6 × 1,000)
7,000	*yeti bïng*	"seven-thousands"	(= 7 × 1,000)
8,000	*säkiz bïng*	"eight-thousands"	(= 8 × 1,000)
9,000	*tokuz bïng*	"nine-thousands"	(= 9 × 1,000)

What is unusual about the ancient Turkish system is the way the numbers from 11 to 99 are expressed. In this range, what is given is first the unit, and then, not the multiple of ten already counted, but the multiple not yet reached, by a kind of "prospective account". This gives, for example:

11	*bir yegirmi*	literally: "one, twenty"
12	*iki yegirmi*	literally: "two, twenty"
13	*üč yegirmi*	literally: "three, twenty"
21	*bir otuz*	literally: "one, thirty"
22	*iki otuz*	literally: "two, thirty"
53	*üč altmïš*	literally: "three, sixty"
65	*beš yetmiš*	literally: "five, seventy"
78	*säkiz säkiz on*	literally: "eight, eighty"
99	*tokuz yüz*	literally: "nine, one hundred"

What is involved is neither a multiplicative nor a subtractive principle but something like an ordinal device, as follows:

11	"the first unit before twenty"
12	"the second unit before twenty"
21	"the first unit before thirty"
23	"the third unit before thirty"
53	"the third unit before sixty"
87	"the seventh unit before ninety"
99	"the ninth unit before a hundred"

This way of counting is reminiscent of the way time is expressed in contemporary German, where, for "a quarter past nine" you say *viertel zehn*, meaning "a quarter of ten" (= "the first quarter before ten"), or for "half past eight" you say *halb neun*, meaning "half nine" (= "the first half before nine").

However, around the tenth century CE, under Chinese influence, which was very strong in the eastern Turkish-speaking areas, this rather special way of counting was "rationalised" (by the Uyghurs, first of all, who always have been close to Chinese civilisation). Using the Turkic stem *artuk*, meaning "overtaken by", the following expressions were created:

11	*on artukï bir*	("ten overtaken by one")
23	*yegirmi artukï üč*	("twenty overtaken by three")
53	*ällig artukï üč*	("fifty overtaken by three")
87	*säkiz on artukï yeti*	("eighty overtaken by seven")

Whence come the simplified versions still in use today:

11	*on bir*	(= 10 + 1)
23	*yegirmi üč*	(= 20 + 3)
53	*ällig üč*	(= 50 + 3)
87	*säkiz on yeti*	(= 80 + 7)

Sanskrit numbering

The numbering system of Sanskrit, the classical language of northern India, is of great importance for several related reasons. First of all, the most ancient written texts that we have of an Indo-European language are the Vedas, written in Sanskrit, from around the fifth century BCE, but with traces going as far back as the second millennium BCE. (All modern European languages with the notable exceptions of Finnish, Hungarian, Basque, and Turkish belong to the Indo-European group: see below). Secondly, Sanskrit, as the sacred language of Brahmanism (Hinduism), was used throughout India and Southeast Asia as a language of literary and scholarly expression, and (rather like Latin in mediaeval Europe) provided a means of communication between scholars belonging to communities and lands speaking widely different languages. The numbering system of Sanskrit, as a part of a written language of great sophistication and precision, played a fundamental role in the development of the sciences in India, and notably in the evolution of a place-value system.

The first ten numbers in Sanskrit are as follows:

1 *eka*
2 *dvau, dva, dve, dvi*
3 *trayas, tisras, tri*
4 *catvaras, catasras, catvari, catur*
5 *pañca*
6 *ṣaṭ*
7 *sapta*
8 *aṣṭau, aṣṭa*
9 *náva*
10 *daśa*

Numbers from 11 to 19 are then formed by juxtaposing the number of units and the number 10:

11	*eka-daśa*	"one-ten"	$(= 1 + 10)$
12	*dva-daśa*	"two-ten"	$(= 2 + 10)$
13	*tri-daśa*	"three-ten"	$(= 3 + 10)$
14	*catvari-daśa*	"four-ten"	$(= 4 + 10)$
15	*pañca-daśa*	"five-ten"	$(= 5 + 10)$
16	*ṣaṭ-daśa*	"six-ten"	$(= 6 + 10)$
17	*sapta-daśa*	"seven-ten"	$(= 7 + 10)$
18	*aṣṭa-daśa*	"eight-ten"	$(= 8 + 10)$
19	*náva-daśa*	"nine-ten"	$(= 9 + 10)$

For the following multiples of 10, Sanskrit has names with particular features:

20	*viṃśati*
30	*triṃśati*
40	*catvāriṃśati*
50	*pañcāśat*
60	*ṣaṣṭi*
70	*sapti*
80	*aśīti*
90	*návati*

Broadly speaking, the names of the tens from 20 upwards are formed from a word derived from the name of the corresponding unit plus a form for the word for 10 in the plural.

One hundred is *śatam* or *śata*, and for multiples of 100 the regular formula is used:

100	*śatam, śata*	
200	*dviśata*	$(= 2 \times 100)$
300	*triśata*	$(= 3 \times 100)$
400	*caturśata*	$(= 4 \times 100)$
500	*pañcaśata*	$(= 5 \times 100)$

For 1,000, the word *sahásram* or *sahásra* is used, in analytical combination with the names of the units, tens and hundreds to form multiples of the thousands, the ten thousands, and hundred thousands:

1,000	*sahásra*	
2,000	*dvisahásra*	$(= 2 \times 1,000)$
3,000	*trisahásra*	$(= 3 \times 1,000)$
. . .	. . .	. . .
10,000	*daśasahásra*	$(= 10 \times 1,000)$
20,000	*viṃsatsahásra*	$(= 20 \times 1,000)$
30,000	*triṃsatsahásra*	$(= 30 \times 1,000)$
. . .	. . .	. . .

100,000	*śatasahásra*	(= 100 × 1,000)
200,000	*dviśatasahásra*	(= 200 × 1,000)
300,000	*triśatasahásra*	(= 300 × 1,000)
. . .	. . .	. . .

This gives the following expressions for intermediate numbers:

4,769:

nava	*ṣaṣti*	*saptaśata*	*ca*	*catursahásra*
("nine	sixty	seven-hundreds	and	four-thousands")
(= 9	+ 60	+ 7 × 100	+	4 × 1,000)

Sanskrit thus has a decimal numbering system, like ours, but with combinations done "in reverse", that is to say starting with the units and then in *ascending* order of the powers of 10.

WHAT IS INDO-EUROPEAN?

"Indo-European" is the name of a huge family of languages spoken nowadays in most of the European land-mass, in much of western Asia, and in the Americas. There has been much speculation about the geographical origins of the peoples who first spoke the language which has split into the many present-day branches of the Indo-European family. Some theories hold that the Indo-Europeans originally came from central Asia (the Pamir mountains, Turkestan); others maintain that they came from the flat lands of northern Germany, between the Elbe and the Vistula, and the Russian steppes, from the Danube to the Ural mountains. The question remains unresolved. All the same, some things are generally agreed. The Indo-European languages derive from dialects of a common "stem" spoken by a wide diversity of tribes who had numerous things in common. The Indo-Europeans were arable farmers, hunters, and breeders of livestock; they were patriarchal and had social ranks or castes of priests, farmers, and warriors; and a religion that involved the cult of ancestors and the worship of the stars. However, we know very little about the origins of these peoples, who acquired writing only in relatively recent times.

The Indo-European tribes began to split into different branches in the second millennium BCE at the latest, and over the following thousand years the following tribes or branches appear in early historical records: *Aryans*, in India, and *Kassites*, *Hittites*, and *Lydians* in Asia Minor; the *Achaeans, Dorians, Minoans,* and *Hellenes*, in Greece; then the *Celts* in central Europe, and the *Italics* in the Italian peninsula. Further migrations from the East occurred towards the end of the Roman Empire in the fourth to sixth centuries CE, bringing the *Germanic* tribes into western Europe.

The Indo-European language family is thus spread over a very wide area and is traditionally classified in the following branches, for each of which the earliest written traces date from different periods, but none from before the second millennium BCE:

- The *Indo-Aryan* branch: Vedic, classical Sanskrit, and their numerous modern descendants, of which there are five main groups:
  - the western group, including Sindhi, Gujurati, Landa, Mahratta, and Rajasthani
  - the central group, including Punjabi, Pahari, and Hindi
  - the eastern group, including Bengali, Bihari, and Oriya
  - the southern group (Singhalese)
  - so-called "Romany" or gypsy languages
- The *Iranian* branch, including ancient Persian (spoken at the time of Darius and Xerxes), Avestan (the language of Zoroaster), Median, Scythian, as well as several mediaeval and modern languages spoken in the area of Iran (Sogdian, Pahlavi, Caspian and Kurdish dialects, Ossetian (spoken in the Caucasus), Afghan, and Baluchi)
- A branch including the *Anatolian* language of the Hittite Empire as well as Lycian and Lydian
- The *Tokharian* branch. This language (with its two dialects, Agnaean and Kutchian) was spoken by an Indo-European population settled in Chinese Turkestan between the fifth and tenth centuries CE, but became extinct in the Middle Ages. As an ancient language related to Hittite as well as to Western branches of the Indo-European family (Greek, Latin, Celtic, Germanic), it is of great importance for historical linguists and is often used in tracing the etymologies of common Indo-European words
- The *Armenian* branch, with two dialects, western (spoken in Turkey) and eastern (spoken in Armenia)
- The *Hellenic* branch, which includes ancient dialects such as Dorian, Achaean, Creto-Minoan, as well as Homeric (classical) Greek, Koiné (the spoken language of ancient Greece), and Modern Greek
- The *Italic* branch, which includes ancient languages such as Oscan, Umbrian and Latin, and all the modern Romance languages (Italian, Spanish, Portuguese, Provençal, Catalan, French, Romanian, Sardinian, Dalmatian, Rhaeto-Romansch, etc.)
- The *Celtic* branch, which has two main groups:
  - "continental" Celtic dialects, including the extinct language of the Gauls
  - "island" Celtic, itself possessing two distinct subgroups, the Brythonic (Breton, Welsh, and Cornish) and the Gaelic (Erse, Manx, and Scots Gaelic)
- The *Germanic* branch, which has three main groups:
  - Eastern Germanic, of which the main representative is Gothic

– the Nordic languages (Old Icelandic, Old Norse, Swedish, Danish)
– Western Germanic languages, including Old High German and its mediaeval and modern descendants (German), Low German, Dutch, Friesian, Old Saxon, Anglo-Saxon, and its mediaeval and modern descendants (Old English, Middle English, contemporary British and American English)

• The *Slavic* branch, of which there are again three main groups:
– Eastern Slavic languages (Russian, Ukrainian, and Belorussian)
– Southern Slavic languages (Slovenian, Serbo-Croatian, Bulgarian)
– Western Slavic languages (Czech, Slovak, Polish, Lekhitian, Sorbian, etc.)

• The *Baltic* branch, comprising Baltic, Latvian, Lithuanian, and Old Prussian

• *Albanian*, a distinct branch of the Indo-European family, with no "close relatives" and two dialects, Gheg and Tosk

• The *Thraco-Phrygian* branch, with traces found in the Balkans (Thracian, Macedonian) and in Asia Minor (Phrygian)

• And finally a few minor dialects with no close relatives, such as Venetian and Illyrian.

INDO-EUROPEAN NUMBER-SYSTEMS

Sanskrit is thus a particular case of a very large "family" of languages (the Indo-European family) all of whose members use decimal numbering systems. The general rule that all these systems have in common is that the numbers from 1 to 9 and each of the powers of 10 have individual names, all other numbers being expressed as analytical combinations of these names.

Nonetheless, some of these languages have additional number-names that seem to have no etymological connection with the basic set of names: for example "eleven" and "twelve" in English, like the German *elf* and *zwölf*, have no obvious connection to the words for "ten" (*zehn*) and "one" (*eins*) or "ten" and "two" (*zwei*) respectively, whereas all the following numbers are formed in regular fashion:

	ENGLISH		GERMAN	
13	thirteen	(= three+ten)	*dreizehn*	(= *drei+zehn*)
14	fourteen	(= four+ten)	*vierzehn*	(= *vier+zehn*)
15	fifteen	(= five+ten)	*fünfzehn*	(= *fünf+zehn*)
16	sixteen	(= six+ten)	*sechszehn*	(= *sechs+zehn*)
17	seventeen	(= seven+ten)	*siebzehn*	(= *sieben+zehn*)
18	eighteen	(= eight+ten)	*achtzehn*	(= *acht+zehn*)
19	nineteen	(= nine+ten)	*neunzehn*	(= *neun+zehn*)

The "additional" number-names in the range 11–19 in the Romance languages, on the other hand, are obvious contractions of the analytical Latin names (with the units in first position) from which they are all derived:

	LATIN		ITALIAN	FRENCH	SPANISH
11	*undecim*	("one-ten")	*undici*	*onze*	*once*
12	*duodecim*	("two-ten")	*dodici*	*douze*	*doce*
13	*tredecim*	("three-ten")	*tredici*	*treize*	*trece*
14	*quattuordecim*	("four-ten")	*quattordici*	*quatorze*	*catorce*
15	*quindecim*	("five-ten")	*quindici*	*quinze*	*quince*
16	*sedecim*	("six-ten")	*sedici*	*seize*	
17	*septendecim*	("seven-ten")			
18	*octodecim*	("eight-ten")			
19	*undeviginti*	("one from twenty")			

The remaining numbers before 20 are constructed analytically: *dix-sept, dix-huit, dix-neuf* (French), *dieci-sette, dieci-otto* (Italian), etc.

In the Germanic languages, the tens are constructed in regular fashion using an ending clearly derived from the word for "ten" on the stem of the word for the corresponding unit : in English, twenty = "two - ty", thirty = "three - ty", and so on, and in German, *zwanzig* = "*zwei - zig*", *dreissig* = *drei + sig*, and so on. In order to avoid confusion between the "teens" and the "tens" in Latin, multiples of 10, which similarly have the unit-name in first position, use the ending "*-ginta*", giving the following contractions in the Romance languages derived from it:

	LATIN	ITALIAN	FRENCH	SPANISH
30	*triginta*	*trenta*	*trente*	*treinta*
40	*quadraginta*	*quaranta*	*quarante*	*cuarenta*
50	*quinquaginta*	*cinquanta*	*cinquante*	*cincuenta*
60	*sexaginta*	*sessanta*	*soixante*	*sessanta*
70	*septuaginta*	*settanta*	*septante**	*setenta*
80	*octoginta*	*ottanta*	*octante**	*ochenta*
90	*nonaginta*	*novanta*	*nonante**	*noventa*

The French numerals marked with an asterisk are the "regular" versions found only in Belgium and French-speaking Switzerland; "standard" French uses irregular expressions for 70 (*soixante-dix*, "sixty-ten"), 80 (*quatre-vingts*, "four-twenty"), and 90 (*quatre-vingt-dix*, "four-twenty-ten"). In addition, of course, we have omitted from the table above the Latin and Romance names for the number 20, which seems to be a problem at first sight. In Latin it is *viginti*, a word with no relation to the words for "ten" (*decem*) or for "two" (*duo*); and its Romance derivatives, with the exception of Romanian, follow the irregularity (*venti* in Italian, *vingt* in French, *veinte* in Spanish). So where does the "Romance twenty" come from?

Roots

The richness of the descendance of the original Indo-European language means that, by comparison and deduction, it is possible to reconstruct the form that many basic words must have had in the "root" or "stem" language, even though no written trace remains of it. Indo-European root words, being hypothetical, are therefore always written with an asterisk. The original number-set is believed to have been this:

1　*oi-no, *oi-ko, *oi-wo
2　*dwō, *dwu, *dwoi
3　*tri (and derivative forms: *treyes, *tisores)
4　*kwetwores, *kwetesres, *kwetwor
5　*pénkwe, *kwenkwe
6　*seks, *sweks
7　*septm
8　*oktṓ, *oktu
9　*néwn̦
10　*dékm̦

This helps us to see that despite their apparent difference, the words for "one" in Sanskrit, Avestan, and Czech, for example (respectively *eka, aeva* and *jeden*) are all derived from the same "root" or prototype, as are the Latin *unus*, German *eins* and Swedish *en*.

All trace has been lost of the concrete meanings that these Indo-European number-names might have had originally. However, Indo-European languages do bear the visible marks of that long-distant time when, in the absence of any number-concept higher than two, the word for all other numbers meant nothing more than "many".

The first piece of evidence of this ancient number-limit is the grammatical distinction made in several Indo-European languages between the *singular*, the *dual*, and the *plural*. In classical Greek, for example, *ho lukos* means "the wolf", *hoi lukoi* means "the wolves", but for "two wolves" a special ending, the mark of the "dual", is used: *tṓ luko*.

Another piece of the puzzle is provided by the various special meanings and uses of words closely associated with the name of the number 3. Anglo-Saxon *thria* (which becomes "three" in modern English) is related to the word *throp*, meaning a pile or heap; and words like *throng* are similarly derived from a common Germanic root having the sense of "many". In the Romance languages there are even more evident connections between the words for "three" and words expressing plurality or intensity: the Latin word *tres* (three) has the same root as the preposition and prefix *trans-* (with meanings related to "up until", "through", "beyond"), and in

French, derivations from this common stem produce words like *très* ("very"), *"trop"* (too much), and even *troupe* ("troop"). It can be deduced from these and many other instances that in the original Indo-European stem language, the name of the number "three" (*tri*) was also the word for plurality, multiplicity, crowds, piles, heaps, and for the beyond, for what was beyond reckoning.

The number-systems of the Indo-European languages, which are all strictly decimal, have remained amazingly stable over many millennia, even whilst most other features of the languages concerned have changed beyond recognition and beyond mutual comprehension. Even the apparent irregularities within the system are for the most part explicable within the logic of the original decimal structure – for example, the problem mentioned above of the "Romance twenty". French *vingt*, Spanish *veinte*, etc. derive from Latin *viginti*, which is itself fairly self-evidently a derivative of the Sanskrit *viṃsati*. And Sanskrit "twenty" is not irregular at all, being a contraction of a strictly decimal *dvi-daśati* ("two-tens") $\Rightarrow$ *visati* $\Rightarrow$ *viṃsati*. Similar derivations can be found in other branches of the Indo-European family of languages. In Avestan, 20 is *visaiti*, formed from *baē*, "two", and *dása* (= 10); and in Tokharian A, where *wu* = 2 and *śäk* = 10, *wi-śäki* (= 2×10) became *wiki*, "twenty".

THE NAMES OF THE NUMBER 1

THE NAMES OF THE NUMBER 2

Indo-European prototypes:	*oi-no, *oi-ko, *oi-wo

Indo-European prototypes:	*dwô, *dwu, *dwoi

SANSKRIT	eka
AVESTAN	aêva
GREEK	hén
EARLY LATIN	oinos, oinom
LATIN	unus, unum
ITALIAN	uno
SPANISH	uno
FRENCH	un
PORTUGUESE	um
ROMANIAN	uno
OLD ERSE	oen
MODERN IRISH	oin
BRETON	eun
SCOTS GAELIC	un
WELSH	un
GOTHIC	ain (-s)
DUTCH	een
OLD ICELANDIC	einn
SWEDISH	en
DANISH	en
OLD SAXON	en
ANGLO-SAXON	an
ENGLISH	one
OLD HIGH GERMAN	ein, eins
GERMAN	ein
CHURCH SLAVONIC	inŭ
RUSSIAN	odin
CZECH	jeden
POLISH	jeden
LITHUANIAN	vienas
BALTIC	vienes

SANSKRIT	dvau, dva, dvi
AVESTAN	baè
HITTITE	tā
TOKHARIAN A	wu, we
ARMENIAN	erku
GREEK	dùô
LATIN	duo, duae
SPANISH	dos
FRENCH	deux
ROMANIAN	doi
OLD ERSE	dáu, dó
MODERN IRISH	da
BRETON	diou
SCOTS GAELIC	dow
WELSH	dwy, dau
GOTHIC	twai, twa
DUTCH	twee
OLD ICELANDIC	tveir
SWEDISH	två
DANISH	to
OLD SAXON	twene
ANGLO-SAXON	twegen
ENGLISH	two
OLD HIGH GERMAN	zwene
GERMAN	zwei
CHURCH SLAVONIC	dŭva, dŭvě
RUSSIAN	dva
POLISH	dwa
LITHUANIAN	dù, dvi
ALBANIAN	dy, dyj

FIG. 2.4A.

FIG. 2.4B.

THE NAMES OF THE NUMBER 3   THE NAMES OF THE NUMBER 4

Indo-European prototypes:	*treyes, *tisores, *tri*
SANSKRIT	*trayas, tisras, tri*
AVESTAN	*thrayŏ, tisrŏ, tri*
HITTITE	*tri*
TOKHARIAN B	*trai*
ARMENIAN	*erekh*
GREEK	*treis*
OSCAN	*trís*
LATIN	*trĕs, tria*
ITALIAN	*tre*
SPANISH	*tres*
FRENCH	*trois*
ROMANIAN	*trei*
OLD ERSE	*téoir, trí*
WELSH	*tri, tair*
GOTHIC	*threis, thrija*
DUTCH	*drie*
OLD ICELANDIC	*thrir*
SWEDISH	*tre*
OLD SAXON	*thria*
ANGLO-SAXON	*thri*
ENGLISH	*three*
OLD HIGH GERMAN	*dri*
GERMAN	*drei*
CHURCH SLAVONIC	*trije, tri*
RUSSIAN	*tri*
POLISH	*trzy*
LITHUANIAN	*trỹs*
ALBANIAN	*tre, tri*

Indo-European prototypes:	*kwetwores, *kwetesres, *kwetwor*
SANSKRIT	*catvaras, catasras, catvari, catur*
AVESTAN	*čathwǎrŏ*
TOKHARIAN A	*śtwar*
TOKHARIAN B	*śtwer*
ARMENIAN	*čorkh*
ANCIENT GREEK	*téttares, téssares, tétores*
OSCAN	*pettiur, petora*
LATIN	*quattuor*
ITALIAN	*quattro*
SPANISH	*cuatro*
FRENCH	*quatre*
ROMANIAN	*patru*
OLD ERSE	*cethir, cethoir*
BRETON	*pevar*
WELSH	*pedwar*
SCOTS GAELIC	*peswar*
GOTHIC	*fidwor*
OLD ICELANDIC	*fjorer*
SWEDISH	*fyra*
OLD SAXON	*fiuwar*
ANGLO-SAXON	*foewer*
ENGLISH	*four*
OLD HIGH GERMAN	*vier*
GERMAN	*vier*
CHURCH SLAVONIC	*četyre*
RUSSIAN	*četyre*
CZECH	*ctyri*
POLISH	*cztery*
LITHUANIAN	*keturi*
BALTIC	*keturi*

FIG. 2.4C.   FIG. 2.4D.

THE NAMES OF THE NUMBER 5

Indo-European prototypes:	*pénkwe, *kwenkwe

SANSKRIT	*pañča*
AVESTAN	*panča*
HITTITE	*panta*
TOKHARIAN A TOKHARIAN B	*päñ* *piš*
ARMENIAN	*hing*
GREEK	*pénte*
LATIN SPANISH FRENCH ROMANIAN	*quinque* *cinco* *cinq* *cinci*
OLD ERSE MODERN IRISH WELSH	*cóic* *coic* *pump*
BRETON	*pemp*
GOTHIC DUTCH OLD ICELANDIC SWEDISH OLD SAXON ANGLO-SAXON ENGLISH OLD HIGH GERMAN GERMAN	*fimf* *viif* *fimm* *fem* *fif* *fif* *five* *finf* *fünf*
CHURCH SLAVONIC RUSSIAN POLISH	*peti̯* *piat'* *piec*
LITHUANIAN	*penki*
ALBANIAN	*pęsë*

FIG. 2.4E.

THE NAMES OF THE NUMBER 6

Indo-European prototypes:	*seks, *sweks

SANSKRIT	*ṣaṭ*
AVESTAN	*xšvaš*
TOKHARIAN A	*ṣäk*
ARMENIAN	*vec*
ANCIENT GREEK MODERN GREEK	*wéks* *héx*
LATIN ITALIAN SPANISH FRENCH ROMANIAN	*sex* *sei* *seis* *six* *shase*
OLD ERSE MODERN IRISH WELSH BRETON	*sé* *se* *chwech* *c'houec'h*
GOTHIC DUTCH OLD ICELANDIC SWEDISH OLD SAXON ANGLO-SAXON ENGLISH OLD HIGH GERMAN GERMAN	*saihs* *zes* *sex* *sex* *sehs* *six* *six* *sehs* *sechs*
CHURCH SLAVONIC RUSSIAN POLISH	*šestĭ* *chest'* *szesc*
LITHUANIAN ALBANIAN	*sesi* *giashtë*

FIG. 2.4F.

THE NAMES OF THE NUMBER 7   THE NAMES OF THE NUMBER 8

Indo-European prototype:	*septṃ

SANSKRIT	sapta
AVESTAN	hapta
HITTITE	sipta
TOKHARIAN A	spät
ARMENIAN	ewhtn
GREEK	heptá
LATIN	septem
SPANISH	siete
FRENCH	sept
ROMANIAN	shapte
OLD ERSE	secht
MODERN IRISH	secht
WELSH	saith
BRETON	seiz
GOTHIC	sibun
DUTCH	zeven
OLD ICELANDIC	siau
SWEDISH	sju
OLD SAXON	sibun
ENGLISH	seven
OLD HIGH GERMAN	siben
GERMAN	sieben
CHURCH SLAVONIC	sedmĭ
RUSSIAN	sem'
POLISH	siedem
LITHUANIAN	septyni

FIG. 2.4G.

Indo-European prototypes:	*októ, *oktu

SANSKRIT	aṣṭ'á, aṣṭau
AVESTAN	asta
TOKHARIAN B	okt
ARMENIAN	uth
GREEK	okto
LATIN	octó
SPANISH	ochó
FRENCH	huit
ROMANIAN	opt
OLD ERSE	ocht
MODERN IRISH	ocht
WELSH	wyth
BRETON	eiz
GOTHIC	ahtau
DUTCH	acht
OLD ICELANDIC	atta
SWEDISH	åtta
OLD SAXON	ahto
ANGLO-SAXON	eahta
ENGLISH	eight
OLD HIGH GERMAN	ahto
GERMAN	acht
CHURCH SLAVONIC	osmi
RUSSIAN	vosem'
POLISH	osiem
LITHUANIAN	aštuoni

FIG. 2.4H.

THE NAMES OF THE NUMBER 9 | THE NAMES OF THE NUMBER 10

Indo-European prototype: *néwn	Indo-European prototype: *dékm̥

SANSKRIT	náva		SANSKRIT	dáśa
AVESTAN	nava		AVESTAN	dasa
TOKHARIAN A	ñu		TOKHARIAN A	śäk
TOKHARIAN B	ñu		TOKHARIAN B	śak
ARMENIAN	inn		ARMENIAN	tasn
GREEK	en-néa		GREEK	déka
LATIN	novem		LATIN	decem
ITALIAN	nove		ITALIAN	dieci
SPANISH	nueve		SPANISH	diez
FRENCH	neuf		FRENCH	dix
ROMANIAN	noue		ROMANIAN	zece
PORTUGUESE	noue		PORTUGUESE	dez
OLD ERSE	nóin		OLD ERSE	deich
MODERN IRISH	nói		MODERN IRISH	deich
WELSH	naw		WELSH	deg
BRETON	nao		BRETON	dek
GOTHIC	nium		GOTHIC	taihun
DUTCH	negon		DUTCH	tien
OLD ICELANDIC	nio		OLD ICELANDIC	tio
SWEDISH	nio		SWEDISH	tio
OLD SAXON	nigun		OLD SAXON	techan
ANGLO-SAXON	nigon		ANGLO-SAXON	tyn
ENGLISH	nine		ENGLISH	ten
OLD HIGH GERMAN	niun		OLD HIGH GERMAN	zehan
GERMAN	neun		GERMAN	zehn
CZECH	devet		CZECH	deset
RUSSIAN	deviat'		RUSSIAN	desiat'
POLISH	dziewiec		POLISH	dziesiec
LITHUANIAN	devyni		LITHUANIAN	desimt
ALBANIAN	nëndë		ALBANIAN	dietë

FIG. 2.4I.   FIG. 2.4J.

	LATIN	ITALIAN	FRENCH	SPANISH	ROMANIAN
1	unus	uno	un	uno	uno
2	duo	due	deux	dos	doi
3	tres	tre	trois	tres	trei
4	quattuor	quattro	quatre	cuatro	patru
5	quinque	cinque	cinq	cinco	cinci
6	sex	sei	six	seis	shase
7	septem	sette	sept	siete	shapte
8	octo	otto	huit	ocho	opt
9	novem	nove	neuf	nueve	noue
10	decem	dieci	dix	diez	zece
11	undecim	undici	onze	once	un spree zece
12	duodecim	dodici	douze	doce	doi spree zece
20	viginti	venti	vingt	veinte	doua-zeci
30	triginta	trenta	trente	treinta	trei-zeci
40	quadraginta	quaranta	quarante	cuarenta	patru-zeci
50	quinquaginta	cinquanta	cinquante	cincuenta	cinci-zeci
60	sexaginta	sessanta	soixante	sesenta	shase-zeci
70	septuaginta	settanta	soixante-dix	setenta	shapte-zeci
80	octoginta	ottanta	quatre-vingts	ochenta	opt-zeci
90	nonaginta	novanta	quatre-vingt-dix	noventa	noua-zeci
100	centum	cento	cent	ciento	o suta
1,000	mille	mille	mille	mil	o mie

	GOTHIC	OLD HIGH GERMAN	GERMAN	ANGLO-SAXON	ENGLISH
1	ains	ein	eins	an	one
2	twa	zwene	zwei	twegen	two
3	preis	dri	drei	pri	three
4	fidwor	vier	vier	feower	four
5	fimf	fünf	fünf	fif	five
6	saíhs	sehs	sechs	six	six
7	sibun	siben	sieben	seofou	seven
8	ahtáu	ahte	acht	eahta	eight
9	niun	niun	neun	nigon	nine
10	taíhun	zehan	zehn	tyn	ten
11	ain-lif	einlif	elf	endleofan	eleven
12	twa-lif	zwelif	zwölf	twelf	twelve
20	twai-tigjus	zwein-zug	zwanzig	twentig	twenty
30	threo-tigjus	driz-zug	dreißig	thritig	thirty
40	fidwor-tigjus	fior-zug	vierzig	feowertig	forty
50	fimf-tigjus	finf-zug	fünfzig	fiftig	fifty
60	saíhs-tigjus	sehs-zug	sechzig	sixtig	sixty
70	sibunt-ehund	sibun-zo	siebzig	hund-seofontig	seventy
80	ahtaút-ehund	ahto-zo	achtzig	hund-eahtatig	eighty
90	niunt-ehund	niun-zo	neunzig	hund-nigontig	ninety
100	taíhun-taíhund	zehan-zo	hundert	hund-teontig	hundred
1,000	thusundi	dusent	tausend	thusund	thousand

FIG. 2.5. *The decimal nature of Indo-European number-names*

OTHER SOLUTIONS TO THE PROBLEM
OF THE BASE

Not all civilisations came up with the same solution to the problem of the base. In other words, base 10 is not the only way of constructing a number-system.

There are many examples of numeration built on a base of 5. For example: Api, a language spoken in the New Hebrides (Oceania), gives individual names to the first five numbers only:

1	*tai*
2	*lua*
3	*tolu*
4	*vari*
5	*luna* (literally, "the hand")

and then uses compounds for the numbers from 6 to 10:

6	*otai*	(literally, "the new one")
7	*olua*	(literally, "the new two")
8	*otolu*	(literally, "the new three")
9	*ovari*	(literally, "the new four")
10	*lualuna*	(literally, "two hands")

The name of 10 then functions as a new base unit:

11	*lualuna tai*	$(= 2 \times 5 + 1)$
12	*lualuna lua*	$(= 2 \times 5 + 2)$
13	*lualuna tolu*	$(= 2 \times 5 + 3)$
14	*lualuna vari*	$(= 2 \times 5 + 4)$
15	*toluluna*	$(= 3 \times 5)$
16	*toluluna tai*	$(= 3 \times 5 + 1)$
17	*toluluna lua*	$(= 3 \times 5 + 2)$

and so on. [Source: T. Dantzig (1930), p. 18]

Languages that use base 5 or have traces of it in their number-systems include Carib and Arawak (N. America); Guarani (S. America); Api and Houailou (Oceania); Fulah, Wolof, Serere (Africa), as well as some other African languages: Dan (in the Mande group), Bete (in the Kroo group), and Kulango (one of the Voltaic languages); and in Asia, Khmer. [See: M. Malherbe (1995); F. A. Pott (1847)].

Other civilisations preferred base 20 – the "vigesimal base" – by which things are counted in packets or groups of twenty. Amongst them we find the Tamanas of the Orinoco (Venezuela), the Eskimos or Inuits (Greenland), the Ainus in Japan and the Zapotecs and Maya of Mexico.

The Mayan calendar consisted of "months" of 20 days, and laid out cycles of 20 years, 400 years (= 20^2) 8,000 years (= 20^3), 160,000 years (= 20^4), 3,200,000 years (= 20^5), and even 64,000,000 years (= 20^6).

Like all the civilisations of pre-Columbian Central America, the Aztecs and Mixtecs measured time and counted things in the same way, as shown in numerous documents seized by the conquistadors. The goods collected by Aztec administrators from subjugated tribes were all quantified in vigesimal terms, as Jacques Soustelle explains:

> For instance, *Toluca* was supposed to provide twice a year 400 loads of cotton cloth, 400 loads of decorated *ixtle* cloaks, 1,200 (3×20^2) loads of white *ixtle* cloth ... *Quahuacan* gave four yearly tributes of 3,600 (9×20^2) beams and planks, two yearly tributes of 800 (2×20^2) loads of cotton cloth and the same number of loads of *ixtle* cloth ... *Quauhnahuac* supplied the Imperial Exchequer with twice-yearly deliveries of 3,200 (8×20^2) loads of cotton cloaks, 400 loads of loin-cloths, 400 loads of women's clothing, 2,000 (5×20^2) ceramic vases, 8,000 (20^3) sheaves of "paper" ...
>
> [From the *Codex Mendoza*]

This is how the Aztec language gives form to a quinary-vigesimal base:

1	*ce*		11	*matlactli-on-ce* (10 + 1)
2	*ome*		12	*matlactli-on-ome* (10 + 2)
3	*yey*		13	*matlactli-on-yey* (10 + 3)
4	*naui*		14	*matlactli-on-naui* (10 + 4)
5	*chica* or *macuilli*		15	*caxtulli*
6	*chica-ce* (5 + 1)		16	*caxtulli-on-ce* (15 + 1)
7	*chica-ome* (5 + 2)		17	*caxtulli-on-ome* (15 + 2)
8	*chica-ey* (5 + 3)		18	*caxtulli-on-yey* (15 + 3)
9	*chica-naui* (5 + 4)		19	*caxtulli-on-naui* (15 + 4)
10	*matlactli*		20	*cem-poualli* (1×20, "a score")

30	*cem-poualli-on-matlactli*	(20 + 10)
40	*ome-poualli*	(2×20)
50	*ome-poualli-on-matlactli*	($2 \times 20 + 10$)
100	*macuil-poualli*	(5×20)
200	*matlactli-poualli*	(10×20)
300	*caxtulli-poualli*	(15×20)
400	*cen-tzuntli*	(1×400, "one four-hundreder")
800	*ome-tzuntli*	(2×400)
1,200	*yey-tzuntli*	(3×400)
8,000	*cen-xiquipilli*	($1 \times 8,000$, "one eight-thousander")

Fig. 2.6.

There are many populations outside of America and Europe (for instance, the Malinke of Upper Senegal and Guinea, the Banda of Central Africa, the Yebu and Yoruba people of Upper Senegal and Nigeria, etc.) who continue to count in this fashion. Yebu numeration is as follows, according to C. Zaslavsky (1973):

1	*otu*	
2	*abuo*	
3	*ato*	
4	*ano*	
5	*iso*	
6	*isii*	
7	*asaa*	
8	*asato*	
9	*toolu*	
10	*iri*	
20	*ohu*	
30	*ohu na iri*	$(= 20 + 10)$
40	*ohu abuo*	$(= 20 \times 2)$
50	*ohu abuo na iri*	$(= 20 \times 2 + 10)$
60	*ohu ato*	$(= 20 \times 3)$
...	...	...
100	*ohu iso*	$(= 20 \times 5)$
200	*ohu iri*	$(= 20 \times 10)$
...	...	...
400	*nnu*	$(= 20^2)$
8,000	*nnu khuru ohu*	$(= 20^3 = $ "400 meets 20")
160,000	*nnu khuru nnu*	$(= 20^4 = $ "400 meets 400")

The Yoruba, however, proceed in a quite special way, using additive and subtractive methods alternately [Zaslavsky (1973)]:

1	*ookan*	
2	*eeji*	
3	*eeta*	
4	*eerin*	
5	*aarun*	
6	*eeta*	
7	*eeje*	
8	*eejo*	
9	*eesan*	
10	*eewaa*	
11	*ookan laa*	$(= 1 + 10$: *laa* from *le ewa*, "added to 10")
12	*eeji laa*	$(= 2 + 10)$
13	*eeta laa*	$(= 3 + 10)$
14	*eerin laa*	$(= 4 + 10)$

15	*eedogun*	(= 20 − 5; from *aarun din ogun*, "5 taken from 20")
16	*erin din logun*	(= 20 − 4)
17	*eeta din logun*	(= 20 − 3)
18	*eeji din logun*	(= 20 − 2)
19	*ookan din logun*	(= 20 − 1)
20	*ogun*	
21	*ookan le loogun*	(= 20 + 1)
25	*eedoogbon*	(= 30 − 5)
30	*ogbon*	
35	*aarun din logoji*	(= (20 × 2) − 5)
40	*logoji*	(= 20 × 2)
50	*aadota*	(= (20 × 3) −10)
60	*ogota*	(= 20 × 3)
. . .	. . .	. . .
100	*ogorun*	(= 20 × 5)
. . .	. . .	. . .
400	*irinwo*	
2,000	*egbewa*	(= (20 × 10) × 10)
4,000	*egbaaji*	(= 2,000 × 2)
20,000	*egbaawaa*	(= 2,000 × 10)
40,000	*egbaawaa lonan meji*	(= (2,000 × 10) × 2)
1,000,000	*egbeegberun*	(literally: "1,000 × 1,000")

The source of this bizarre vigesimal system lies in the Yorubas' traditional use of cowrie shells as money: the shells are always gathered in "packets" of 5, 20, 200 and so on.

According to Mann (JAI, 16), Yoruba number-names have two meanings – the number itself, and also the things that the Yoruba count most of all, namely cowries. "Other objects are always reckoned against an equivalent number of cowries . . . " he explains. In other words, Yoruba numbering retains within it the ancient tradition of purely cardinal numeration based on matching sets.

Various other languages around the world retain obvious traces of a 20-based (vigesimal) number-system. For example, Khmer (spoken in Cambodia) has some combinations based on an obsolete word for 20, and, according to F. A. Pott (1847), used to have a special word (*slik*) for 400 (= 20 × 20). Such features are of course also to be found in European languages, and nowhere more clearly than in the English word *score*. "Four score and seven years ago . . . " is the famous opening sentence of Abraham Lincoln's Gettysburg Address. Since *to score* also means to scratch, mark, or incise (wood, stone or paper), we can see the very ancient origin of its use for the number 20: a *score* was originally a counting stick "scored" with twenty notches.

French also has many traces of vigesimal counting. The number 80 is "four-twenties" (*quatre-vingts*) in modern French, and until the seventeenth century other multiples of twenty were in regular use. *Six-vingts* (6 x 20 = 120) can be found in Molière's *Le Bourgeois Gentilhomme* (Act III, scene iv); the seventeenth-century corps of the sergeants of the city of Paris, who numbered 220 in all, was known as the *Corps des Onze-Vingts* (11 x 20), and the hospice, originally built by Louis IX to house 300 blind veteran soldiers, was and still is called the *Hôpital des Quinze-Vingts* (15 x 20 = 300).

Danish also has a curious vigesimal feature. The numbers 60 and 80 are expressed as "three times twenty" (*tresindstyve*) and "four times twenty" (*firsindstyve*); 50, 70 and 90, moreover, are *halvtresindstyve, halvfirsindstyve*, and *halvfemsindstyve*, literally "half three times twenty", "half four times twenty", and "half five times twenty", respectively. The prefix "half" means that only half of the last of the multiples of 20 should be counted. This accords with the kind of "prospective account" that we observed in ancient Turkish numeration (see above, p. 50):

$$50 = 3 \times 20 \ \textit{minus} \ \text{half of the third twenty} = 3 \times 20 - 10$$
$$70 = 4 \times 20 \ \textit{minus} \ \text{half of the third twenty} = 4 \times 20 - 10$$
$$90 = 5 \times 20 \ \textit{minus} \ \text{half of the third twenty} = 5 \times 20 - 10$$

Even clearer evidence of vigesimal reckoning is found in Celtic languages (Breton, Welsh, Irish). In modern Irish, for example, despite the fact that 100 and 1,000 have their own names by virtue of the decimality that is common to all Indo-European languages, the tens from 20 to 50 are expressed as follows:

20	*fiche*	("twenty")
30	*deich ar fiche*	("ten and twenty")
40	*da fiche*	("two-twenty")
50	*deich ar da fiche*	("ten and two-twenty")

We can only presume that the Indo-European peoples who settled long ago in regions stretching from Scandinavia to the north of Spain, including the British Isles and parts of what is now France, found earlier inhabitants whose number-system used base 20, which they adopted for the common-est numbers up to 99, integrating these particular vigesimal expressions into their own Indo-European decimal system. Since all trace of the languages of the pre-Indo-European inhabitants of Western Europe has disappeared, this explanation, though plausible, is only speculation, but it is supported, if not confirmed, by the use of base 20 in the numbering system of the Basques, one of the few non-Indo-European languages spoken in Western Europe and whose presence is not accounted for by any recorded invasion or conquest.

	IRISH		WELSH		BRETON	
1	*oin*		*un*		*eun*	
2	*da*		*dau*		*diou*	
3	*tri*		*tri*		*tri*	
4	*cethir*		*pedwar*		*pevar*	
5	*coic*		*pump*		*pemp*	
6	*se*		*chwe*		*chouech*	
7	*secht*		*saith*		*seiz*	
8	*ocht*		*wyth*		*eiz*	
9	*nói*		*naw*		*nao*	
10	*deich*		*dec, deg*		*dek*	
11	*oin deec*	$1 + 10$	*un ar dec*	$1 + 10$	*unnek*	$1 + 10$
12	*da deec*	$2 + 10$	*dou ar dec*	$2 + 10$	*daou-zek*	$2 + 10$
13	*tri deec*	$3 + 10$	*tri ar dec*	$3 + 10$	*tri-zek*	$3 + 10$
14	*cethir deec*	$4 + 10$	*pedwar ac dec*	$4 + 10$	*pevar-zek*	$4 + 10$
15	*coic deec*	$5 + 10$	*hymthec*	$5 + 10$	*pem-zek*	$5 + 10$
16	*se deec*	$6 + 10$	*un ar hymthec*	$1 + 15$	*choue-zek*	$6 + 10$
17	*secht deec*	$7 + 10$	*dou ar hymthec*	$2 + 15$	*seit-zek*	$7 + 10$
18	*ocht deec*	$8 + 10$	*tri ar hymthec*[1]	$3 + 15$	*eiz-zek*[2]	$8 + 10$
19	*noi deec*	$9 + 10$	*pedwar ar hymthec*	$4 + 15$	*naou-zek*	$9 + 10$
20	*fiche*	20	*ugeint*	20	*ugent*	20
30	*deich ar fiche*	$10 + 20$	*dec ar ugeint*	$10 + 20$	*tregont*	
40	*da fiche*	2×20	*de-ugeint*	2×20	*daou-ugent*	2×20
50	*deich ar dafiche*	$10 + (2 \times 20)$	*dec ar de-ugeint*	$10 + (2 \times 20)$	*hanter-kant*	half-100
60	*tri fiche*	3×20	*tri-ugeint*	3×20	*tri-ugent*	3×20
70	*dech ar tri fiche*	$10 + (3 \times 20)$	*dec ar tri-ugeint*	$10 + (3 \times 20)$	*dek ha tri-ugent*	$10 + (3 \times 20)$
80	*ceithri fiche*	4×20	*pedwar-ugeint*	4×20	*pevar-ugent*	4×20
90	*deich ar ceithri fiche*	$10 + (4 \times 20)$	*dec ar pedwar-ugeint*	$10 + (4 \times 20)$	*dek ha pevar-ugent*	$10 + (4 \times 20)$
100	*cet*		*cant*		*kant*	
1,000	*mile*		*mil*		*mil*	

[1]alternatively, *deu naw* (= 2×9) [2]alternatively, *tri-ouech* (= 3×6)

FIG. 2.7. *Celtic number-names*

Basque numbers are as follows:

1	*bat*		16	*hamasei*	$= 10 + 6$
2	*bi, biga, bida*		17	*hamazazpi*	$= 10 + 7$
3	*hiru, hirur*		18	*hamazortzi*	$= 10 + 8$
4	*lau, laur*		19	*hemeretzi*	$= 10 + 9$
5	*bost, bortz*		20	*hogei*	$= 20$
6	*sei*		30	*hogeitabat*	$= 10 + 20$
7	*zazpi*		40	*berrogei*	$= 2 \times 20$
8	*zortzi*		50	*berrogei-tamar*	$= (2 \times 20) + 10$
9	*bederatzi*		60	*hirurogei*	$= 3 \times 20$
10	*hamar*		70	*hirurogei-tamar*	$= (3 \times 20) + 10$
11	*hamaika*	irregular	80	*laurogei*	$= 4 \times 20$
12	*hamabi*	$= 10 + 2$	90	*laurogei-tamar*	$= (4 \times 20) + 10$
13	*hamahiru*	$= 10 + 3$	100	*ehun*	
14	*hamalau*	$= 10 + 4$	1,000	*mila*	
15	*hamabost*	$= 10 + 5$			

The mystery of Basque remains entire. As can be seen, it is a decimal system for numbers up to 19, then a vigesimal system for numbers from 20 to 99, and it then reverts to a decimal system for larger numbers. It may be that, like the Indo-European examples given above (Danish, French, and Celtic), it was originally a decimal system which was then "contaminated" by contact with populations using base 20; or, on the contrary, Basque may have been originally vigesimal, and subsequently "reformed" by contact with Indo-European decimal systems. The latter seems to be supported by the obviously Indo-European root of the words for 100 (not unlike "hundred") and 1,000 (almost identical to Romance words for "thousand"); but neither hypothesis about the origins of Basque numbering can be proven.

THE COMMONEST BASE IN HISTORY: 10

Base 20, although quite widespread, has never been predominant in the history of numeration. Base 10, on the other hand, has always been by far the commonest means of establishing the rule of position. Here is a (non-exhaustive) alphabetical listing of the languages and peoples who have used or still use a numbering system built on base 10:

AMORITES:	Northwestern Mesopotamia, founders of Babylon c. 1900 BCE, and of the first Babylonian dynasty
ARABS:	before and after the birth of Islam
ARAMAEANS:	Syria and northern Mesopotamia, second half of second millennium BCE

ASSYRIANS:	Mesopotamia, from the start of the second millennium BCE to c. 500 BCE
BAMOUNS :	Cameroon
BAOULE :	Ivory Coast
BERBERS :	Fair-skinned people settled in North Africa since at least Classical times
SHAN :	Indo-China, from second century CE
CHINESE :	from the origins
EGYPTIANS :	from the origins
ELAMITES:	Khuzestan, southwestern Iran, from fourth century BCE
ETRUSCANS :	probably from Asia Minor, settled in Tuscany from the late seventh century BCE
GOURMANCHES :	Upper Volta
GREEKS :	from the Homeric period
HEBREWS :	before and after the Exile
HITTITES:	Anatolia, from second millennium BCE
INCAS :	Peru, Ecuador, Bolivia, twelfth to sixteenth centuries CE
INDIA :	All civilisations of northern and southern India
INDUS CIVILISATION:	River Indus area, c. 2200 BCE
LYCIANS :	Asia Minor, first half of first millennium BCE
MALAYSIANS	
MALAGASY :	Madagascar
MANCHUS	
MINOANS :	Crete, second millennium BCE
MONGOLIANS	
NUBIANS :	Northeast Africa, since Pharaonic times
PERSIANS	
PHOENICIANS	
ROMANS	
TIBETANS	
UGARITIC PEOPLE :	Syria, second millennium BCE
URARTIANS :	Armenia, seventh century BCE

In the world today, base 10 is used by a multitude of languages, including:

Albanian; the Altaic languages (Turkish, Mongolian, Manchu); Armenian; Bamoun (Cameroon); Baoule (Ivory Coast); Batak; Chinese; the Dravidian languages (Tamil, Malayalam, Telugu); the Germanic languages (German, Dutch, Norwegian, Danish, Swedish, Icelandic, English); Gourmanche (Upper Volta); Greek; Indo-Aryan languages (Sindhi, Gujurati, Mahratta, Hindi, Punjabi, Bengali, Oriya, Singhalese); Indonesian; Iranian languages (Persian, Pahlavi, Kurdish, Afghan); Japanese; Javanese; Korean; Malagasy; Malay; Mon-Khmer languages (Cambodian, Kha); Nubian (Sudan); Polynesian languages (Hawaiian, Samoan, Tahitian, Marquesan); the Romance languages (French, Spanish, Italian, Portuguese, Romanian, Catalan, Provençal, Dalmatian); Semitic languages (Hebrew, Arabic, Amharic, Berber); the Slavic languages (Russian, Slovene, Serbo-Croat, Polish, Czech, Slovak); Thai languages (Laotian, Thai, Vietnamese); Tibeto-Burmese languages (Tibetan, Burmese, Himalayan dialects); Uralian (Finno-Ugrian) languages (Finnish, Hungarian).

These lists show, if it needed to be shown, just how successful base 10 has been and ever remains.

ADVANTAGES AND DRAWBACKS OF BASE 10

The ethnic, geographical, and historical spread of base 10 is enormous, and we can say that it has become a virtually universal counting system. Is that because of its inherent practical or mathematical properties? Certainly not!

To be sure, base 10 has a distinct advantage over larger counting units such as 60, 30, or even 20: its magnitude is easily managed by the human mind, since the number of distinct names or symbols that it requires is quite limited, and as a result addition and multiplication tables using base 10 can be learned by rote without too much difficulty. It is far, far harder to learn the sixty distinct symbols of a base 60 system, even if large numbers can then be written with far fewer symbols; and the multiplication tables for even very simple Babylonian arithmetic require considerable feats of memorisation (sixty tables, each with sixty lines.)

At the other extreme, small bases such as 2 and 3 produce very small multiplication and addition tables to learn by heart; but they require very lengthy strings to express even relatively small numbers in speech or writing, a difficulty that base 10 avoids.

Let us look at a concrete alternative system, an English oral numbering system using base 2. Initially such a system would have only two number-names: "one" to express the unit, and "two" (let us call it "twosome") to express the base.

1   2

one   *twosome*

It would then acquire special names for each of the powers of the base: let us say "foursome" for 2^2, "eightsome" for 2^3, "sixteensome" for 2^4, and so on. Analytical combinations would therefore produce a set of number-names something like this:

1	one	10	eightsome twosome
2	twosome	11	eightsome twosome-one
3	twosome-one	12	eightsome foursome
4	foursome	13	eightsome foursome-one
5	foursome one	14	eightsome foursome twosome
6	foursome twosome	15	eightsome foursome twosome-one
7	foursome twosome-one	16	sixteensome
8	eightsome	17	sixteensome-one
9	eightsome one		and so on.

If our written number-system, using the rule of position, were constructed on base 2, then we would need only two digits, 0 and 1. The number two ("twosome"), which constitutes the base of the system, would be written 10, just like the present base "ten", but meaning "one twosome and no units"; three would be written 11 ("one twosome and one unit"), and so on:

1	would be written	1	
2	would be written	10	
3	would be written	11	$= 1 \times 2 + 1$
4	...	100	$= 1 \times 2^2 + 0 \times 2 + 0 \times 1$
5		101	$= 1 \times 2^2 + 0 \times 2 + 1 \times 1$
6		110	$= 1 \times 2^2 + 1 \times 2 + 0 \times 1$
7		111	$= 1 \times 2^2 + 1 \times 2 + 1 \times 1$
8		1000	$= 1 \times 2^3 + 0 \times 2^2 + 0 \times 2 + 0 \times 1$
9		1001	$= 1 \times 2^3 + 0 \times 2^2 + 0 \times 2 + 1 \times 1$
10		1010	$= 1 \times 2^3 + 0 \times 2^2 + 1 \times 2 + 0 \times 1$
11		1011	$= 1 \times 2^3 + 0 \times 2^2 + 1 \times 2 + 1 \times 1$
12		1100	$= 1 \times 2^3 + 1 \times 2^2 + 0 \times 2 + 0 \times 1$
13		1101	$= 1 \times 2^3 + 1 \times 2^2 + 0 \times 2 + 1 \times 1$
14		1110	$= 1 \times 2^3 + 1 \times 2^2 + 1 \times 2 + 0 \times 1$
15		1111	$= 1 \times 2^3 + 1 \times 2^2 + 1 \times 2 + 1 \times 1$
16		10000	$= 1 \times 2^4 + 0 \times 2^3 + 0 \times 2^2 + 0 \times 2 + 0$
17		10001	$= 1 \times 2^4 + 0 \times 2^3 + 0 \times 2^2 + 0 \times 2 + 1$

FIG.2.8.

Now, whilst we now require only four digits to express the number two thousand four hundred and forty-eight (2,448) in a base 10 number system, a base 2 or binary system (which is in fact the system used by computers) requires no fewer than twelve digits:

$$100110010000$$
$$(= 1 \times 2^{11} + 0 \times 2^{10} + 0 \times 2^9 + 1 \times 2^8 + 1 \times 2^7 + 0 \times 2^6 + 0 \times 2^5 + 1 \times 2^4 + 0 \times 2^3 + 0 \times 2^2 + 0 \times 2 + 0)$$

Using these kinds of expressions would produce real practical problems in daily life: cheques would need to be the size of a sheet of A3 paper in order to be used to pay the deposit on a new house, for example; and it would take quite a few minutes just to say how much you think a second-hand Ferrari might be worth.

Nonetheless, there are several other numbers that could serve as base just as well as 10, and in some senses would be preferable to it.

There is nothing impossible or impracticable about changing the "steps on the ladder" and counting to a different base. Bases such as *7*, *11*, *12*, or even *13* would provide orders of magnitude that would be just as satisfactory as base 10 in terms of the human capacity for memorisation. As for arithmetical operations, they could be carried out just as well in these other bases, and in exactly the same way as we do in our present decimal system. However, we would have to lose our mental habit of giving a special status to 10 and the powers of 10, since the corresponding names and symbols would be just as useless in a 12-based system as they would in one based on 11.

If we were to decide one day on a complete reform of the number-system, and to entrust the task of designing the new system to a panel of experts, we would probably see a great battle engaged, as is often the case, between the "pragmatists" and the "theoreticians". "What we need nowadays is a system that is mathematically satisfactory," one of them would assert. "The best systems are those with a base that has the largest number of divisors," the pragmatist would propose. "And of all such bases, *12* seems to me to be by far the most suitable, given the limits of human memory. I don't need to remind you how serviceable base 12 was found to be by traders in former times – nor that we still have plenty of traces of the business systems of yore, such as the *dozen* and the *gross* (12×12), and that we still count eggs, oysters, screws and suchlike in that way. Base 10 can only be divided by 2 and 5; but 12 has 2, 3, 4, and 6 as factors, and that's precisely why a duodecimal system would be really effective. Just think how useful it would be to arithmeticians and traders, who would much more easily be able to compute halves, thirds, quarters, and even sixths of every quantity or sum. Such fractions are so natural and so common that they crop up all the time even without our noticing. And that's not the whole story! Just think how handy it would be for calculations of time: the number of months in the year would be equal to the base of the system; a day would be twice the base in terms of hours; an hour would be five times the base in minutes; and a minute the same number of seconds. It would be enormously helpful as well for geometry, since arcs and angles would be measured in degrees equal to five times the base in minutes, and minutes would be the same number of seconds. The full circle would be thirty times the base 12, and a straight line just fifteen times the base. Astronomers too would find it more than handy ..."

"But those are not the most important considerations in our day and age," the theoretician would argue. "I've no historical example to support what I'm going to propose, but enough time has passed for my ideas to stand up on their own. The main purpose of a written number system – I'm sure everyone will agree – is to allow its users to represent all numbers simply and unambiguously. And I do mean *all* numbers – integers, fractions, rational and irrational numbers, the whole lot. So what we are looking for is a numbering system with a base that has no factor other than itself, in other words, a number system having a prime number as its base. The only example I'll give is base 11. This would be much more useful than base 10 or 12, since under base 11 most fractions are irreducible: they would therefore have one and only one possible representation in a system with base 11. For instance: the number which in our present decimal system is written 0.68 corresponds in fact to several other fractions – 68/100, 34/50, and 17/25. Admittedly, these expressions all refer to the same fraction, but there is an ambiguity all the same in representing it in so many different ways. Such ambiguities would vanish completely in a system using base 11 or 7 (or indeed, any system with a prime number as its base), since the irreducibility of fractions would mean that any number had one and only one representation. Just think of the mathematical advantages that would flow from such a reform ..."

So, since it has only two factors and is not a prime number, base 10 would have no supporters on such a committee of experts!

Base 12 really has had serious supporters, even in recent times. British readers may recall the rearguard defence of the old currency – 12 pence (d) to the shilling, 20 shillings to the pound sterling – at the time it was abandoned in 1971: the benefits of teaching children to multiply and divide by 2, 3, 4, and 6 (for the smaller-value coins of 3d and 6d) and by 8 (for the "half-crown", worth 2s 6d) were vigorously asserted, and many older people in Britain continue to maintain that youngsters brought up on decimal coinage no longer "know how to count". In France, a civil servant by the name of Essig proposed a duodecimal system for weights and measures in 1955, but failed to persuade the nation that first universalised the metric system to all forms of measurement.

It seems quite unrealistic to imagine that we could turn the clock back now and modify the base number of both spoken and written number-systems. The habit of counting in tens and powers of 10 is so deeply ingrained in our traditions and minds as to be well-nigh indestructible. The best thing to do was to reform the bizarre divisions of older systems of weights and measures and to replace them with a unified system founded on the all-powerful base of 10. That is precisely what was done in France in the Revolutionary period: the Convention (a form of parliament) created the metric system and imposed it on the nation by the Laws of 18 Germinal Year III, in the revolutionary calendar (8 April 1795) and 19 Frimaire Year VIII (19 December 1799).

A BRIEF HISTORY OF THE METRIC SYSTEM

Until the late eighteenth century, European systems of weights and measures were diverse, complicated, and varied considerably from one area to another. Standards were fixed with utter whimsicality by local rulers, and quite arbitrary objects were used to represent lengths, volumes, etc. From the late seventeenth century onwards, as the experimental sciences advanced and the general properties of the physical world became better understood, scholars strove to devise stable and coherent measuring systems based on permanent, universal and unmodifiable standards. The growth of trade throughout the eighteenth century also created a need for common measurements at least within each country, and a uniform system of weights and measures. Thus the metric system emerged towards the end of the eighteenth century. It is a fully consistent and coherent measurement system using base 10 (and therefore fully compatible with the place-value system of written numbering that the Arabs had brought to Europe in the Middle Ages, having themselves learned it from the Indians), which the French Revolution offered "to all ages and to all peoples, for their greater benefit". It produced astounding progress in applied areas, since it is perfectly adapted to numerical calculation and is extremely simple to operate in fields of every kind.

Around 1660: In order to harmonise measurement of time and length and also so as to compare the various standards used for measuring length around the world, the Royal Society of London proposed to establish as the unit of length the length of a pendulum that beats once per second. The idea was taken up by Abbé Jean Picard in *La Mesure de la Terre* ("The Measurement of the Earth") in 1671, by Christian Huygens in 1673, and by La Condamine in France, John Miller in England, and Jefferson in America.

1670: Abbé Gabriel Mouton suggested using the sexagesimal minute of the meridian (= 1/1000 of the nautical mile) as the unit of length. But this unit, of roughly 1.85 metres, was too long to be any practical use.

1672: Richer discovered that the length of a pendulum that beats once per second is less at Cayenne (near the Equator) than in Paris. The consequence of this discovery was that, because of the variation in length of the pendulum caused by the variation in gravity at different points on the globe, the choice of the location of the standard pendulum would be politically very tricky. As a result the idea of using the one-second pendulum as a unit of length was eventually abandoned.

1758: In *Observations sur les principes métaphysiques de la géometrie* ("Observations on the Metaphysical Principles of Geometry"), Louis Dupuy suggested unifying measurements of length and weight by fixing the unit of weight as that of a volume of water defined by units of length.

1790: 8 May: Talleyrand proposed, and the *Assemblée constituante* (Constituent Assembly) approved the creation of a stable, simple and uniform system of weights and measures. The task of defining the system was entrusted to a committee of the Academy of Sciences, with a membership consisting of Lagrange, Laplace, and Monge (astronomical and calendrical measurements), Borda (physical and navigational measurement), and Lavoisier (chemistry). The base unit initially chosen was the length of the pendulum beating once per second.

1791: 26 March: The committee decided to abandon the pendulum as the base unit and persuaded the Constituent Assembly to choose as the unit of length the ten-millionth part of one quarter of the earth's meridian, which can be measured exactly as a fraction of the distance from the pole to the Equator. At Borda's suggestion, this unit would be called the *metre* (Greek for "measure").

What the committee then had to do was to produce conventional equivalencies between the various units chosen so that all of them (except units of time) could be derived from the metre. So, for measuring surface area, the unit chosen was the *are*, a square with a side of 10 metres; for measuring weight, the *kilogram* was defined as the weight of a unit of volume (1 litre) of pure water at the temperature of melting ice, corrected for the effects of latitude and air pressure. All that now had to be done to set up the entire metric system was to make the key measurement, the distance from the pole to the Equator – a measurement that was all the more interesting at that time as Isaac Newton had speculated that the globe was an ellipsoid with flattened ends (contradicting Descartes, who believed it was a sphere with elongated or pointed ends).

1792: The "meridian expedition" began. A line was drawn from Dunkirk to Barcelona and measured out by triangulation points located thanks to

Borda's goniometer, with some base stretches measured out with greater precision on the ground. Under the direction of Méchain and Delambre, one team was in charge of triangulation, one was responsible for the standard length in platinum, and one for drafting the users' manuals of the new system. Physicists such as Coulomb, Haüy, Hassenfrantz, and Borda, and the mathematicians Monge, Lagrange, and Laplace were amongst the many scientists who collaborated on this project which was not fully completed until 1799.

1793: 1 August: The French government promulgated a decree requiring all measures of money, length, area, volume, and weight to be expressed in decimal terms : all the units of measure would henceforth be hierarchised according to the powers of 10. As it overturned all the measures in current use (most of them using base 12), the decimalisation decree required new words to be invented, but also created the opportunity for much greater coherence and accuracy in counting and calculation.

1795: 7 April: Law of 18 Germinal, Year III, which organised the metric system, gave the first definition of the metre as a fraction of the terrestrial meridian, and fixed the present nomenclature of the units (decimetre, centimetre, millimetre; are, deciare, centiare, hectare; gram, decigram, centigram, kilogram; franc, centime; etc.)

1795: 9 June: Lenoir fabricated the first legal metric standard, on the basis of the calculation made by La Caille of the distance between the pole and the Equator at 5,129,070 *toises de Paris* (in 1799, Delambre and Méchain obtained a different, but actually less accurate figure of 5,130,740 *toises de Paris*).

1795: 25 June: Establishment of the *Bureau des Longitudes* (Longitude Office) in Paris.

1799: First meeting in Paris of an international conference to discuss universal adoption of the metric system. The system was considered "too revolutionary" to persuade other nations to "think metric" at that time.

1799: 22 June: The definitive standard metre and kilogram, made of platinum, were deposited in the French National Archives.

1799: 10 December: Law of 10 Frimaire, Year VIII, which confirmed the legal status of the definitive standards, gave the second definition of the metre (the length of the platinum standard in the National Archives, namely 3 feet and 11.296 "lines" of the *toise de Paris*), and in theory made the use of the metric system obligatory. (In fact, old habits of using pre-metric units of measurement persisted for many years and were tolerated.)

1840: 1 January: With the growing spread of primary education in France, the law was amended to make the use of the metric system genuinely obligatory on all.

1875: Establishment of the International Bureau of Weights and

Measures at Sèvres (near Paris). The Bureau created the new international standard metre, made of iridoplatinum.

1876: 22 April: The new international standard metre was deposited in the Pavillon de Breteuil, at Sèvres, which was then ceded by the nation to the International Weights and Measures Committee and granted the status of "international territory".

1899: The General Conference on Weights and Measures met and provided the third definition of the metre. The length of the meridian was abandoned as a basis of calculation. Henceforth, the metre was defined as the distance at 0°C of the axis of the three median lines scored on the international standard iridoplatinum metre.

1950s: The invention of the laser allowed significant advances in optics, atomic physics, and measurement sciences. Moreover, quartz and atomic clocks resulted in the discovery of variations in the length of the day, and put an end to the definition of units of time in terms of the earth's rotation on its axis.

1960: 14 October: Fourth definition of the metre as an optical standard (one hundred times more accurate than the metre of 1899): the metre now becomes equal to 1,650,763.73 wave-lengths of orange radiation in a void of krypton 86 (krypton 86 being one of the isotopes of natural krypton).

1983: 20 October: The XVIIth General Conference on Weights and Measures gives the fifth definition of the metre, based on the speed of light in space (299,792,458 metres per second): a metre is henceforth the distance travelled by light in space in 1/299,792,458 of a second. As for the second, it is defined as the duration of 9,192,631,770 periods of radiation corresponding to the transition between the two superfine levels of the fundamental state of an atom of caesium 133. At the same conference, definitions of the five other basic units (kilogram, amp, kelvin, mole, and candela) were also adopted, as well as the standards that constitute the current International Standards system (IS).*

THE ORIGIN OF BASE 10
Well, then: where *does* base 10 come from?

In the second century CE, Nicomachus of Gerasa, a neo-Pythagorean from Judaea, wrote an *Arithmetical Introduction* which, in its many translations, influenced Western mathematical thinking throughout the Middle Ages. For Nicomachus, the number 10 was a "perfect" number, the number of the divinity, who used it in his creation, notably for human toes and fingers,

---

* For information contained in this section on the metric system I am indebted to Jean Dhombres, President of the French Association for the History of Science.

and inspired all peoples to base their counting systems on it. For many centuries, indeed, numbers were thought to have mystical properties; in Pythagorean thinking, 10 was held to be "the first-born of the numbers, the mother of them all, the one that never wavers and gives the key to all things".

Such attitudes to numbers, which had their place in a world-view which was itself mystical through and through, now seem as circular and self-defeating as the observation that God had the wisdom to cause rivers to flow through the middle of towns.

In fact, the almost universal preference for base 10 comes from nothing more obscure than the fact that we learn to count on our fingers, and that we happen to have ten of them. We would use base 10 even if we had no language, or were bound to a vow of total silence: for just like the North African shepherd and his shells and straps discussed on p. 24–25 above, we could use our raised fingers to count out the first ten in silence, a colleague could then raise one finger to keep count of the tens, and so on to 99, when (for numbers of 100 and more) the fingers of a third colleague would be needed. Fig. 2.9 shows the position of the three silent colleagues' hands at number 627.

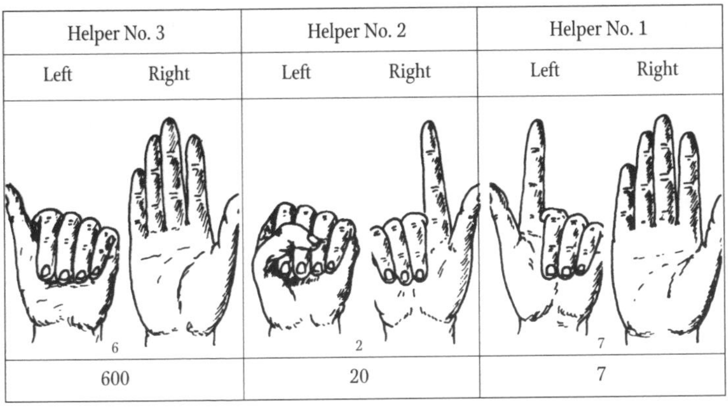

Helper No. 3		Helper No. 2		Helper No. 1	
Left	Right	Left	Right	Left	Right
6		2		7	
600		20		7	

Fig. 2.9.

The obvious practicality of such a non-linguistic counting system using only our own bodies shows that the idea of grouping numbers into packets of ten and powers of ten is based on the "accident of nature" that is the physiology of the human hand. Since that physiology is universal, base 10 necessarily occupies a dominant, not to say inexpugnable position in counting systems.

If nature had given us six fingers, then the majority of counting systems would have used base 12. If on the other hand evolution had brought us down to four fingers on each hand (as it has for the frog), then we would doubtless have long-standing habits and traditions of counting on base 8.

THE ORIGINS OF THE OTHER BASES

The reason for the adoption of vigesimal (base 20) systems in some cultures can be seen by the basic idea of Aztec numbering as laid out in Fig. 2.6 above. In the language of the Aztecs

> • the names of the first five numbers can be associated with the fingers of one hand;
>
> • the following five numbers can be associated with the fingers of the other hand;
>
> • the next five numbers can be associated with the toes of one foot;
>
> • and the last five numbers can be associated with the toes on the other foot.

And so 20 is reached with the last toe of the second foot (see Fig. 2.10).

This is no coincidence. It is simply that some communities, because they realised that by leaning forward a little they could count toes as well as fingers, ended up using base 20.

One remarkable fact is that both the Inuit (Greenland) and the Tamanas (in the Orinoco basin) used the same expression for the number 53, literally meaning: "of the third man, three on the first foot".

According to C. Zaslavsky (1973), the Banda people in Central Africa express the number 20 by saying something like "a hanged man": presumably because when you hang a man you can see straight away all his fingers and toes. In some Mayan dialects, the expression *hun uinic*, which means 20, also means "one man". The Malinke (Senegal) express 20 and 40 by saying respectively "a whole man" and "a bed" – in other words, two bodies in a bed!

In the light of all this there can be no doubt at all that the origin of vigesimal systems lies in the habit of counting on ten fingers *and* ten toes ...

The origin of base 5 is similarly anthropomorphic. Quinary reckoning is founded on learning to count using the fingers of one hand only.

The following finger-counting technique, which is found in various parts of Oceania and is also currently used by many Bombay traders for various specific purposes, is a good example of how a primitive one-hand counting system can give rise to more elaborate numbering. You use the five fingers of the left hand to count the first five units. Then, once this number is reached, you extend the thumb of the right hand, and go on counting to 10 with the fingers of the left hand; then you extend the index finger of the right hand and count again on the left hand from 11 to 15; and so on, up to 25. The series can be extended to 30 since the fingers of the left hand are usable six times over in all.

However, this obviously fails to resolve the basic mystery: why did base 5 – which must be considered the most natural base by far, since it is

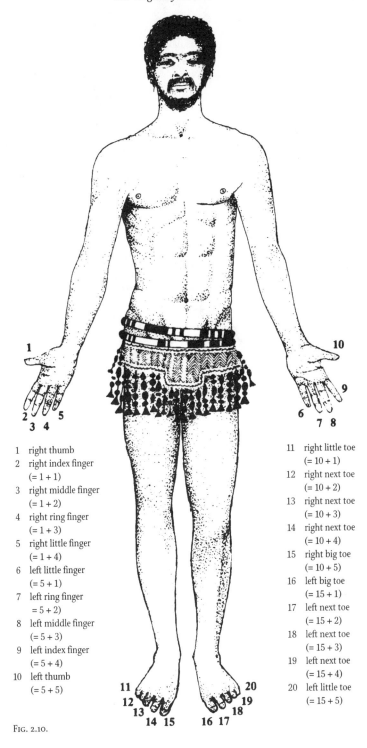

1 right thumb
2 right index finger
   (= 1 + 1)
3 right middle finger
   (= 1 + 2)
4 right ring finger
   (= 1 + 3)
5 right little finger
   (= 1 + 4)
6 left little finger
   (= 5 + 1)
7 left ring finger
   = 5 + 2)
8 left middle finger
   (= 5 + 3)
9 left index finger
   (= 5 + 4)
10 left thumb
   (= 5 + 5)

11 right little toe
   (= 10 + 1)
12 right next toe
   (= 10 + 2)
13 right next toe
   (= 10 + 3)
14 right next toe
   (= 10 + 4)
15 right big toe
   (= 10 + 5)
16 left big toe
   (= 15 + 1)
17 left next toe
   (= 15 + 2)
18 left next toe
   (= 15 + 3)
19 left next toe
   (= 15 + 4)
20 left little toe
   (= 15 + 5)

Fig. 2.10.

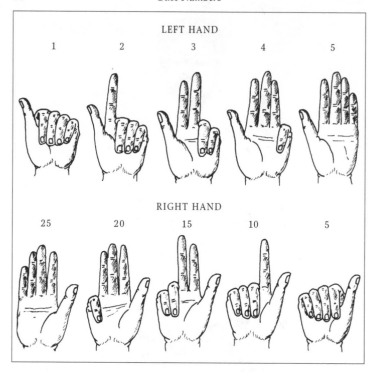

LEFT HAND

| 1 | 2 | 3 | 4 | 5 |

RIGHT HAND

| 25 | 20 | 15 | 10 | 5 |

Fig. 2.11.

virtually dictated by the basic features of the human body and must be self-evident from the very moment of learning to count – why did base 5 not become adopted as the universal human counting tool? Why, in other words, was the apparently inevitable construction of quinary counting generally avoided? Why did so many cultures go up to 10, to 20 or, in the case of the Sumerians, whom we will discuss again, as far as 60? Even more mysterious are those cultures which possessed a concept of number and knew how to count, but went back down to 4 for their numerical base.

L. L. Conant (1923) tackled the whole problem in detail without claiming to have found the final answer. The anthropologist Lévy-Bruhl, on the other hand, thought it was a false problem. In his view, we should not suppose that people ever invented number-systems *in order* to carry out arithmetical operations or devised systems that were intended to be best suited to operations that, prior to the devising of the system, could not be imagined. "Numbering systems, like languages, from which they can hardly be distinguished, are in the first place social phenomena, closely dependent on collective mentalities," he claimed. "The mentality of any society is completely bound up with its internal functioning and its institutions."

To conclude this chapter we shall return with Lévy-Bruhl to very primitive counting systems which do not yet clearly distinguish between the cardinal and ordinal aspects of number. In the kind of "body-counting" explained above and demonstrated in Fig. 1.30, 1.31 and 1.32, there are no "privileged" points or numbers, and therefore no concept of a base at all. Using Petitot's dictionary of the language of Dene-Dindjie Indians (Canada), Lévy-Bruhl explains how things are counted out in a system with no base:

You hold out your left hand (always the left hand) with the palm turned towards your face, and bend your little finger, saying

for 1 :   *the end is bent*

or   *on the end*

Then you bend your ring finger and say:

for 2:   *it's bent again*

Then you bend your middle finger and say

for 3:   *the middle one is bent*

Then you bend your index finger, leaving the thumb stretched out, and say:

for 4:   *there's only that left*

Then you open out your whole hand and say:

for 5:   *it's OK on my hand*

or   *on a hand*

or   *my hand*

Then you fold back three fingers together on your left hand, keeping the thumb and index stretched out, and touch the left thumb with the right thumb, saying:

for 6:   *there's three on each side*

or: *three by three*

Then you bend down four fingers on your left hand and touch your left thumb (still stretched out) with the thumb and index finger of your right hand, and say:

for 7:   *on one side there are four*

or   *there are still three bent*

or   *three on each side and one in the middle*

Then you stretch out three fingers of your right hand and touch the outstretched thumb of your left hand, creating two groups of four fingers (bent and extended), and say:

for 8:   *four on four*

or   *four on each side*

Then you show the little finger of your right hand, the only one now bent, and say:

for 9:   *there's still one down*

or   *one still short*

or   *the little finger's lying low*

Then you start the gestures over again, saying "*one full plus one*", or "*one counted plus one*", "*one counted plus two*", "*one counted plus three*", and so on.

Lévy-Bruhl argues that in this system, which does not prevent the Dene-Dindjie from counting properly, there is no concept of a quinary base: 6 is not "a second one", 7 not "a new two", as we find in so many other numbering systems. On the contrary, he says, 6 here is "three and three" – which shows that finishing the count on one hand is in no way a "marker" or a "privileged number" in this system. The periodicity of numbers is not derived from the physical manner of counting, does not come from the series of movements made to indicate the sequence of the numbers.

In this view, numbering systems relate much more directly to the "mental world" of the culture or civilisation, which may be mythical rather than practical, attributing more significance to the four cardinal points of the compass, or to the four legs of an animal, than to the five fingers of the hand. We do not have to try and guess why this base rather than another was "chosen" by a given people for their numbering system, even if they do effectively use the five fingers of their hand for counting things out. Where a numbering system has a base, the base was never "chosen", Lévy-Bruhl asserts. It is a mistake to think of "the human mind" constructing a number system in order to count: on the contrary, people began to count, slowly and with great difficulty, long before they acquired the concept of number.

However, it is clear that the adoption of base 5 is related to the way we count on the fingers of our hands. But why did those cultures that adopted base 5 not extend it, like so many others, to the base 10 that corresponds to the fingers of both hands? Dantzig has speculated that it may have to do with the conditions of life in warrior societies, in which men rarely go about unarmed. If they want to count, they tuck their weapon under their left arm and count on the left hand, using the right hand as a check-off. The right hand remains free to seize the weapon if needed. This may explain why the left hand is almost universally used by right-handed people for counting, and vice versa [T. Dantzig (1930), p. 13].

However this may be, base numbers arise for many reasons, many of which have nothing at all to do with their suitability for counting or for arithmetical operations; and they may indeed have arisen long before any kind of abstract arithmetic was invented.

CHAPTER 3

THE EARLIEST CALCULATING MACHINE – THE HAND

That uniquely flexible and useful tool, the human hand, has also been the tool most widely used at all times as an aid to counting and calculation.

Greek writers from Aristophanes to Plutarch mention it, and Cicero tells us that its use was as common in Rome: *tuos digitos novi* – "I well know your skill at calculating on your fingers" (*Epistulae ad Atticum*, V, 21, 13); Seneca says much the same: "The geometer was my teacher of arithmetic: I learned to make my fingers the servants of my desires" (*Epistles*, LXXXVII); and, later, Tertullian said: "Meanwhile, I have to sit surrounded by piles of papers, bending and unbending my fingers to keep track of numbers." The famous orator Quintilian stressed the importance of calculating on the fingers, especially in the context of pleading at law: "Skill with numbers is needed not only by the Orator, but also by the pleader at the Bar. An Advocate who stumbles over a multiplication, or who merely exhibits hesitation or clumsiness in calculating on his fingers, immediately gives a bad impression of his abilities;" and the digital techniques he referred to, which were in common use by the inhabitants of Rome, required very considerable dexterity (see Fig. 3.13). Pliny the Elder, in his *Natural History* (XXXIV), described how King Numa offered up to the God Janus (the god of the Year, of Age, and of Time) a statue whose fingers displayed the number of days in a year. Such practices were by no means confined to the Greeks and Romans. Archaeologists, historians, ethnologists, and philologists have come upon them at all times and in all regions of the world, in Polynesia, Oceania, Africa, Europe, Ancient Mesopotamia, Egypt under the Pharaohs, the Islamic world, China, India, the Americas before Columbus, and the Western world in the Middle Ages. We can conclude, therefore, that the human hand is the original "calculating machine". In the following we shall show how, once people had grasped the principle of the base, over the ages they developed the arithmetical potential of their fingers to an amazing degree. Indeed, certain details of this are evidence of contacts and influences between different peoples, which could never have been inferred in any other way.

EARLY WAYS OF COUNTING ON FINGERS

The simplest method of counting on the fingers consists of associating an integer with each finger, in a natural order. This may be done in many ways.

One may start with the fingers all bent closed, and count by successively straightening them; or with the fingers open, and successively close them. One may count from the left thumb along the hands to the right little finger, or from the little finger of the left hand through to the thumb of the right, or from the index finger to the little finger and finally the thumb (see Fig. 3.3). The last method was especially used in North Africa. It seems likely that at the time of Mohammed the Arabs used this method. One of the *Ḥadiths* tells how the Prophet showed his disciples that a month could have 29 days, showing "his open hands three times, but with one finger bent the third time". Also, a Muslim believer always raises the index finger when asserting the unity of Allah and expressing his faith in Islam, in performing the prayer of *Shahādah* ("witness").

FIG. 3.1. *Finger-counting among the Aztecs (Pre-Columbian Mexico). Detail of a mural by Diego Rivera. National Museum of Mexico*

FIG. 3.2. *Boethius (480–524 CE), the philosopher and mathematician, counting on his fingers. From a painting by Justus of Ghent (15th C.). See P. Dedron and J. Itard (1959).*

A STRANGE WAY OF BARGAINING

There is a similar method, of very ancient origin, which persisted late in the East and was common in Asia in the first half of the twentieth century. It is a special way of finger-counting used by oriental dealers and their clients in negotiating their terms. Their very curious procedure was described by the celebrated Danish traveller Carsten Niebuhr in the eighteenth century, as follows:

> I have somewhere read, I think, that the Orientals have a special way of settling a deal in the presence of onlookers, which ensures that none of these becomes aware of the agreed price, and they still regularly make use of it. I dreaded having someone buy something on my behalf in this way, for it allows the agent to deceive the person for whom he

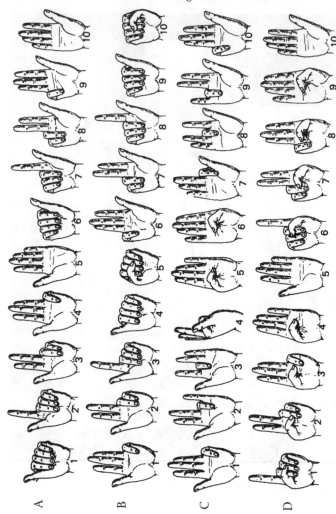

FIG. 3·3. *Variants of basic counting with the fingers*

is acting, even when he is watching. The two parties indicate what price is asked, and what they are willing to pay, by touching fingers or knuckles. In doing so, they conceal the hand in a corner of their dress, not in order to conceal the mystery of their art, but simply in order to hide their dealings from onlookers.... (*Beschreibung von Arabien*, 1772)

To indicate the number 1, one of the negotiators takes hold of the index finger of the other; to indicate 2, 3 or 4, he takes hold of index and middle fingers, index, middle and fourth, or all four fingers. To indicate 5, he grasps the whole hand. For 6, he twice grasps the fingers for 3 (2×3), for 7, the fingers for 4 then the fingers for 3 ($4 + 3$), for 8, he twice grasps the fingers for

4 (2 × 4), and for 9, he grasps the whole hand, and then the fingers for 4 (5 + 4). For 10, 100, 1,000 or 10,000 he again takes hold of the index finger (as for 1); for 20, 200, 2,000 or 20,000 the index and middle fingers (as for 2), and so on (see Fig. 3.4). This does not, in fact, lead to confusion because the two negotiators will have agreed beforehand on roughly what the price will be (whether about 40 dinars, or 400, for example). Niebuhr does not tell that he himself saw such a deal take place, but J. G. Lemoine found traces of the method in Bahrain, a place famous for its pearl fishery, when he made a study of this topic at the beginning of this century. He gathered information from pearl dealers in Paris, who had often visited Bahrain and had occasion to employ this procedure in dealing with the Bahrainis. He states:

> The two dealers, seated face to face, bring their right hands together and, with the left hand, hold a cloth over them so that their right hands are concealed. The negotiation, with all its "discussions", takes place without a word being spoken, and their faces remain totally impassive. Those who have observed this find it extremely interesting, for the slightest visible sign could be taken to the disadvantage of one or other of the dealers.

Similar methods of negotiation have been reported from the borders of the Red Sea, from Syria, Iraq and Arabia, from India and Bengal, China and Mongolia, and – from the opposite end of the world – Algeria. P. J. Dols (1918), reporting on "Chinese life in Gansu province", describes how dealings were still being conducted in China and Mongolia in the early twentieth century.

> The buyer puts his hands into the sleeves of the seller. While talking, he takes hold of the seller's index finger, thereby indicating that he is offering 10, 100 or 1,000 francs. "No!" says the other. The buyer then takes the index and middle fingers together. "Done!" says the seller. The deal has been struck, and the object is sold for 20, or for 200, francs. Three fingers together means 30 (or 300 or 3,000), four fingers 40 (or 400 or 4,000). When the buyer takes the whole hand of the seller, it is 50 (or 500 or 5,000). Thumb and little finger signify 60 (note the difference from the Middle Eastern system described above). Placing the thumb in the vendor's palm means 70, thumb and index together 80. When the buyer, using his thumb and index finger together, touches the index finger of the seller, this indicates 90.

COUNTING ALONG THE FINGERS

There is more to fingers than a single digit: they have a knuckle, two joints, and three bones (but one joint and two bones for the thumbs). Amongst many Asiatic peoples, this more detailed anatomy has been

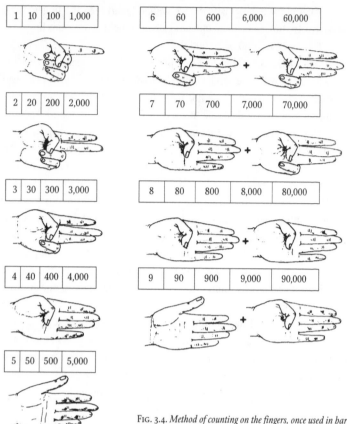

FIG. 3.4. *Method of counting on the fingers, once used in bargaining between oriental dealers*

exploited for counting. In southern China, Indo-China and India, for example, people have counted one for each joint, including the knuckle, working from base to tip of finger (or in reverse) and from little finger to thumb, pointing with a finger of the other hand. Thus each hand can count up to 14, and both hands up to 28 (see Fig. 3.5). A Chinaman from Canton once told me of a singular application of this method. Since a woman's monthly cycle lasts 28 days, his mother used to tie a thread around each joint as above for each day of her cycle, to detect early or late menstruation. The Venerable Bede (673–735 CE, a monk in the Monastery of Saints Peter and Paul at Wearmouth and Jarrow and author of the influential *De ratione temporum*, "Of the Division of Time"), applied similar counting methods for his calculations of time. To count the twenty-eight years of the solar cycle, beginning with a leap year, he started from the tip of the little finger and counted across the four fingers, winding back and forth and working down to the base of the fingers to count up to 12, then moving to the other hand to count up to 24, finally using the two thumbs to count up to 28 (see Fig. 3.6).

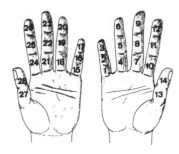

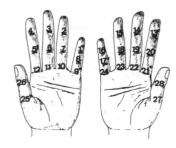

FIG. 3.5. *The method used in China, Indo-China, and India, using the fourteen finger-joints of each hand*

FIG. 3.6. *The Venerable Bede's method of counting the twenty-eight years of the solar cycle on the knuckles (7th C.). Leap years are marked with asterisks*

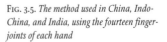

He used a similar method to count the nineteen years of the lunar cycle, counting up to 14 on the knuckles of the left hand, going on up to 19 by also counting the fingernails (see Fig. 3.7). The objective was to determine the date of Easter, the subject of complicated controversies in the early Church. In particular there was dispute between the British and Irish Churches, on the one hand, and the Roman Church on the other, regarding which lunar cycle to adopt for the date of Easter. Bede's calculations brought together the solar year and the solar and lunar cycles of the Julian calendar, and its leap years.

A different method of counting on the knuckles was long used in northeast India, and is still found in the regions of Calcutta and Dacca. It was reported by seventeenth- and eighteenth-century travellers, especially the Frenchman J. B. Tavernier (1712). According to N. Halhed (1778), the Bengalis counted along the knuckles from base to tip, starting with the little finger and ending with the thumb, using the ball of the thumb as well and thus counting up to 15 (Fig. 3.8).

This method of counting on the knuckles has given rise to the practice, common among Indian traders, of fixing a price by offering the hand under cover of a cloth; they then touch knuckles to raise or lower their proposed prices (Halhed).

There are 15 days in the Hindu month, the same number as can be counted on a hand; and, according to Lemoine, this is no coincidence. The Hindu year (360 days) consists of 12 seasons (*Nitus*) each of two "months" (*Masas*). One month of 15 days corresponds to one phase of the moon, and the following month to another phase. The first, waxing, phase is called *Rahu*, and the second, waning, phase is called *Ketu*. In this connection we may refer to the legend which tells how, before the raising up of the oceans, these two "faces" of the moon formed a single being, subsequently cut in two by Mohini (*Vishnu*). Such a system for counting on the hands is also

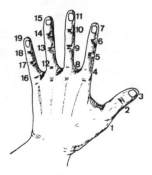
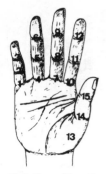

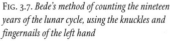

FIG. 3.7. *Bede's method of counting the nineteen years of the lunar cycle, using the knuckles and fingernails of the left hand*

FIG. 3.8. *The Indian and Bengali method, using the knuckles of fingers and thumb, and the ball of the thumb*

found throughout the Islamic world, but mainly for religious purposes in this case. Muslims use it when reciting the 99 incomparable attributes of Allah or for counting in the litany of *subḥa* (which consists of the 33-fold repetition of each of the three "formulas"), which is recited following the obligatory prayer. To do either of these conveniently, a count of 33 must be achieved. This is done by counting the knuckles, from base to tip, of each finger and the thumb (including the ball of the thumb), first on the left hand and then on the right. In this way a count of 30 is attained, which is brought up to 33 by further counting on the tips of the little, ring, and middle fingers of the right hand.

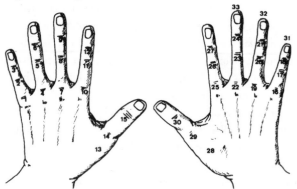

FIG. 3.9. *The Muslim method of counting up to 33, used for reciting the 99 (= 3 times 33) attributes of Allah, and for the 33-fold repetitions of the* subḥa

Nowadays, Muslims commonly adopt a rosary of prayer-beads for this purpose, but the method just described may still be adopted if the beads are not to hand. However, the hand-counting method is extremely ancient and undoubtedly pre-dates the use of the beads. Indeed, it finds mention in the oral tradition, in which the Prophet is described as admonishing women believers against the use of pearls or pebbles, and encouraging them to use their fingers to count the Praises of Allah. I. Godziher (1890) finds in this tradition some disapproval, by the Islamic authorities, of the use of the rosary subsequent to its emergence in the ninth century CE, which persisted until the fifteenth century CE. Abū Dawūd al Tirmidhī tells it as follows: "The Prophet of Allah has said to us, the women of Medina: Recite the *tasbīḥ*, the *tahlīl* and the *taqdīs*; and count these Praises on your fingers, for your fingers are for counting." This parallelism between Far-Eastern commercial practices and the ancient and widespread customs of Islamic religious tradition is extremely interesting.

THE GAME OF MORRA

For light relief, let us consider the game of Morra, a simple, ancient and well-known game usually played between two players. It grew out of finger-counting. The two players stand face to face, each holding out a closed fist. Simultaneously, the two players open their fists; each extends as many fingers as he chooses, and at the same time calls out a number from 1 to 10. If the number called by a player equals the sum of the numbers of fingers shown by both players, then that player wins a point. (The players may also use both hands, in which case the call would be between 1 and 20.) The game depends not only on chance, but also on the quickness, concentration, judgment and anticipation of the players. Because the game is so well defined, and also of apparently ancient origin, it is very interesting for our purpose to follow its traces back into history, and across various peoples; and we shall come upon many signs of contacts and influences which will be important for us. It is still popular in Italy (where it is called *morra*), and is also played in southeast France (*la mourre*), in the Basque region of Spain, in Portugal, in Morocco and perhaps elsewhere in North Africa. As a child I played a form of it myself in Marrakesh, with friends, as a way of choosing "It". We would stand face to face, hands behind our backs. One of the two would bring forward a hand with a number of fingers extended. The other would call out a number from 1 to 5, and if this was the same as the number of fingers then he was "It"; otherwise the first player was "It". In China and Mongolia the game is called *hua quan* (approximately, "fist quarrel"). According to Joseph Needham, it is a popular entertainment in good circles. P. Perny (1873) says: "If the guests know each other well, their host will propose *qing hua quan* ('let us have a fist quarrel'). One of the guests is appointed umpire. For reasons of politeness, the host and one of the guests commence, but the host will soon give way to someone else. The one who loses pays the 'forfeit' of having to drink a cup of tea." The game of Morra was very popular, in Renaissance times, in France and Italy, amongst valets, pages and other servants to while away their idle hours. "The pages would play Morra at a flick of the fingers" (Rabelais, *Pantagruel* Book IV, Ch. 4); "Sauntering along the path like the servants sent to get wine, wasting their time playing at Morra" (Malherbe, *Lettres* Vol. II p. 10). Fifteen hundred years earlier, the Roman plebs took great delight in the game, which they called *micatio* (Fig. 3.10). Cicero's phrase for a man one could trust was: "You could play *micatio* with him in the dark." He says it was a common turn of phrase, which indicated the prevalence of the game in the popular culture.

FIG. 3.10. *Mural painting showing the game of Morra. Farnesina, Rome*

The game also served in the settling of disputes, legal or mercantile, when no other means prevailed, much as in "drawing the short straw", and was even forbidden by law in public markets (G. Lafaye, 1890). Vases and other Ancient Greek relics depict the game (Fig. 3.11). According to legend it was Helen who invented the game, to amuse her lover Paris.

Much earlier, the Egyptians had a similar game, as shown in the two funerary paintings reproduced in Fig. 3.12. The top is from a tomb at Beni

FIG. 3.11. *The game of Morra in Greek times. (Left) Painted vase in the Lambert Collection, Paris. (Right) Painted vase, Munich Museum. (DAGR, pp. 1889–90)*

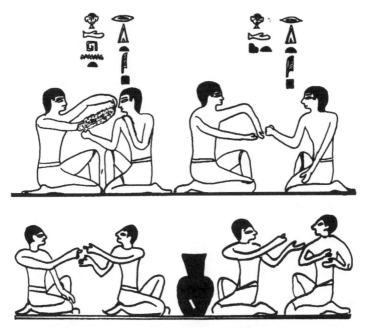

FIG. 3.12. *Two Egyptian funeral paintings showing the game of Morra. (Above) Tomb no. 9 of Beni-Hassan (Middle Kingdom). See Newberry, ASE vol. 2 (1893), plate 7. (Below) Theban tomb no. 36 (Aba's tomb, XXVIth dynasty). See Wilkinson (1837), vol. 2, p. 55 (Fig. 307). See also photo no. 9037 by Schott at the Göttingen Institute of Egyptology.*

Hassan dating from the Middle Kingdom (21st–17th centuries BCE), and it shows two scenes. In the first scene, one man holds his hand towards the eyes of the other, hiding the fingers with his other hand. The other man holds his closed fist towards the first. The lower scene depicts similar gestures, but directed towards the hand. According to J. Yoyotte (in G. Posener, 1970), the hieroglyphic inscriptions on these paintings mean: *Left legend*: Holding the *íp* towards the forehead; *Right legend*: Holding the *íp* towards the hand. The Egyptian word *íp* means "count" or "calculate", so these paintings must refer to a game like Morra. The lower painting, from Thebes, is from the time of King Psammetichus I (seventh century BCE) and was (according to Leclant) copied from an original from the Middle Kingdom. This too shows two pairs of men, showing each other various combinations of open and closed fingers.

We may therefore conclude that the game of Morra, in one form or another, goes back at least to the Middle Kingdom of Pharaonic Egypt. In the world of Islam, Morra is called *mukhāraja* ("making it stick out"). At the start of the present century it was played in its classical form in remote areas of Arabia, Syria and Iraq. *Mukhāraja* was above all, however, a divination ritual amongst the Muslims and was therefore forbidden to the faithful (fortune-telling is proscribed by both Bible and Koran); so it was a much more serious matter than a mere game. An Arabian fortune-telling manual shows circular maps of the universe (*Zā' irjat al 'alam*), divided into sectors corresponding to the stars, where each star has a number. There are also columns of numbers which give possible "answers" to questions which might be asked. The *mukhāraja* was then used to establish a relationship between the two sets of numbers.

COUNTING AND SIGN-LANGUAGE

There is a much more elaborate way of counting with the hand which, from ancient times until the present day, has been used by the Latins and can also be found in the Middle East where, apparently, it may go back even further in time. It is rather like the sign language used by the deaf and dumb. Using one or both hands at need, counting up to 9,999 is possible by this method. From two different descriptions we can reconstruct it in its entirety. These are given in parallel to each other in Fig. 3.13.

The first was written in Latin in the seventh century by the English monk Bede ("The Venerable") in his *De ratione temporum*, in the chapter *De computo vel loquela digitorum* ("Counting and talking with the fingers"). The other is to be found in the sixteenth-century Persian dictionary *Farhangi Djihangiri*. There is a most striking coincidence between these two descriptions written nine centuries apart and in such widely separated places.

With one hand (the left in the West and the right in the East), the little finger, fourth and middle fingers represented units, and either the thumb or the index finger (or both) was used for tens. With the other hand, hundreds and thousands were represented in the same way as the units and tens.

Both accounts also describe how to show numbers from 10,000 upwards. In the Eastern description: "for 10,000 bring the whole top joint of the thumb in contact with the top joint of the index finger and part of its second joint, so that the thumbnail is beside the nail of the index finger and the tip of the thumb is beside the tip of the index finger." For his part, Bede says: "For 10,000 place your left hand, palm outwards, on your breast, with the fingers extended backwards and towards your neck." Therefore the two descriptions diverge at this point.

Let us however follow Bede a little further.

> For 20,000 spread your left hand wide over your breast. For 30,000 the left hand should be placed towards the right and palm downwards, with the thumb towards the breastbone. For 50,000 similarly place the left hand at the navel. For 60,000 bring your left hand to your left thigh, inclining it downwards. For 70,000 bring your left hand to the

WESTERN DESCRIPTION	EASTERN DESCRIPTION
From the Latin of the Venerable Bede, seventh century	*From a sixteenth-century Persian dictionary*

A. UNITS

When you say "one", bend your left little finger so as to touch the central fold of your palm

1

For 1, bend down your little finger

1

For "two", bend your next finger to touch the same spot

2

For 2, your ring finger must join your little finger

2

When you say "three", bend your third finger in the same way

3

For 3, bring your middle finger to join the other two

3

When you say "four", raise up your little finger from its place

4

For 4, raise the little finger (the other fingers should stay where they were before)

4

Saying "five", raise your second finger in the same way

5

For the number 5, also raise your ring finger

5

When you say "six", you also raise your third finger, but you must keep your ring finger in the middle of your palm

6

For the number 6, raise the middle finger, keeping your ring finger down (so that its tip is in the centre of the palm)

6

FIG. 3.13.

A. UNITS (continued)		
 7	Saying "seven", raise all your other fingers except the little finger, which should be bent onto the edge of the palm	For 7, the ring finger is also raised, but the little finger is lowered so that its tip points towards the wrist 7
 8	To say "eight", do the same with the ring finger	For 8, do the same with the ring finger 8
 9	Saying "nine", you place the middle finger also in the same place	For 9, do just the same with the middle finger 9
B. TENS		
 10	When you say "ten", place the nail of the index finger into the middle joint of the thumb	For 10, the nail of the right index finger is placed on the first joint (counting from the tip) of the thumb, so that the space between the fingers is like a circle 10
20	For "twenty" put the tip of the thumb between the index and the middle fingers	For 20, place the middle-finger side of the lower joint of the index finger over the face of the thumbnail, so that it appears that the tip of the thumb is gripped between the index and middle fingers. But the middle finger must not take part in this gesture, for by varying the position of this one also you may obtain other numbers. The number 20 is expressed solely by the contact between the thumbnail and the lower joint of the index finger 20

Fig. 3.13. *(continued)*

B. TENS (continued)		
30	For "thirty", touch thumb and index in a gentle kiss	For 30, hold the thumb straight and touch the tip of the thumbnail with the index finger, so that together they resemble the arc of a circle with its chord (if you need to bend the thumb somewhat, the number will be equally well indicated and no confusion should result) 30
40	For "forty", place the inside of the thumb against the side or the back of the index finger, keeping both of them straight	For 40, place the inside of the tip of the thumb on the back of the index finger, so that there is no space between the thumb and the edge of the palm 40
50	For "fifty", bend the thumb across the palm of the hand, with the top joint bent over, like the Greek letter Γ	For 50, hold the index finger straight up, but bend the thumb and place it in the palm of the hand, in front of the index finger 50
60	For "sixty", with the thumb bent as for fifty, the index finger is brought down to cover the face of the thumbnail	For 60, bend the thumb and place the second phalanx of the index finger on the face of the thumbnail 60
70	For "seventy", with the index finger as before, that is closely covering the thumbnail, raise the thumbnail across the middle joint of the index finger	For 70, raise the thumb and place the underside of the first joint of the index finger on the tip of the thumbnail so that the face of the thumbnail remains uncovered 70

FIG. 3.13. *(continued)*

B. TENS (continued)			

 80

For "eighty", with the index raised as above, and the thumb straight, place the thumbnail within the bent middle phalanx of the index finger

For 80, hold the thumb straight and place the tip of the index finger on the curve of its top joint. (Note the discrepancy between the two accounts here)

 80

 90

For "ninety", press the nail of the index finger against the root of the thumb

For 90, put the nail of the index finger over the joint of the second phalanx of the thumb (just as, for 10, you place it over the joint of the first phalanx)

 90

C. HUNDREDS AND THOUSANDS

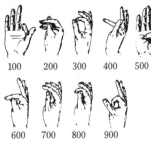

100	200	300	400	500

| 600 | 700 | 800 | 900 | |

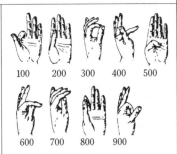

100	200	300	400	500

| 600 | 700 | 800 | 900 | |

When you say "a hundred", on your right hand do as for ten on the left hand; "two hundred" on the right hand is like twenty on the left; "three hundred" on the right like thirty on the left; and so on up to "nine hundred"

When you say "a thousand", with your right hand you do as for one with the left; "two thousand", on the right is like two on the left; "three thousand" on the right like three on the left, and so on up to "nine thousand"

Once you have mastered these eighteen numbers, the nine combinations of the little, ring and middle fingers as well as the nine combinations of the thumb and the index finger then you can readily understand that what serves on the right hand to show the units from 1 to 9 will on the left hand show from 1,000 to 9,000; and that what on the right hand shows the tens, on the left hand shows the hundreds from 100 to 900

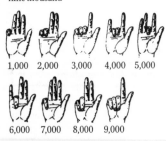

1,000	2,000	3,000	4,000	5,000

| 6,000 | 7,000 | 8,000 | 9,000 | |

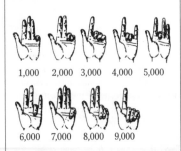

1,000	2,000	3,000	4,000	5,000

| 6,000 | 7,000 | 8,000 | 9,000 | |

FIG. 3.13. *(continued)*

same place, but palm outwards. For 80,000 grasp your thigh with your hand. For 90,000 grasp your loins with the left hand, the thumb towards the genitals.

Bede continues by describing how, by using the same signs on the right-hand side of the body, and with the right hand, the numbers from 100,000 to 900,000 may be represented. Finally he explains that one million may be indicated by crossing the two hands, with the fingers intertwined.

FINGER-COUNTING THROUGHOUT HISTORY

The method described above is extremely ancient. It is likely that it goes back to the most extreme antiquity, and it remained prominent until recent times in both the Western and Eastern worlds and, in the latter, persisted until recent times. In the Egypt of the Pharaohs it was in use from the Old Kingdom (2800–2300 BCE), as it would seem from a number of funeral paintings of the period. For example, Fig. 3.14 shows, from right to left, three men displaying numbers on their fingers according to the method just described. The first figure seems to be indicating 10 or 100, the fourth 6 or 6,000 and the sixth 7 or 7,000. According to traditions which have been repeated by various authors, Egypt clearly appears to have been the source of this system.

FIG. 3.14. *Finger-counting shown on a Egyptian monument of the Old Kingdom (Fifth Dynasty, 26th century BCE). Mastaba D2 at Saqqara. See Borchardt (1937), no. 1534A, plate 48.*

C. Pellat (1977) quotes two Arab manuscripts. One of these is at the University of Tunis (no. 6403) and the other is in the library of the Waqfs in Baghdad (*Majami‘* 7071/9). The counting system in question is, in the first manuscript, attributed to "the Copts of Egypt"; the title of the second clearly suggests that it is of Egyptian origin. (*Treatise on the Coptic manner of counting with the hands*).

A *qasida* (poem in praise of a potential patron) attributed to Mawsili al Ḥanbali describes "the sign language of the Copts of Egypt, which expresses numbers by arranging the fingers in special ways". Ibn al Maghribi states, "See! I follow in the steps of every learned man. The spirit moves me to write something of this art and to compose a *Ragaz*, to be called *The Table of Memory*, which shall include the art of counting of the Copts." Finally, Juan Perez de Moya (*Alcala de Henares*, 1573) comes to the following conclusion: "No one knows who invented this method of counting, but since the Egyptians loved to be sparing of words (as Théodoret has said), it must be from them that it has come."

There is also evidence for its use in ancient Greece. Plutarch (*Lives of Famous Men*) has it that Orontes, son-in-law of Artaxerxes King of Persia, said: "Just as the fingers of one who counts are sometimes worth ten thousand and sometimes merely one, so also the favourites of the King may count for everything, or for nothing."

The method was also used by the Romans, as we know in the first instance from "number-tiles" discovered in archaeological excavations from several parts of the Empire, above all from Egypt, which date mostly from the beginning of the Christian era (Fig. 3.15). These are small counters or tokens, in bone or ivory, each representing a certain sum of money. The Roman tax collectors gave these as "receipts". On one side there was a representation of one of the numbers according to the sign system described above, and on the other side was the corresponding Roman numeral. (It would seem, however, that these numbers never went above 15 in these counters from the Roman Empire).

FIG. 3.15. *Roman numbered tokens (tesserae) from the first century CE. The token on the left shows on one face the number 9 according to a particular method of finger-counting; on the reverse face, the same value is shown in Roman numerals. British Museum. The token on the right shows a man making the sign for 15, according to the same system, on the fingers of his left hand. Bibliothèque nationale (Paris). Tessera no. 316. See Frohner (1884).*

We also know about this from the writings of numerous Latin authors. Juvenal (c. 55–135 CE) speaks thus of Nestor, King of Pylos, who lived, it is said, for more than a hundred years: "Fortunate Nestor who, having attained one hundred years of age, henceforth shall count his years on his right hand!" This tells us that the Romans counted tens and units on the left hand, and hundreds and thousands on the right hand. Apuleus (c. 125–170 CE) describes in his *Apologia* how, having married a rich widow, a certain Aemilia Pudentilla, he was accused of resorting to magic means to win her heart. He defended himself before the Pro-Consul

Claudius Maximus in the presence of his chief accuser Emilianus. Emilianus had ungallantly declared that Aemilia was sixty years old, whereas she was really only forty. Here is how Apuleus challenges Emilianus.

> How dare you, Emilianus, increase her true number of years by one half again? If you said thirty for ten, we might think that you had ill-expressed it on your fingers, holding them out straight instead of curved (Fig. 3.16). But forty, now that is easily shown: it is the open hand! So when you increase it by half again this is not a mistake, unless you allow her to be thirty years old and have doubled the consular years by virtue of the two consuls.

FIG. 3.16.

And we may cite Saint Jerome, Latin philologist of the time of Saint Augustine:

> One hundred, sixty, and thirty are the fruits of the same seed in the same earth. Thirty is for marriage, since the joining of the two fingers as in a tender kiss represents the husband and the wife. Sixty depicts the widow in sadness and tribulation. And the sign for one hundred (pay close attention, gentle reader), copied from the left to right with the same fingers, shows the crown of virginity (Fig. 3.17).

FIG. 3.17.

Again, the patriarch Saint Cyril of Alexandria (376–444) gives us the oldest known description of this system (*Liber de computo*, Chapter CXXXVIII: *De Flexibus digitorum*, III, 135). The description exactly matches a passage in a sixth-century Spanish encyclopaedia, *Liber etymologiarum*, which was the outcome of an enormous compilation instituted by Bishop Isidor of Seville (570–636). The Venerable Bede, in his turn, drew inspiration from it in the seventh century for his chapter *De computo vel loquela digitorum*.

One of the many reasons why this system remained popular was its secret, even mysterious, aspect. J. G. Lemoine (1932) says: "What a splendid method for a spy to use, from the enemy camp, to inform his general at a distance of the numerical force of the enemy, by a simple, apparently casual, gesture or pose." Bede also gives an example of such silent communication: "A kind of manual speech [*manualis loquela*] can be expressed by the system which I have explained, as a mental exercise or as an amusement." Having established a correspondence between the Latin letters and the integers, he says: "To say *Caute age* ('look out!') to a friend amongst doubtful or dangerous people, show him (the following finger gestures)" (Fig. 3.18).

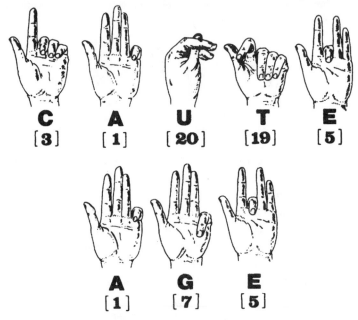

C [3] **A** [1] **U** [20] **T** [19] **E** [5]

A [1] **G** [7] **E** [5]

F*IG.* 3.18.

Following the fall of the Roman Empire, the same manual counting remained extraordinarily in vogue until the end of the Middle Ages (Fig. 3.19 to 3.21), and played a most important part in mediaeval education. The finger counting described in Bede's *De computo vel loquela digitorum* (cited above) was extensively used in the teaching of the *Trivium* of grammar, rhetoric and logic during the undergraduate years leading to the B.A. degree, which, with the *Quadrivium* (literally "crossroads", the meeting of the Four Ways of arithmetic, geometry, astronomy, and music) studied in the following years leading to the M.A. degree, made up the Seven Liberal Arts of the scholarly curriculum, from the sixth to the fifteenth centuries. Barely four hundred years ago, a textbook of arithmetic was not considered complete without detailed explanations of this system (Fig. 3.22). Only when written arithmetic became widespread, with the adoption of the use of Arabic numerals, did the practice of arithmetic on hands and fingers finally decline.

FIG. 3.19. *The system described in Fig. 3.13 illustrated in a manuscript by the German theologian Rabanus Maurus (780–856). Codex Alcobacense 394, folio 152 V. National Library of Lisbon. See Burnam (1912–1925), vol. 1, plate XIV.*

FIG. 3.20. *The same system again, in a Spanish manuscript of 1130. Detail of a codex from Catalonia (probably from Santa Maria de Ripoll). National Library of Madrid, Codex matritensis A19, folio 3V. See Burnam (1912–1925), vol. 3, plate XLIII.*

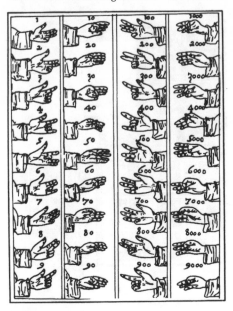

FIG. 3.21. *The same system yet again in a mathematical work published in Vienna in 1494. Extract from the work by Luca Pacioli:* Summa de Arithmetica, Geometrica, proportioni e proportionalita

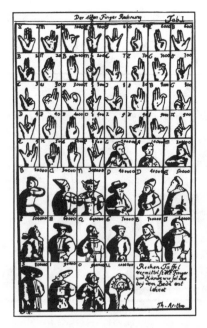

FIG. 3.22. *The same system of signs in a work on arithmetic published in Germany in 1727: Jacob Leupold,* Theatrum Arithmetico-Geometricum

FIG. 3.23. *In the Arab-Persian system of number gestures, the number 93 is shown by placing the nail of the index finger right on the joint of the second phalanx of the thumb (which represents 90), and then bending the middle, ring and little fingers (which represents 3); and this, nearly enough, gives rise to a closed fist.*

93

In the Islamic world, the system was at least as widely spread as in the West, as recounted by many Arab and Persian writers from the earliest times. From the beginning of the Hegira, or Mohammedan era (dated from the flight of Mohammed from Mecca to Medina on 15 July 622 CE), we find an oblique allusion among these poets when they say that a mean or ungenerous person's hand "makes 93" (see the corresponding closed hand, symbol of avarice, in Fig. 3.23). One of them, Yaḥyā Ibn Nawfal al Yamānī (seventh century) says: "Ninety and three, which a man may show as a fist closed to strike, is not more niggardly than thy gifts, Oh Yazid." Another, Khalîl Ibn Aḥmad (died 786), grammarian and one of the founders of Arab poetry, writes: "Your hands were not made for giving, and their greed is notorious: one of them makes 3,900 (the mirror image of 93) and the other makes 100 less 7."

One of the greatest Persian poets, Abu'l Kassim Firdūsi, dedicated *Shah Naméh* (The Book of Kings) to Sultan Mahmūd le Ghaznavide but found himself poorly rewarded. In a satire on the Sultan's gross avarice, he wrote: "The hand of King Mahmūd, of noble descent, is nine times nine and three times four."

A *qasîda* of the Persian poet Anwari (died 1189 or 1191) praises the Grand Vizir Nizam al Mūlk for his precocity in arithmetic: "At the age when most children suck their thumbs, you were bending the little finger of your left hand" (implying that the Vizir could already count to a thousand) (Fig. 3.13C).

A dictum of the Persian poet Abu'l Majîd Sanāyî (died 1160) reminds us that by twice doing the same thing in one's life, one may take away from its value: "What counts for 200 on the left hand, on the right hand is worth no more than 20" (Fig. 3.24). The poet Khāqānī (1106–1200) exclaims: "If I could count the turns of the wheel of the skies, I would number them on my left hand!" and: "Thou slayest thy lover with the glaive of thy glances, so many as thou canst count on thy left hand" (the left hand counts the hundreds and the thousands).

Another quotation from Anwari: "One night, when the service I rendered thee did wash the face of my fortune with the water of kindness, you did give to me that number (50) which thy right thumb forms when it tries to hide its back under thy hand" (Fig. 3.25).

And some verses of Al Farazdaq (died 728) refer to forming the number 30 by opposing thumb and forefinger, in a description of crushing pubic lice (Fig. 3.26).

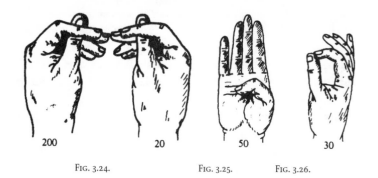

FIG. 3.24.   FIG. 3.25.   FIG. 3.26.

According to Levi della Vida (1920), one of the earliest datable references from the Islamic world to this numerical system can be found in the following quotation from Ibn Sa'ad (died c. 850): "Hudaifa Ibn al Yamān, companion of the Prophet, signalled the murder of Khalif 'Otman as one shows the number 10 and sighed: 'This will leave a void [forming a round between finger and thumb, Fig. 3.27] in Islam which even a mountain could hardly fill.'"

A poem attributed to Al Mawsilī al Ḥanbalī says: "If you place the thumb against the forefinger like – listen carefully – someone who takes hold of an arrow, then it means 60" (Fig. 3.28); and, in verses attributed to Abūl Ḥassan 'Alī: "For 60, bend your forefinger over your thumb, as a bowman grasps an arrow [Fig. 3.28] and for 70 do like someone who flicks a dinar to test it" (Fig. 3.29).

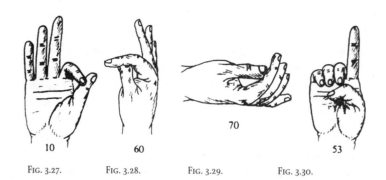

FIG. 3.27.   FIG. 3.28.   FIG. 3.29.   FIG. 3.30.

Aḥmad al Barbīr al Tarābulusī (a writer on secular Arab and Persian texts), talking of what he calls counting by bending the fingers, says: "We know the traditionalists use it, because we find references; and it is the same with the *fuqahā*,* for these lawyers refer to it in relation to prayer in connection with the Confession of Faith;† they say that, according to the rule of tradition, he who prays should place his right hand on his thigh when he squats for the Tashahud, forming the number 53" (Fig. 3.30).

From the poet Khāqānī we have: "What struggle is this between Rustem and Bahrām? What fury and dispute is it that perturbs these two sons of noble lines? Why, they fight day and night to decide which army shall do a 20 on the other's 90."

This may seem obscure to the modern reader, unfamiliar with the finger signs in question. But look closely at the gestures that correspond to the numbers (Fig. 3.31): "90" undoubtedly represents the anus (and, by extension, the backside), as it commonly did in vulgar speech; while "to do a 20 on someone" is undoubtedly an insulting reference to the sexual act (apparently expressed as "to make a thumb" in Persian) and therefore (by extension in this military context) to "get on top of".

More obscenely, Aḥmād al Barbīr al Tarābulusī could not resist offering his pupils the following mnemonic for the gestures representing 30 and 90:

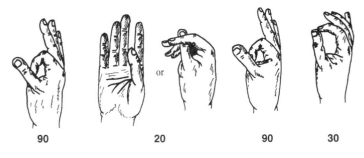

or

90   **20**   **90**   **30**

FIG. 3.31.   FIG. 3.32.

"A poet most elegantly said, of a handsome young man: Khālid set out with a fortune of 90 dirhams, but had only one third of it left when he returned!" plainly asserting that Khālid was homosexual (Fig. 3.32), having started "narrow" (90) but finished "wide" (30).

These many examples amply show how numbers formed by the fingers served as figures of speech, no doubt much appreciated by the readers of the time. These ancient origins find etymological echo today, as in *digital computing*. There is no longer any question of literally counting on the fingers, but the Latin words *digiti* ("fingers") and *articuli* ("joints") came to represent "units" and "tens", respectively, in the Middle Ages, whence *digiti*

* Islamic lawyers who concern themselves with every kind of social or personal matter, with the order of worship and with ritual requirements
† Asserting that Allah is One, affirming belief in Mohammed, at the same time raising the index finger and closing the others

in turn came to mean the signs used to represent the units of the decimal system. The English word *digit*, meaning a single decimal numeral, is derived directly from this. In turn, this became applied to computation, hence the term *digital* computing in the sense of "computing by numbers". With the development and recent enormous spread of "computers" (*digital* computers), the meaning of "digital computation" has been extended to include every aspect of the processing of information by machine in which any entity, numerical or not, and whether or not representing a variable physical quantity, is given a *discrete* representation (by which is meant that distinct representations correspond to different values or entities, there is a finite – though typically enormous – number of possible distinct representations, and different repesentations are encoded as sequences of symbols taken from a finite set of available symbols). In the modern digital computer, the primitive symbols are two in number and denoted by "0" and "1" (the *binary* system) and realised in the machine in terms of distinct physical states which are reliably distinguishable.

HOW TO CALCULATE ON YOUR FINGERS

After this glance at the modern state of the art of digital information processing, let us see how the ancients coped with their "manual informatics".

The hand can be used not only for counting, but also for systematically performing arithmetical calculations. I used to know a peasant from the Saint-Flour region, in the Auvergne, who could multiply on his fingers, with no other aid, any two numbers he was given. In so doing, he was following in a very ancient tradition.

For example, *to multiply 8 by 9*, he closed on one hand as many fingers as the excess of 8 over 5, namely 3, keeping the other two fingers extended. On the other hand, he closed as many fingers as the excess of 9 over 5, namely 4, leaving the fifth finger extended (Fig. 3.33). He would then (mentally) multiply by 10 the total number (7) of closed fingers (70), multiply together the numbers of extended fingers on the two hands ($2 \times 1 = 2$), and finally add these two results together to get the answer (72). That is to say:

$$8 \times 9 = (3 + 4) \times 10 + (2 \times 1) = 72$$

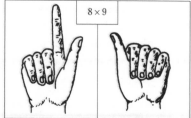

8×9

Fig. 3.33.

Similarly, *to multiply 9 by 7*, he closed on one hand the excess of 9 over 5, namely, 4, and on the other the excess of 7 over 5, namely 2, in total 6; leaving extended 1 and 3 respectively, so that by his method the result is obtained as

$$9 \times 7 = (4 + 2) \times 10 + (1 \times 3) = 63$$

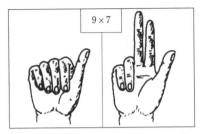

9×7

FIG. 3.34.

Although undoubtedly discovered by trial and error by the ancients, this method is infallible for the multiplication of any two whole numbers between 5 and 10, as the following proves by elementary (but modern) algebra. To multiply together two numbers x and y each between 5 and 10, close on one hand the excess $(x - 5)$ of x over 5, and on the other the excess $(y - 5)$ of y over 5; the total of these two is $(x - 5) + (y - 5)$, and 10 times this is $$((x - 5) + (y - 5)) \times 10 = 10\,x + 10\,y - 100$$

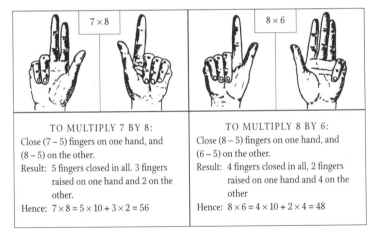

7×8 8×6

TO MULTIPLY 7 BY 8:	TO MULTIPLY 8 BY 6:
Close $(7 - 5)$ fingers on one hand, and $(8 - 5)$ on the other.	Close $(8 - 5)$ fingers on one hand, and $(6 - 5)$ on the other.
Result: 5 fingers closed in all. 3 fingers raised on one hand and 2 on the other.	Result: 4 fingers closed in all, 2 fingers raised on one hand and 4 on the other
Hence: $7 \times 8 = 5 \times 10 + 3 \times 2 = 56$	Hence: $8 \times 6 = 4 \times 10 + 2 \times 4 = 48$

FIG. 3.35.

The number of fingers remaining extended on the first hand is $5 - (x - 5) = 10 - x$, and on the other, similarly, $10 - y$. The product of these two is

$$(10 - x) \times (10 - y) = 100 - 10\,x - 10\,y + x\,y$$

Adding these two together, according to the method, therefore results in

$$(10\,x + 10\,y - 100) + 100 - 10\,x - 10\,y + xy = x\,y$$

namely the desired result of multiplying x by y.

He had a similar way of multiplying numbers exceeding 9. For example, *to multiply 14 by 13*, he closed on one hand as many fingers as the excess of 14 over 10, namely 4, and on the other, similarly, 3, making in all 7. Then he mentally multiplied this total (7) by 10 to get 70, adding to this the product ($4 \times 3 = 12$) to obtain 82, finally adding to this result $10 \times 10 = 100$ to obtain 182 which is the correct result.

By similar methods, he was able to multiply numbers between 15 and 20, between 20 and 25, and so on. It is necessary to know the squares of 10, 15, 20, 25 and so on, and their multiplication tables. The mathematical justifications of some of these methods are as follows.

To multiply two numbers x and y between 10 and 15:

$$10\,[(x-10)+(y-10)] + (x-10) \times (y-10) + 10^2 = x\,y$$

> MULTIPLYING NUMBERS BETWEEN 10
> AND 15
> ON THE FINGERS
>
> (It must be known by heart that 100 is the square of 10)
>
> Example: 12×13
> Close: $(12 - 10)$ fingers on one hand, and
> $(12 - 10)$ on the other.
> Result: 2 fingers closed on one hand, and 3 on the other.
> Hence: $12 \times 13 = 10 \times (2 + 3) + (2 \times 3) + 10 \times 10$
> $= 156$

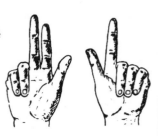

FIG. 3.36.

To multiply two numbers x and y between 15 and 20:

$$15\,[(x-15)+(y-15)] + (x-15) \times (y-15) + 15^2 = xy$$

> MULTIPLYING NUMBERS
> BETWEEN 15 AND 20
> ON THE FINGERS
>
> (It must be known by heart that 225 is the square of 15)
>
> Example: 18×16
> Close: $(18 - 15)$ fingers on one hand, and
> $(16 - 15)$ on the other.
> Result: 3 fingers closed on one hand, and 1 on the other
> Hence: $15 \times (3 + 1) + (3 \times 1) + 15 \times 15$
> $= 288$

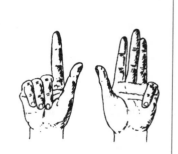

FIG. 3.37.

To multiply two numbers x and y between 20 and 25:

$$20\,[(x-20)+(y-20)]+(x-20)\times(y-20)+20^2 = xy$$

and so on.*

It can well be imagined, therefore, how people who did not enjoy the facility in calculation which our "Arabic" numerals allow us were none the less able to devise, by a combination of memory and a most resourceful ingenuity in the use of the fingers, ways of overcoming their difficulties and obtaining the results of quite difficult calculations.

FIG. 3.38. *Calculating by the fingers shown in an Egyptian funeral painting from the New Kingdom. This is a fragment of a mural on the tomb of Prince Menna at Thebes, who lived at the time of the 18th Dynasty, in the reign of King Thutmosis, at the end of the 15th century BCE. We see six scribes checking while four workers measure out grain and pour bushels of corn from one heap to another. On the right, on one of the piles of grain, the chief scribe is doing arithmetic on his fingers and calling out the results to the three scribes on the left who are noting them down. Later they will copy the details onto papyrus in the Pharaoh's archives. (Theban tomb no. 69)*

COUNTING TO THOUSANDS USING THE FINGERS

The method to be described is a much more developed and mathematically more interesting procedure than the preceding one. There is evidence of its use in China at any rate since the sixteenth century, in the arithmetical textbook *Suan fa tong zong* published in 1593. E. C. Bayley (1847) attests that it was in use in the nineteenth century, and Chinese friends of mine from Canton and Peking have confirmed that it is still in use.

In this method, each knuckle is considered to be divided into three parts: left knuckle, middle knuckle and right knuckle. There being three knuckles to a finger, there is a place for each of the nine digits from 1 to 9. Those on the

---

* The general rule being:
$N((x-N)+(y-N))+(x-N)(y-N)+N^2 = Nx + Ny - 2N^2 + xy - xN - yN + N^2 + N^2 = xy.$

little finger of the right hand correspond to the units, those on the fourth finger to the tens, on the middle finger to the hundreds, the forefinger to the thousands, and finally the right thumb corresponds to the tens of thousands. Similarly on the left hand, the left thumb corresponds to the hundreds of thousands, the forefinger to the millions, the middle finger to the tens of millions, and so on (Fig. 3.39); finally, therefore, on the little finger of the left hand we count by steps of thousands of millions, i.e. by billions.

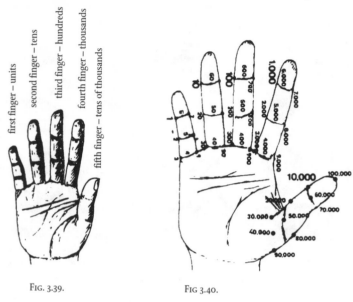

FIG. 3.39.   FIG 3.40.

With the right hand palm upwards (Fig. 3.40), we count on the little finger from 1 to 3 by touching the "left knuckles" from tip to base; then from 4 to 6 by touching the "centre knuckles" from base to tip; and finally, from 7 to 9 by touching the "right knuckles" from tip to base. We count the tens similarly on the fourth finger, the hundreds on the middle finger, and so on.

In this way it is, in theory, possible to count up to 99,999 on one hand, and up to 9,999,999,999 with both: a remarkable testimony to human ingenuity.

CHAPTER 4

HOW CRO-MAGNON MAN COUNTED

Among the oldest and most widely found methods of counting is the use of marked bones. People must have made use of this long before they were able to count in any abstract way.

The earliest archaeological evidence dates from the so-called Aurignacian era (35,000–20,000 BCE), and are therefore approximately contemporary with Cro-Magnon Man. It consists of several bones, each bearing regularly spaced markings, which have been mostly found in Western Europe (Fig. 4.1).

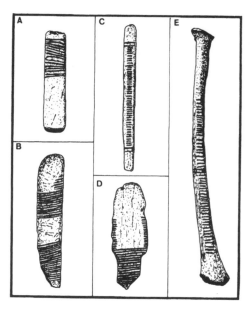

FIG. 4.1. *Notched bones from the Upper Palaeolithic age. A and C: Aurignacian. Musée des Antiquités nationales, St-Germain-en-Laye. Bone C is from Saint-Marcel (Indre, France). B and D: Aurignacian. From the Kulna cave (Czech Republic). E: Magdalenian (19,000 – 12,000 BCE). From the Pekarna cave (Czech Republic). See Jelinek (1975), pp. 435–453.*

Amongst these is the radius bone of a wolf, marked with 55 notches in two series of groups of five. This was discovered by archaeologists in 1937, at Dolní Věstonice in Czechoslovakia, in sediments which have been dated as approximately 30,000 years old. The purpose of these notches remains mysterious, but this bone (whose markings are systematic, and not artistically motivated) is one of the most ancient arithmetic documents to have come down to us. It clearly demonstrates that at that time human beings were not only able to conceive number in the abstract sense, but also to represent number with respect to a base. For otherwise, why would the notches have been grouped in so regular a pattern, rather than in a simple unbroken series?

The man who made use of this bone may have been a mighty hunter. Each time he made a kill, perhaps he made a notch on his bone. Maybe he had a different bone for each kind of animal: one for bears, another for deer, another for bison, and so on, and so he could keep the tally of the larder. But, to avoid having to re-count every single notch later, he took to grouping them in fives, like the fingers of the hand. He would therefore have established a true graphical representation of the first few whole numbers, in base 5 (Fig. 4.2).

I I I I I	I I I I I	I I I I I	I I I I I	...
1 2 3 4 5	6 7 8 9 10	11 ... 15	16 ... 20	
1 hand	2 hands	3 hands	4 hands	

Fig. 4.2.

Also of great interest is the object shown in Fig. 4.3, a point from a reindeer's antler found some decades ago in deposits at Brassempouy in the Landes, dating from the Magdalenian era. This has a longitudinal groove which separates two series of transverse notches, each divided into distinct groups (3 and 7 on one side, 5 and 9 on the other). The longitudinal notch, which is much closer to the 9–5 series than to the 3–7 series, seems to form a kind of link or *vinculum* (as is sometimes used in Mathematics) joining the group of nine to the group of five.

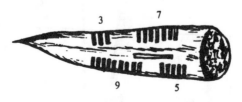

Fig. 4.3. *Notched bone from the Magdalenian era (19,000–12,000 BCE), found at Brassempouy (Landes, France). Bordeaux, Museum of Aquitaine*

Now what could this be for? Was it perhaps a simple tool, or a weapon, which had been grooved to stop it slipping in the hand? Unlikely. Anyway, what purpose would the longitudinal groove then serve? And even if this were the case, why do we not find such markings on similar prehistoric implements?

In fact, this object also bears witness of some activity with arithmetical connotations. The way the numbers 3, 5, 7, and 9 are arranged, and the frequency with which these numbers occur in many artefacts from the same period, suggest a possible explanation.

Let us suppose that the longitudinal groove represents unity, and that the transverse lines represent other odd numbers (which are all prime except for 9 which is the square of 3).

This spike from an antler with its grooves then makes a kind of arithmetical tool, showing a graphical representation of the first few odd numbers arranged in such a way that some of their simpler properties are exhibited (Fig. 4.4).

$$3 \bullet \quad \bullet 7$$
$$9 \bullet \quad \bullet 5$$
$$9 - 7 = 5 - 3 = 2$$

$$3 \bullet \leftarrow - \bullet 7$$
$$9 \bullet - \rightarrow \bullet 5$$
$$7 - 3 = 9 - 5 = (9 + 5) - (7 + 3) = 4$$

$$3 \bullet \quad \bullet 7$$
$$9 \bullet \quad \bullet 5$$
$$3 \times 3 = 9$$

$$3 \bullet \quad \bullet 7$$
$$9 \bullet \quad \bullet 5$$
$$3 + 9 = 5 + 7 = 12$$

As well as giving us concrete evidence for the memorisation and recording of numbers, the practice of making tally marks such as described is also a precursor of counting and book-keeping. We are therefore led to suppositions such as the following.

Our distant forefathers possibly used this piece of antler for taking count of people, things or beasts. It could perhaps have served a tool-maker to keep account of his own tools:

> 3 graters and 7 knives (in stone)
> 9 scrapers and 5 needles (in bone)

where the longitudinal groove linking the 5 and the 9 may, in this man's mind, have denoted the common material (bone) from which they were made.

Or perhaps a warrior might similarly keep count of his weapons:

> 3 knives and 7 daggers
> 9 spears with plain blade, and 5 with split blade

Or the hunter might record the numbers of different types of game brought back for the benefit of his people:

> 3 bison and 7 buffalo
> 9 reindeer and 5 stags

We can also imagine how a herdsman could count the beasts in his keeping, sheep and goats on the one hand, cattle on the other.

A messenger could use an antler engraved in this way to carry a promissory note to a neighbouring tribe:

> In 3 moons and 7 days we will bring
> 9 baskets of food and 5 fur animals

We can also imagine it being used as a receipt for goods, or a delivery manifest, or for accounting for an exchange or distribution of goods.

Of course, these are only suppositions, since the true meaning has eluded the scholars. And in fact the true purpose of these markings will remain unknown for ever, because with this kind of symbolism the things themselves to which the operations apply are represented only by their quantity, and not by specific signs which depict the nature of the things.

Human kind was still unable to write. But, by representing as we have described the enumeration of this or that kind of unit, the owner of the antler, and his contemporaries, had nonetheless achieved the inventions of written number: in truth, they wrote figures in the most primitive notation known to history.

CHAPTER 5

TALLY STICKS
ACCOUNTING FOR BEGINNERS

Notched sticks – tally sticks – were first used at least forty thousand years ago. They might seem to be a primitive method of accounting, but they have certainly proved their value. The technique has remained much the same through many centuries of evolutionary, historical, and cultural change, right down to the present day. Although our ancestors could not have known it, their invention of the notched stick has turned out to be amongst the most permanent of human discoveries. Not even the wheel is as old; for sheer longevity, only fire could possibly rival it.

Notch-marks found on numerous prehistoric cave-wall paintings alongside outlines of animals leave no doubt about the accounting function of the notches. In the present-day world the technique has barely changed at all.

For instance, in the very recent past, native American labourers in the Los Angeles area used to keep a tally of hours worked by scoring a fine line in a piece of wood for each day worked, with a deeper or thicker line to mark each week, and a cross for each fortnight completed.

More colourful users of the device in modern times include cowboys, who made notches in the barrels of their guns for each bison killed, and the fearsome bounty hunters who kept a tally in the same manner for every outlaw that they gunned down. And Calamity Jane's father also used the device for keeping a reckoning of the number of marriageable girls in his town.

On the other side of the world, the technique was in daily use in the nineteenth century, as we learn from explorer's tales:

> On the road, just before a junction with a smaller track, I came upon a
> heavy gate made of bamboo and felled tree trunks, and decorated with
> hexagonal designs and sheaves. Over the track itself was hung a small
> plank with a set of regularly-sized notches, some large, some small, on
> each side. On the right were twelve small notches, then four large ones,
> then another set of twelve small ones. This meant: *Twelve days march
> from here, any man who crosses our boundary will be our prisoner or will
> pay a ransom of four water-buffalo and twelve ticals* (rupees). On the left,
> eight large notches, eleven middle-sized ones, and nine small ones,
> meaning: *There are eight men, eleven women and nine children in our
> village.* [J. Harmand (1879): Laos]

In Sumatra, the Lutsu declared war by sending a piece of wood scored with lines together with a feather, a scrap of tinder and a fish. Translation: they will attack with as many hundreds (or thousands) of men as there are scored lines; they will be as swift as a bird (the feather), will lay everything waste (the tinder = fire), and will drown their enemies (the fish). [J. G. Février (1959)]

Only a few generations ago, shepherds in the Alps and in Hungary, as well as Celtic, Tuscan, and Dalmatian herders, used to keep a tally of the number of head in their flocks by making an equivalent number of notches or crosses on wooden sticks or planks. Some of them, however, had a particularly developed and subtle version of the technique as L. Gerschel describes:

> On one tally-board from the Moravian part of Walachia, dating from 1832, the shepherd used a special form of notation to separate the milk-bearing sheep from the others, and within these, a special mark indicated those that only gave half the normal amount. In some parts of the Swiss Alps, shepherds used carefully crafted and decorated wooden boards to record various kinds of information, particularly the number of head in their flock, but they also kept separate account of sterile animals, and distinguished between sheep and goats ....

We can suppose that shepherds of all lands cope with much the same realities, and that only the form of the notation varies (using, variously, knotted string or *quipu* [see Chapter 6 below], primitive notched sticks, or a board which may include (in German-speaking areas) words like *Küo* (cows), *Gallier* (sterile animals), *Geis* (goats) alongside their tallies. There is one constant: the shepherd must know how many animals he has to care for and feed; but he also has to know how many of them fall into the various categories – those that give milk and those that don't, young and old, male and female. Thus the counts kept are not simple ones, but threefold, fourfold or more *parallel tallies* made simultaneously and entered side by side on the counting tool.

FIG. 5.1. *Swiss shepherd's tally stick (Late eighteenth century, Saanen, Canton of Bern). From the Museum für Völkerkunde, Basel; reproduced from Gmür (1917)*

In short, shepherds such as these had devised a genuine system of accounting.

Another recent survival of ancient methods of counting can be found in the name that was given to one of the taxes levied on serfs and commoners in France prior to 1789: it was called *la taille*, meaning "tally" or "cut", for the simple reason that the tax-collectors totted up what each taxpayer had paid on a wooden tally stick.

In England, a very similar device was used to record payments of tax and to keep account of income and expenditure. Larger and smaller notches on wooden batons stood for one, ten, one hundred, etc., pounds sterling (see Fig. 5.2). Even in Dickens's day, the Treasury still clung on to this antiquated system! And this is what the author of *David Copperfield* thought of it:

> Ages ago, a savage mode of keeping accounts on notched sticks was introduced into the Court of Exchequer; the accounts were kept, much as Robinson Crusoe kept his calendar on the desert island. In course of considerable revolutions of time ... a multitude of accountants, book-keepers, actuaries and mathematicians, were born and died; and still official routine clung to these notched sticks, as if they were pillars of the constitution, and still the Exchequer accounts continued to be kept on certain splints of elm wood called "tallies". Late in the reign of George III, some restless and revolutionary spirit originated the suggestion, whether, in a land where there were pens, ink and paper, slates and pencils, and systems of accounts, this rigid adherence to a barbarous usage might not border on the ridiculous? All the red tape in the public offices turned redder at the bare mention of this bold and original conception, and it took till 1826 to get these sticks abolished.
>
> [Charles Dickens (1855)]

Britain may be a conservative country, but it was no more backward than many other European nations at that time. In the early nineteenth century, tally sticks were in use in various roles in France, Germany, and Switzerland, and throughout Scandinavia. Indeed, I myself saw tally sticks in use as credit tokens in a country bakery near Dijon in the early 1970s. This is how it is done: two small planks of wood, called *tailles*, are both marked with a notch each time the customer takes a loaf. One plank stays with the baker, the other is taken by the customer. The number of loaves is totted up and payment is made on a fixed date (for example, once a week). No dispute over the amount owed is possible: both planks have the same number of notches, in the same places. The customer could not have removed any, and there's an easy way to make sure the baker hasn't added any either, since the two planks have to match (see Fig. 5.3).

The French baker's tally stick was described thus in 1869 by André Philippe, in a novel called *Michel Rondet:*

> The women each held out a piece of wood with file-marks on it. Each piece of wood was different – some were just branches, others were

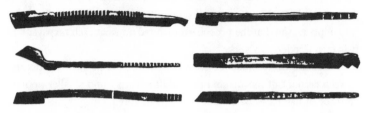

FIG. 5.2. *English accounting tally sticks, thirteenth century. London, Society of Antiquaries Museum*

FIG. 5.3. *French country bakers' tally sticks, as used in small country towns*

planed square. The baker had identical ones threaded onto a strap. He looked out for the one with the woman's name on it on his strap, and the file-marks tallied exactly. The notches matched, with Roman numerals – I, V, X – signifying the weight of the loaves that had been supplied.

René Jonglet relates a very similar scene that took place in Hainault (French-speaking Belgium) around 1900:

> The baker went from door to door in his wagon, calling the housewives out. Each would bring her "tally" – a long and narrow piece of wood, shaped like a scissor-blade. The baker had a duplicate of it, put the two side by side, and marked them both with a saw, once for each six-pound loaf that was bought. It was therefore very easy to check what was owed, since the number of notches on the baker's and house-wife's tally stick was the same. The housewife couldn't remove any from both sticks, nor could the baker add any to both.

The tally stick therefore served not just as a curious form of bill and receipt, but also as a wooden credit card, almost as efficient and reliable as the plastic ones with magnetic strips that we use nowadays.

French bakers, however, did not have a monopoly on the device: the use of twin tally sticks to keep a record of sums owing and to be settled can be found in every period and almost everywhere in the world.

The technique was in use by the Khâs Boloven in Indo-China, for example, in the nineteenth century:

> For market purchases, they used a system similar to that of country bakers: twin planks of wood, notched together, so that both pieces held the same record. But their version of this memory-jogger is much more complicated than the bakers', and it is hard to understand how they coped with it. Everything went onto the planks – the names of the

sellers, the names of the buyer or buyers, the witnesses, the date of delivery, the nature of the goods and the price. [J. Harmand (1880)]

As Gerschel explains, the use of the tally stick is, in the first place, to keep track of partial and successive numbers involved in a transaction. However, once this use is fully established, other functions can be added: the tally stick becomes a form of memory, for it can hold a record not just of the intermediate stages of a transaction, but also of its final result. And it was in that new role, as the record of a completed transaction, that it acquired an economic function, beyond the merely arithmetic function of its first role.

The *mark of ownership* was the indispensable additional device that allowed tally sticks to become economic tools. The mark symbolised the name of its owner: it was his or her "character" and represented him or her legally in any situation, much like a signature. Improper use of the mark of ownership was severely punished by the law, and references to it are found in French law as late as the seventeenth century.

The mark of ownership thus took the notched stick into a different domain. Originally, notched sticks had only notches on them: but now they also carried signs representing not numbers but names.

FIG. 5.4. *Examples of marks of ownership used over the ages. The signs were allocated to specific members of the community and could not be exchanged or altered.*

Here is how they were used amongst the Kabyles, in Algeria:

Each head of cattle slaughtered by the community is divided equally between the members, or groups of members. To achieve this, each member gives the chief a stick that bears a mark; the chief shuffles the sticks and then passes them to his assistant, who puts a piece of meat on each one. Each member then looks for his own stick and thus obtains his share of the meat. This custom is obviously intended to ensure a fair share for everyone. [J. G. Février (1959)]

The mark of ownership probably goes back to the time before writing was invented, and it is the obvious ancestor of what we call a signature (the Latin verb *signare* actually means "to make a cross or mark"). So the mark, the "signature" of the illiterate, can be associated with the tally stick, the accounting device for people who cannot count.

But once you have signatures, you have contracts: which is how tally sticks with marks of ownership came to be used to certify all sorts of commitments and obligations. One instance is provided by the way the Cheremiss and Chuvash tribesfolk (central Russia) recorded loans of money in the nineteenth century. A tally stick was split in half lengthways, each half therefore bearing the same number of notches, corresponding to the amount of money involved. Each party to the contract took one of the halves and inscribed his personal mark on it (see Fig. 5.4), and then a witness made his or her mark on both halves to certify the validity and completeness of the transaction. Each party then took and kept the half with the *other's* signature or mark. Each thus retained a certified, legally enforceable and unalterable token of the amount of capital involved (indicated by the notched numbers on both tally sticks). The creditor could not alter the sum, since the debtor had the tally stick with the creditor's mark; nor could the debtor deny his debt, since the creditor had the tally stick with the debtor's mark on it.

According to A. Conrady (1920), notched sticks similarly constituted the original means of establishing pacts, agreements and transactions in pre-literate China. They gave way to written formula only after the development of Chinese writing, which itself contains a trace of the original system: the ideogram signifying *contract* in Chinese is composed of two signs meaning, respectively, "notched stick" and "knife".

FIG. 5.5.

The Arabs (or their ancestors) probably had a similar custom, since a similar derivation can be found in Arabic. The verb-root *farada* means both "to make a notch" and "to assign one's share (of a contract or inheritance) to someone".

In France, tally sticks were in regular use up to the nineteenth century as waybills, to certify the delivery of goods to a customer. Article 1333 of the *Code Napoléon*, the foundation stone of the modern French legal system, makes explicit reference to tally sticks as the means of guaranteeing that deliveries of goods had been made.

In many parts of Switzerland and Austria, tally sticks constituted until recently a genuine social and legal institution. There were, first of all, the *capital tallies* (not unlike the tokens used by the Chuvash), which recorded loans made to citizens by church foundations and by local authorities. Then there were the *milk tallies*. According to L. Gerschel, they worked in the following ways:

> At Ulrichen, there was a single tally stick of some size on which was inscribed the mark of ownership of each farmer delivering milk, and opposite his mark, the quantity of milk delivered. At Tavetsch (according to Gmür), each farmer had his own tally stick, and marked on it the amount of milk he owed to each person whose mark of ownership was

on the stick; reciprocally, what was owed to him appeared under his mark of ownership on others' tally sticks. When the sticks were compared, the amount outstanding could be computed.

There were also *mole tallies*: in some areas, the local authorities held tallies for each citizen, marked with that citizen's mark of ownership, and would make a notch for each mole, or mole's tail, surrendered. At the year's end, the mole count was totted up and rewards paid out according to the number caught.

Tallies were also used in the Alpine areas for recording pasture rights (an example of such a tally, dated 1624, is said by M. Gmür to be in the Swiss Folklore Museum in Basel) and for water rights. It must be remembered that water was scarce and precious, and that it almost always belonged to a feudal overlord. That ownership could be rented out, sold and bequeathed. Notched planks were used to record the sign of ownership of the family, and to indicate how many hours (per day) of a given water right it possessed.

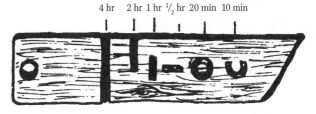

Fig. 5.6. *A water tally from Wallis (Switzerland). Basel, Museum für Völkerkunde. See Gmür, plate XXVI*

Finally, the Alpine areas also used *Kehrtesseln* or "turn tallies", which provided a practical way of fixing and respecting a duty roster within a guild or corporation (night watchmen, standard-bearers, gamekeepers, churchwardens, etc.).

In the modern world there are a few surviving uses of the notched-stick technique. Brewers and wine-dealers still mark their barrels with Xs, which have a numerical meaning; publicans still use chalk-marks on slate to keep a tally of drinks yet to be settled. Air Force pilots also still keep tallies of enemy aircraft shot down, or of bombing raids completed, by "notching" silhouettes of aircraft or bombs on the fuselage of their aircraft.

The techniques used to keep tallies of numbers thus form a remarkably unbroken chain over the millennia.

CHAPTER 6

NUMBERS ON STRINGS

Although it was certainly the first physical prop to help our ancestors when they at last learned to count, the hand could never provide more than a fleeting image of numerical concepts. It works well enough for representing numbers visually and immediately: but by its very nature, finger-counting cannot serve as a recording device.

As crafts and trade developed within different communities and cultures, and as communication between them grew, people who had not yet imagined the tool of writing nonetheless needed to keep account of the things that they owned and of the state of their exchanges. But how could they retain a durable record of acts of counting, short of inventing written numerals? There was nothing in the natural world that would do this for them. So they had to invent something else.

In the early years of the sixteenth century, Pizarro and his Spanish conquistadors landed on the coast of South America. There they found a huge empire controlling a territory more than 4,000 km long, covering an area as large as Western Europe, in what is now Bolivia, Peru, and Ecuador. The Inca civilisation, which went back as far as the twelfth century CE, was then at the height of its power and glory. Its prosperity and cultural sophistication seemed at first sight all the more amazing for the absence amongst these people of knowledge of the wheel, of draught animals, and even of writing in the strict sense of the word.

However, the Incas' success can be explained by their ingenious method of keeping accurate records by means of a highly elaborate and fairly complex system of knotted string. The device, called a *quipu* (an Inca word meaning "knot") consisted of a main piece of cord about two feet long onto which thinner coloured strings were knotted in groups, these pendant strings themselves being knotted in various ways at regular intervals (see Fig. 6.1).

Quipus, sometimes incorrectly described as "abacuses", were actually recording devices that met the various needs of the very efficient Inca administration. They provided a means for representing liturgical, chronological, and statistical

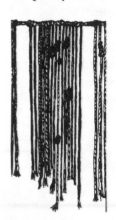

FIG. 6.1. *A Peruvian* quipu

records, and could occasionally also serve as calendars and as messages. Some string colours had conventional meanings, including both tangible objects and abstract notions: white, for instance, meant either "silver" or "peace"; yellow signified "gold"; red stood for "blood" or "war"; and so on. *Quipus* were used primarily for book-keeping, or, more precisely, as a concrete enumerating tool. The string colours, the number and relative positions of the knots, the size and the spacing of the corresponding groups of strings all had quite precise numerical meanings (see Fig. 6.2, 6.3 and 6.4). *Quipus* were used to represent the results of counting (in a decimal verbal counting system, as previously stated) all sorts of things, from military matters to taxes, from harvest reckonings to accounts of animals slain in the enormous annual culls that were held, from delivery notes (see Fig. 6.5) to population censuses, and including calculations of base values for levies and taxes for this or that administrative unit of the Inca Empire, inventories of resources in men and equipment, financial records, etc.

FIG. 6.2. *The first nine numbers represented on a string in the manner of the Inca* quipu

thousands	3	**3**	3,000
hundreds	6	**6**	600
tens	4	**4**	40
units	3	**3**	3

FIG. 6.3. *The number 3,643 as it would be represented on a string in the manner of a Peruvian* quipu

A = 38 B = 273 C = 258 D = 89
E = A + B + C + D = 658

FIG. 6.4. *Numerical reading of a bunch of knotted strings, from an Inca* quipu, *American Museum of Natural History, New York, exhibit B 8713, quoted in Locke (1924): the number 658 on string E equals the sum of the numbers represented on strings A, B, C and D.*

Quipus were based on a fairly simple, strictly decimal system of positions. Units were represented by the string being knotted a corresponding number of times around the first fixed position-point (counting from the end or bottom of the string), tens were represented similarly by the number of times the string was knotted around the second position-point, the third point served for recording hundreds, the fourth for thousands, etc. So to "write" the number 3,643 on Inca string (as shown in Fig. 6.3), you knot the string three times at the first point, four times at the second, six times at the third, and three times at the fourth position-point.

Officers of the king, called *quipucamayocs* ("keepers of the knots"), were appointed to each town, village and district of the Inca Empire with responsibility for making and reading *quipus* as required, and also for supplying the central government with whatever information it deemed important (see Fig. 6.5). It was they who made annual inventories of the region's produce and censuses of population by social class, recorded the results on string with quite surprising regularity and detail, and sent the records to the capital.

FIG. 6.5. *An Inca* quipucamayoc *delivering his accounts to an imperial official and describing the results of an inventory recorded on the quipu. From the Peruvian Codex of the sixteenth-century chronicler Guaman Poma de Ayala (in the Royal Library, Copenhagen), reproduced from* Le Quipucamayoc, *p. 335*

One of the *quipucamayocs* was responsible for the revenue accounts, and kept records of the quantities of raw materials parcelled out to the workers, of the amount and quality of the objects each made, and of the total amount of raw materials and finished goods in the royal stores. Another kept the register of births, marriages and deaths, of men fit for combat, and other details of the population in the kingdom. Such records were sent in to the capital every year where they were read by officers learned in the art of deciphering these devices. The Inca government thus had at its disposal a valuable mass of statistical information: and these carefully stored collections of skeins of coloured string constituted what might have been called the Inca National Archives. [Adapted from W. H. Prescott (1970)]

Quipus are so simple and so valuable that they continued to be used for many centuries in Peru, Bolivia and Ecuador. In the mid-nineteenth century, for example, herdsmen, particularly in the Peruvian Altiplano, used *quipus* to keep tallies of their flocks [M. E. de Rivero & J. D. Tschudi (1859)]. They used bunches of white strings to record the numbers of their sheep and goats, usually putting sheep on the first pendant string, lambs on the second, goats on the third, kids on the fourth, ewes on the fifth, and so on; and bunches of green string to count cattle, putting the bulls on the first pendant string, dairy cows on the second, heifers on the third, and then calves, by age and sex, and so on (see Fig. 6.6).

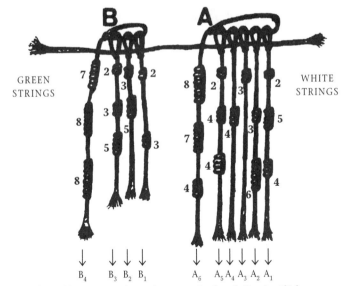

Fɪɢ. 6.6. *A livestock inventory on a nineteenth-century* quipu *from the Peruvian Altiplano. On bunch A (white string), small livestock: 254 sheep (string A_1), 36 lambs (A_2), 300 goats (A_3), 40 kids (A_4), 244 ewes (A_5), total = 874 sheep and goats (A_6). On bunch B (green string), cattle: 203 bulls (B_1), 350 dairy cows (B_2), 235 sterile cows (B_3), total = 788 head of cattle (B_4).*

Even today native Americans in Bolivia and Peru use a very similar device, the *chimpu*, a direct descendant of the *quipu*. A single string is used to represent units up to 9, with each knot on it indicating one unit, as on a *quipu*; tens are figured by the corresponding number of knots tied on two strings held together; hundreds in like manner on three strings, thousands on four strings, and so on. On *chimpus*, therefore, the magnitude of a number in powers of 10 is represented by the number of strings included in the knot – six knots may have the value of 6, 60, 600 or 6,000 according to whether it is tied on one, two, three, or four strings together.

5	knots on four strings together	→ 5,000
4	knots on three strings together	→ 400
7	knots on two string together	→ 70
7	knots on one string	→ 7

FIG. 6.7. *A chimpu (Bolivia and Peru)*

These remarkable systems are not however uniquely found in Inca or indeed South American civilisations. The use of knotted string is attested since classical times, and in various regions of the world.

Herodotus (485–425 BCE) recounts how, in the course of one of his expeditions, Darius, King of Persia (522–486 BCE) entrusted the rearguard defence of a strategically vital bridge to Greek soldiers, who were his allies. He gave them a leather strap tied into sixty knots, and ordered them to undo one knot each day, saying:

> "If I have not returned by the time all the knots are undone, take to your boats and return to your homes!"

In Palestine, in the second century CE, Roman tax-collectors used a "great cable", probably made up of a collection of strings, as their register. In addition, receipts for taxes paid took the form of a piece of string knotted in a particular way.

Arabs also used knotted string over a long period of time not only as a concrete counting device, but also for making contracts, for giving receipts, and for administrative book-keeping. In Arabic, moreover, the word *aqd*, meaning "knot", also means "contract", as well as any class of numbers constituted by the products of the nine units to any power of ten (several Arabic mathematicians refer to the *aqd* of the hundreds, the *aqd* of the thousands, and so on).

The Chinese were also probably familiar with knotted-string numbers in ancient times before writing was invented or widespread. The semi-legendary Shen Nong, one of the three emperors traditionally credited with founding Chinese civilisation, is supposed to have had a role in developing a counting system based on knots and in propagating its use for book-keeping and for chronicles of events. References to a system reminiscent of Peruvian *quipus* can be found in the *I Ching* (around 500 BCE) and in the *Tao Te Ching*, traditionally attributed to Lao Tse.

The practice is still extant in the Far East, notably in the Ryû-Kyû Islands. On Okinawa, workers in some of the more mountainous areas use plaited straw to keep a record of days worked, money owed to them, etc. At Shuri, moneylenders keep their accounts by means of a long piece of reed or bark to which another string is tied at the middle. Knots made in the upper half of the main "string" signify the date of the loan, and on the lower half, the amount. On Yaeyama, harvest tallies were kept in similar fashion; and taxpayers received, in lieu of a written "notice to pay", a piece of string so knotted as to indicate the amount due [J. G. Février (1959)].

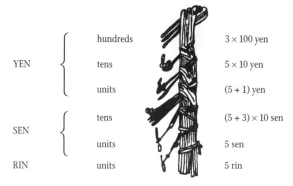

FIG. 6.8. *A sum of money as expressed in knotted string in the style used by workers on Okinawa and tax-collectors on Yaeyama. The figure shows 356 yen, 85 sen and 5 rin (1yen = 100 sen, 1 sen = 10 rin). The number 5 is represented by a knot at the end of the trailing straw. See also Chapter 25, Fig. 25.9.*

The same general device can be found in the Caroline Islands, in Hawaii, in West Africa (specifically amongst the Yebus, who live in the hinterland of Lagos (Nigeria)), and also at the other ends of the world, amongst native Americans such as the Yakima (eastern Washington State), the Walapai and the Havasupai (Arizona), the Miwok and Maidu (North & South Carolina), and of course amongst the Apache and Zuñi Indians of New Mexico.

A bizarre survival of the formerly wide role of knotted string was to be found as late as the end of the last century amongst German flour-millers, who kept records of their dealings with bakers by means of rope (see Fig. 6.9 below). Similarly, knotted-string rosaries (like their beaded and notched counterparts), for keeping count of prayers, are common to many religions. Tibetan monks, for example, count out the *one hundred and eight unities* (the number 108 is considered a sacred number) on a bunch of 108 knotted strings (or a string of 108 beads) whose colour varies with the deity to be invoked: yellow string (or beads) for prayers to *Buddha*; white string (or white beads made from shells) for *Bodhisattva*; red strings (or coral beads) for *the one who converted Tibet*; etc. A very similar practice was current only a few decades ago amongst various Siberian tribes (Voguls, Ostyaks, Tungus, Yakuts, etc.); and there is also a Muslim tradition, handed down by Ibn Sa'ad, according to which Fatima, Mohammed's daughter, counted out the 99 attributes of Allah and the supererogatory laudations on a piece of knotted string, not on a bead rosary.

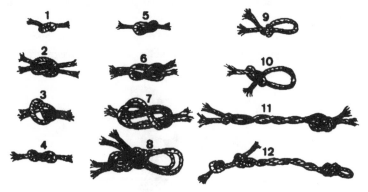

FIG. 6.9. *German millers' counting device using knotted rope (the system in force at Baden in the nineteenth century is illustrated)*

FIG. 6.10. *The bands and fringe of prayer-shawl*

For morning prayers (*Shaḥrit*) and other services in the synagogue, Jews wear a prayer-shawl (*talit*) adorned with fringes (*tsitsit*). Now, the four corner-threads of the fringe are always tied into a quite precise number of knots: 26 amongst "Eastern" (Sephardic) Jews, and 39 amongst "Western" (Ashkenazi) Jews. The number 26 corresponds to the numerical value of the Hebrew letters which make up the name of God, YHWH (see below, Chapters 17 and 20, for more detail on letter-counting systems), and 39 is the total of the number-values of the letters in the expression "God is One", YHWH EHD (see below). 39 is also the "value" of the Hebrew word meaning "morning dew" (*tal*), and rabbis have often commented that at prayer the religious Jew is able to hear the word of God "which falls from his mouth as morning dew falls on the grass".

	ה ו ה י	
YHWH	5 + 6 + 5 + 10	= 26
Yahwe, "the Lord"		
	ד ח א ה ו ה י	
YḤWH EHD	4 + 8 + 1 + 5 + 6 + 5 + 10	= 39
Yahwe ehad, "the Lord is One"		
	ל ט	
ṬL	30 + 9	= 39
tal, "morning dew"		

FIG. 6.11.

Knotted string has thus served not only as a device for concrete numeration, but also as a mnemotechnic tool (for recording numbers, maintaining administrative archives, keeping count of contracts, calendars, etc.). Although knotted string does not constitute a form of writing in the strict sense, it has performed all of writing's main functions – to preserve the past and to ensure the survival of contracts between members of the same society. Numbers on strings can therefore be considered for our purposes as a special form of written numbers.

CHAPTER 7

NUMBER, VALUE AND MONEY

At a time when people lived in small groups, and could find what they needed in the nature around them, there would have been little need for different communities to communicate with each other. However, once some sort of culture developed, and people began to craft objects of use or desire, then, because the raw resources of nature are unequally distributed, trade and exchange became necessary.

The earliest form of commercial exchange was barter, in which people exchange one sort of foodstuff or goods directly for another, without making use of anything resembling our modern notion of "money". On occasion, if the two parties to the exchange were not on friendly terms, these exchanges took the form of silent barter. One side would go to an agreed place, and leave there the goods on offer. Next day, in their place or beside them, would be found the goods offered in exchange by the other side. Take it or leave it: if the exchange was considered acceptable, the goods offered in exchange would be taken away and the deal was done. However, if the offer was not acceptable then the first side would go away, and come back next day hoping to find a better offer. This could go on for several days, or even end without a settlement.

Among the Aranda of Australia, the Vedda of Ceylon, Bushmen and Pygmies of Africa, the Botocoudos in Brazil, in Siberia, in Polynesia – such transactions have been observed. But with growth in communication, and the increasing importance of trade, barter became increasingly inconvenient, depending as it does on the whims of individuals or on interminable negotiations.

The need grew, therefore, for a stable system of equivalences of value. This would be defined (much as numbers are expressed in terms of a base) in terms of certain fixed units or standards of exchange. With such a system it is not only possible to evaluate the transactions of trade and commerce, but also to settle social matters – such as "bride price" or "blood money" – so that, for instance, a woman would be worth so many of a certain good as a bride, the reparation for a robbery so many. In pre-Hellenic Greece, the earliest unit of exchange that we find is the ox. According to Homer's *Iliad* (XXIII, 705, 749–751; VI, 236; eighth century BCE), a "woman good for a thousand tasks" was worth four oxen, the bronze armour of Glaucos was worth nine, and that of Diomedes (in gold) was worth 100. And, in decreasing order of value, are given: a chased silver cup, an ox, and half a golden talent. The Latin word *pecunia* (money), from which we get "pecuniary", comes from *pecus*, meaning "cattle"; and the related word *peculium* means "personal property", from which we also get "peculiar". In fact the strict sense of *pecunia* is "stock of cattle". The English word *fee* has come to us partly from Old English *feoh* meaning both "cattle" and "property" which itself is believed to be derived via a Germanic root from *pecus* (compare modern German *Vieh*, "livestock"), and partly from Anglo-French *fee* which is probably also of similar Germanic derivation. Like the Sanskrit *rupa* (whence "rupee"), these words remind us of a time when property, recompense, offerings, and ritual sacrifices were evaluated in heads of cattle. In some parts of East Africa, the dowry of a bride is counted in cattle. The Latin *capita* ("head") has given us "capital". In Hebrew, *keseph* means both "sheep" and "money"; and the root-word made of the letters GML stands for both "camel" and "wages".

In ancient times, however, barter was a far from simple affair. It was surrounded by complicated formalities, which were probably associated with mysticism and magical practices, as is confirmed by ethnological study of contemporary "primitive" societies and by archaeological findings. We may imagine, therefore, that in pastoral societies the concept of the "ox standard" grew out of the "ox for the sacrifice" which itself depended on the intrinsic value attributed to the animal.

L. Hambis (1963–64), describing certain parts of Siberia, says "Buying and selling was still done by barter, using animal pelts as a sort of monetary unit; this system was employed by the Russian government until 1917 as a means of levying taxes on the people of these parts." In the Pacific islands, on the other hand, goods were valued not in terms of livestock but in terms of pearl or sea-shell necklaces. The Iroquois, Algonquin, and other northeast American Indians used strings of shells called *wampum*. Until recently, the Dogon of Mali used cowrie shells. One Ogotemmêli, interviewed by M. Griaule (1966), says "a chicken is worth three times eighty cowries, a goat or a sheep three times eight hundred, a donkey forty times eight hundred, a horse eighty times eight hundred, an ox one hundred and twenty times eight hundred." "But", continues Griaule, "in earlier times the unit of exchange was not the cowrie. At first, people bartered strips of cloth for animals or goods. The cloth was their money. The unit was the 'palm' of a strip of cloth twice eighty threads wide. So a sheep was worth eight cubits of three 'palms' ... Subsequently, values were laid down in terms of cowries by Nommo the Seventh, Master of the Word."

With some differences of detail, practices were similar in pre-Columbian Central America. The Maya used also cotton, cocoa, bitumen, jade, pots, pearls, stones, jewels, and gold. For the Aztecs, according to J. Soustelle,

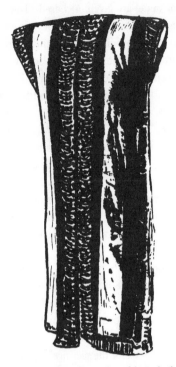

Fig. 7.1. *Tunic worn in the nineteenth century by members of the Tyal tribe in Formosa. More than 2,500 precious stones are attached in bands, at the edges and on either side of the centre line. Such tunics were used as "money" in buying livestock and in the trade in young women. The bands of precious stones could be detached separately to serve as pocket money for everyday purchases. New York, Chase Manhattan Bank Museum of Money*

"certain foodstuffs, goods or objects were employed as standards of value and as tokens of exchange: the *quachtli* (a piece of cloth) and 'the load' (20 *quachtlis*); the cocoa bean used as 'small change' and the *xiquipilli* (a bag of 8,000 beans); little T-shaped axes of copper; feather quills filled with gold." The same kind of economy was practised in China prior to the adoption of money in the modern sense. In the beginning, foodstuffs and goods were exchanged, their value being expressed in terms of certain raw materials, or certain necessities of life, which were adopted as standards. These might include the teeth and horns of animals, tortoise shells, sea-shells, hides, or fur pelts. Later, weapons and utensils were adopted as tokens of value: knives, shovels, etc. These would at first have been made of stone, but later, from the Shang Dynasty, of bronze (sixteenth to eleventh century BCE).

However, regular use of such kinds of items was cumbersome and not always easy. As a result, metal played an increasingly important role, in the form of blocks or ingots, or fashioned into tools, ornaments or weapons, until finally metal tokens were adopted as money in preference to other forms, for the purposes of buying and selling. The value of a merchandise was measured in terms of weight, with reference to a standard weight of one metal or another.

FIG. 7.3. *Lance-head once used as money by Central Congo tribes. One of these would possibly buy a fowl; five or six would be the price of a slave. New York, Chase Manhattan Bank Museum of Money*

FIG. 7.2. *Bronze "knife" from the Zhou period, used as a unit of barter in China; approximately 1000 BCE. Beijing Museum*

Thus it was that "When Abraham purchased the Makpelah Cave, he weighed out four hundred silver shekels for Ephron the Hittite."* Later on Saul, seeking his father's she-asses, sought the help of a seer for which he gave one quarter of a shekel of silver (I Samuel IX, 8). Similarly, the fines laid down in the Code of the Alliance were stipulated in shekels of silver, as also was the poll tax (Exodus XXX, 12–15) [A. Negev (1970)].

In the Egypt of the Pharaohs likewise, foodstuffs and goods were often valued, and paid for, with metal (copper, bronze, sometimes gold or silver) measured out in nuggets or in flakes, or given in the form of bars or rings which were measured by weight. The principal standard of weight was the *deben*, equivalent to 91 grams of our measure. For certain purchases, value was determined in certain fractions of the *deben*. For example, in the Old Kingdom (2780–2280 BCE) the *shât*, one twelfth of a *deben*, was used (equivalent, therefore, to 7.6 grams). In the New Kingdom (1552–1070 BCE) the *shât* gave way to the *qat*, one tenth of the *deben* or 9.1 grams.

In a contract from the Old Kingdom we can see how value was expressed in terms of the *shât*. According to this, the rent of a servant was to be paid as follows. the values being in *shâts* of bronze:

8	bags of grain	value	5	*shâts*
6	goats	value	3	*shâts*
	silver	value	5	*shâts*
	Total	value	13	*shâts*

* The Old Testament shekel is equivalent to 11.4 grams of our measure.

As another example, the following account from the New Kingdom shows *debens* of copper being used as a standard of value.

> Sold to Hay by Nebsman the Brigadier:
> 1 ox, worth 120 *debens* of copper
> Received in exchange:
> 2 pots of fat, value 60 *debens*
> 5 loin-cloths in fine cloth, worth 25 *debens*
> 1 vestment of southern flax, worth 20 *debens*
> 1 hide, worth 15 *debens*

In this example we can see how goods could be used in payment as well as metal tokens in the marketplace of ancient times. That ox, for instance, cost 120 *debens* of copper, but not one piece of real metal had changed hands: 60 of the *debens* owing had been settled by handing over 2 pots of fat, 25 more with 5 loin-cloths, and so on.

Although goods had been exchanged for goods, therefore, this was not a straightforward barter. It in fact reflected a real monetary system. Thenceforth, by virtue of the metal standard, goods were no longer bartered at the whim of the dealers or according to arbitrary established practice, but in terms of their "market price".

There is a letter dating from around 1800 BCE which gives a vivid illustration of these matters. It comes from the Royal Archives of the town of Mari, and was sent by Iškhi-Addu, King of Qatna, to Išme-Dagan, King of Êkallâtim. Iškhi-Addu roundly reproaches his "brother" for sending a meagre "sum" in pewter, in payment for two horses worth several times that amount.

> Thus [speaks] Iškhi-Addu thy brother:
> This should not have to be said! But speak I must, to console my heart. ... Thou hast asked of me the two horses that thou didst desire, and I did have them sent to thee. And see! how thou hast sent to me merely twenty rods of pewter! Didst thou not gain thy whole desire from me without demur? And yet thou dare'st send me so little pewter! ... Know thou that here in Qatna, these horses are worth six hundred shekels of silver. And see, how thou hast sent me but twenty rods of pewter! What will they say of this, when they hear of it?

An understandable indignation, since a shekel of silver was worth three or four rods of pewter at the time.

It should not be thought, though, that "money", in the modern sense of the word, was used in payment in those times. It was not a "coinage" in the sense of pieces of metal, die-cast in a mint which is the prerogative of the State, and guaranteed in weight and value. The idea of a coinage sound in weight and alloy did not come about until the first millennium

BCE, most probably with the Lydians. Until that time, only a kind of "base-weight" played a role in transactions and in legal deeds, acting as a unit of value in terms of which the prices of individual items of merchandise, or individual deeds, could be expressed. On this basis, this or that metal was first counted out in ingots, rings, or other objects, and then its weight, in units of the "base-weight", was determined, and in this way could be used as "salary", "fine", or "exchange".

Let us go back a few thousand years and, in the description of Maspero, observe a market from Egypt of the Pharaohs.

Early in the morning endless streams of peasants come in from the surrounding country, and set up their stalls in the spots reserved for them as long as anyone can remember. Sheep, geese, goats and wide-horned oxen are gathered in the centre to await buyers. Market gardeners, fishermen, fowlers and gazelle hunters, potters and craftsmen squat at the roadside and beside the houses, their goods heaped in wicker baskets or on low tables, fruits and vegetables, fresh-baked bread and cakes, meats raw or variously prepared, cloths, perfumes, jewels, the necessities and the frivolities of life, all set out before the curious eyes of their customers. Low and middle class alike can provide for themselves at lower cost than in the regular shops, and take advantage of it according to their means.

The buyers have brought with them various products of their own labours, new tools, shoes, mats, pots of lotion, flasks of drink, strings of cowrie shells or little boxes of copper or silver or even golden rings each weighing one *deben*,* which they will offer to exchange for the things they need.

For purchase of a large beast, or of objects of great value, loud, bitter and protracted arguments take place. Not only the price, but in what species the price shall be paid, must be settled, so they draw up lists whereon beds, rods, honey, oil, pick-axes or items of clothing may make up the value of a bull or a she-ass.

FIG. 7.4. *Brass ingot formerly used as monetary standard in the black slave market of the West African coast. New York, Chase Manhattan Bank Museum of Money*

* Maspero uses *tabnou*, here replaced by the the more precise term *deben*.

The retail trading does not involve so much complicated reckoning. Two townsmen have stopped at the same moment in front of a fellah with onions and corn displayed in his basket.* The first's liquid assets are two necklaces of glass pearls or coloured enamelled clay; the second one has a round fan with a wooden handle, and also one of those triangular fans which cooks use to boost the fire.

FIG. 7.5. *Market scenes in an Egyptian funeral painting of the Old Kingdom, Fifth or Sixth Dynasty (around 2500 BCE). The painting adorns the tomb of Feteka at the northern end of the necropolis of Saqqara (between Abusîr and Saqqara). See Lepsius (1854–59), vol. II, page 96 (Tomb no. 1), and Porter & Moss (1927–51), vol. 3 part 1, page 351.*

"This necklace would really suit you," calls the first, "it's just your style!"

"Here is a fan for your lady and a fan for your fire," says the other.

Still, the fellah calmly and methodically takes one of the necklaces to examine it:

"Let's have a look, I'll tell you what it's worth."

With one side offering too little, and the other asking for too much, they proceed by giving here and taking there, and finally agree on the number of onions or the amount of grain which will just match the value of the necklace or the fan.

Further along, a shopper wants some perfume in exchange for a pair of sandals and cries his wares heartily:

"Look, fine solid shoes for your feet!"

---

* Some of the scenes described can be seen on an Egyptian funeral painting from the Old Kingdom, reproduced here in Fig. 7.5.

But the merchant is not short of footwear just now, so he asks for a string of cowries for his little jars:

"See how sweet it smells when you put a few drops around!" he says winningly.

A woman passes two earthen pots, probably of ointment she has made, beneath the nose of a squatting man.

"This lovely scent will catch your fancy!"

Behind this group, two men argue the relative worth of a bracelet and a packet of fish-hooks; and a women with a small box in her hand is negotiating with a man selling necklaces; another woman is trying to get a lower price on a fish which the seller is trimming for her.

Barter against metal requires two or three more stages than simple barter. The rings or the folded sheets which represent *debens* do not always have the standard content of gold or silver, and may be of short weight. So they must be weighed for each transaction to establish their real value, which offers the perfect opportunity for those concerned to enter into heated dispute. After they have passed a quarter of an hour yelling that the scales do not work, that the weighing has been messed up, that they have to start all over again, they finally weary of the struggle and come to a settlement which roughly satisfies both sides.

However, sometimes someone cunning or unscrupulous will adulterate the rings by mixing their precious metal with as much false metal as possible short of making their trickery apparent. An honest trader who is under the impression that he received a payment of eight gold *debens*, who was in fact paid in metal which was one third silver, has unwittingly lost almost one third of his part. Fear of being cheated in this way held back the common use of *debens* for a long time, and caused the use of produce and artisanal objects in barter to be maintained.

At the end of the day, the use of *money* (in the modern sense of the term) became established once the metal was cast into small blocks or coins, which could be easily handled, had constant weight, and were marked with the official stamp of a public authority who had the sole right to certify good weight and sound metal.

This ideal system of exchange in commercial transactions was invented in Greece and Anatolia during the seventh century BCE. (In China the earliest similar usage occurred also at about the same time, apparently, around 600–700 BCE, during the Chow Dynasty.) Who might have first thought of it? Some consider that Pheidon, king of Argos in the Peloponnese, introduced the system in his own city and in Ægina, around 650 BCE. However, the majority of scholars agree that the honour of the invention should go to Asia Minor under the Greeks, most probably to Lydia.

Be that as it may, the many advantages of the use of coins led to its rapid adoption in Greece and Rome, and amongst many other peoples. The rest is another story.

Fig. 7.6. *Greek coins.*
Left: silver tetradrachma from Agrigento, around 415 BCE.
Right: tetradrachma from Syracuse, around 310 BCE.
Agrigento Museum

By learning how to count in the abstract, grouping every kind of thing according to the principle of numerical base, people also learned how to *estimate*, *evaluate* and *measure* all sorts of magnitudes – weights, lengths, areas, volumes, capacities and so on. They likewise managed to conceive ever larger numbers, though they could not yet attain the concept of infinity. They worked out many technical procedures (mental, material and later written), and laid the early foundations of arithmetic which, at first, was purely practical and only later became abstract and led on to algebra.

The way also opened up for the devising of a calendar, for a systemisation of astronomy, and for the development of a geometry which was at first based straightforwardly on measurement of length, area and volume, before becoming theoretical and axiomatic. In short, the grasp of these fundamental data allowed the human race to attempt the measurement of its world, little by little to understand it better and better, to press into humanity's service some of their world's innumerable secrets, and to organise and to develop their economy.

CHAPTER 8

NUMBERS OF SUMER

WRITING: THE INVENTION OF SUMER

Writing, as a system enabling articulated speech to be recorded, is beyond all doubt among the most potent intellectual tools of modern man. Writing perfectly meets the need (which every person in any advanced social group feels) for visual representation and the preservation of thought (which of its nature would otherwise evanesce). It also offers a remarkable method of expression and of preservation of communication, so that anyone can keep a permanent record of words long since spoken and flown. However, it is much more than a mere instrument.

> By recording speech in silent form, writing does not merely conserve it, but also stimulates thought such as, otherwise, would have remained latent. The simplest of marks made on stone or paper are not just a tool: they entomb old thoughts, but also bring them back to life. As well as fixing language, writing is also a new language, silent perhaps, which lays a discipline on thought and, in transcribing it, organises it. ... Writing is not only a means of durable expression: it also gives direct access to the world of ideas. It faithfully represents the spoken word, but it also facilitates the understanding of thought and gives thought the means to traverse both space and time. [C. Higounet (1969)]

Writing, therefore, in revolutionising human life, is one of the greatest of all inventions. The earliest known writing appeared around 3000 BCE, not far from the Persian Gulf, in the land of Sumer, which lay in Lower Mesopotamia between the Tigris and Euphrates rivers. Here also were developed the earliest agriculture, the earliest technology, the first towns and cities, by the Sumerians, a non-Semitic people of still obscure origins.

As evidence of this we have numerous documents known as "tablets" which were used as a kind of "paper" by the inhabitants of this region. The oldest of these (which also carry the most archaic form of the writing) were discovered at the site of Uruk,* more precisely at the archaeological level designated as *Uruk IVa*.†

These tablets are, in fact, small plaques of dry clay, roughly rectangular

---

* The royal city of Uruk was situated south of Lower Mesopotamia on the Iraqi site of Warka (now about twenty km north of the Euphrates). It has given its name to the epoch in which, it is presumed, the Sumerian people first appeared in the region and in which writing was invented in Mesopotamia.

(A)

Numbers

Writing

Print of a cylindrical seal

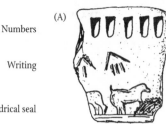

ATU 565

(B)

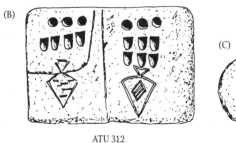

ATU 312

(C)

ATU 111

(E)

(D)

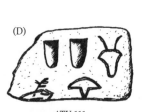

ATU 111   ATU 264

FIG. 8.1. *Archaic Sumerian tablets, discovered at Uruk (level IVa). They are among the earliest known instances of Sumerian writing. Several of these tablets are divided by horizontal and vertical lines into panels which contain numbers and signs representing writing (which already seem to follow a standard pattern). These indicate a degree of precise analytical thought, composed of separate elements brought together, as in articulate speech. The Iraqi Museum, Baghdad*

in outline and convex on their two faces (see Fig. 8.1). On one side, sometimes on both, they bear hollowed-out markings of various shapes and sizes. These marks were made on the clay while still soft by the pressure of a particular tool. As well as these hollow markings we may also find outline drawings made with a pointed tool, representing all kinds of things or

† The best known of the Sumerian archaeological sites, and the first to be excavated, Uruk has served to establish a "time scale" for this civilisation. In certain sectors deep excavation has revealed a series of strata to which archaeologists refer to determine approximate dates for their finds: the ordering of the different layers, from top to bottom, corresponds to the different stages in the history of the civilisation.

beings. The hollow markings correspond to the different units in the Sumerian sequence of enumeration (in the archaic graphology); they are, therefore, the most ancient "figures" known in history (see Fig. 8.2). The drawings are simply the characters in the archaic writing system of Sumeria (Fig. 8.3).

Some of these tablets also have symbolic motifs in relief, made by rolling cylindrical seals over the surface of the tablet, from one end to the other.

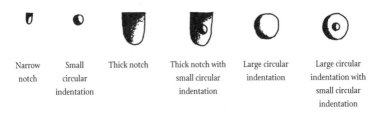

| Narrow notch | Small circular indentation | Thick notch | Thick notch with small circular indentation | Large circular indentation | Large circular indentation with small circular indentation |

FIG. 8.2. *The shapes of archaic Sumerian numbers*

These tablets seem to have served as records of various quantities associated with different kinds of goods – invoices, as it were, for supplies, deliveries, inventories, or exchanges. Let us have a closer look at the drawings on these tablets, and try to discern the principal character of this writing system. Some of these drawings are very realistic and show the essential outlines of material objects, which may be quite complex (Fig. 8.3).

On occasion, the drawings are much simplified, but still strongly evoke their subject. For example, the heads of the ox, the ass, the pig, and the dog are drawn in a concrete though very stylised way, and the drawing of the animal's head stands for the animal itself.

More often, however, the original object is no longer directly recognisable; the part stands for the whole, and effect represents cause, in a stylised and condensed symbolism. A woman, for instance, is represented by a schematic drawing of the triangle of pubic hair (Fig. 8.3 F), and the verb to impregnate by a drawing of a penis (Fig. 8.3 E).

Generally speaking, as a result of these abbreviations and the subtly simplified relation between representation and object represented, the latter mostly eludes us. The symbols are simple geometric drawings, and the represented objects (where we can determine what they are, by semantic or palaeographic means) have little apparently in common with their representations. Consider the sign for a sheep, for example (Fig. 8.3 U): what might this drawing possibly represent, a circle surrounding a cross? A sheep-pen? A brand? We have no idea.

What is striking about these drawings is their constant and definite character* in which each particular symbol exhibits little variation of form.

---

* This means that the design has been finalised once and for all, so that "writing" implies choosing and setting up a repertoire of generally accepted and recognised symbols.

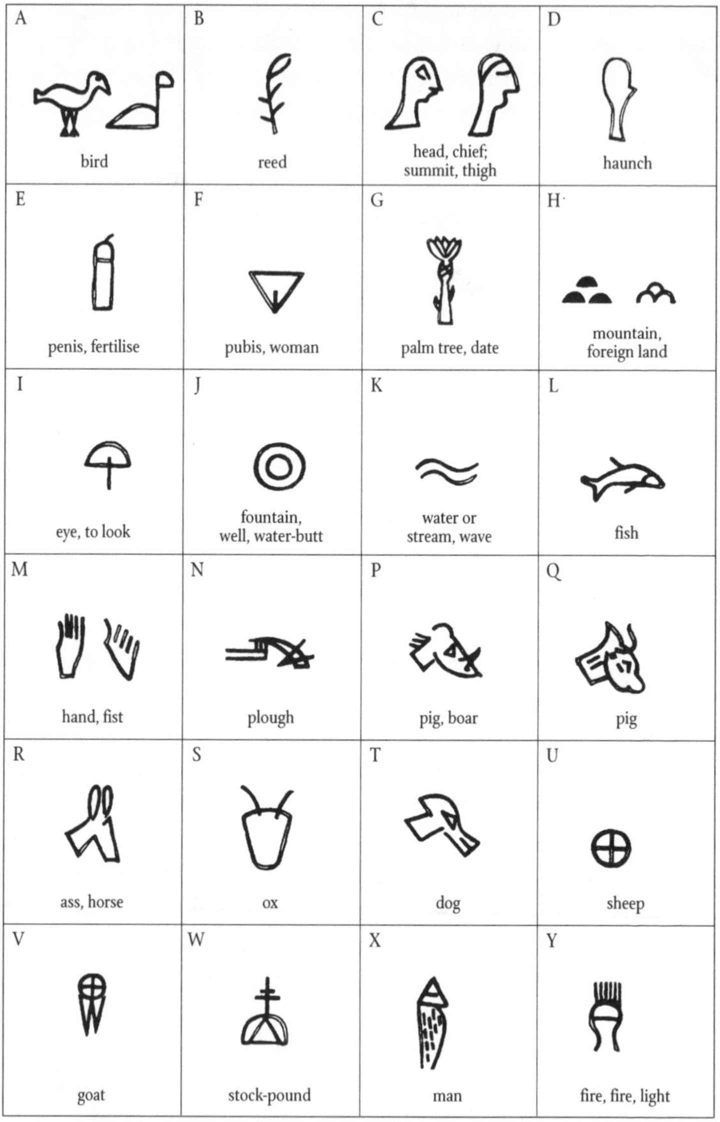

A	B	C	D
bird	reed	head, chief; summit, thigh	haunch
E	**F**	**G**	**H·**
penis, fertilise	pubis, woman	palm tree, date	mountain, foreign land
I	**J**	**K**	**L**
eye, to look	fountain, well, water-butt	water or stream, wave	fish
M	**N**	**P**	**Q**
hand, fist	plough	pig, boar	pig
R	**S**	**T**	**U**
ass, horse	ox	dog	sheep
V	**W**	**X**	**Y**
goat	stock-pound	man	fire, fire, light

Fig. 8.3. *Pictograms from archaic Sumerian writing*

Comparing this with the number of variations which will emerge in subsequent periods, we are obliged to see in this constancy and regularity the mark of true writing – in the sense of a fully worked-out system which everyone has adopted – and therefore to consider that we are seeing the very origin of writing or, at any rate, its earliest stages, based no doubt on earlier usages but bearing this essential new feature of being a generally accepted uniform practice.

We find ourselves contemplating, therefore, a system of graphical

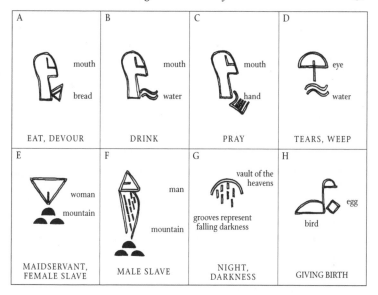

FIG. 8.4. *Some examples of the evocative "logical aggregates" used in archaic Sumerian writing*

symbols intended to express the precise thoughts which occur in speech. However, it is still not writing in the strict and full sense:* we are still in the "prehistory", or rather the "protohistory", of the development of writing (that is, in the pictographic stage).

All of these symbols, whether we know what they mean or not, are graphical representations of material objects.

But we should still not conclude that they can only represent material objects. Each object can be used to symbolise not only the activities or actions directly implied by the object, but also related concepts. The leg, for example, can also represent "walking", "going" or "standing up"; the hand can stand for "taking", "giving", "receiving" (Fig. 8.3 M); the rising sun for "day", "light" or "brightness"; the plough for "ploughing", "sowing", "digging" (Fig. 8.3 N) and, by extension, "ploughman", "farmer", and so on.

The scope of each ideogram can also be extended by a device which had already, at this time, been long applied to symbolism. Two parallel lines can represent the idea of "friend" and "friendship"; two lines crossing each other, the idea of "enemy" or "hostility". The Sumerians gave great place to this idea of enlarging the possible range of meanings of their drawings by combining two or more together to represent new ideas, or aspects of

---

* In the strict sense of the word, a mere visual representation of thought by means of symbols of material objects cannot be considered true writing, since it is more closely related to spoken words than to thought itself. For it to be considered true writing, it would in addition need to be a systematic representation of spoken language, since writing, like language, is a *system* and not a random sequence of items. Février (1959) says: "Writing is a system for human communication using well-defined conventional symbols for the representation of language, which can be transmitted and received, which can be equally well understood by both parties, and which are associated with the words of the spoken language."

reality otherwise hard to express. The combination *mouth* + *bread* thereby expresses "eat", "devour"; *mouth* + *water* expresses "drink"; *mouth* + *hand* expresses "prayer" (in accordance with Sumerian ritual); and *eye* + *water* denotes "tears", "weeping".

In the same way, an *egg* beside *fowl* suggests the act of "giving birth", strokes underneath a semi-circle suggest darkness falling from the heavenly vault, "night", "the dark". In that flat, lowland country, where "mountain" was synonymous with foreign lands and enemy country (Fig. 8.3 H), the juxtaposition *woman* + *mountain* meant "foreign woman" (literally "woman from the mountains") and therefore, by extension, "female slave" or "maid-servant" (since women were brought to Sumer, bought or captured, to serve as slaves). The same association of ideas gave rise to the combination *man* + *mountain* to denote a male slave (Fig. 8.4 F).

Human thought could therefore be better expressed by this system of pictograms and ideograms than by a purely representational visual art. This system was a systematic attempt to express the whole of thought in the same way as it was represented and dissected in spoken language. But it was still far from perfect, being a long way yet from being able to denote with precision, and without ambiguity, everything that could be expressed in spoken language. Because it depended excessively on the material world of objects which could be drawn as pictures, it required a very large number of different symbols. In fact, the total number of symbols used in this first age of writing in Mesopotamia has been estimated to be about two thousand.

Furthermore, not only was this writing system difficult to manipulate, it was also seriously ambiguous. If, for example, *plough* can also mean "ploughman", how are we supposed to know which one is meant? Even for one and the same word, how can its various nuances be distinguished – nuances which language can meticulously encapsulate and which are essential to complete understanding of the thought (including such qualities as gender, singular and plural, quality, and the countless relationships between things in time and in space)? How can one distinguish the many ways in which actions vary with time?

This writing was certainly a step towards representing spoken language, but it was limited to what could be expressed in images, that is to say to the immediately representable aspects of objects and actions, or to their imme-diately cognate extensions. For such reasons the original Sumerian writing remains, and will no doubt always remain, undecipherable. Consider the *bull's head* in Fig. 8.1 D. Is it really "the head of a bull"? Or is it – more plausibly – "a bull", a unit of livestock ("one ox"), one of the many products one can obtain from cattle (leather, milk, horn, meat)? Or does it represent some person who may have had a name on the lines of "Mr Bull" (thus being the equivalent of a signature)? Only the few people immediately implicated would be in a position to know what exactly was intended by the bull's head on this particular tablet.

In these circumstances, Sumerian writing at this stage of development is better thought of as an aide-mémoire than as a written record in the proper sense of the term: something which served to help people recall what they already exactly knew (possibly missing out some essential detail), rather than something which could exactly express this to someone who had never known it directly.

Such a scheme answered the purposes of the time well enough. Apart from a few "lists of symbols", all of the known archaic Sumerian tablets carry summaries of administrative actions or of exchanges, as we can see from the totalled numbers which can be found at the end of the document (or on its other side). All of these tablets are, therefore, accounts (in the financial sense of the term). Pure economic necessity therefore played, beyond doubt, a leading role in the story:* the emergence of this writing system was undoubtedly inspired by the necessities of accounting and stock-taking, which caused the Sumerians to become aware of the fact that the old order, which was still based on a purely oral tradition, was running out of steam and that a completely new approach to the organisation of work was called for.

As P. Amiet explains, "Writing was invented by accountants faced with the task of noting economic transactions which, in the rapidly developing Sumerian society, had become too numerous and too complex to be merely entrusted to memory. Writing bears witness to a radical transformation of the traditional way of life, in a novel social and political environment already heralded by the great constructions of the preceding era." At that time the temples were solely responsible for the economy of all Sumer, where continual over-production required a very centralised system of redistribution which became increasingly complicated, a situation which undoubtedly gave rise to the invention of writing. But accounting is simply the recording, by rote or by writing, of operations which have already taken place, and which concern solely the displacements of objects and of people. According to J. Bottero, archaic Sumerian writing is perfectly adapted to this function, which is the reason why its earliest form – which had a profound effect on later developments – was such as to serve above all as an aide-mémoire.

In order, however, to become completely intelligible, and above all in order to attain the status of "writing" in the true sense of the word

---

* Does the development of this writing have solely an economic explanation? Did not different needs (religious, divinatory, even literary) also play a part? Did people not communicate with each other at a distance in writing, for instance? There are those who think so; but so far no archaeological find has lent support to such possibilities.

(i.e. capable of recording unambiguously whatever could be expressed in language), this archaic picto-ideography was therefore obliged to make great advances not only in clarity and precision, but also in universality of reference.

This transition began to occur around 2800–2700 BCE, at which time Sumerian writing became allied to spoken language (which is the most developed way of analysing and communicating reality).

The idea at the root of this development was to use the picture-signs, no longer merely pictorially or ideographically, but phonetically, by relating them to spoken Sumerian, somewhat as in our picture-puzzles, where a phrase is punningly represented by objects whose names form parts of the sounds in the spoken phrase. For example, a picture showing a needle and thread being used to sew a bunch of thyme, a goalkeeper blocking a goal-kick, and the digit "9", could (in English) represent the saying "A stitch in time saves nine".

Thus a picture of an oven is at this time (2800–2700 BCE) no longer used to represent the object, but rather to represent the sound *ne*, which is the Sumerian word for "oven". Likewise, a picture of an arrow (in Sumerian *ti* stands for the sound *ti*; and since the word for "life" in Sumerian is also pronounced *ti* the arrow picture also stands for this word. As Bottero explains: "Using the pictogram of the arrow (*ti*) to denote something quite different which is also pronounced *ti* ('life') completely breaks the primary relation of the image to the object (arrow) and transfers it to the phoneme (*ti*); to something, that is, which is not situated in the material world but is inherent solely in spoken language, and has a more universal nature. For while the arrow, purely as a pictogram, can only refer to the object 'arrow' and possibly to a limited group of related things (weapon, shooting, hunting, etc.), the sound *ti* denotes precisely the phoneme, no matter where it may be encountered in speech and without reference to any material object whatever, and corresponds solely to this word, or to this part of a word (as in *ti-bi-ra*, 'blacksmith'). The sign of the arrow is therefore no longer a pictogram (it depicts nothing) but a *phonogram* (evoking a *phoneme*). The graphical system no longer serves to write things, but to write words, and it no longer communicates one single idea, but the whole of speech and language."

This represents an enormous advance, because such a system is now capable of representing the various grammatical parts of speech: pronouns, articles, prefixes, suffixes, nouns, verbs, and phrases, together with all the nuances and qualifications which can hardly, if at all, be represented in any other way. "As such," adds Bottero, "even if this now means that the reader must know the language of the writer in order to understand, the system can record whatever the spoken language expresses, exactly as it is expressed: the system no longer serves merely as a record to assist memory and recall, but can also inform and instruct."

It is not our business to go into the specific details of the language for which the Sumerians developed their graphical system, once they had reached the phonetic stage in the above way. But we may echo Bottero in saying that Sumerian writing (enormous advance though it was), because it was born of a pictography designed to aid and extend the memory, remained fundamentally a way of writing words: an aide-mémoire developed into a system, enhanced by the extension into phonetics, but not essentially transformed by it. (After the entry of the phonetic aspect, the Sumerians in fact kept many of their archaic ideograms of which each one continued to denote a word designating a specific entity or object, or even several words connected by more or less subtle relations of meaning, causality or symbolism.)

THE SUMERIANS
(Adapted from G. Rachet's *Dictionnaire de l'archéologie*)

The geographical origins of the Sumerians remain a topic of controversy. Though some would have them originate from Asia Minor, it seems rather that they arrived in Lower Mesopotamia from Iran, having come from central Asia.

Their language, which remains imperfectly known, was agglutinative, like the early Asiatic (pre-Semitic and pre-Indo-European) languages, and the Caucasian and Turco-Mongolian languages of today. In any case, wherever they came from was mountainous, as is shown by two things which they brought with them to south Mesopotamia: the *ziggurat*, a relic of ancient mountain religions, and stone-carving; whereas the Mesopotamian region is bare of stone.

Their most likely date of arrival in Mesopotamia can be placed in the so-called *Uruk* period, during the second half of the fourth millennium BCE, either during the *Uruk IV* period, or that of *Uruk V*. Quite possibly they arrived gradually, in minor waves, thereby leaving no archaeological traces for the whole of the Uruk period. It certainly seems that this city, home of the epic hero Gilgamesh, had been the primordial centre of the culture they bore. And it is certain the so-called *Jemdet Nasr* period began under their initiative, at the end of the fourth millennium BCE, to be followed by the *pre-Sargonic* era or *Ancient Dynasty* which saw the first culmination of Sumerian civilisation.

These periods were marked by three cultural manifestations: the development of *glyptics* (where cylinders engraved with parades of animals, and various scenes of a religious nature are dominant among the tablets);

the development of sculpture with relief on stone vases, animals and personages in the round, themes treated with great mastery and with a force which did not exclude elegance, the masterpiece of this period being the mask, known as the Lady of Warka, imbued with a delicate realism; finally, the emergence of writing which, if it has not given us annals, allows us to identify the gods to whom the temples were dedicated and to learn the names of certain personages, in particular those which have been found in the royal tombs of Ur.

The towns of the land of Sumer: Ur, Uruk, Lagaš, Umna, Adab, Mari, Kiš, Awan, Akšak, were constituted as city-states or, as Falkenstein has said, city-temples, which fought incessantly to exert a hegemony which they exercised more or less by turns. Up to the Archaic Dynasty II, we nowhere find a palace, since the king was in reality a priest, vicar of the god, who lived in the precincts of the temple, the Gir-Par, of which it seems we have an example in the edifice of Nippur.

The priest-king bore the title of EN, "Lord"; it is only during the Archaic Dynasty II that the title of king, *Lugal*, emerges, and at the same time the palace, witness to the separation of State and priesthood, and the emergence of a military monarchy. The earliest known palace is that of Tell A at Kiš, and the first personage who bore the title of Lugal was in fact a king of Kiš, Mebaragesi (around 2700 BCE). The furnishings of the tombs of Ur, which date from subsequent centuries, reveal the high level of material civilisation which the Sumerians had attained. The metallurgists had acquired a great mastery of their art and the sculptors had produced fine in-the-round works. We see a parallel development of urbanisation and of monumental building: the oval temple of Khafaje, the square temple of Tell Asmar, the temple of Ishtar at Mari, the temple of Inanna at Nippur. The expansion of the Sumerian cities was brusquely arrested in the twenty-fourth century BCE by the formation of the Semitic empire of Akkad. But the Akkadians assimilated the Sumerian culture and spread it beyond the land of Sumer. Savage tribes from the neighbouring mountains, Lullubi and Guti, put an end to the Akkadian Empire and ravaged the countryside until the king of Uruk, Utu-Hegal, overthrew the power of the Guti and captured their king, Tiriqan. Now an age of Sumerian renaissance began, with the hegemony of Lagaš and above all of Ur.

At the beginning of the second millennium BCE, the Sumerians were once again dominant with the dynasties of Isin and of Larsa, but after the triumph of Babylon, under Hammurabi, Sumer disappeared politically; but nevertheless the Sumerian language remained a language of priests, and many features of their civilisation, assimilated by the Babylonian Semites, were to survive across the Mesopotamian culture of Babylon.

THE SEXAGESIMAL SYSTEM

Let us now pass to the numbers themselves. The Sumerians did not count in tens, hundreds and thousands, but adopted instead the numerical base 60, grouping things by sixties and by powers of 60.

We ourselves have vestiges of this base, visible in the ways we express time in hours, minutes and seconds, and circular measure in degrees, minutes and seconds. For instance, if we have to set a digital timepiece to

$$9; 08; 43$$

then we know that this corresponds to 9 hours, 8 minutes and 43 seconds, being time elapsed since midnight; and this can be expressed in seconds as follows:

$$9 \times 60^2 + 8 \times 60 + 43 = 32{,}923 \text{ seconds.}$$

Likewise, when a ship's officer determines the latitude of a position he will express it as, for instance: $25°$; $36'$; $07''$, and everyone then knows that the position is

$$25 \times 60^2 + 36 \times 60 + 7 = 92{,}167''$$

north of the Equator.

With the Greeks, and later the Arabs, this was used as a scientific number-system, adopted by astronomers. Since the Greeks, however, with few and belated exceptions, this system has been used solely to express fractions (e.g. minutes and seconds as subdivisions of an hour). But in more distant times, as excavations in Mesopotamia have revealed, it gave rise to two quite separate number-systems which were used for whole numbers as well as fractions. One was the system used solely for scientific purposes by the Babylonian mathematicians and astronomers, later inherited by the Greeks who in turn passed it down to us by way of the Arabs. The other, more ancient yet and which we are about to discuss, was the number- system in common use amongst the Sumerians, predecessors of the Babylonians, and exclusively amongst them.

THE SUMERIAN ORAL COUNTING METHOD

60 is certainly a large number to use as base for a number-system, placing considerable demands on the memory since – in principle at least – it requires knowledge of sixty different signs or words to stand for the numbers from 1 to 60. But the Sumerians overcame this difficulty by using 10 as an intermediary to lighten the burden on the memory, as a kind of stepping-stone between the different sexagesimal orders of magnitude (1, 60, 60^2, 60^3, etc.).

Ignoring sundry variants, the Sumerian names for the first ten numbers, according to Deimel, Falkenstein and Powell, were

1	geš (or aš or dič)	6	àš
2	min	7	imin
3	eš	8	ussu
4	limmu	9	ilimmu
5	iá	10	u

FIG. 8.5A.

They also gave a name to each multiple of 10 below 60 (so, up this point, it was a decimal system):

10	u
20	niš
30	ušu
40	nišmin (or nimin or nin)
50	ninnû
60	geš (or gešta)

FIG. 8.5B.

Apart from the case of 20 (*niš* eems to be independent of *min* = 2 and of *u* = 10), these names are in fact compound words. The word for 30, there-fore, is formed by combining the word for 3 with the word for 10:

$$30 = ušu < {}^*eš.u = 3 \times 10$$

(where the asterisk indicates that an intermediate word has been restored).

In the same way, the word for 40 is derived by combining the word for 20 with the word for 2:

$$40 = nišmin = niš.min = 20 \times 2 \ .$$

The variants of this are simply contractions of *nišmin*:

$$40 = nin < ni.(-m).in = ni.(-š).min < nišmin.$$

The word for 50 comes from the following combination:

$$50 = ninnû < {}^*nimnu = niminu = nimin.u = 40 + 10.$$

In the words of F. Thureau-Dangin, the Sumerian names for the numbers 20, 40 and 50 seem like a sort of "vigesimal enclave" in this system. Note, by the way, that the word for 60 (*geš*) is the same as the word for unity. No doubt this was because the Sumerians thought of 60 as a large unity. Nevertheless, to avoid ambiguity, it was sometimes called *gešta*.

The number 60 represents a certain level, above which, in this oral numeration system, multiples of 60 up to 600 were expressed by using 60 as a new unit:

60	*geš*		360	*geš-àš*	$(= 60 \times 6)$
120	*geš-min*	$(= 60 \times 2)$	420	*geš-imin*	$(= 60 \times 7)$
180	*geš-eš*	$(= 60 \times 3)$	480	*geš-ussu*	$(= 60 \times 8)$
240	*geš-limmu*	$(= 60 \times 4)$	540	*geš-ilimmu*	$(= 60 \times 9)$
300	*geš-iá*	$(= 60 \times 5)$	600	*geš-u*	$(= 60 \times 10)$

FIG. 8.5C.

The next level is reached at 600, which is now treated as another new unit whose multiples were used up to 3,000:

600	*geš-ú*		2,400	*geš-u-limmu*	$(= 600 \times 4)$
1,200	*geš-u-min*	$(= 600 \times 2)$	3,000	*geš-u-iá*	$(= 600 \times 5)$
1,800	*geš-u-eš*	$(= 600 \times 3)$	3,600	*šàr*	$(= 60^2)$

FIG. 8.5D.

The number 3,600 (sixty sixties) is the next level, and it is given a new name (*šàr*) and in turn becomes yet another new unit:

šàr	3,600	$(= 60^2)$	*šàr-àš*	21,600	$(= 3,600 \times 6)$
šàr-min	7,200	$(= 3,600 \times 2)$	*šàr-imin*	25,200	$(= 3,600 \times 7)$
šàr-eš	10,800	$(= 3,600 \times 3)$	*šàr-ussu*	28,800	$(= 3,600 \times 8)$
šàr-limmu	14,400	$(= 3,600 \times 4)$	*šàr-ilimu*	32,400	$(= 3,600 \times 9)$
šàr-iá	18,000	$(= 3,600 \times 5)$	*šàr-u*	36,000	$(= 3,600 \times 10)$

FIG. 8.5E.

The following levels correspond to the numbers 36,000, 216,000, 12,960,000, and so on, proceeding in the same sort of way as above:

36,000	*šàr-u*	$(= 60^2 \times 10)$	144,000	*šàr-u-limmu*	$(= 36,000 \times 4)$
72,000	*šàr-u-min*	$(= 36,000 \times 2)$	180,000	*šàr-u-iá*	$(= 36,000 \times 5)$
108,000	*šàr-u-eš*	$(= 36,000 \times 3)$	216,000	*šàrgal*	$(= 60^3)$
				(literally: "big 3,600")	

FIG. 8.5F.

216,000	*šàrgal*	$(= 60^3)$	1,296,000	*šàrgal-aš*	$(= 216,000 \times 6)$
432,000	*šàrgal-min*	$(= 216,000 \times 2)$	1,512,000	*šàrgal-imin*	$(= 216,000 \times 7)$
....					
1,080,000	*šàrgal-iá*	$(= 216,000 \times 5)$	2,160,000	*šàrgal-u*	$(= 216,000 \times 10)$

FIG. 8.5G.

2,160,000	*šàrgal-u*	(= 60³× 10)	8,640,000	*šàrgal-u-limmu*	(= 2,160,000 × 4)
4,320,000	*šàrgal-u-min*	(= 2,160,000 × 2)	10,800,000	*šàrgal-u-iá*	(= 2,160,000 × 5)
6,480,000	*šàrgal-u-eš*	(= 2,160,000 × 3)			

12,960,000   *šàrgal-šu-nu-tag*   (= 60⁴)

("Unit greater than big šàr")

FIG. 8.5H.

FROM THE ORAL TO THE WRITTEN NUMBER-SYSTEM

When, around 3200 BCE, the Sumerians devised a numerical notation, they gave a special graphical symbol to each of the units 1; 10; 60; 600 (= 60 × 10); 3,600 (= 60^2); 36,000 (= 60^2 × 10), that is to say to each term in the sequence generated by the following schema:

1

10

10×6

$(10 \times 6) \times 10$

$(10 \times 6 \times 10) \times 6$

$(10 \times 6 \times 10 \times 6) \times 10$

They therefore mimicked the names of the different units in their oral system which, as we have seen, used base 60 and proceeded by a system of levels constructed alternately on auxiliary bases of 6 and of 10 (Fig. 8.6).

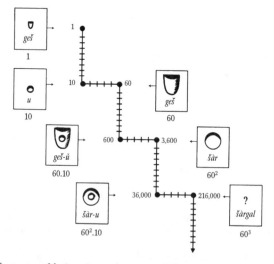

FIG. 8.6. *The structure of the Sumerian number-system, which was a sexagesimal system constructed upon a base of 10 alternating with a base of 6 (thus activating in turn two divisors of the base 60: 10 × 6 = 60)*

THE VARIOUS FORMS OF SUMERIAN NUMBERS

In the archaic epochs, unity was represented by a small notch (sometimes elongated), 10 by a circular indentation of small diameter, 600 (= 60 × 10) by a combination of these two, 3,600 (= 60²) by a circular indentation of large diameter, and 36,000 (= 3,600 × 10) by the smaller circular indentation within the larger circular indentation (Fig. 8.2 and 8.6).

To start with, these symbols were impressed on the tablets in the following orientation:

| 1 | 10 | 60 | 600 | 3,600 | 36,000 |

FIG. 8.7.

However, starting in the twenty-seventh century BCE, these became rotated anticlockwise through 90°. Thus the non-circular symbols thenceforth no longer pointed from top to bottom but from left to right:

| 1 | 10 | 60 | 600 | 3,600 | 36,000 |

FIG. 8.8.

After the development of the cuneiform script, these number-symbols took on a completely new form, angular and with much sharper outlines.

  • The number 1 was thereafter represented by a small vertical wedge (instead of a small cylindrical notch);

  • the number 10 was represented by a chevron (instead of the small circular impression);

  • the number 60 was represented by a larger vertical wedge (instead of a wide notch);

  • the number 600 by this larger vertical wedge combined with the chevron of the number 10;

  • the number 3,600 by a polygon formed by joining up four wedges (instead of the larger circle);

  • the number 36,000 by the polygon for 3,600, with the wedge for 10 in its centre;

and, finally, the number 216,000 (the cube of 60, for which a special symbol was introduced into cuneiform script) was represented by combining the polygon for 3,600 with the wedge for 60 (see Fig. 8.9).

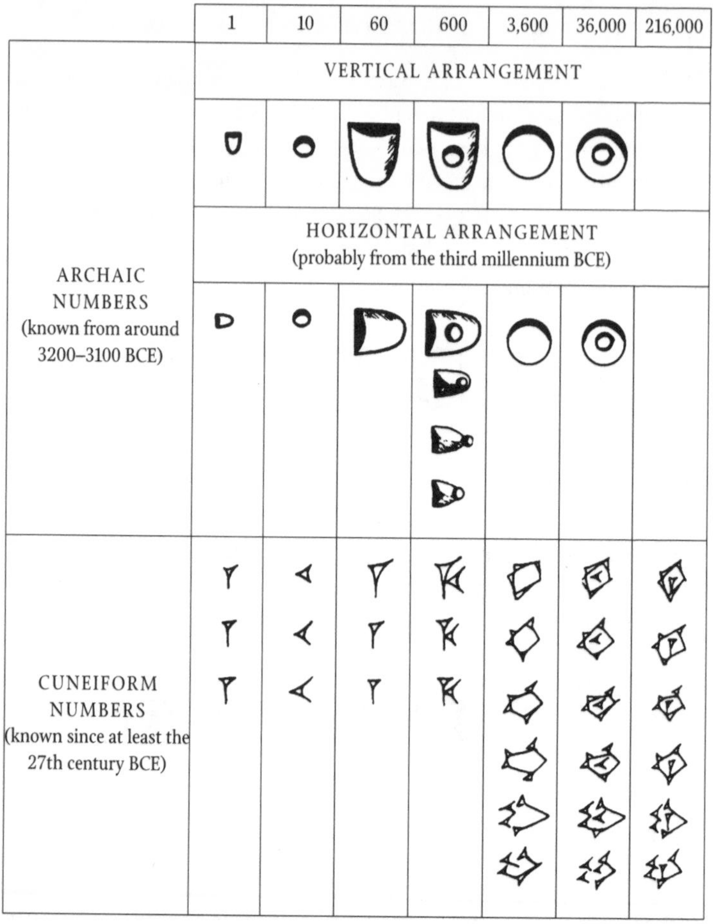

	1	10	60	600	3,600	36,000	216,000
ARCHAIC NUMBERS (known from around 3200–3100 BCE)	VERTICAL ARRANGEMENT						
	HORIZONTAL ARRANGEMENT (probably from the third millennium BCE)						
CUNEIFORM NUMBERS (known since at least the 27th century BCE)							

FIG. 8.9. *The development of the shapes of numbers originating in Sumeria. The change from the archaic to the cuneiform shapes resulted from the replacement of the "old stylus", which was cylindrical at one end, and pointed at the other, by the "flat" stylus shaped something like a modern ruler. This new writing instrument conduced its users to break the curves into a series of wedges or chevrons. See Deimel (1924, 1947) and Labat (1976 and in EPP).*

Clay as Mesopotamian "paper" and how to write on it

In Mesopotamia, stone is rare; wood, leather and parchment are difficult to preserve, and the soil consists of alluvial deposits. The inhabitants of this region therefore took what came to hand for the purpose of expressing their thoughts or for recording the spoken word, and what they had to hand was clay. They had used this raw material since very early times for modelling figurines, for sculpture, and for glyptics*, and later most ingeniously put it

---

* Close examination of archaeological finds leads one to believe that the usages of clay held no secrets for the Mesopotamians, four thousand years BCE. This is an important consideration for the history of writing in this region, since it effectively implies that they were fully aware of the possibilities of the medium. The

to diverse uses, especially for the purpose of writing, for more than three thousand years, in more than a dozen languages†. To borrow a phrase from J. Nougayrol (1945), you might say that these people created "civilisations of clay".

The originality of Mesopotamian graphics directly reflects the nature of this material and the techniques available to work with it, and we have an interest in devoting some attention to this; what follows will allow us to better trace the evolution of the forms of figures and written characters which originated in Sumer.

We have seen how Sumerian figures were hollow marks of different shapes and sizes (Fig. 8.2), while the written characters were real drawings representing beings and objects of every kind (Fig. 8.3). Originally, there-fore, there were fundamental differences of technique between the production of the one and the production of the other. The number-signs, like the motifs created using cylindrical or stamp-like seals, were produced by *impression*; the written characters on the other hand were *traced*.

For these purposes the Sumerians used a reed stem (or possibly a rod made of bone or ivory), which at one end was shaped into a cylindrical stylus, while the other end was sharpened to a point somewhat like a modern pen (Fig. 8.10).

The pictograms were made by pressing the pointed end quite deeply into the clay, still fresh, of the tablets (Fig. 8.11). To draw a line, the same

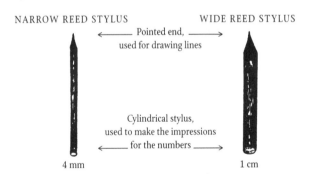

FIG. 8.10. *Reconstruction of the writing instruments of the Sumerian scribes (archaic era)*

character, possibly religious but certainly symbolic, of the motifs appearing on these vases and jugs, their repeated occurrence and their systematic stylisation, must not only have accustomed their creators to express a number of thoughts and ideas in this way but also to subsume these into ever simpler and more concise designs.

† At the dawn of the second millennium BCE, at the time when writing emerged in Sumer, the use of clay for "tablets" intended to bear conventional signs was already widespread in the region. This point also is very important, for it clarifies one reason why Sumerian writing moved on to a systematic phase: consid-ering the difficulty of sculpture, carving, and painting, and the fact that they demand time for their execu-tion, the universal adoption of clay throughout Mesopotamia is readily explained by the ease with which it can be worked (compared with wood, bone, or stone, whether for engraving, embossing, impression, moulding, or cutting).

pointed end was pressed in as before, and then drawn parallel to the surface through the required distance. Of course, this would often result in a wavy line, and could give rise to spillover on either side, because of the softness of the material.

Fig. 8.11. *How the archaic Sumerian pictograms were drawn on soft clay tablets*

For the numbers, on the other hand, the Sumerians made these by making an imprint on the soft damp clay with the other end of the instrument, the end shaped into a circular stylus. This was done with the stylus held at a certain angle to the surface of the clay. They had two styluses of different diameters: one about 4 mm, the other about 1 cm (Fig. 8.10). According to the angle at which they held the stylus, either a circular imprint, or a notch, would be obtained, and its size would depend on the diameter of the stylus (Fig. 8.12):

- a circular imprint of smaller or larger diameter if the stylus was held perpendicular to the surface of the clay;
- a notch, narrow or wide, if the stylus was held at an angle of 30° – 45° to the surface; the imprint would be more elongated if the angle was small.

ACTION		RESULT
I ⟋ II 45°	Narrow stylus applied at an angle of 45°	▽ Small notch
I II 90°	Narrow stylus applied perpendicularly	○ Small circular indentation
I ⟍ II 45°	Wide stylus applied at an angle of 45°	▽ Wide notch
I II 45° III 90°	(I): Wide stylus applied at 45° (II): Narrow stylus applied perpendicularly	Wide notch with small indentation
I II 90°	Wide stylus applied perpendicularly	◯ Large circular indentation
I II 90° III 90°	(I): Wide stylus applied perpendicularly (II): Narrow calamus applied perpendicularly	◎ Large circular indentation with small circular indentation

Fig. 8.12. *How the archaic numbers were impressed on soft clay tablets*

Why Sumerian writing changed direction

In the very earliest times, the signs used in Sumerian writing were drawn on the clay tablets in the natural orientation of whatever they were supposed to represent: vases stood upright, plants grew upwards, living things were vertical, etc. Similarly, the non-circular figures for numbers were also vertical (the stylus being held sloping towards the bottom of the tablet).

These signs and figures were generally arranged on the tablets in horizontal rows which, in turn, were subdivided into several compartments or boxes (Fig. 8.1, tablet E). Within each box, the figures were generally at the top, starting from the right, while the drawings used for writing were at the very bottom, like this:

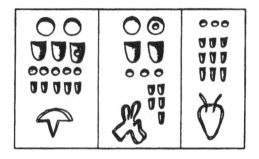

FIG. 8.13.

Now, if we examine the arrangement of figures and drawings on one of the tablets of the so-called Uruk period (around 3100 BCE), we find that where one of the boxes is not completely full the empty space is always on the left of the box (see the second box from the right in the top row of the tablet in Fig. 8.14).

This proves that the scribes of the earliest times wrote from right to left and from the top to the bottom. The non-circular figures were vertical, and

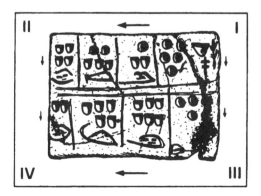

FIG. 8.14. *Sumerian tablet from Uruk, from around 3100 BCE. Iraqi Museum, Baghdad*

the drawings had their natural orientations. In short, in the beginning Sumerian writing was read from right to left and from top to bottom.

This arrangement long persisted on Mesopotamian stone inscriptions. It can be seen especially on the Stele of the Vultures (where the text is arranged in horizontal bands, and the boxes succeed each other from right to left and from top to bottom), in the celebrated Code of Hammurabi (whose inscription, which is read from right to left and from left to right, is arranged in vertical columns), and in several legends later than the seventeenth century BCE.

It went quite differently in the case of clay tablets, however: that is, in the case of everyday writings. Starting around the twenty-seventh century BCE, the signs used for writing, and the figures used for numbers, underwent a rotation through 90° anticlockwise.

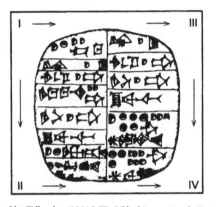

FIG. 8.15. *Sumerian tablet (Tello, about 3500 BCE). Bibliothèque nationale, Paris, Cabinet des Médailles (CMH 870 F). See de Genouillac (1909), plate IX*

To verify this, consider the tablet in Fig. 8.15, and look at in the direction I → II indicated by the long arrow in the Fig., after turning it 90° clockwise so that I → II is from right to left and at the top. Then we can see that if a compartment is not full up, the empty space is at the bottom, and not at the left. Likewise, in the original position of the tablet, the empty space is at the right.

According to C. Higounet (1969), this would be due to a change in the orientation with which the tablets were held.

With the small tablets of the earliest times, holding the tablet obliquely in the hand made it easier to trace drawings in columns from top to bottom. But, when the tablets became larger, the scribes had to place them upright in front of them, and the signs became horizontal, and the writing went in lines from left to right.

Be that as it may, thenceforth the drawings and the non-circular figures had an orientation 90° anticlockwise from their original one (Fig. 8.16);

"turned sideways, they became less pictorial, and therefore more liable to undergo a certain systematisation." [R. Labat]

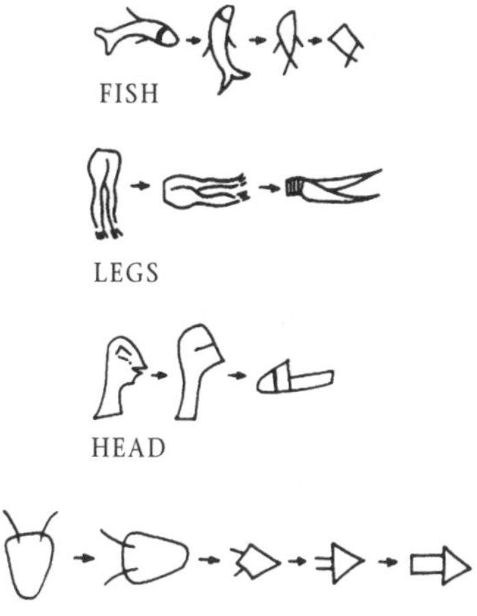

FISH

LEGS

HEAD

OX

FIG. 8.16. *Anticlockwise rotation, through a quarter turn, of the Sumerian signs and numbers*

The emergence of the cuneiform signs

The radical transformation which the Sumerian characters underwent after the Pre-Sargonic era (2700–2600 BCE) is due simply to a change of implement.

While the drawings used in writing had originally been traced out with the pointed end of the stylus, this changed when they had the idea of using instead, for this, the method which had always been used for the figures denoting numbers, namely impressing the marks on the clay. Instead of using a pointed stylus for tracing lines, they preferred to use a reed stem (or a rod of bone or ivory) whose end was trimmed in such a way that its tip formed a straight edge, and no longer a circle or a point. This edge was then pressed into the clay, to achieve cleanly, at one stroke, a line segment of a certain length; this clearly was much more rapid than drawing it with a pointed tool.

Of course this new type of stylus made characters of quite a different shape, with sharper lines and an angular appearance; these signs are called *cuneiform* (from the Latin *cuneus*, "a wedge") (Fig. 8.17).

The angular shapes of the imprints made by such a stylus on the clay naturally led to greater stylisation of the shapes of the various signs. Curves were broken up, and where necessary were replaced by a series of line segments, so that a picture was reduced to a collection of broken lines. In this new form of Sumerian writing, a circle, for example, became a polygon, and curves were replaced by polygonal lines (Fig. 8.18).

This did not occur all at once, however. It is not seen at all around 2850 BCE. It begins to appear in the archaic tablets of Ur (2700–2600 BCE), and in those of Fara (Šuruppak), where the majority of the signs are made up of impressed lines, while many other tablets of the same period continue to show the curved lines traced by the older method.

SHAPE OF THE STYLUS

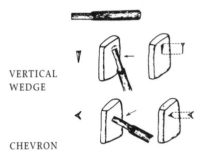

VERTICAL
WEDGE

CHEVRON

FIG. 8.17. *Impressing cuneiform signs on soft clay. The vertical wedge was made by pressing lightly on the clay with one of the corners of the "beak" of the stylus (the heavier the pressure, the larger the wedge).*

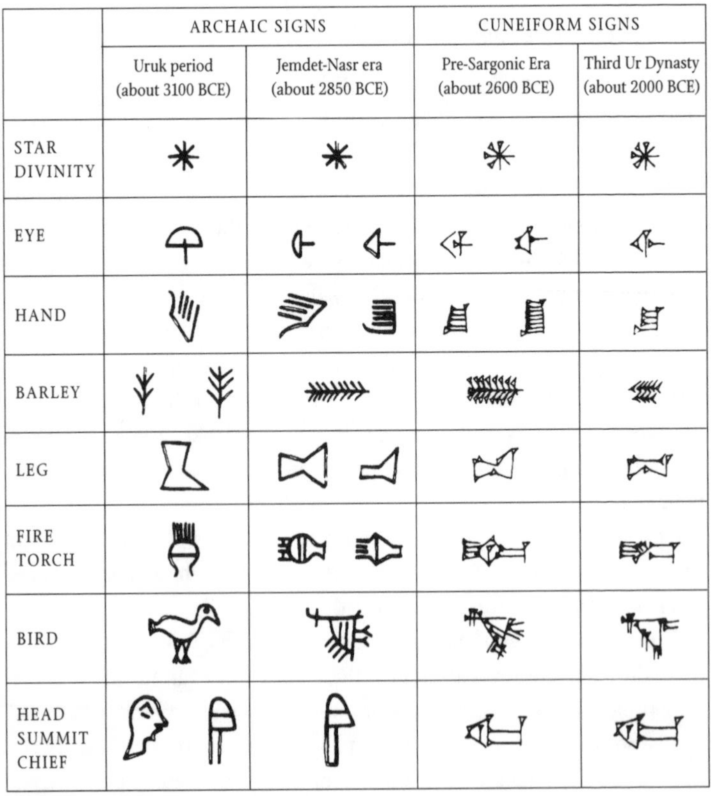

	ARCHAIC SIGNS		CUNEIFORM SIGNS	
	Uruk period (about 3100 BCE)	Jemdet-Nasr era (about 2850 BCE)	Pre-Sargonic Era (about 2600 BCE)	Third Ur Dynasty (about 2000 BCE)
STAR DIVINITY				
EYE				
HAND				
BARLEY				
LEG				
FIRE TORCH				
BIRD				
HEAD SUMMIT CHIEF				

FIG. 8.18.

At the beginning of this change in form, the signs nevertheless remained very complex, since people wished to preserve as much as possible of the detail of the original drawings, and because in the majority of cases the objective was still to achieve the outline of a concrete object. But, after a long period of adaptation, from the end of the third millennium BCE the scribes only kept what was essential and therefore made their marks much more rapidly than before.

And this is how the signs in Sumerian writing finally lost all resemblance to the real objects which they were meant to represent in the first place.

THE SUMERIAN WRITTEN COUNTING METHOD

Starting with these basic symbols, the first nine whole numbers were represented by repeating the sign for unity as often as required; the numbers 20, 30, 40, and 50 by repeating the sign for 10 as often as required, the numbers 120, 180, 240, etc. by repeating the symbol for 60, and so on.

Generally, since the system was based on the additive principle, a number was represented by repeating, at the level of each order of magnitude, the requisite symbol as often as required.

For example, a tablet dating from the fourth millennium BCE (Fig. 8.1, tablet C) carries the representation of the number 691 in the following form:

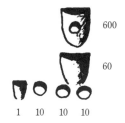

FIG. 8.19.

Likewise, on a tablet from Šuruppak, from around 2650 BCE, the number 164,571 is represented as follows (Fig. 8.20 and 12.1):

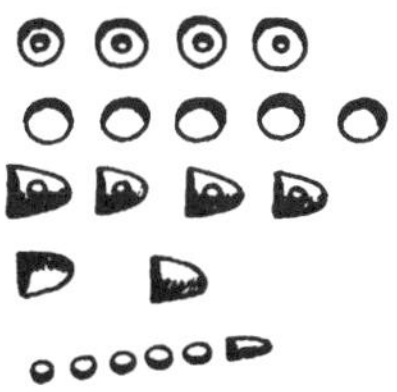

◉ ◉ ◉ ◉	36,000 drawn 4 times over =	36,000 × 4 =	144,000
○ ○ ○ ○ ○	3,000 drawn 5 times over =	3,600 × 5 =	18,000
▷ ▷ ▷ ▷	600 drawn 4 times over =	600 × 4 =	2,400
▷ ▷	60 drawn 2 times over =	60 × 2 =	120
○ ○ ○ ○ ○ ▷	10 drawn 5 times over =	10 × 5 =	50
	1 drawn 1 time =	=	1
			164,571

FIG. 8.20.

𒐀	𒌍𒐋	𒐕𒐏𒐋	𒐏𒐜𒐕	𒐏𒐜𒐕	𒐕𒌋𒐋
	30 8	60 50 7	180 40 1	240 40 1	120 10 9
4	38	117	221	281	139

FIG. 8.21A.

TRANSLATION

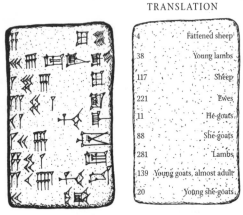

4	Fattened sheep
38	Young lambs
117	Sheep
221	Ewes
11	He-goats
88	She-goats
281	Lambs
139	Young goats, almost adult
20	Young she-goats

FIG. 8.21B. *Sumerian tablet from about 2000 BCE, giving a tally of livestock by means of cuneiform signs and numbers. Translation by Dominique Charpin. See de Genouillac (1911), plate V, no. 4691 F*

Similarly, for the cuneiform representation, on a tablet dating from the third dynasty of Ur (about 2000 BCE), found in a warehouse at Drehem (Ašnunak Patesi), various numbers are represented as shown on Figs. 8.21 A and B.

Finally, and in the same way, on a tablet contemporary with this last one, but from a clandestine excavation at Tello, we find the numbers 54,492 and 199,539 also expressed in cuneiform symbols:

54,492	36,000 drawn 1 time = 36,000
	3,600 drawn 5 times over = 18,000
	60 drawn 8 times over = 480
	10 drawn 1 time = 10
	1 drawn 2 times over = 2
	54,492
199,539	36,000 drawn 5 times over = 180,000
	3,600 drawn 5 times over = 18,000
	600 drawn 2 times over = 1,200
	60 drawn 5 times over = 300
	10 drawn 3 times over = 30
	1 drawn 9 times over = 9
	199,539

FIG. 8.22. *Barton (1918), Table Hlb 24, no. 16*

We may observe in passing that the Sumerians grouped the identical repeated symbols in such a way as to facilitate the grasp, in one glance, of the values of the assemblages within each order of magnitude. Considering just the representations of the first nine numbers, these groupings were initially made according to a dyadic or binary principle (Fig. 8.23) and later according to a ternary principle in which the number 3 played a special role (Fig. 8.24).

ARCHAIC NUMBERS

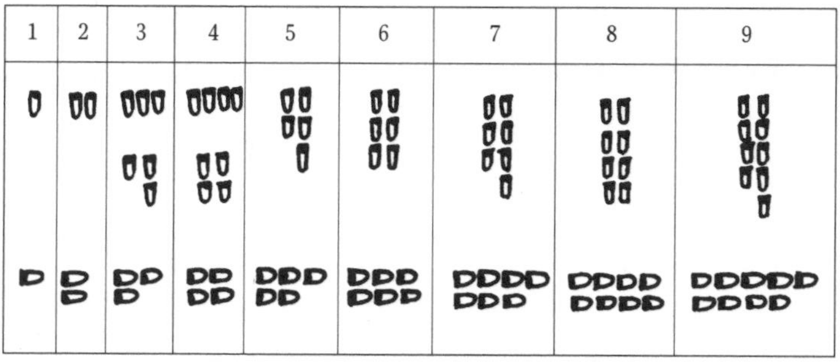

CUNEIFORM NUMBERS

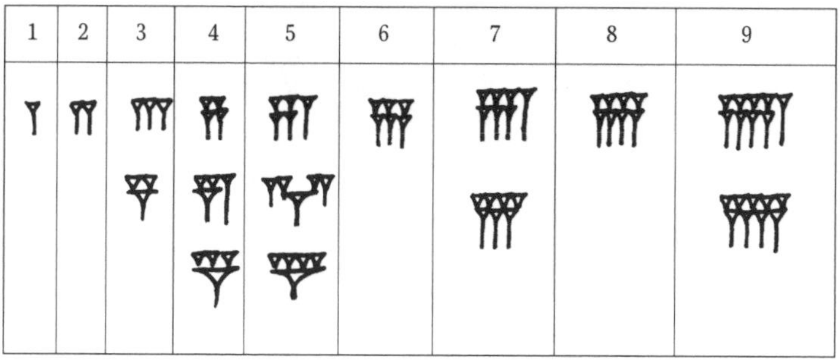

FIG. 8.23. *The dyadic (binary) principle of representing the nine units*

CUNEIFORM NUMBERS

FIG. 8.24. *The ternary principle of representing the nine units*

Thus the Sumerian numbering system sometimes required inordinate repetitions of identical marks, since it placed symbols side by side to represent addition of their values. For example, the number 3,599 required a total of twenty-six symbols!

For this reason, the Sumerian scribes would seek simplification by often using a subtractive convention, writing numbers such as 9, 18, 38, 57, 2,360 and 3,110 in the form:

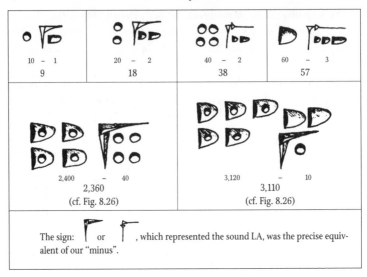

10 – 1 9	20 – 2 18	40 – 2 38	60 – 3 57

2,400 – 40 2,360 (cf. Fig. 8.26)	3,120 – 10 3,110 (cf. Fig. 8.26)

The sign: ⌐ or ⌐ , which represented the sound LA, was the precise equivalent of our "minus".

FIG. 8.25.

FIG. 8.26. *Sumerian tablet from Šuruppak (Fara), 2650 BCE. Istanbul Museum. See Jestin (1937), plate LXXXIV, 242 F*

From the pre-Sargonic era (about 2500 BCE), certain irregularities start to appear in the cuneiform representation of numbers. As well as the subtractive convention just described, the multiples of 36,000 can be found represented as shown in Fig. 8.27, instead of simply repeating the symbol for 36,000 once, twice, or three, four, or five times.

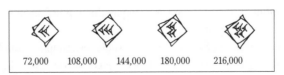

72,000	108,000	144,000	180,000	216,000

FIG. 8.27. *See Deimel*

These forms evidently correspond to the arithmetical formulae

72,000 = 3,600 × 20 (instead of 36,000 + 36,000)
108,000 = 3,600 × 30 (instead of 36,000 + 36,000 + 36,000)
144,000 = 3,600 × 40 (instead of 36,000 + 36,000 + 36,000 + 36,000)
180,000 = 3,600 × 50
(instead of 36,000 + 36,000 + 36,000 + 36,000 + 36,000).

In this, the Sumerians were doing nothing other than what we would today refer to as "expressing in terms of a common factor". Observing that the symbol for 3,600 is itself made up of the symbol for 360 with the symbol for 10, they also, after their fashion, made the number 3,600 a common factor so that, for instance, instead of representing 144,000 in the form

$$(3,600 \times 10) + (3,600 \times 10) + (3,600 \times 10) + (3,600 \times 10)$$

they used instead the simpler form

$$3,600 \times (10 + 10 + 10 + 10).$$

Another special point arising in the cuneiform notation concerned the two numbers 70 (= 60 + 10) and 600 (= 60 × 10), since both involved juxtaposing the symbol for 60 and the symbol for 10. This can clearly lead to ambiguity, since for 70 they are combined additively, and for 600 multiplicatively. This ambiguity was not present, however, in the archaic notation:

60 + 10	60 × 10 or	
70		60 × 10
Fig. 8.28a.	Fig. 8.28b.	600

They were, however, able to eliminate any possible confusion. In the case of 70, they placed a clear separation between the wedge (for 10) and the chevron (for 60) so as to indicate addition (Fig. 8.29 A), while for 600 they put them in contact so as to form an indivisible group, to represent multiplication (Fig. 8.29 B).

| 60 + 10 | 60 × 10 | |
| 70 | Fig. 8.29a. | 600 | Fig. 8.29b. |

A different problem arose with the representation of the numbers 61, 62, 63, etc. In the beginning, the number 1 was represented by a small wedge, and the number 60 by a larger wedge, and so there was no ambiguity:

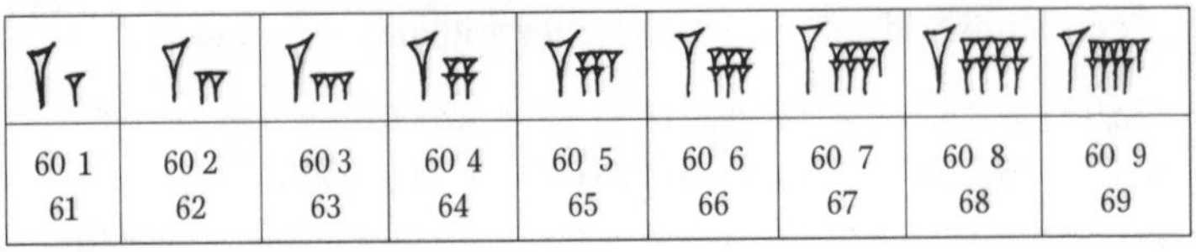

60 1	60 2	60 3	60 4	60 5	60 6	60 7	60 8	60 9
61	62	63	64	65	66	67	68	69

FIG. 8.30.

Later, however, 1 and 60 came to be represented by the same size of vertical wedge, and it was very difficult to distinguish between 2 and 61, or between 3 and 62, for example:

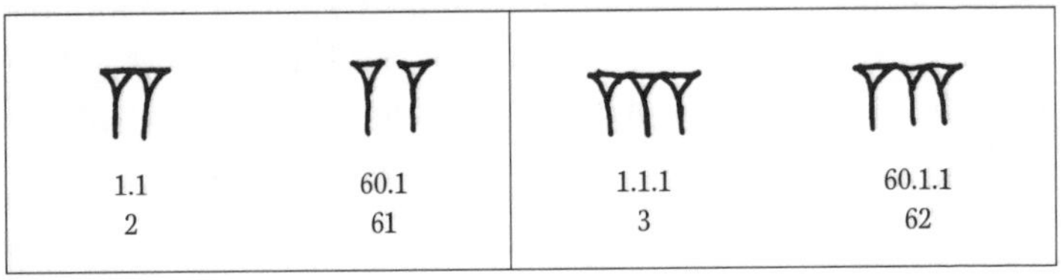

1.1	60.1	1.1.1	60.1.1
2	61	3	62

FIG. 8.31.

Therefore they had the idea of distinctly separating the unit symbols for the sixties from those for the units.

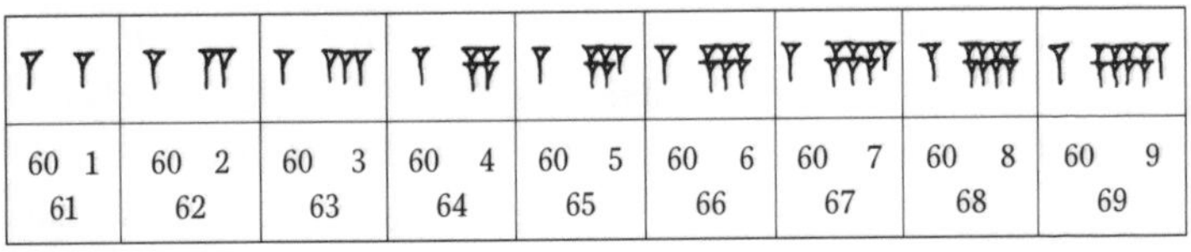

60 1	60 2	60 3	60 4	60 5	60 6	60 7	60 8	60 9
61	62	63	64	65	66	67	68	69

FIG. 8.32.

This particular problem with the cuneiform sexagesimal notation was the root of a most interesting simplification to which we shall return in Chapter 13.

For a long time, the cuneiform characters (known since at least twenty-seven centuries BCE) coexisted with the archaic numeral signs (Fig. 8.9). On certain tablets contemporary with the kings of the Akkad Dynasty (second half of the third millennium BCE), we see the cuneiform numbers side by side with their archaic counterparts. The intention, it seems, was to mark a distinction of rank between the people being enumerated: the cuneiform figures were for people of higher social standing, and the others for slaves or common people [M. Lambert, *personal communication*]. The cuneiform number-symbols did not definitively supplant the archaic ones until the third dynasty of Ur (2100–2000 BCE).

CHAPTER 9

THE ENIGMA OF THE SEXAGESIMAL BASE

In all of human history the Sumerians alone invented and made use of a *sexagesimal system* – that is to say, a system of numbers using 60 as a base. This invention is without doubt one of the great triumphs of Sumerian civilisation from a technical point of view, but it is nonetheless one of the greatest unresolved enigmas in the history of arithmetic. Although there have been many attempts to make sense of it since the time of the Greeks, we do not know the reasons which led the Sumerians to choose such a high base. Let us begin with a review of the explanations that have been put forward so far.

THEON OF ALEXANDRIA'S HYPOTHESIS

Theon of Alexandria, a Greek editor of Ptolemaic texts, suggested in the fourth century CE that the Sumerians chose base 60 because it was the "easiest to use" as well as the lowest of "all the numbers that had the greatest number of divisors". The same argument also cropped up 1,300 years later in *Opera mathematica*, by John Wallis (1616–1703), and again, in a slightly different form, in 1910, when Löfler argued that the system arose "in priestly schools where it was realised that 60 has the property of having all of the first six integers as factors".

FORMALEONI'S AND CANTOR'S HYPOTHESES

In 1789 a different approach was suggested by the Venetian scholar Formaleoni, and then repeated in 1880 by Moritz Cantor. They held that the Sumerian system derived from exclusively "natural" considerations: on this view, the number of days in a year, rounded down to 360, was the reason for the circle being divided into 360 degrees, and the fact that the chord of a sextant (one sixth of a circle) is equal to the radius gave rise to the division of the circle into six equal parts. This would have made 60 a natural unit of counting.

LEHMANN-HAUPT'S HYPOTHESIS

In 1889, Lehmann-Haupt believed he had identified the origin of base 60 in the relationship between the Sumerian "hour" (*danna*), equivalent to two of our current hours, and the visible diameter of the sun expressed in units of time equivalent to two minutes by current reckoning.

NEUGEBAUER'S HYPOTHESIS

In 1927 O. Neugebauer proposed a new solution which located the source of base 60 in terms of systems of weights and measures. This is how the proposal was explained by O. Becker and J. E. Hoffmann (1951):

> It arose from the combination of originally quite separate measurement units using base 10 and having (as in spoken language, and like the Egyptian systems) different symbols for 1, 10, and 100 as well as for the "natural fractions", 1/2, 1/3, and 2/3. The need to combine the systems arose particularly for measures of weight corresponding to measures of price or value. The systems were too disparate to be harmonised by simple equivalence tables, and so they were combined to give a continuous series such that the elements in the set of higher values (B) became whole multiples of elements in the set of lower values (A). Since both sets of values had the structure 1/1, 1/2, 2/3, 1, 2, 3 ... 10, the relationship between the two sets A and B had to allow for division by 2 and by 3, which introduced factor 6. So from the decimal structure of the original number-system, the Sumerians ended up with 60 as the base element of the new (combined) system.

On the other hand, F. Thureau-Dangin (1929) took the view that this entirely theoretical explanation cannot be a correct account of the origin of Sumerian numbering, because it is "undoubtedly the case that base 60 only occurs in Sumerian weights and measures because it was already available in the number-system".

OTHER SPECULATIONS

The Mesopotamians, according to D. J. Boorstin (1986), got to 60 by multiplying the number of planets (Mercury, Venus, Mars, Jupiter, and Saturn) by the number of months in the year: 5×12 is also a multiple of 6.

In 1910, E. Hoppe tried to refute, then to adapt Neugebauer's hypothesis: in this view, the Sumerians would have seen that base 30 provided for most of their needs, but chose the higher base of 60 because it was also divisible by 4. He subsequently proposed another explanation, based on geometry: the sexagesimal system, he argued, must have been in some relationship to the division of the circle into six equal parts instead of into four right angles, which made the equilateral triangle, instead of the square, the fundamental figure of Sumerian geometry. If the angle of an equilateral triangle is divided into 10 "degrees", in a decimal numbering system, then the circle would have 60 degrees, thus giving the origin of base 60 for the developed numbering system.

However, as was pointed out by the Assyriologist G. Kewitsch (1904), neither astronomy nor geometry can actually explain the origin of a number-system. Hoppe's and Neugebauer's speculations are far too theoretical, presupposing as they do that abstract considerations preceded concrete applications. They require us to believe that geometry and astronomy existed as fully-developed sciences before any of their practical applications. The historical record tells a very different story!

I once knew a professor of mathematics who likewise tried to persuade his students that abstract geometry was historically prior to its practical applications, and that the pyramids and buildings of ancient Egypt "proved" that their architects were highly sophisticated mathematicians. But the first gardener in history to lay out a perfect ellipse with three stakes and a length of string certainly held no degree in the theory of cones! Nor did Egyptian architects have anything more than simple devices – "tricks", "knacks" and methods of an entirely empirical kind, no doubt discovered by trial and error – for laying out their ground plans. They knew, for example, that if you took three pieces of string measuring respectively three, four, and five units in length, tied them together, and drove stakes into the ground at the knotted points, you got a perfect right angle. This "trick" *demonstrates* Pythagoras's theorem (that in a right-angled triangle the square on the hypotenuse equals the sum of the squares on the other two sides) with a particular instance in whole numbers $((3 \times 3) + (4 \times 4) = 5 \times 5)$, but it does not presuppose *knowledge* of the abstract formulation, which the Egyptians most certainly did not have.

All the same, the Sumerians' mysterious base 60 has survived to the present day in measurements of time, arcs, and angles. Whatever its origins, its survival may well be due to the specific arithmetical, geometrical and astronomical properties of the number.

KEWITSCH'S HYPOTHESIS

Kewitsch speculated in 1904 that the sexagesimal system of the Sumerians resulted from the fusion of two civilisations, one of which used a decimal number-system, and the other base 6, deriving from a special form of finger-counting. This is not easily acceptable as an explanation, since there is no historical record of a base 6 numbering system anywhere in the world [F. Thureau-Dangin (1929)].

BASE 12

On the other hand, *duodecimal systems* (counting to base 12) are widely attested, not least in Western Europe. We still use it for counting eggs and oysters, we have the words *dozen* and *gross* (= 12 × 12), and measurements of length and weight based on 12 were current in France prior the Revolution of 1789, in Britain until only a few years ago, and still are in the United States.

The Romans had a unit of weight, money, and arithmetic called the *as*, divided into 12 *ounces*. Similarly, one of the monetary units of pre-Revolutionary France was the *sol*, divided into 12 *deniers*. In the so-called Imperial system of weights and measures, in use in continental Europe prior to the introduction of the metric system (see above, pp. 42–3), length is measured in *feet* divided into 12 *inches* (and each inch into 12 *lines* and each line into 12 *points*, in the obsolete French version).

The Sumerians, Assyrians, and Babylonians used base 12 and its multiples and divisors very widely indeed in their measurements, as the following table shows:

LENGTH

1 *ninda*		12 cubits
1 *ninni*	"perch"	10 × 12 ells
1 *šu*		2/12ths of a cubit

WEIGHT

1 *gín*	"shekel"	3 × 12 *šu*	(8.416 grams)

AREA

1 *bùr*	150 × 12 *sar*	
1 *sar*	12 × 12 square cubits	(35.29 centiares)

VOLUME

1 *gur*	25 × 12 *sìla*	
1 *pi*	3 × 12 *sìla*	
1 *baneš*	3 × 6 *sìla*	
1 *bán*	6 *sìla*	
1 *sìla*		(842 ml)

The Mesopotamian day was also divided into twelve equal parts (called *danna*), and they divided the circle, the ecliptic, and the zodiac into twelve equal sectors of 30 degrees.

Moreover, there is clear evidence on tablets from the ancient city of Uruk [see Green & Nissen (1985); Damerov & Englund (1985)] of several different Sumerian numerical notations, which must have been used concurrently with the classical system (see Fig. 8.9, recapitulated in Fig. 9.1 below), amongst which there are the measures of length shown in Fig. 9.2.

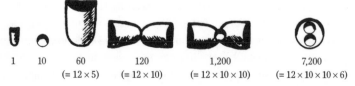

1	10	60	120	1,200	7,200
		(= 12 × 5)	(= 12 × 10)	(= 12 × 10 × 10)	(= 12 × 10 × 10 × 6)

FIG. 9.1.

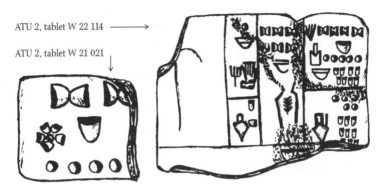

ATU 2, tablet W 22 114 ⟶

ATU 2, tablet W 21 021 ↓

FIG. 9.2. *Archaic Sumerian tablets from Uruk, showing a numerical notation that is different from the standard one. (Numerous tablets of this kind prove that the Sumerians had several parallel systems). Date: c. 3000 BCE. Baghdad, Iraqi Museum. Source: Damerov & Englund (1985)*

To sum up, base 12 could well have played a major role in shaping the Sumerian number-system.

AN ATTRACTIVE HYPOTHESIS

The major role given to base 10 in Sumerian arithmetic is similarly well-attested: as we saw in Chapter 8, it was used as an auxiliary unit to circumvent the main difficulty of the sexagesimal system, which in theory requires sixty different number-names or signs to be memorised. This is all the more interesting because the Sumerian word for "ten", pronounced *u*, means "fingers", strongly suggesting that we have a trace of an earlier finger-counting system of numerals.

This makes it possible to go back to Kewitsch's hypothesis and to give it a different cast: to suppose that the choice of base 60 was a learned solution to the union between two peoples, one of which possessed a decimal system and the other a system using base 12. For 60 is the lowest common multiple of 10 and 12, as well as being the lowest number of which all the first six integers are divisors.

Our hypothesis is therefore this: that Sumerian society had to begin with both decimal and duodecimal number-systems; and that its mathematicians, who reached a fairly advanced degree of sophistication (as we can see from the record of their achievements), subsequently devised a learned system that combined the two bases according to the principle of the LCM (lowest common multiple), producing a sexagesimal base, which had the added advantage of convenience for numerous types of calculation.

This is a very attractive and quite plausible hypothesis: but it fails as a historical explanation of origins on the obvious grounds that it presupposes too much intellectual sophistication. For we must not forget that most historically and ethnographically attested base numbers arose for reasons quite independent of arithmetical convenience, and that they were chosen very often without reference to a structure or even to the concepts of abstract numbers.

ARE THERE MYSTICAL REASONS FOR BASE 60?

Sacred numbers played a major role in Mesopotamian civilisations; Sumerian mathematics developed in the context of number-mysticism; and so it is tempting to see some kind of religious or mystical basis for the sexagesimal system.

> Sumerian mathematics, like astrology, cannot be disentangled from numerology, with which it has reciprocal relations. From the dawn of the third millennium BCE, the number 50 was attributed to the temple of Lagaš, son of the earth-god, and this shows that from the earliest times numbers had "speculative" meanings. The Akkadians brought number-symbolism into Babylonian thought, making it an essential element of the Name, the Individual and the Work. Alongside their scientific or intellectual functions, numbers became part of the way the Mesopotamians conceived the structure of the world. For example, the numeral *šar* or *šaros* (= 3,600) is written in cuneiform as a sign which is clearly a deformation of the *circle* [see Fig. 8.9], and it also means "everything", "totality", "cosmos". In Sumerian cosmogony, two primordial entities, the "Upper Totality" or *An-Šar* and the "Lower Totality" or *Ki-Šar* came together to give birth to the first gods. Moreover, the full circle of 360° is divided into degrees, whose basic unit of 1/360 is called *Geš* – and the symbol for *Geš* is precisely what is used to signify "man" and thus for elaborating the names of masculine properties. The higher base unit or *sosse* (= 60) is also pronounced *Geš* [see Fig. 8.5], and its sign (with an added asterisk or star) is the figure of the "Upper God", or heaven, whose name is pronounced *An(u)*, by virtue of the ideogram that defines it as a divinity and as heaven. So the celestial god, 60, is the father of the earth-god, 50; the god of the Abyss is 40, two thirds of 60. The moon-god is 30 (it has been suggested, without any evidence, that the moon-god has this number in virtue of the number of days in the lunar cycle); and the sun-god has the number 20, which is also the determining number of "king" ...
> [Adapted from M. Rutten (1970)]

It seems plausible, in this context, to think that base 60 commended itself to the mystic minds of Sumerians because of their cult of the "Upper God" *Anu*, whose number it was.

There are many attested examples, in Australia, Africa, the Americas, and Asia, of number-systems with a base (most often, base 4) that has mystical ramifications. However, the Sumerian system is much more developed than any of these, and presupposes complete familiarity with abstract number-concepts. For this reason it does not seems right to consider Sumerian mysticism as the origin of the Sumerian base 60. Things should rather be looked on the other way round: it is far more probable that 60 was the "number" of the Upper God Anu precisely because it was *already* the larger of the units of Sumerian arithmetic.

THE PROBABLE ORIGIN OF THE SEXAGESIMAL SYSTEM

So where does base 60 come from? Here is what I believe to be the solution to this enigma.

It is necessary to suppose (without a great deal of material evidence) that the Mesopotamian basin had one or more *indigenous populations* prior to Sumerian domination. A second essential premise (but one that is not at all controversial) is that the Sumerians were immigrants, that they *came from somewhere else,* more than probably in the fourth millennium BCE. Though we know very little about the indigenous population, and almost nothing about the prior cultural connections of the Sumerians, who seem to have broken all ties with their previous environment, we can speculate with a fair degree of confidence that these two cultures possessed different counting systems, one of which was duodecimal, and the other quinary.

Let us look again at Sumerian number-names.

1	2	3	4	5	6	7	8	9	10
geš	min	eš	limmu	iá	àš	imin	ussu	ilimmu	u

Geš (1) is a word that also means "man", "male" and "erect phallus"; *min* (2) also means "woman"; and *eš* (3) is also the plural suffix in Sumerian (rather like -*s* in English). The symbolism of these number-names is both apparent and very ancient indeed, taking us back to "primitive" perceptions of man as vertical (in distinction to all other animals) and alone, of woman as the "complement" of a pair (man and woman, or woman and child), and of "the many" beginning at three. (In Pharaonic Egypt as in the Hittite Empire, plurals were indicated by writing the same hieroglyph three times over, or by adding three vertical bars after the sign; in classical Chinese, the ideogram for "forest" consists of three ideograms for "tree", whereas the concept "crowd" was represented by a triple repetition of the ideogram for "man".) So the semantic meanings of the names of the first three numbers of Sumerian is a trace of those lost ages when people had only the most rudimentary concepts of number, counting only "one, two, and many".

More importantly, however, the names of the numbers in spoken Sumerian also carry unmistakable traces of a quinary system. *Às*, six, looks like an elision of *iá* and *geš*, "five (and) one"; *imin*, seven, is more certainly a contraction of *iá* and *min*, "five and two"; *ilimmu* is clearly related to *iá* and *limmu*, "five and four". In other words, Sumerian number-names derive from a vanished system using base 5. We speculate therefore that one of the two populations involved had a quinary counting system, and that in contact with a civilisation using base 12, the sexagesimal system was invented or chosen, since $5 \times 12 = 60$.

As we have already seen, the quinary base is anthropomorphic and derives from learning to count on the fingers of one hand and using the other hand as a "marker" when counting beyond 5. However, the origin of base 12 is far less obvious. My own view is that it was probably also based on the human hand.

Each finger has three articulations (or phalanxes): and if you leave out the thumb (as you have to, since you use it to check off the phalanxes counted), you can get to 12 using only the fingers of one hand, as in Fig. 9.3 below:

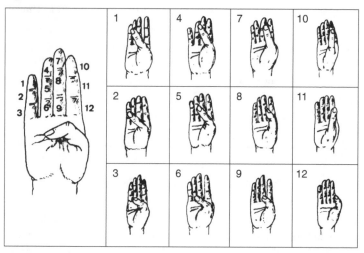

FIG. 9.3.

Repeating the device once over, you get from 13 to 24, then from 25 to 36, and so on. In other words, with a finger-counting device of this kind, base 12 seems the most natural for a numbering system.

This hypothesis is difficult to prove, but phalanx-counting of this type does exist and is in use today in Egypt, Syria, Iraq, Iran, Afghanistan, Pakistan, and some parts of India. Sumerians could therefore easily have used it at the dawn of their civilisation.

How then can we explain the fact that *u*, the Sumerian word for "10", means "fingers", and that there is no trace of a duodecimal system in spoken numbers, and no special word for the dozen ("12" is *u-min*, meaning "ten-two")?

My view is that spoken Sumerian numbers carry no trace of either base 12 or base 10: in other words, the name of the number 10 is not evidence of a lost decimal number-system, but merely the metaphoric expression of a universal human perception of human anatomy, the fact that there are ten fingers in all on the two hands.

At all events, my hypothesis has the advantage over all other speculations of giving a concrete explanation for the mysterious origin of base 60. As we saw in Chapter 3, basic finger-counting techniques, supplemented by mental effort (which quickly becomes quite "natural" once the principle of the base has been grasped), has often opened the way to arithmetical elaborations far superior to the original rudimentary system involved. From this, we can assert that the origin of base 60 could well have been connected to the finger-counting scheme shown in Fig. 9.4, currently in use across a broad band stretching from the Middle East to Indo-China.

This particular device makes 60 the principal base, with 5 and 12 serving as auxiliaries. This is how it is done:

Using your right hand, you count from 1 through 12 by pressing the tip of your thumb onto each of the three phalanxes (articulations) of the four opposing fingers. When you reach a dozen on the right hand, you check it off by folding the little finger of your left hand. You return to the right hand and count from 13 through 24 in similar fashion, then fold down the ring finger of your left hand, then count from 25 through 36 again on the right hand. The middle finger of the left hand is folded down to mark off 36, and you proceed to count from 37 to 48 on the right, then folding down the left index finger. Repeating the operation once more, you get to 60, and fold down the last remaining finger of the left hand (the thumb). As you can't count any higher numbers with this system, 60 is the obvious base.

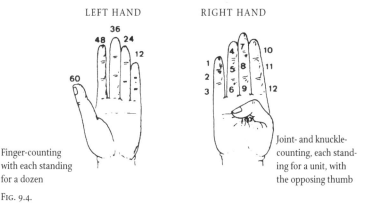

LEFT HAND   RIGHT HAND

Finger-counting with each standing for a dozen

Joint- and knuckle-counting, each standing for a unit, with the opposing thumb

FIG. 9.4.

My hypothesis can therefore be told as a story. As a result of the symbiosis of two different cultures, one of which used a quinary finger-counting method, and the other a duodecimal base deriving from a system of counting the phalanxes with the opposing thumb, 60 was chosen as the new higher unit of counting as it represented the combination of the two prior bases.

Since 60 was a pretty large number to use as a base, arithmeticians looked for an intermediate number to use, so as to mitigate the difficulties that arise from people's limited capacity to memorise number-names. Base 5 was too small compared to 60 – it would have required very long number-strings to express intermediate numbers; so 10 was chosen, a number provided by nature, so to speak, and of an ideal magnitude for the task in hand. Why not base 12? It has many advantages over 10, but it would probably have disoriented those accustomed to the quinary base, for whom 10, being twice the number of fingers on one hand, must have seemed more natural.

Since 6 is the coefficient required to turn 10 into 60, the Sumerian system, by its own dynamic, or rather, because of the inherent properties of the numbers involved, became a kind of compromise between 6 and 10, which served as the alternate and auxiliary bases of the sexagesimal system. Only subsequently could it have been observed that the resulting base had very valuable arithmetical properties as well as advantages for astronomy and geometry, which could only have been discovered as mastery of the counting tool and of the applied sciences progressed. Those properties and advantages came to seem so considerable and numerous that the Sumerians gave the main units the names of their own gods.

That is, in my view, the most plausible explanation of base 60. All the same, it should only be taken as a story – a story for which no archaeological proof or even evidence exists, as far as I know. However, if it were the true story of the origin of the sexagesimal system, then it would give added support to the anthropomorphic origin of the other common and historically-attested bases (5, 10 and 20), and thus underline the huge importance of human fingers in the history of numbers and counting.

CHAPTER 10

THE DEVELOPMENT OF
WRITTEN NUMERALS IN ELAM
AND MESOPOTAMIA

As we have seen, by the fourth millennium BCE clay was already a traditional material, not only for building work, but also and above all as the *basic medium for the expression of human thought*. In this period, the Mesopotamian peoples were entirely at ease with clay in a wide range of applications, and they used it to throw earthenware and ceramic vessels and figurines, to mix mortar, to mould bricks, and to shape seals, beads, jewels, and so on. It is therefore not unreasonable to suppose that the inhabitants of Sumer, long before they devised their written numerals and writing system, made diverse kinds of clay or earthenware objects or tokens with conventional values in order to symbolise and to manipulate numbers.

FROM PEBBLES TO ARITHMETIC

Concrete arithmetic (which as we shall see most certainly existed in the region of Mesopotamia) necessarily derived from the archaic "heap of pebbles" counting method, put to numerical use. The "pebbles" method is attested in every corner of the globe and clearly played a major role in the history of arithmetic – for pebbles first allowed people to learn how to perform arithmetical operations.

The English word *calculus*, like the French *calcul* (which has a more general meaning of "arithmetic", "counting operation", or "calculation") comes directly from the Latin *calculus*, which means ... a pebble, and, by extension, a ball, token, or counter. The Latin word is related to *calx, calcis*, which, like the similar-sounding Greek word *khaliks*, means "rock" or "limestone", and which has numerous etymological derivations in modern European languages, from German *Kalkstein*, "limestone", to English *calcium* and French *calcul*, in the sense of "kidney stones".

Because the Greeks and Romans taught their children to count and to perform arithmetical operations with the help of pebbles, balls, tokens, and counters made of stone (probably limestone, which is lighter and easier to fashion than marble or granite) their word for "doing pebbles" (*calculation*) has come to refer to all and any of the elementary arithmetical operations – addition, subtraction, multiplication, and division.

Greek and Arabic both have their own independent etymological proofs of the origins of arithmetic in the manipulation of stones. Greek *pséphos* means both "stone" and "number"; Arabic *ḥaswa*, meaning "pebble", has the same root as *iḥsā'*, which means "a count (of things)" and "statistics".

At its simplest, the pebble method is extremely primitive: even more than the basic forms of notched-stick counting described in Chapter 1, it represents the "absolute zero" of number-techniques. It can only supply cardinal numbers, requires no memorisation and no abstraction, and uses exclusively the principle of one-for-one correspondence.

However, once abstract counting has been mastered, the pebble method is sufficiently adaptable to allow great strides to be made. In some African villages, accounts of marriageable girls (and of boys of military age) were kept until quite recently by this method. On reaching puberty each village girl gave a metal bangle to the local matchmaker, who threaded it onto a strap alongside other similar bangles; when her marriage was imminent, the girl would take back her bangle, leaving the matchmaker with an accurate and immediate account of the number of "matches" left to make. It is a most convenient way of performing a subtraction in the absence of any knowledge of arithmetic as such.

In Abyssinia (now Ethiopia) tribal warriors had a similar device. On leaving for a foray each warrior placed a pebble on a heap, and on return to the village, each survivor removed his pebble. The number of unclaimed pebbles provided the precise total of losses in the skirmish. Exactly the same device is portrayed in the opening sequence of Eisenstein's film, *Ivan The Terrible* (part I): we see each soldier in the army of Ivan IV Vassilievich, Tsar of all the Russias, placing a metal token on a tray before setting off for the siege of Kazan.

In the course of time it became apparent that the device could not be taken very far, and did not satisfy many perfectly ordinary requirements. For instance, you need to collect a thousand pebbles just to count up to 1,000! But once the principle of a base in a numbering system had arisen, pebbles could be used more imaginatively.

In some cultures, the idea arose of replacing natural pebbles with pieces of stone of various sizes and attributing a conventional value of a different order of magnitude to each size. So, in a decimal numbering system, a unit of the first order might be represented by a small pebble, a unit of the second order (tens) by a slightly larger pebble, a unit of the third order (hundreds) by a larger stone, a unit of the fourth order (thousands) by an even bigger one, and so on. To represent the other numbers in the series by this method, you just needed an appropriate number of pebbles of the appropriate size.

It was a practical device, but not yet quite serviceable, because it was hard to find a sufficiency of pebbles or stones of identical sizes and shapes.

For some societies there was an additional and quite crucial obstacle if they inhabited regions where stone was uncommon.

The pebble method was therefore perfected by recourse to malleable earth, a material far better suited to making regular counting tokens. That is what happened in Elam and Mesopotamia in prehistoric times, prior to the invention of writing and of written numerals.

MESOLITHIC AND NEOLITHIC TOKENS IN THE MIDDLE EAST

In several archaeological sites in the Middle East, in places as far from each other as Anatolia, the Indus Valley, the shores of the Caspian Sea, and the Sudan (Fig. 10.2), researchers have unearthed thousands upon thousands of small objects in a wide variety of sizes and regular geometrical shapes, such as cones, discs, spheres, pellets, sticks, tetrahedrons, cylinders, and so on (Fig. 10.1). These are the objects to which we shall apply the generic term of *calculi*.

Some of these *calculi* are inscribed with parallel lines, crosses, and other similar patterns (Fig. 10.1 B, C, D, E, M, O). Others are decorated with carved or moulded figurines that are visible representations of different kinds of beings or things (jars, cattle, dogs, etc.). Finally, there are some that have neither pattern nor figurine (Fig. 10.1 A, F, G, H, I, J, L, N, P, Q, R, S, T, U).

The oldest *calculi* found so far, dating from the ninth to the seventh millennium BCE, come from Beldibi (Anatolia), Tepe Asiab (Mesopotamia), Ganj Dareh Tepe (Iran), Khartoum (Sudan), Jericho (West Bank)) and Abu Hureyra (Syria). The most recent, dating from the second millennium BCE, were found at Tepe Hissar (Iran), Megiddo (Israel), and Nuzi (Mesopotamia).

Most of the *calculi* were found scattered around at ground level. Others, however, were found inside or next to egg-shaped or spherical hollow clay balls or *bullae*. However, although *bullae* are not found prior to the fourth millennium BCE, hundreds of them have been unearthed at Tepe Yahya (Iran), Habuba Kabira (Syria), Uruk (Mesopotamia), Susa (Iran), Chogha Miš (Iran), Nineveh (Mesopotamia), Tall-i-Malyan (Iran) and Nuzi (Mesopotamia).

WHAT DO THE COUNTING TOKENS MEAN?

Denise Schmandt-Besserat has put together all that is known about these tokens and argues that for the Middle Eastern civilisations of the ninth to the second millennium BCE they constituted three-dimensional *pictograms* of the specific goods or produce which they served to account for in commercial exchanges.

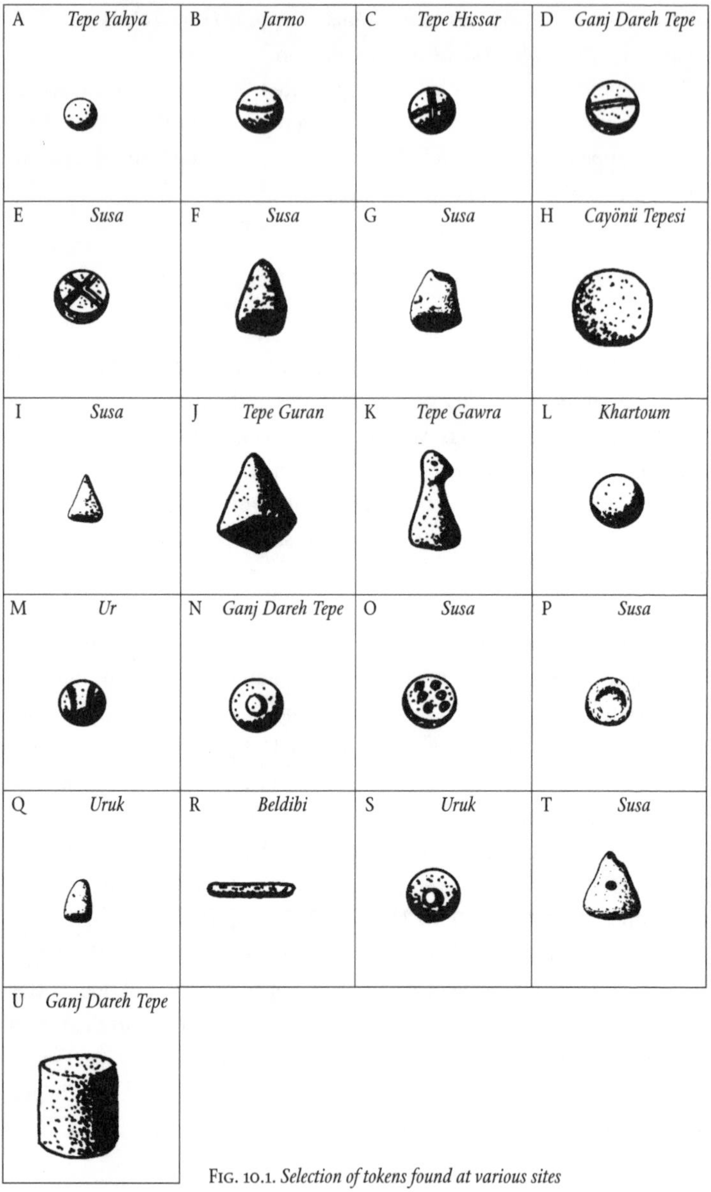

A	Tepe Yahya	B	Jarmo	C	Tepe Hissar	D	Ganj Dareh Tepe
E	Susa	F	Susa	G	Susa	H	Cayönü Tepesi
I	Susa	J	Tepe Guran	K	Tepe Gawra	L	Khartoum
M	Ur	N	Ganj Dareh Tepe	O	Susa	P	Susa
Q	Uruk	R	Beldibi	S	Uruk	T	Susa
U	Ganj Dareh Tepe						

FIG. 10.1. *Selection of tokens found at various sites*

In other words, she believes that the tokens were shaped so as to represent or symbolise the very things that they "counted", using actual or schematic images of, for instance, pots, heads of cattle, and so on, and in some cases were marked with dots or lines to indicate their places in a numbering series (one rectangular plaque has 2×5 dots on it, for example, and one cow's head figurine has 2×3 dots marked on it).

TYPE OF TOKEN

MILLENNIUM BCE	SITE	REGION	cylinders	discs	spheres and pellets	cones (various sizes)	sticks	hollow clay balls or bullae
9th	Beldibi	Anatolia	★	★	★	★	★	
	Tepe Asiab	Mesopotamia	★	★	★	★	★	
9th–8th	Ganj Dareh Tepe	Iran	★	★	★	★	★	
8th	Khartoum	Sudan	★	★	★			
8th–7th	Cayönü Tepesi	Anatolia	★	★	★	★	★	
7th–6th	Jericho	West Bank			★	★		
	Tell Ramad	Syria	★	★	★	★		
	Ghoraife	Syria		★	★	★		
	Suberde	Anatolia	★		★	★	★	
	Jarmo	Mesopotamia	★	★	★	★	★	
	Tepe Guran	Iran		★	★	★		
	Anau	Iran		★	★	★		
6th	Tell As Sawwan	Mesopotamia			★	★		
	Can Hasan	Anatolia		★	★	★		
	Tell Arpichiya	Mesopotamia			★	★		
6th–5th	Chaga Sefid	Iran	★	★	★	★	★	
	Tal-i-Iblis	Iran		★	★	★		
4th	Tepe Yahya	Iran		★	★	★		★
	Habuba Kabira	Iran		★	★	★		★
	Warka (Uruk)	Mesopotamia	★	★	★	★		★
	Susa	Iran	★	★	★	★	★	★
	Chogha Miš	Iran	★	★	★	★		★
	Nineveh	Mesopotamia						★
4th–3rd	Tall-i-Malyan	Iran	★		★	★		★
	Tepe Gawra	Iran				★		
3rd	Jemdet Nasr	Mesopotamia		★		★		
	Kiš	Mesopotamia		★	★	★		
	Tello	Mesopotamia	★	★	★	★		
	Fara	Mesopotamia		★	★	★		
3rd–2nd	Tepe Hissar	Iran		★	★	★		
2nd	Megiddo	Israel	★					
	Nuzi	Mesopotamia						★

FIG. 10.2. *Middle Eastern archaeological sites with finds of clay objects of various shapes and sizes, some of which are known to have been used for arithmetical operations and for accounts*

It is an attractive idea, and if it could be proved it would show that there was a very sophisticated accounting system in the Middle East in the earliest periods of the prehistoric record in that area.

However, it is only a hypothesis, and there is no solid evidence to support it. It presupposes the existence of a sufficiently complex market economy to have created the need felt for such an elaborate counting system. Schmandt-Besserat nonetheless takes her argument even further, and claims that this "three-dimensional symbolic system" is the origin of Sumerian pictograms and ideograms, that is to say is the source of the earliest writing system in the world.

Her conclusions derive from the discovery of a very large number of objects of various shapes (discs, spheres, cones, cylinders, and triangles) inscribed with exactly the same motifs – parallel lines, concentric circles, crossed lines – as are found on Sumerian tablets of the Uruk period, where a cross inside a circle stands for "sheep", three parallel lines inside a circle stands for "clothes", and so on (Fig. 10.3). The signs of Sumerian writing, she says, are simply two-dimensional reproductions of the three-dimensional tokens.

This important claim is nonetheless somewhat specious because it presupposes that there was a completely common and standardised set of traditions and conventions over a huge geographical area throughout a period of several thousand years – and what we know of the area and period suggests on the contrary that its cultures were very diverse. It is quite wrong to "explain" Sumerian pictograms by the shape of tokens found as far afield as Beldibi, Jericho, Khartoum, or Tepe Asiab, dating from eras as varied as the fourth, sixth, and ninth millennia BCE, since the cultures of these places in those periods probably had nothing whatsoever to do with developments in Sumer itself. (S. J. Liebermann gives a full critique.)

However, Schmandt-Besserat's general idea is not unacceptable, provided it is handled more methodically, by studying, not the entire collection of tokens in existence, but each subset of them in the context of its particular culture, in its specific location, at a particular period.

There must be major reservations about the overall conclusion concerning the origins of Sumerian writing. If it is ever demonstrated satisfactorily that there was a proper "system" of three-dimensional representation in these early Middle Eastern cultures, we will certainly find not one, but many different "systems" in the area. If we do establish a derivation from one such system to Sumerian writing, we are unlikely to establish it for more than a small number of individual signs.

All the same, these three-dimensional tokens must have meant *something* for their inventors and users, even if they did not form part of a system

Token shape	Sumerian sign	Known meaning of the sign
		jar, pot, vase
		oil, grease, fat
		sheep
		bread, food
		leather
		clothing

Fig. 10.3. *Comparison of tokens and their allegedly corresponding pictograms in early Sumerian writing (from D. Schmandt-Besserat, 1977)*

in the proper sense. They are obviously connected to ancient practices of symbolisation, which we can see in use on painted ceramics and in glyptics. One conclusion that might well come out of these speculations if sufficient evidence is found is that these tokens represent perhaps the final intermediate stage in the evolution of purely symbolic expressions of thought into formal notations of articulated language.

MULTIFUNCTION OBJECTS

The variety of the tokens is so great, their geographical locations so diverse, and their chronological origins are so widely separated that they could not possibly have belonged to a single system.

Even within a single period and place, they did not all serve the same purposes.

We can all the same make a few plausible guesses about their meaning if we bear in mind the specific nature of the cultures to which they belong. Some of the tokens, for example, that have holes in the middle and were found threaded on string, were probably objects of personal decoration. Such "necklaces" may also have served as counting beads, much like rosaries, allowing priests to count out gods or prayers. Other tokens decorated with the heads of animals may have been amulets, invoking the spirits of the animals represented, in terms of superstitions about the protective values of the different species (warding off the evil eye, illnesses, accidents, etc.). And since clay was plentiful and easy to shape, we can suppose that a fair number of these tokens were *playing pieces*, for ancient games like fives, draughts, chess, and so on.

FROM TOKENS TO *CALCULI*

However, the most interesting tokens from our point of view, and whose function is not in any doubt, are those small clay objects of varying shapes and sizes found inside the hollow clay balls called *bullae*. They were in use in Sumer and Elam (a region contiguous to Mesopotamia, covering the western part of the Iranian plateau and the plain to the east of Mesopotamia proper) from the second half of the fourth millennium BCE, and they served both as concrete accounting tools and also, as we shall see, as *calculi* which permitted the performance of the various arithmetical operations of addition, subtraction, multiplication, and even division. The Assyrians and the Babylonians called these counting tokens *abnū* (plural: *abnāti*), a word used to mean: 1. *stone*, 2. *stone object*, 3. *stone* (of a fruit), 4. *hailstone*, 5. *coin* [from R. Labat (1976) item 229]. Long before that, the Sumerians had called them *imna*, meaning "clay stone" (S. J. Liebermann in *AJA*). We will call them *calculi*, remembering that by this term we refer exclusively to the tokens found inside or close to hollow clay balls, the *bullae*.

THE FORMAL ORIGINS OF SUMERIAN NUMERALS

Archaic Sumerian numerals suggest very strongly at first glance that they derive from a pre-existing concrete number- and counting system, but they also seem to have obviously formal origins. The various symbols used (Fig. 8.2, repeated in Fig. 10.4 below) look very much like some of the *calculi* "copied" onto clay tablets once writing had been invented: specifically, the little cone, the pellet, the large cone, the perforated large cone, the sphere, and the perforated sphere. To put things the other way round (Fig. 10.4):

- the fine line representing the unit in archaic Sumerian numerals looks a two-dimensional representation of the small cone token;
- the small circular imprint representing the tens looks like a pellet-shaped token;
- the thick indentation for 60 looks like a large cone;
- the thick dotted indentation for 600 looks like a large perforated cone;
- the large circular imprint (3,600) looks like a sphere;
- the large dotted circular imprint (36,000) looks like a perforated sphere.

These resemblances are so obvious that the relationship would have to be accepted even if there were no other proof. But as we shall see, the archaeological record contains more than adequate confirmation of these identifications.

SPOKEN NUMERALS		CALCULI		WRITTEN NUMERALS		
	Number-names			Archaic	Cuneiform	Mathematical structure
1	*geš*	◮	small cone	◖	Y	1
10	*u*	●	pellet	●	◀	10
60	*geš*	◭	large cone	◗	Y	10.6 (= 60)
600	*geš u*	◭	perforated large cone	◖	⥟	10.6.10 (= 60.10)
3,600	*šàr*	◓	sphere	○	✧	10.6.10.6 (= 60²)
36,000	*šàr-u*	◉	perforated sphere	◉	✦	10.6.10.6.10 (= 60².10)
216,000	*šàrgal*	?		?	✦	10.6.10.6.10.6 (= 60³)
Archaeological date (BCE)		From mid-4th millennium		From c. 3200	From c. 2650	

FIG. 10.4. *Number-names, numerals and* calculi *of Sumerian civilisation. The* calculi *come from several Mesopotamian sites (Uruk, Nineveh, Jemdet Nasr, Kiš, Ur, Tello, Šurrupak, etc.)*

THE HOLLOW CLAY BALLS FROM
THE PALACE OF NUZI

It was in 1928–29 that Mesopotamian *calculi* were first properly identified, when the American archaeologists from the Oriental Research Institute in Baghdad excavating the Palace of Nuzi (a second-millennium BCE site near Kirkuk, in Iraq) came across a hollow clay ball clearly containing "something else", inscribed with cuneiform writing in Akkadian (Fig. 10.5) which in translation reads as follows:

> *Abnāti* ("things") about sheep and goats:
>
> | 21 | ewes which have lambed | 6 | she-goats that |
> |---|---|---|---|
> | 6 | female lambs | | have had kids |
> | 8 | adult rams | 1 | he-goat |
> | 4 | male lambs | [2] | kids |

The sum of the count is 48 animals. When the clay ball was opened, it was found to contain precisely 48 small, pellet-shaped, unbaked clay objects (which were subsequently mislaid). It seemed logical to assume that these tokens had previously been used to count out the livestock, despite the difficulty of distinguishing between the different categories by this system of reckoning.

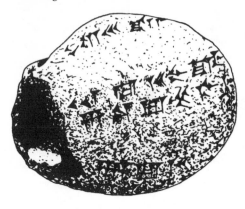

FIG. 10.5. *Hollow clay ball or bulla found at the Palace of Nuzi, 48mm × 62mm x 50mm. Fifteenth century BCE. From the Harvard Semitic Museum, Cambridge, MA (inventory no. SMN 1854)*

The archaeologists might have thought nothing of their discovery without a chance occurrence that suddenly explained the original purpose of the find. One of the expedition porters had been sent to market to buy chickens, and by mistake he let them loose in the yard before they had been counted. Since he was uneducated and did not know how to count, the porter could not say how many chickens he had bought, and it would have been impossible to know how much to pay him for his purchases had he not come up with a bunch of pebbles, which he had set aside, he said, "one for each chicken". So an uneducated local hand had, without knowing it, repeated the very same procedures that herdsman had used at the same site over 3,500 years before.

Thirty years later, A. L. Oppenheim at the University of Chicago carried out a detailed study of all the archaeological finds at Nuzi, and discovered that the Palace kept a double system of accounting. The cuneiform tablets of the Palace revealed the existence of various objects called *abnu* ("stones") that were used to make calculations and to keep a record of the results. The texts written on the tablets make clear reference to the "deposit" of *abnu*, to "transfers" of the same, and to "withdrawals". The meticulous cuneiform accounts made by the Palace scribes were "doubled", as Schmandt-Besserat explains, by a tangible or concrete system. One set of *calculi* may for instance have represented the palace livestock. In spring, the season of lambing, the appropriate number of new *calculi* would have been added; *calculi* representing dead animals would have been withdrawn; perhaps *calculi* were even moved from one shelf to another when animals were moved between flocks, or when flocks moved to new pasture, or when they were shorn.

The hollow clay ball was therefore probably made by a Palace accountant for recording how many head of livestock had been taken to pasture by local shepherds. The shepherds were illiterate, to be sure, but the accountant must have known how to count, read, and write: he was probably a priest, as he possessed the great privilege of Knowledge, and must have been one of the managers of the Nuzi Palace's goods and chattels. The proof of this lies in the Akkadian word *sangu*, which means both "priest" and "manager of the Temple's wealth"; it is written in cuneiform in exactly the same way as the verb *manû*, which means "to count".

When shepherds left for pasture, the functionary would make as many unbaked clay pellets as there were sheep, and then put them inside the clay "purse". Then he would seal the purse and mark on it, in cuneiform, an account of the size of the flock, which he then signed with his mark. When the shepherd came back the purse could be broken open and the flock checked off against the pellets inside. There could be no disputing the numbers, since the signed account on the outside certified the size of the flock as far as the masters of the Palace were concerned, and the *calculi* provided the shepherd with his own kind of certified account.

The later discovery of an oblong accounting tablet shaped like the base of the hollow clay ball in the ruins of the same palace, but from a higher (and therefore more recent) stratum, gave further support to Oppenheim's views.

The story now moves to Paris, where, at the Musée du Louvre, there are about sixty of these hollow clay balls brought back c. 1880 by the French Archaeological Mission to Iran, which had been excavating the city of Susa (about 300 km east of Sumer, in present-day southwestern Iran, Susa was the capital of Elam and then of the Persian Empire under Darius). Up until recently the only interest that had been shown in them concerned the imprints of cylinder-seals with which most of them are decorated (Fig. 10.10). Several of the *bullae* had been broken during shipment to Paris, other had been found broken. All the same, some of them were intact, and sounded like rattles when shaken. X-ray photography showed that they contained *calculi* – but not all of the same uniform type. When some of them were very carefully opened, they were found to contain clay discs, cones, pellets, and sticks (Fig. 10.6)

As P. Amiet then argued, these "documents", since they came from a site dated about 3300 BCE, proved that Elam had an accounting system far more elaborate than that of Nuzi with its plain "unit counters", and had it 2,000 years earlier. In other words, this counting system had survived for two millennia, but had regressed over that period, losing the use of a base, and retreating to a rudimentary and purely cardinal method.

It was therefore correctly assumed that the counting system of Susa consisted of giving tangible form to numbers by the means of various *calculi* which symbolised numerical values both by their own number and by their respective shapes and sizes, which corresponded to some order of magnitude within a given number-system (for example, a stick was a unit of the first order of magnitude, a pellet for a unit of the second order, a disc for a unit of the third order, and so on).

More recent finds in Iran (Tepe Yahya, Chogha Miš, Tall-i-Malyan, Šahdad, etc.), in Iraq (Uruk, Nineveh, Jemdet Nasr, Kiš, Tello, Fara, etc.), and in Syria (Habuba Kabira) have proved Oppenheim and Amiet to be correct. What they have also shown is that the system was not restricted to Elam, but that similar accounting methods were used throughout the neighbouring region, including Mesopotamia. These methods are thus even more ancient than the one used for the accounting tablets of the Uruk period.

FROM CLAY BALLS TO ACCOUNTING TABLETS

It then seemed very likely that the archaic accounting tablets of Sumer were directly descended from the clay *calculi*-and-*bulla* accounting system. The archaic Sumerian figures obviously were related to the *calculi*; and, unlike the later, perfectly rectangular tablets that were made to a standard pattern, the archaic counting tablets are just crude oblong or roughly oval slabs (Fig. 8.1 C above). So there really had been a point in time when the stones were supplanted by their own images in two-dimensional form, and the hollow clay balls replaced by these flat clay slabs. But this remained only a conjecture in the absence of all the archaeological evidence needed to reconstitute the intermediate stages of the supposed development and of evidence to allow firm datings.

In the 1970s, the French Archaeological Delegation to Iran (DAFI), under the direction of Alain Le Brun, excavated the Acropolis of Susa, and established a far more accurate and substantiated stratigraphy of Elamite civilisation than had previously been possible, and, in 1977–78, important finds were made which make the transition comprehensible in archaeological terms. A word of warning, however: the development we describe below is attested only at Susa. Nonetheless there are good reasons for believing that much the same thing happened at Sumer.

FIG. 10.6. *Sketch of the contents of an unbroken* bulla, *as revealed by X-rays*

The first reason is that Elamite civilisation is pretty much contemporary with Sumer, and flourished in very similar fashion in precisely similar circumstances in the second half of the fourth millennium BCE. For that reason various aspects of Elamite civilisation are used as reference points (or as potentially applicable models) for the civilisation of Uruk. All the same the Elamites retained many features that are distinct from those of their Mesopotamian neighbours.

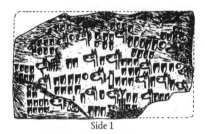

Side 1

Side 2

FIG. 10.7. *Proto-Elamite tablet (Susa, level unknown), c. 3000 BCE. J. Schell (1905) identified this as an inventory of stallions (erect manes), mares (flat manes), and colts (no manes), with the numbers of each indicated by various indentation-marks. Side 2 bears the imprint of a cylinder-seal representing standing and resting goats. Paris, Musée du Louvre, Sb 6310*

The second reason is that the Elamites, like the Sumerians, were fully conversant with the use of clay for expressing human thought visually and symbolically, and later on in using it to represent articulated language.

For we know that the Elamites acquired a writing system around 3000 BCE, the earliest traces of which are the clay "tablets" (Fig. 10.7) found at several Iranian sites, mainly at Susa, at archaeological level XVI. Like archaic Sumerian tablets, they bear on one side (sometimes both sides) a number of numerical signs alongside more or less schematic drawings, and occasionally the imprint of a cylinder-seal.

And finally, as we have seen, the system of *calculi* and *bullae* was used in Elam as well as Sumer since at least 3500–3300 BCE.

Such manifest analogies between the two civilisations lead us to hope that new archaeological finds at Sumerian sites will one day establish once and for all the relationship between Sumer and Elam.

WHO WERE THE ELAMITES?

The oldest Iranian civilisation arose in the area now called Khuzestan. Its people called themselves *Haltami*, which the Bible transforms into *Elam*.

The origins of Elam are as ill understood as its language, despite the efforts of many linguists to decipher it. We know only that the name of Elam means "land of God". Elamite appears to be an agglutinative language, like Sumerian and other Asianic languages; some linguists think it belongs to the Dravidian group (southern India) and is related to Brahaoui, which is currently spoken in Baluchistan. It should be noted that from the beginning of the third millennium BCE there appear to have been close relations between Elam and Tepe Yahya (Kirman), which is located on a possible migration route from India. The Elamite tablets found there have been dated as late fourth millennium BCE.

It seems most likely that the Elamites arrived and settled in the area that was to bear their name in the fifth millennium BCE, joining a farming culture of which the earliest traces date from the eighth millennium BCE. The earliest pieces of Susan art are decorated ceramics, showing archers and beasts of prey (Tepe Djowzi), and horned snakes (Tepe Bouhallan), and Susa, which became a full-blown city in the fourth millennium, seems to have been the most important Elamite town. Painted ceramics were abandoned during what Amiet calls the earlier period of "proto-urban" Elamite civilisation.

Throughout its history, Mesopotamia had relations with Elam, from which it imported wood, copper, lead, silver, tin, building stone, and rare stones such as alabaster, diorite, and obsidian, but from the start of the third millennium BCE relations were intense. The periods are divided as follows: from 3000 to 2800 BCE, the palaeo-Elamite period; from 2800 to 2500 BCE, the Sumero-Elamite period (subdivided into early and late, during which Sumerian influence is very noticeable); from 2500 to 1850 BCE, the Awan Dynasty, interrupted by an Akkadian conquest, was replaced by the dynasty of Shimash.

Susa became the central city in the second millennium BCE, and Elamite civilisation reaches its apogee in the middle of the thirteenth century BCE under the reign of Untash Gal who built Tchoga-Zanbil. During the first millennium BCE, Elam is closely connected to the Kingdom of Anshan which, from the sixth century BCE, became one of the key points in the Achaemenian Persian Empire.

THE STAGES OF ELAMITE ACCOUNTING

With the help of the latest discoveries made by DAFI, we can now reconstruct the stages in the development of accounting systems in Elam. We begin in the second half of the fourth millennium BCE, in an advanced urban society where trading is increasing every day. And with an active economy, there is a pressing need to keep durable records of sales and purchases, stock lists and tallies, income and expenditure ...

First stage: 3500 – 3300 BCE

Levels: Susa XVIII; Uruk IVb. For sources, see Fig. 10.4, 10.8, and 10.10

Susan officials have an accounting system through which they can represent any given number (for example, a price or a cost) by a given number of unbaked clay *calculi* each of which is associated with an order of magnitude according to the following conventions:

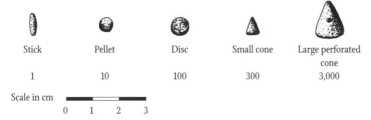

Stick	Pellet	Disc	Small cone	Large perforated cone
1	10	100	300	3,000

Scale in cm

0   1   2   3

FIG. 10.8. *The only* calculi *found in or very near hollow clay balls at the Acropolis of Susa. The values shown derive from the decipherment explained in Chapter 11 below. From DAFI 8, plate 1 (Susa, level XVIII)*

Intermediate numbers are represented by using as many of each type of *calculus* as required. For example, the number 297 calls for 2 discs, 9 pellets, and 7 sticks:

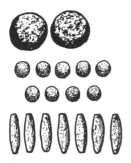

FIG. 10.9.

FIG. 10.10. *Exterior of a* bulla *marked with a cylinder-seal. Susa, c. 3300 BCE. From the Musée du Louvre, item Sb 1943*

You then place these objects with conventional values (whose use is not entirely dissimilar to our current use of coins or standard weights) into a hollow ball, spherical or ovoid in shape (Fig. 10.10), the outside of which is then marked by a cylinder-seal, so as to authenticate its origin and to guarantee its accuracy. For in Elam, as in Sumer, men of substance each had their own individual seal – a kind of tube of more or less precious stone on which a reversed symbolic image was carved. The cylinder-seal, invented around 3500 BCE, was its owner's representative mark. The owner used it to mark any clay object as his own, or to confer his authority on it, by rolling the cylinder on its axis over the still-soft surface (Fig. 10.11).

Let us imagine we are at the Elamite capital of Susa. A shepherd is about to set off for a few months to a distant pasture to graze a flock of 297 sheep that a wealthy local owner has entrusted to him. The shepherd and the owner call on one of the city's counting men to record the size of the flock. After checking the actual number of sheep, the counting master makes a hollow clay ball with his hands, about 7 cm in diameter, that is to say hardly bigger than a tennis ball. Then, through the thumb-hole left in the ball, he puts inside it 2 clay discs each standing for 100 sheep, 9 pellets that each stand for 10 sheep, and 7 little sticks, each one representing a single animal. Total contents: 297 heads (Fig. 10.9).

When that is done, the official closes up the thumb-hole, and, to certify the authenticity of the item he has just made up, rolls the owner's cylinder-seal over the outside of the ball, making it into the Elamite equivalent of a signed document. Then to guarantee the whole thing he rolls his own cylinder-seal over the ball. This makes it unique and entirely distinct from all other similar-looking objects.

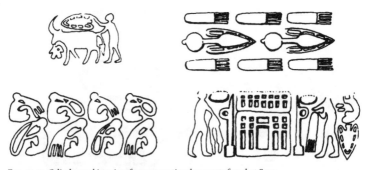

FIG. 10.11. *Cylinder-seal imprints from accounting documents found at Susa*

The counting master then lets the *bulla* dry and stores it with other documents of the same kind. With its tokens or *calculi* inside it, the *bulla* is now the official certification of the count of sheep that has taken place, and serves as a record for both the shepherd and the owner. On the shepherd's return from the pastures, they will both be able to check whether or not the right number of sheep have come back – all they need to do is break open the ball, and check off the returning sheep against the tokens that it contains.

At about the same period, the Sumerians used a very similar system: hollow clay balls have been found at Warka at the level of Uruk IVb, at Nineveh and Habuba Kabira (Fig. 10.4). The Sumerians, however, were accustomed to counting to base 60, using tens only as a supplementary system to reduce the need for memorisation (Fig. 8.5, 8.6 above), and the tokens that they used were also shaped rather differently. At Sumer, the small cone stood for 1, the pellet for 10, the large cone for 60, the perforated large cone stood for 600, the sphere represented 3,600, and the perforated sphere meant 36,000 (Fig. 10.4).

It was a sophisticated system for the period, since values were regularly multiplied by 10 by means of the perforation. By pushing a small circular stylus through the cone signifying 60, or through the sphere signifying 3,600, the values of 600 (60 × 10) and 36,000 (3,600 × 10) were obtained. The hole or circle was thus already a virtual graphic sign for the pellet, with a value of 10.

Let us now imagine ourselves in the market of the royal city of Uruk, capital of Sumer. A cattle farmer and an arable farmer have made a deal to exchange 15 head of cattle against 795 bags of wheat. However, the livestock dealer has only got 8 head of cattle at the market, and the grain seller has only 500 bags of wheat immediately available. The deal is done nonetheless, but to keep things above board there has to be a contract. The cattle man agrees to deliver a further 7 cattle by the end of the month, and the arable farmer promises to supply 295 bags of grain after that year's harvest. To make a firm record of the agreement, the cattle man makes a clay ball and puts in 7 small cones, one for each beast due, then closes the ball and marks its surface with his own cylinder-seal, as a signature. The arable farmer, for his part, makes another clay ball and puts in it 4 large cones, each one standing for 60 bags of wheat, 5 pellets each standing for 10 bags, and 5 small cones for the 5 remaining bags due, then seals and signs the clay ball in like manner. Then a witness puts his own "signature" on the two documents, to guarantee the completeness and accuracy of the transaction. Finally, the two traders exchange their *bullae* and go their separate ways.

So although this remains an illiterate society, it possesses a means of recording transactions that has exactly the same force and value as written contracts do for us today.

At a time when cities were still relatively small, and where trade was still relatively simple, business relations were conducted by people who knew each other, and whose cylinder-seals were unambiguously identifiable. For that reason, the nature of a transaction recorded in a *bulla* is implicit in the identity of the seal(s) upon it: the symbolic shapes on the outside of the clay ball tell you whether you are dealing with this farmer or that miller, with a particular craftsman or a specific potter. As for the numbers involved in the transaction, they are unambiguously recorded by the nature and number of the *calculi* inside.

Cheating is therefore ruled out. Each party to the deal possesses the record of what his partner owes him, a record certified by his business partner's own identity, in the form of his seal.

Second stage: 3300 BCE

Level: Susa XVIII. For sources see Fig. 10.13

The great defect of the system in place was that the hollow clay balls had to be broken in order to verify that settlements conformed to the contracts. To overcome this, the idea arose of making various imprints on the outer surface of the *bullae* (alongside the imprints of the necessary cylinder-seals) to symbolise the various tokens or *calculi* that are inside them. Technically, the device harks back to the more ancient practice of notching, but it is quite altered in its significance by the new context.

The corresponding marks are: a long, narrow notch, made by a stylus with its point held sideways on to the surface, to represent the stick; a small circular imprint, made by the same stylus pressed in vertically, to represent the pellet; a large circular imprint, made by a larger stylus or just by pressing in a finger-tip, to represent a disc; a thick notch, made by a large stylus held obliquely, to represent a cone; and a thick notch with a circular imprint to represent a perforated cone.

| long narrow notch | small circular imprint | large circular imprint | thick notch | perforated thick notch |

Fig. 10.12. *Numerical markings on* bullae *found at Susa*

This constitutes a kind of résumé of the contract, or rather a graphic symbolisation of the contents of each accounting "document".

Henceforth, an Elamite *bulla* containing (let us say) 3 discs and 4 sticks (making a total of $3 \times 100 + 4 = 304$ units) carries on its outer face, alongside the cylinder-seal imprints, 3 large circular indentations and 4 narrow lines. No longer is it necessary to break open the clay balls simply to check a sum or to make an inventory – because the information can now be "read" on the outside of the *bullae*.

The cylinder-seal imprint or imprints show the *bulla's* origin and guarantee it as a genuine document, and the indentations specify the quantities of beings or things involved in the accounting operation.

Third stage: c. 3250 BCE

Level: Susa XVIII. See Fig. 10.15 below

These indentations thus constitute real numerical symbols, since each of them is a graphic sign representing a number. Together they make up a genuine numbering system (Fig. 10.14). So why carry on using *calculi* and putting them in *bullae*, when it's much simpler to represent the corresponding values by making indentations on slabs of clay? Mesopotamian and Elamite accountants very quickly realised that of the two available systems, one was redundant, and the *calculi* were rapidly abandoned. The spherical or ovoid *bullae* came to be replaced by crudely rounded or oblong clay slabs, on which the same information as was formerly put on the casing of the *bullae* was recorded, but on one side only.

The cylinder-seal imprint remained the mark of authenticity on these new types of accounting records, whose shape, at the start, roughly imitates that of a *bulla*. The sums involved in the transaction are represented on the soft clay by graphic images of the *calculi* that would previously have been

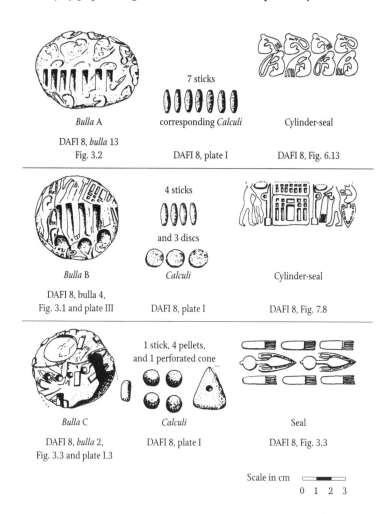

Bulla A	7 sticks corresponding *Calculi*	Cylinder-seal
DAFI 8, *bulla* 13 Fig. 3.2	DAFI 8, plate I	DAFI 8, Fig. 6.13
Bulla B	4 sticks and 3 discs *Calculi*	Cylinder-seal
DAFI 8, bulla 4, Fig. 3.1 and plate III	DAFI 8, plate I	DAFI 8, Fig. 7.8
Bulla C	1 stick, 4 pellets, and 1 perforated cone *Calculi*	Seal
DAFI 8, *bulla* 2, Fig. 3.3 and plate I.3	DAFI 8, plate I	DAFI 8, Fig. 3.3

Scale in cm 0 1 2 3

FIG. 10.13. Bullae *containing the same number of* calculi *as are symbolised on the outer surface by indentations next to the cylinder-seal imprints. Susa, level XVIII (approx. BCE 3300), excavated by DAFI in 1977–1978. Similar* bullae *have been found at Tepe Yahya and Habuba Kabira, but none so far at Uruk.*

Written Numerals in Elam and Mesopotamia

enclosed in a *bulla*. This stage therefore marks the appearance of the first "accounting tablets" in Elam.

It should be noted that the three stages laid out above occurred in a relatively short period of time, since all the evidence for them is attested at the same archaeological level (Susa XVIII), in the same room, and on the same floor level. The imprint of the same cylinder-seal on one *bulla* and two tablets (see *bulla* C in Fig. 10.13 and tablet B in Fig. 10.15 below, for example) seems to confirm that both systems existed side by side at least for a time.

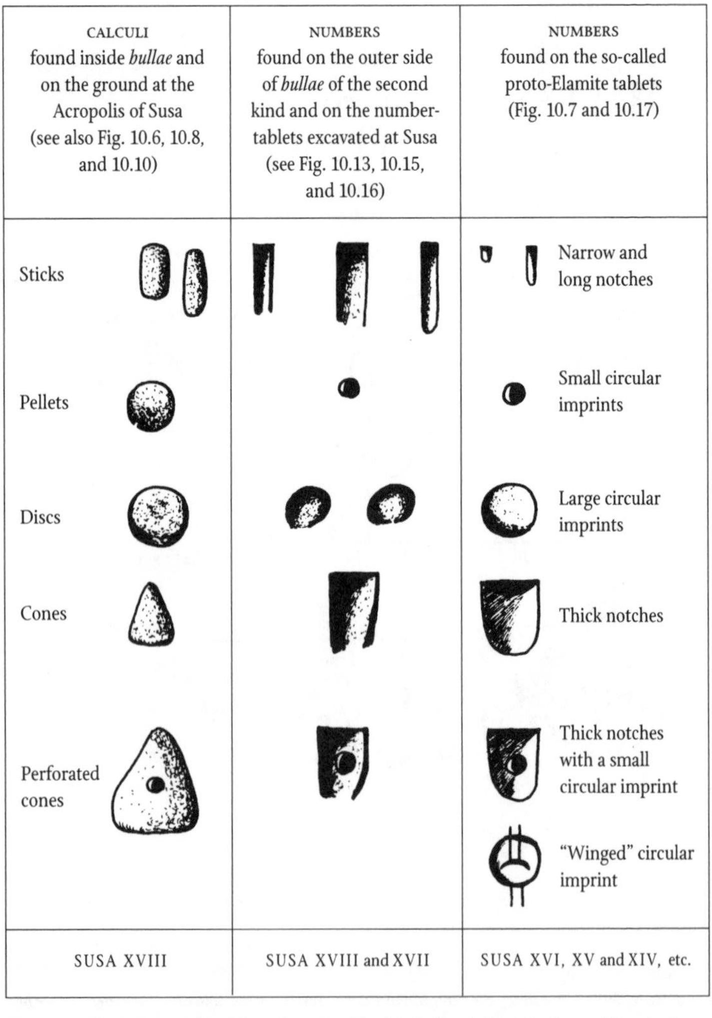

CALCULI found inside *bullae* and on the ground at the Acropolis of Susa (see also Fig. 10.6, 10.8, and 10.10)	NUMBERS found on the outer side of *bullae* of the second kind and on the number-tablets excavated at Susa (see Fig. 10.13, 10.15, and 10.16)	NUMBERS found on the so-called proto-Elamite tablets (Fig. 10.7 and 10.17)
Sticks		Narrow and long notches
Pellets		Small circular imprints
Discs		Large circular imprints
Cones		Thick notches
Perforated cones		Thick notches with a small circular imprint
		"Winged" circular imprint
SUSA XVIII	SUSA XVIII and XVII	SUSA XVI, XV and XIV, etc.

FIG. 10.14. *The indentations made on the outer side of the* bullae *imitate the shape of the* calculi *that are enclosed. Moreover, these marks resemble not only the number-tablets found at Susa but also the figures on the proto-Elamite tablets of later periods.*

Fourth stage: 3200–3000 BCE

Levels: Susa XVII; Uruk IVa. See Fig. 8.1 above and 10.16 below

This stage sees only a slow refinement of the system in place already: exactly the same types of information are included on the accounting tablets of the fourth period as on those of the third. However, the tablets themselves become less crudely shaped, the numbers are less deeply indented in the clay, and their shapes become more regular. In addition, the cylinder-seals are now imprinted on both sides of the tablet, and not just on the "top".

However, like the earlier *bullae* and crude tablets, this stage of development is still not "writing" in the proper sense. The notation records only numerical and symbolic information, and the things involved are described only in terms of their quantity, not by signs specifying their nature. Nor is the nature of the operation indicated by any of these documents: we have

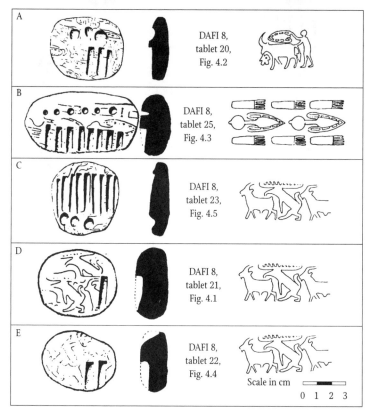

FIG. 10.15. *Roughly circular or oblong tablets containing indented numerical marks (similar to those found on* bullae*) alongside one or two cylinder-seal imprints. Items dated c. 3250 BCE, from Susa level XVIII, excavated dy DAFI in 1977–1978.*

no idea if they are records of a sale, a purchase, or an allocation, nor can we know the names, the numbers, the functions, or the locations of any of the parties to the transaction. We have already made the assumption that the cylinder-seals, since they indicate the identities of the contracting parties, would also have indicated the type of transaction in a society where

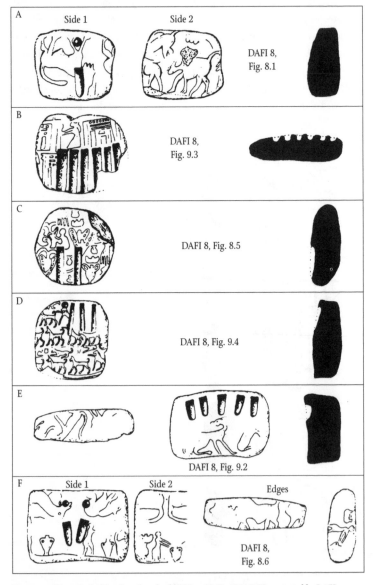

FIG. 10.16. *Numerical tablets from Susa level XVII , c. 3200– 3000 BCE, excavated by DAFI in 1972*

people were known to one another. This makes very clear just how concise, but also how imprecise are the purely symbolic visual notations of these documents, which constitute the trace of the very last stage in the prehistory of writing. Cylinder-seal imprints do in fact disappear from the tablets as soon as pictograms and ideograms make their appearance.

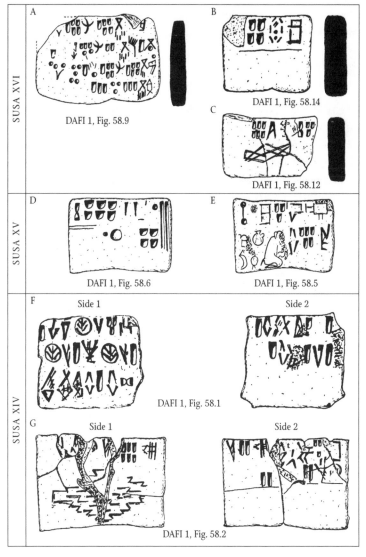

FIG. 10.17. *The first proto-Elamite tablets. They are less crude, rectangular tablets giving written name-signs alongside the corresponding numbers. From Susa, c. 3000–2800 BCE; excavated by DAFI in 1969–1971. (Cf. A. Le Brun)*

At Sumer, writing emerged at the same time as these regular tablets from Elam. The first Uruk tablets date from 3200–3100 BCE (Fig. 8.1 above) and, although they remain exclusively economic documents, they use a notation (archaic Sumerian numerals) which is founded not on making a "picture" of a vague idea, but on something much more precise, analytical and articulated. In tablet E of Fig. 8.1, for example, you can see how the document is divided into horizontal and vertical lines, marking out squares in which pictograms are placed beside groups of numbers. Sumerian tablets are thus ahead of the Susan ones of the same period: Sumer has something like writing, and Susa has only symbols.

Fifth stage: 3200–2900 BCE
Level: Susa XVI. See also Fig. 10.17, tablets A, B, C

The tablets from this period are thinner and more regularly rectangular (standardised), but most significantly they carry the first signs of "proto-Elamite" script alongside numerical indentations. The purpose of the signs is to specify the nature of the objects involved in the transaction associated with the tablet. On several tablets found at Susa XVI, there are no cylinder-seal imprints.

Sixth stage: 2900–2800 BCE
Level: Susa XV and XIV. See Fig. 10.17, D, E and F

In this period, the proto-Elamite script on the tablets grows to cover more of the surface than the number-signs. Could this mean that the script might hold the key to the grammar of the language? Is proto-Elamite the earliest alphabetic script? We do not know, as it remains to be deciphered.

THE PROBLEMS OF SO-CALLED PROTO-ELAMITE SCRIPT

This script appeared at the dawn of the third millennium BCE and spread from the area around Susa to the centre of the Iranian plateau. It remained in use in Elam until around 2500 BCE, when it was supplanted by cuneiform writing systems from Mesopotamia, whence derived Elamite script proper, whose final form was neo-Elamite.

How did proto-Elamite arise? Some scholars believe the Elamites invented it, independently of the Sumerians. This presupposes that it resulted from a similar set of steps, starting from identical circumstances, and following the same generic idea based on earlier rudimentary trials in the area. That is not implausible, especially in the light of the developments we have just charted.

Other scholars take the opposite view, namely that proto-Elamite script was inspired by Sumerian. This is also quite plausible, even if the nature of the "inspiration" must have been quite a distant one. Some of the proto-Elamite signs look as if they might be related to specific Sumerian pictograms and ideograms, but most of the signs are too different to allow any systematic comparison of the two scripts. On the other hand, it may well be that the Sumerian invention of writing inspired their neighbours the Elamites (Uruk and Susa are less than two hundred miles apart) to invent a writing of their own. Sumerian accounting tablets are one or two centuries older than their Elamite equivalents, and there is no doubt in which direction the invention flowed.

It seems probable that writing would have been invented in Susa even without the example or inspiration of Sumer, since all the social and economic dynamics that led to the invention of writing elsewhere were present amongst the Elamites. For as the history of numbers shows, people in similar circumstances and faced with similar needs often do make very similar inventions, even when separated by centuries and continents.

Be that as it may, proto-Elamite script remains a mystery. The signs almost certainly represented beings and things of various kinds, but the forms used are simplified and conventionalised to a point where guessing their meaning is impossible. We also know next to nothing about the language which this script represents.

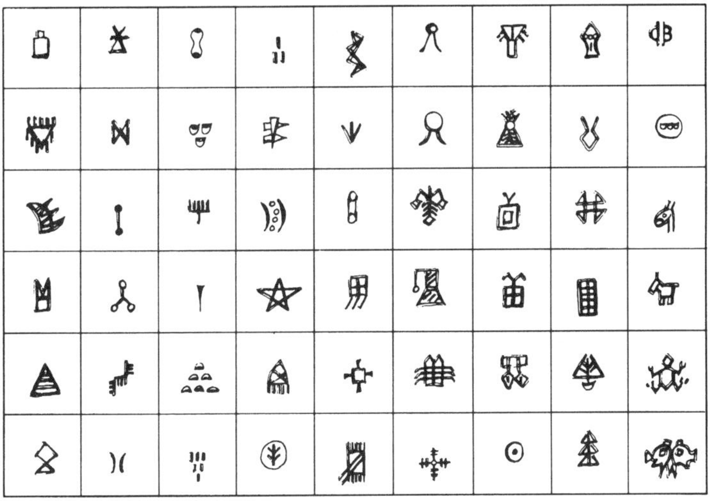

FIG. 10.18. *The signs of proto-Elamite script. References: Mecquenen; Scheil; Meriggi*

CHAPTER 11

THE DECIPHERMENT OF A FIVE-THOUSAND-YEAR-OLD SYSTEM

In 1981, when I published the first edition of *The Universal History of Numbers*, the number-signs in the proto-Elamite script (Fig. 11.1) still presented major problems.

A table drawn up by W. C. Brice (1962), and later also referred to by A. Le Brun and F. Vallat (1978), clearly shows how these number-symbols received very varied, indeed contradictory, interpretations over the years on the part of the majority of epigraphists and specialists in these questions.

Despite the great difficulties, I decided to apply myself to the task. In 1979 I began my researches which, one year later, culminated in the complete decipherment of these number-signs, after close examination of a large number of invoice tablets which had been discovered by the French Archaeological Mission to Iran at the end of the last century. These documents may be found in the collections of the Louvre and the Museum of Teheran.

We shall come shortly to the method which I followed. But, in order to appreciate it, we must first make yet another visit to the land of Sumer ...

THE INVENTION OF THE BALANCE SHEET
IN SUMERIA

The period from 3200 to 3100 BCE saw, as we have observed, the beginnings of written business accounts.

At first, however, the system was primitive. The documents held only one kind of numerical record at a time: one tablet for 691 jugs, for example (Fig. 8.1 C above), another tablet for 120 cattle (Fig. 8.1 D), another for 567 sacks of corn, another for 23 chickens, yet another for 89 female slaves imported from abroad, and so on.

But from around 3100 BCE as business transactions and distributions of goods became increasingly numerous and varied, the inventories and the accounts for each transaction also grew more complex and voluminous, and the accountants found they had to cut down on the cost of clay. From this time on the pictures and the numbers took up increasing amounts of space on the tablets. Onto a single rectangular sheet of clay, divided into boxes by horizontal and vertical lines, were recorded inventories of livestock in all their different kinds (sheep, fat sheep, lambs, lambkins, ewes, goats, kids male and female or half-grown, etc.) in all necessary detail. A single tablet, too, was used to summarise an agricultural audit in which all the different kinds of species were distinguished.

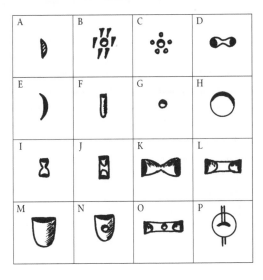

FIG. 11.1. *The proto-Elamite number-signs*

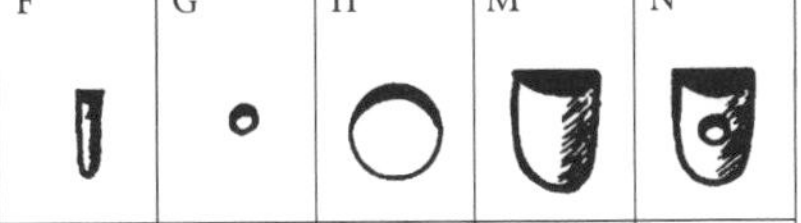

	F	G	H	M	N
System proposed by Scheil See MDP VI (1905)	1	10	100	1,000	10,000
System proposed by Scheil See MDP XVII (1923)	1	10	100	60	600
System proposed by Scheil See MDP XVII (1923)	1	10	100	600	6,000
System proposed by Langdon See JRAS (1925)	1	?	100	1,000	10,000
System proposed by Scheil See MDP XXVI (1935)	1	10	100	1,000	10,000
System proposed by de Mecquenem See MDP XXXI (1949)	1	10	100	300	1,000

FIG. 11.2. *Various contradictory conclusions drawn over the years concerning the values of the proto-Elamite number-signs*

But then: the *balance sheet* was invented. Now people wrote on both sides of the tablet: the "recto" side bore the details of a transaction, the "verso" the totals under the various headings.

The idea took hold, and with refinement proved to be of the greatest usefulness. At Uruk, in 2850 BCE, a proposal of marriage has been made. The girl's father and the father of the future spouse have just agreed on the "bride price". When the ceremony has taken place, the bride's father will receive from the other 15 sacks of barley, 30 sacks of corn, 60 sacks of beans, 40 sacks of lentils, and 15 hens. But, in view of the frailties of human memory and in order to avoid any quarrels later, the two men betake themselves to one of the religious leaders of the town in order to draw up the contract in due form and give the force of law to the engagement.

Having taken note of all the elements of the marriage contract, the notary then fashions a roughly rectangular tablet of clay, and takes up his "tracing tools".

For writing, he uses two ivory sticks of different cross-section, pointed at one end and, at the other, fashioned into a kind of cylindrical stylus (Fig. 8.10 above). The pointed ends are used to draw lines or to trace pictograms on the soft clay (Fig. 8.11 above), and the cylindrical styluses are used to mark numbers by pressing at a given angle on the surface of the tablet. According to the angle between stylus and the tablet, the impression made on the soft clay will be either a notch or a circular imprint, whose size will depend on the diameter of the stylus which is used. As in Fig. 8.12 above, this will be a narrow or a wide notch, according as the wide or narrow stylus is used, if the angle is 30°–45°; or it will be a circular imprint of small or large diameter, according to the stylus, if it is applied perpendicular to the surface of the tablet.

Then, holding the tablet with its long side horizontal, the scribe draws four vertical lines, thereby dividing it into five sections, one for each item in the contract. At the bottom of the rightmost division he draws a "sack of barley", in the next a "sack of corn", then a "sack of beans", then a "sack of lentils", then finally in the leftmost division he draws a "hen". Then he places the corresponding numerical quantities: in the first division, a small circular imprint for the number 10, and 5 small notches each worth 1, thus making up the total of 15 sacks of barley; in the second, three imprints of 10 for the number 30; in the third, he marks the number 60 with a large imprint, and so on.

On the back of the tablet, he makes the summary, that is, the totals of the inventory according to the numbers on the front, namely "145 sacks (various)" and "15 hens".

This done, the two men append their signatures to the bottom of the tablet, but not as used to be done by rolling a cylinder-seal over it. Instead, they use the pointed end of the stylus to trace conventional signs which represent them. Then, having given the document into the safekeeping of the notary, they part.

HOW THE SUMERIAN NUMBERS WERE DECIPHERED

The story reconstituted in the preceding section was not imaginary: it was achieved on the basis of the document shown in Fig. 11.3, which provides detailed evidence of how the Sumerian scribes used to note on one side of the tablets the details of the accounting, and on the other side a kind of summary of the transaction in the form of totals under different headings.

		Translation	
Side 1	Side 2	Side 1	Side 2
		15 sacks of barley	145 sacks (various)
		30 sacksof corn	
		60 sacks of ?	15 hens
		40 sacks of ?	signature (?)
		15 hens	(?)
			(?)

FIG. 11.3. *Sumerian "invoice" discovered at Uruk, said to be from the Jemdet Nasr era (c. 2850 BCE). Iraqi Museum, Baghdad. ATU 637*

But it is precisely this feature which has enabled the experts to decipher various ancient number-systems such as Sumerian, hieroglyphic or linear Cretan, and so on. The values of the numbers could therefore be determined with certainty by virtue of applying a large number of checks and verifications to these totals.

Observing, for example, that on the front of some tablet there were ten narrow notches here and there, while on the back there was a single small circular imprint, and then finding this correspondence confirmed in a sufficient number of similar cases, they can conclude that the narrow notch denotes unity and the small circular imprint denotes 10.

$$\text{narrow notch} = 1 \qquad \bullet = 10$$

Now suppose that we are trying to discover the unknown value, which we shall denote by x, of the wide notch:

 $= x \,?$

Of course, lacking any other indication, and in the absence of a bilingual "parallel text" (linguistic or mathematical), the value of this number would have long remained a mystery. But a happy chance has placed into our hands the tablet shown in Fig. 11.3, which bears the three numbers described above of which two have already been deciphered, which will indeed be our "Rosetta Stone".

We begin, of course, by ignoring the count of the 15 hens (one small circular imprint and 5 narrow notches, together with the pictogram of the bird), since this is reproduced exactly on the reverse of the document. So we shall only bother with the details of the inventory of sacks (goods denoted by the same writing sign throughout). Adding up the numbers on side 1, we therefore obtain

$$10 \quad + \quad 5 \quad + \quad 30 \quad + \quad x \quad + \quad 40 \quad = x + 85$$

while on side 2 we find

$$2x \quad + \quad 20 \quad + \quad 5 \quad = 2x + 25$$

On equating these two results, we obtain the equation

$$x + 85 = 2x + 25$$

which, on reduction, finally gives the result we are seeking, namely

$$= x = 60$$

However, we are only entitled to draw this conclusion as to the value of the sign in question if the value so determined gives consistent results for several other tablets of similar kind. And this turns out to be the case.

SIMILAR PRACTICE OF THE ELAMITE SCRIBES

It was precisely by observing similar practice on the part of the Elamite scribes, and carrying out systematic verifications of the same kind on a multitude of proto-Elamite tablets (some of the most important of which will be shown below) that I was able, myself, to arrive at the solution of this thorny problem.

Some of these tablets can lead us to it, even though the values of the proto-Elamite numbers may remain unknown. Consider for example the tablet in Fig. 11.4 A which refers to a similar accounting operation. The goods in question are represented by writing signs (whose meaning, in many cases, still eludes us). But the numbers associated with the various goods are clearly indicated by groups of number-signs. The subsequent diagram (Fig. 11.4 B) shows what we shall from now on call the "rationalised transcription" of the original tablet.

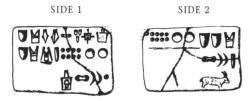

SIDE 1   SIDE 2

Fig. 11.4A. *Accounting tablet from Susa. Louvre. See MDP, VI, diagram 358*

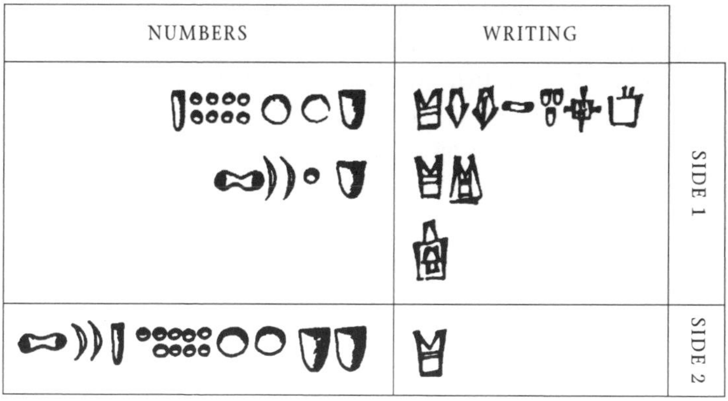

NUMBERS	WRITING	
▯⦂⦂⦂ ○ ◖ ▯ ⦿ ⟩⟩ ∘ ▯	▯◈◈⟜᛭⬧▯ ▯▯ ◈	SIDE 1
⦿⟩⟩▯ ⦂⦂⦂⦂ ○○ ▯▯	▯	SIDE 2

Fig. 11.4B.

Now we see, on the front of the tablet:
- the wide notch twice;
- the large circular impression twice;
- the small circular impression 9 times;
- the narrow, lengthened notch once;
- a circular arc twice;
- and a peculiar number (Fig. 11.1 D) once only.

This, moreover, is exactly what we also find on the reverse of the tablet. The number which is shown on side B therefore corresponds to the grand total of the inventory on the front.

In the same way, on the tablet shown in Fig. 11.5, the front and the reverse both show six narrow notches.

SIDE 1   SIDE 2

FIG. 11.5. *Tablets from Susa. Teheran Museum. See MDP, XXVI, diagram 437*

DETERMINING THE VALUES OF THE PROTO-ELAMITE NUMBERS

Now consider the tablet shown in Fig. 11.6. In the present state of the tablet, on the front side the narrow notch occurs only 18 times, and the smaller circular impression occurs 3 times, while on the reverse the narrow notch occurs 9 times and the circular impression 4 times.

If we proceed by analogy with the Sumerian numbers of similar form, attributing value 1 to the narrow notch and value 10 to the circular imprint, then the total from the front of the tablet ($18 + 3 \times 10 = 48$) and the total from the reverse ($9 + 4 \times 10 = 49$) differ by 1. We may conjecture that this difference is the result of a missing piece broken off from its left-hand side, which would have damaged the numerical representation in the last line of the top face.

Since, moreover, there are similar tablets* on which we find exactly equal totals on the two sides, we may conclude that this explanation for the discrepancy is in fact correct.

Therefore we may definitively fix the value of the narrow notch as 1, and the value of the small circular impression as 10.

SIDE 1

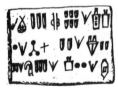

FIG. 11.6A. *Accounting tablet from Susa. Teheran Museum. See MDP, XXVI, diagram 297*

SIDE 2

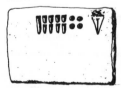

	SIDE 1

FIG. 11.6B.

	SIDE 2

* See, for example, tablet 353 of *MDP*, VI (Louvre; Sb 3046).

SIDE 1 SIDE 2

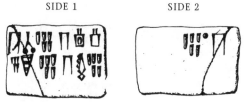

FIG. 11.7. *Accounting tablet from Susa. Louvre. See MDP, XVII, diagram 3*

Now we must take account of the fact that the Elamites set their numbers down from right to left (in the same direction as their writing), starting with the highest-order units and proceeding left towards the lower-order units. Furthermore, close examination of the tablets shows that the Elamite scribes used two different systems for writing numbers, both of which were based on the notion of juxtaposition to represent addition. These two systems made use, in general, of different symbols (Fig. 11.10 and 11.11).

For the first of these two proto-Elamite systems, it is pretty clear that the number-signs were always written in the following order, from right to left and from highest value to lowest value (Fig. 11.8).

FIG. 11.8.

The number-signs of the second system always occur as follows, again from right to left and in decreasing order (Fig. 11.9).

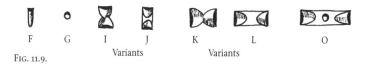

FIG. 11.9.

The above shows, therefore, that

- on the one hand, the numbers labelled A, B, C, D, and E (which always occur to the left of the narrow notch which represents 1) correspond to orders of magnitude below 1, that is to say to fractions;
- on the other hand, H, M, N, and P, and also I (or J), K (or L) and O correspond to orders of magnitude above 10 (since they always occur to the right of the small circular impression representing 10) (Fig. 11.10 and 11.11).

In the end, therefore, by working out the totals on many other tablets, I was able to obtain the following results which, as we shall see below, can be confirmed in other ways.

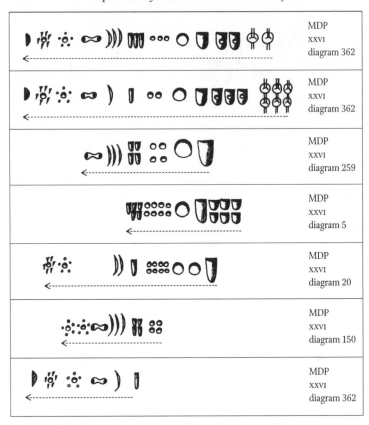

FIG. 11.10. *Instances of number taken from accounting tablets, which show how the earliest proto-Elamite number-system worked*

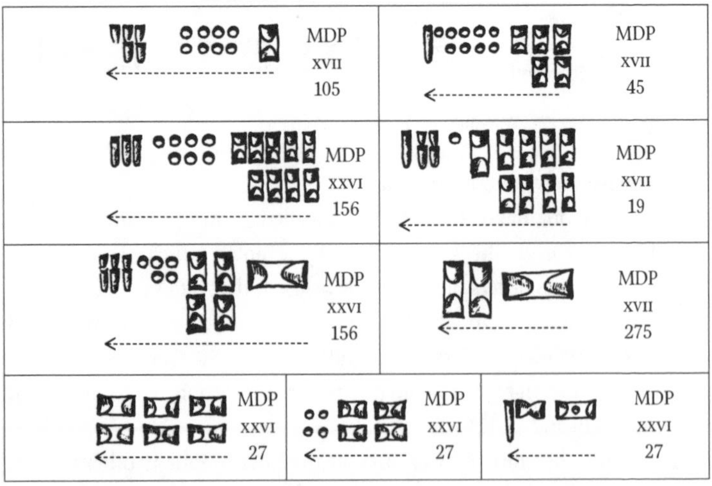

FIG. 11.11. *Instances from accounting tablets which illustrate the second proto-Elamite number-system*

$$A = \frac{1}{120} \quad B = \frac{1}{60} \quad C = \frac{1}{30} \quad D = \frac{1}{10} \quad E = \frac{1}{5}$$

For the number E (the circular arc), for example, I considered the tablet shown in Fig. 11.12 which, as can be seen from its rationalised transcription, bears two kinds of inventory:

UPPER SIDE

LOWER SIDE

FIG. 11.12A. *Accounting tablet from Susa. Louvre. Ref. MDP, XVII, diagram 17*

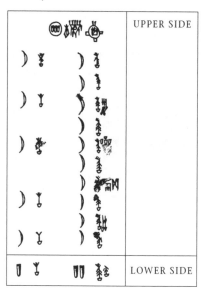

FIG. 11.12B.

• one, associated with the script character 🜚, which has 10 circular arcs on the top face and 2 narrow notches on the reverse;
• the other, associated with the ideogram 🜛, which has 5 circular arcs on the top face and 1 narrow notch on the reverse.

Therefore, denoting by x the unknown value of the number (E) in question, these two inventories give, according to the totals of the two sides, the two equations

$$x + x + x + x + x + x + x + x + x + x = 2$$
$$x + x + x + x + x = 1$$

namely

$$10x = 2$$
$$5x = 1$$

which is precisely how it was possible to determine the value $\frac{1}{5}$ for the circular arc.

Now let us try to evaluate the large circular imprint and the wide notch (H and M in Fig. 11.1). Because they look just like the Sumerian signs associated with 60 and 3,600 respectively (Fig. 8.7 and 9.15 above), we are at first tempted to conclude that the same values should be attributed to them in the present case. But when we examine the proto-Elamite tablets we find that this cannot be true. As we have seen, the Elamites set their numbers down from right to left, in decreasing order of magnitude and always commencing with the highest. Therefore, if these signs had the Sumerian values, the large circular impression should come before the wide notch in writing numbers. But this is not the case, as can be seen from Fig. 11.10 for example.

The document shown in Fig. 11.13 leads without difficulty to the ascertainment of the value of the proto-Elamite large circular impression.

UPPER SIDE

LOWER SIDE

FIG. 11.13A. *Accounting tablet from Susa. Teheran Museum. Ref. MDP, XXXI, diagram 3*

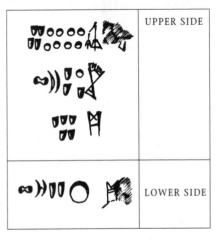

FIG. 11.13B.

Ignoring the two circular arcs and the doubled round imprint which are on both sides of the tablet, we find

• 9 small circular impressions and 12 narrow notches on the upper face;
• 1 large circular impression and 2 narrow notches on the lower.

Therefore, if we now evaluate these numerical elements on the two faces of the tablet, bearing in mind what we have already found out, we obtain the following:

$$\text{Upper} \quad 9 \times 10 + 12 = 102$$
$$\text{Lower} \quad 1 \times x + 2 = x + 2$$

Since these must be equal, we find the equation $x + 2 = 102$, whose solution is that $x = 100$.

Now consider the tablet shown in Fig. 11.14, on which we find

• 20 small circular impressions, and 2 large ones, on the upper face;
• 1 wide notch and one large circular impression, on the lower.

Let us now give the value 100 to the large circular impression, as we have just determined, and denote by y the value of the wide notch. We then obtain the following totals:

Upper $20 \times 10 + 2 \times 100 = 400$
Lower $1 \times y + 100 = y + 100$

Since these also, as before, must be equal, we obtain the equation $y + 100 = 400$, whose solution is that $y = 300$.

From the preceding arguments, therefore, we attribute the value 100 to the large circular impression, and the value 300 to the wide notch.

UPPER SIDE   LOWER SIDE

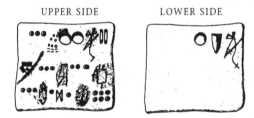

FIG. 11.14. *Tablet from Susa. Teheran Museum. Ref. MDP, XXVI, digram 118*

Of course, this would not allow us to conclude that these values correspond to a general reality unless we also find at least one other tablet which gives completely concordant results. This is, however, precisely the case for the tablets shown in Fig. 11.15 and 11.16.

UPPER SIDE   LOWER SIDE

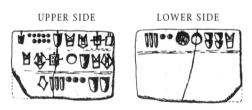

FIG. 11.15A. *Tablet from Susa. Louvre, Ref. MDP, VI, diagram 220*

UPPER SIDE	
	$300 + 9 \times 10$ 390
	$300 + 100$ 400
	$2 \times 300 + 3 \times 10 + 3$ 633
	1,423
LOWER SIDE	
	$4 \times 300 + 2 \times 100 + 2 \times 10 + 3 \ldots 1,423$

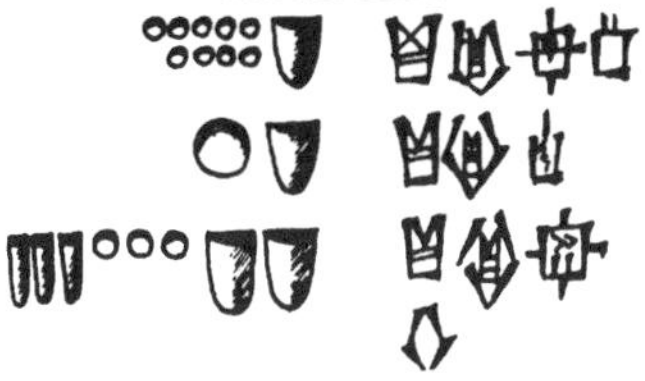

FIG. 11.15B.

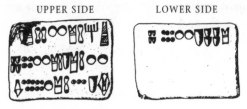

FIG. 11.16A. *Tablet from Susa. Teheran Museum, Ref. MDP, XXVI, diagram 439*

UPPER SIDE	
○○ 𒐽𒐏𒐊	2×100 200
○○ ▽ 𒐽𒐊	$300 + 2 \times 100$ 500
▽ 𒐽𒐊	300 300
𒐊𒐊○○ 𒐽𒐊	$2 \times 100 + 4 \times 10 + 4$ 244
●●●▽ 𒀭	$300 + 3 \times 10$ 330
●●●●●○ 𒐽𒐊	$100 + 9 \times 10$ 190
𒀸	$\overline{1{,}764}$
LOWER SIDE	
𒐊 ●●●○○ 𒐽𒐽𒐽 𒐽	$5 \times 300 + 2 \times 100 + 6 \times 10 + 4 \dots 1{,}764$

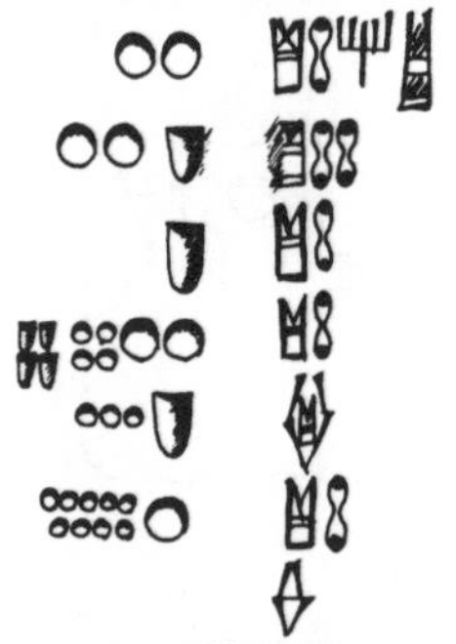

FIG. 11.16B.

In conclusion, the results established so far (which from now on will be considered definitive) are the following:

◗	⦚⦚⦚	⦂∴	∞	⟩	▮	○	◯	▽
$\dfrac{1}{120}$	$\dfrac{1}{60}$	$\dfrac{1}{30}$	$\dfrac{1}{10}$	$\dfrac{1}{5}$	1	10	100	300

FIG. 11.17.

Therefore, of the eleven number-signs of the proto-Elamite system, nine have been deciphered.

Now let us consider the delicate problem of the following two number-signs:

N   P

FIG. 11.18.

As we have already shown in Fig. 11.2, these two numbers have been interpreted in the most diverse ways since the beginning of this century (the number labelled N, for example, has been assigned to 600, to 6,000, to 10,000, or even to 1,000). To try to have a better understanding of the situation, we shall consider the tablet shown in Fig. 11.19 A. According to V. Scheil, this is "an important example of an exercise in agricultural accounting". As far as I know, this is the only preserved intact proto-Elamite document which contains both the entire set of number-signs of the first system and also a grand summary total.

On this tablet, we find:

- on the top face, a series of twenty numerical entries (corresponding to an inventory of twenty lots of the same kind denoted, it would seem, by the script character at the right of the top line);
- on the reverse, the corresponding grand total (itself preceded by the same written character).

UPPER SIDE   LOWER SIDE

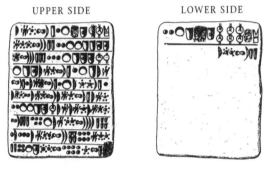

FIG. 11.19A. *Accounting tablet from Susa. Ref. MDP, XXVI, diagram 362*

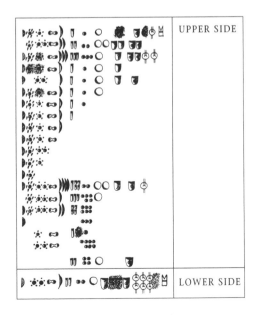

FIG. 11.19B.

Considering the results we have already obtained, we shall make various attempts to reconcile the totals of the numbers on this tablet, by trying various different possible values for the numbers labelled N and P, and making use of the numbers of occurrences of the different signs as shown in Fig. 11.19 C.

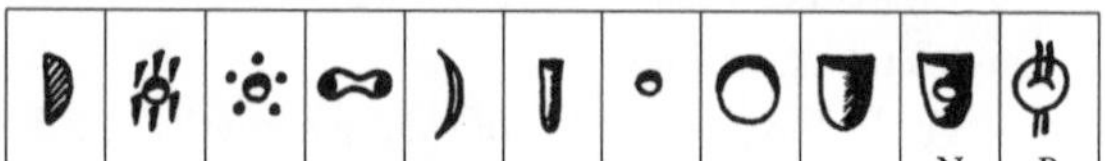

											N	P
Number of times each sign occurs	on the upper side	15	15	24	14	19	26	39	11	7	8	5
	on the lower side	1	0	2	1	1	2	2	1	1	3	6

FIG. 11.19C. *Complete listing of all the numerical signs on the tablet*

First attempt:

Following Scheil (1935, see MDP, XXVI), let us assign the value 10,000 to the wide notch with the circular impression (N), and the value 100,000 to the circle with the little wings (P). On the upper face of the tablet, we then obtain the following total for the numbers which appear there (Fig. 11.19 C):

$$15 \times \frac{1}{120} + 15 \times \frac{1}{60} + 24 \times \frac{1}{30} + 14 \times \frac{1}{10} + 19 \times \frac{1}{5}$$
$$+ 26 + 39 \times 10 + 11 \times 100 + 7 \times 300 + 8 \times 10{,}000 + 5 \times 100{,}000$$

namely $583{,}622 + \dfrac{45}{120}$

On the lower, similarly (Fig. 11.19 C):

$$1 \times \frac{1}{120} + 0 \times \frac{1}{60} + 2 \times \frac{1}{30} + 1 \times \frac{1}{10} + 1 \times \frac{1}{5}$$
$$+ 2 + 2 \times 10 + 1 \times 100 + 1 \times 300 + 3 \times 10{,}000 + 6 \times 100{,}000$$

namely $630{,}422 + \dfrac{45}{120}$

The difference between these two results is 46,800, far too great to allow this attempt to be considered correct, if we attribute the discrepancy to an error on the part of the scribe.

Second attempt:

Now consider the possibilities of assigning the values:

N = 6,000 [V. Scheil (1923)], P = 100,000 [V. Scheil (1935)]

By a similar calculation, we obtain (Fig. 11.19 C):

$$\text{Upper side } 551,622 + \frac{45}{120} \qquad \text{Lower side B } 618,000 + \frac{45}{120}$$

This attempt also must be considered to fail, since the discrepancy between the two faces is again too large.

Third attempt:

Now let us try:

$$N = 6,000 \text{ [V. Scheil (1923)], } P = 10,000 \text{ [S. Langdon (1925)]}$$

This again fails, since we obtain (Fig. 11.19 C):

$$\text{Upper side } 101,622 + \frac{45}{120} \qquad \text{Lower side } 78,422 + \frac{45}{120}$$

Fourth attempt:

Now let us consider the values proposed by R. de Mecquenem in 1949:

$$N = 1,000, \text{ and } P = 10,000$$

Again from Fig. 11.19 C, we obtain the results

$$\text{Upper side } 61,622 + \frac{45}{120} \qquad \text{Lower side } 63,422 + \frac{45}{120}$$

This possibility seemed to me for a long time to be the most likely solution. The results it gives are relatively satisfactory, since the discrepancy between the totals for the two faces of the tablet is only 1,800. On this belief, I had therefore supposed that the scribe had made some error in calculation, or had omitted to inscribe on the tablet the numbers corresponding to this difference. This, after all, could be likely enough, considering the many number-signs crowded onto the tablet – *errare humanum est*! Let us not forget that, just as in our own day, the scribes of old were capable of making mistakes in arithmetic.

Nonetheless, on reflection, it seemed to me that there was something illogical in attributing the value 1,000 to the number N, for two reasons.

Consider, first of all, the following two numerical entries taken from proto-Elamite tablets:

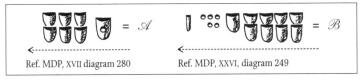

Ref. MDP, XVII diagram 280   Ref. MDP, XXVI, diagram 249

Fig. 11.20.

On Mecquenem's hypothesis, these would respectively have values

$$\mathscr{A} = 1 \times 1,000 + 6 \times 300 \quad = 2,800$$
$$\mathscr{B} = 9 \times 300 + 5 \times 10 + 1 \quad = 2,751$$

Now, still adopting this hypothesis, the following numbers would be units of consecutive orders of magnitude:

$$1 \quad 10 \quad 100 \quad 300 \quad 1,000 \quad 10,000$$

Therefore, in the first place, the question arises: if the notch with the circular impression really corresponded to the value 1,000, why should the scribes have adopted the above representations of the numbers 2,800 and 2,751, and not the more regular forms in Fig. 11.21 following?

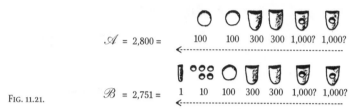

$$\mathscr{A} = 2,800 = \quad \underset{100}{} \quad \underset{100}{} \quad \underset{300}{} \quad \underset{300}{} \quad \underset{1,000?}{} \quad \underset{1,000?}{}$$

$$\mathscr{B} = 2,751 = \quad \underset{1}{} \quad \underset{10}{} \quad \underset{100}{} \quad \underset{300}{} \quad \underset{300}{} \quad \underset{1,000?}{} \quad \underset{1,000?}{}$$

FIG. 11.21.

On the other hand, we know that for the Sumerians the small circular impression had value 10, the wide notch 60, and the combination of the latter including the former had value 600:

FIG. 11.22.   10   60 $60 \times 10 = 600$

in other words, that the last figure follows the multiplicative principle.

But for the Elamites the small circular impression had value 10 while the wide notch had value 300. By analogy with the Sumerian system, the value $300 \times 10 = 3,000$ should be assigned to the wide notch compounded with the small circle:

FIG. 11.23.   10   300 $300 \times 10 = 3,000?$

For these reasons I was led to reject Mecquenem's hypothesis.

Fifth attempt:

We are therefore now led to consider the proposed values:

$$N = 3,000 \text{ and } P = 10,000$$

[the latter from S. Langdon (1925) and R. de Mecquenem (1949), the former from the above reasoning]. Again comparing the totals from the two faces of the tablet, this hypothesis gives the following results:

$$\text{Upper side } 77,622 + \frac{45}{120} \quad \text{Lower side B } 69,422 + \frac{45}{120}$$

This hypothesis therefore does not work either. But, if we wish to keep the value of 3,000 for the number N, we must seek a different value for the number P.

Now, close examination of the mathematical structure which can be inferred from the values so far determined in the proto-Elamite number-system caused me to suppose that the following three values could be possible for the number P:

$$9,000, 18,000 \text{ and } 36,000$$

I was led to this supposition by postulating that the proto-Elamite system of fractions was developed on the same lines as the notation for the whole numbers, namely that there had to be a certain correspondence between a scale of increasing values, and a scale of decreasing values, relative to a given base number.

This, however, is exactly what one observes if one expresses the different values determined so far in terms of the number M = 300 (Fig. 11.24).

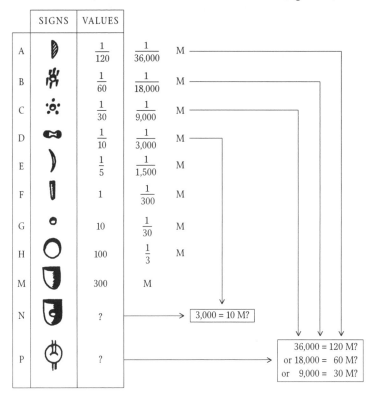

FIG. 11.24.

Sixth attempt:

This now leads us to contemplate the possibilities based on these three possible values for P, of which the first is (Fig. 11.19 C):

$$N = 3,000, P = 9,000$$

But on comparing the totals which result, we find a serious discrepancy:

Upper side $72,622 + \dfrac{45}{120}$ Lower side $63,422 + \dfrac{45}{120}$

Difference 9,200

Therefore this suggestion must be rejected.

Seventh attempt:

The same results from trying the second possibility inferred above, since the values:

$$N = 3,000, P = 36,000$$

also lead to implausible results (Fig. 11.19 C):

Upper side $207,622 + \dfrac{45}{120}$ Lower side $225,422 + \dfrac{45}{120}$

Difference 17,800

Final attempt, and the solution of the problem:

Now consider the final possibility, with the following values:

$$N = 3,000, P = 18,000$$

This system, which is compatible with a coherent mathematical structure, also gives satisfyingly close agreement:

Upper side $117,622 + \dfrac{45}{120}$ $(117,622 + \dfrac{1}{5} + \dfrac{1}{10} + \dfrac{2}{30} + \dfrac{1}{120})$

Lower side $117,422 + \dfrac{45}{120}$ $(117,422 + \dfrac{1}{5} + \dfrac{1}{10} + \dfrac{2}{30} + \dfrac{1}{120})$

Whence, however, comes this discrepancy of 200 which exists between the two faces if we adopt this hypothesis? Quite simply, I believe, from a "typographical error".

Instead of inscribing on the lower side the grand total corresponding to the inventory on upper side, which should be in the form:

$\dfrac{1}{120} + \dfrac{1}{30} + \dfrac{1}{30} + \dfrac{1}{10} + \dfrac{1}{5} + 1 + 1 + 10 + 10 + 300 + 300 + 3,000 + 3,000 + 3,000 + 18,000 \times 6$

$117,622 + \dfrac{1}{5} + \dfrac{1}{10} + \dfrac{1}{30} + \dfrac{1}{30} + \dfrac{1}{120}$

Fig. 11.25A.

the scribe in fact made a large circular impression in the place of one of the two wide notches:

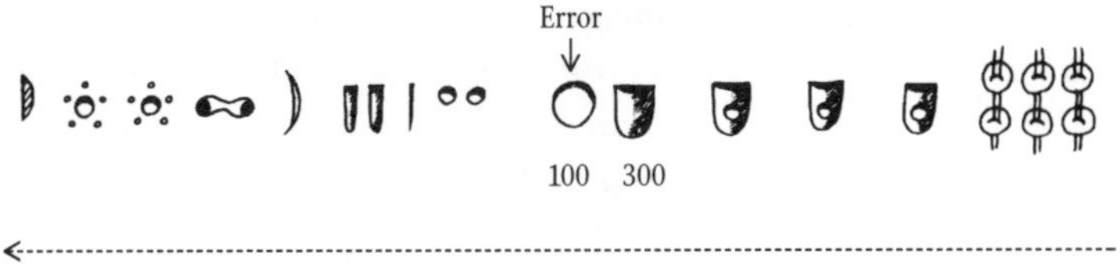

Error
↓

100   300

$$117{,}422 + \frac{1}{5} + \frac{1}{10} + \frac{1}{30} + \frac{1}{30} + \frac{1}{120}$$

FIG. 11.25B.

It is easy to see how this could happen. The scribe held his stylus with large circular cross-section in the wrong position (See Fig. 8.10 and 8.12 above): instead of pressing the stylus at an angle of 30°–45° to the surface of the soft clay, which would have given him a wedge, he held it perpendicular to the surface thereby obtaining the circle.

That is, instead of doing this:

FIG. 11.26A.   Result

he did this:

FIG. 11.26B.   Result

Therefore, in all probability, we may conclude that the wide notch with a small circular imprint corresponds to the value 3,000, and the circle with the little wings corresponds to the value 18,000.

All the numbers in the proto-Elamite system have, therefore, been definitively deciphered.

We have good reason to suppose that this system is the more ancient of the two since the following numerals appear on the proto-Elamite accounting tablets from the archaic epoch onwards.

1   10   100   300   3,000

FIG. 11.27.

The same set of numerals appears on the earliest numerical tablets, as well as on the outside of the counting balls recently discovered on the site of the Acropolis of Susa. Finally, the numerals also are those which, according to their respective shapes, correspond to the archaic *calculi* which were formerly enclosed in the counting balls, in fact to the number-tokens of various shapes and sizes which stood for these numbers (and whose values, in turn, have themselves now been determined as a result of the decipherment described above; see also Fig. 10.8 and 10.14 above):

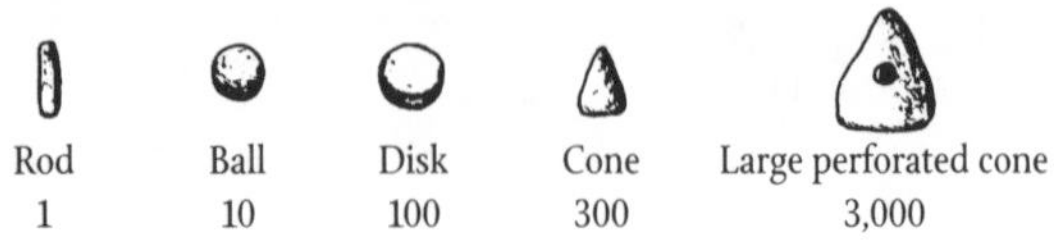

Rod	Ball	Disk	Cone	Large perforated cone
1	10	100	300	3,000

FIG. 11.28.

As to the second system of writing numbers, I believe that the Elamites constructed it – maybe in a relatively recent era – for the purpose of recording quantities of objects or of goods, or magnitudes, of a different kind from those for which the symbols of the first system were used.

I base this hypothesis on an analogy with Sumerian usage. During the third millennium BCE, the scribes of Lower Mesopotamia in fact used three different numerical notations:

- the first, the commonest and oldest, which we have studied in Chapter 8, was used for numbers of men, beasts, or objects, or for expressing measures of weight and length;
- the second was used for measures of volume;
- the third was used for measures of area.

This hypothesis is in fact confirmed by the tablet shown in Fig. 11.29, which carries two inventories which have been very clearly differentiated.

UPPER SIDE   LOWER SIDE

FIG. 11.29A. *Accounting tablet from Susa. Teheran Museum. Ref. MDP, XXVI, diagram 156*

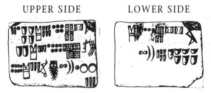

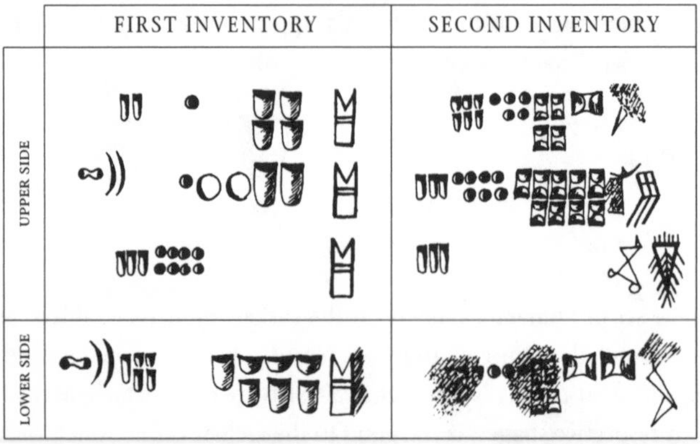

	FIRST INVENTORY	SECOND INVENTORY
UPPER SIDE		
LOWER SIDE		

FIG. 11.29B.

The first of these inventories is indicated by a characteristic script character, and the corresponding quantities are expressed in the numerals of the first proto-Elamite system (Fig. 11.29 B). The second inventory is indicated by the signs (which have not yet been deciphered):

and the corresponding quantities are expressed in the numerals of the second proto-Elamite system (Fig. 11.9).

The numbers given on the reverse of this tablet correspond respectively to the total of the first inventory and to the total of the second. Using the values we have already obtained, we can make the totals for the first inventory:

a) on upper side:

$$6 \times 300 + 2 \times 100 + 10 \times 10 + 5 + \frac{2}{5} + \frac{1}{10} = 2{,}105 + \frac{2}{5} + \frac{1}{10}$$

b) on lower side:

$$7 \times 300 + 5 + \frac{2}{5} + \frac{1}{10} = 2{,}105 + \frac{2}{5} + \frac{1}{10}$$

(which, by the way, is a further confirmation of the validity of our earlier result).

Now let us consider the different numerals on the second inventory, and let us give value 1 to the narrow notch, 10 to the small circular impression, 100 to the double vertical notch and 1,000 to the double horizontal notch. Then the totals come out as follows:

a) on upper side:

$$1{,}000 + 13 \times 100 + 12 \times 10 + 12 = 2{,}432$$

b) on lower side:

$$2 \times 1{,}000 + 4 \times 100 + 3 \times 10 + 2 = 2{,}432$$

We may therefore fix the values of the following numerals as shown:

 or

100   1,000

FIG. 11.30.

(where the former of these values, for example, is confirmed by the tablet in Fig. 11.31, since the totals come to 591 on both sides).

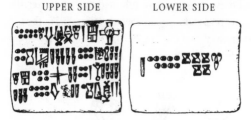

UPPER SIDE   LOWER SIDE

FIG. 11.31A. *Accounting tablet from Susa. Louvre. Ref. MDP, XVII, diagram 45*

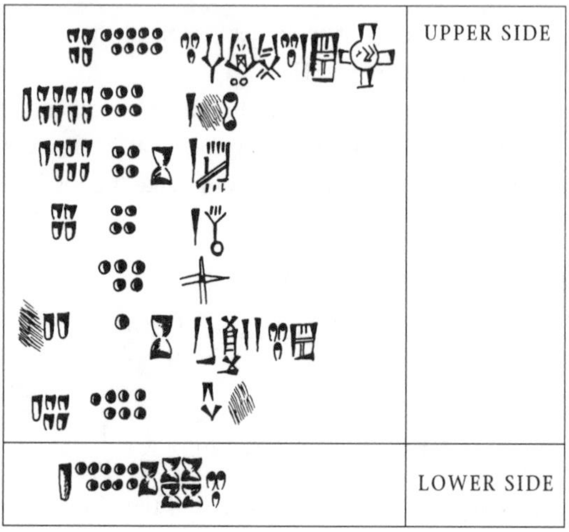

FIG. 11.31B.

So there we see pretty well all of the proto-Elamite numerals deciphered. At the same time, we have discovered that at Susa two different number-writing systems were in use, probably corresponding to two different systems of expressing numbers:

- one, strictly decimal* (Fig. 11.32);
- the other, visibly "contaminated" by the base 60 (Fig. 11.33).

1	10	100	100	1,000	1,000	10,000?
F	G	I	J	K	L	O

FIG. 11.32. *The values of the number-signs of the second proto-Elamite number-system*

---

* A question remains for the numeral formed from a double horizontal notch with a small circular impression in its centre (Fig. 11.32, sign O). Is this the numeral representing 10,000 = 1,000 × 10? It seems likely. But this could not be stated with certainty, since we lack documents better preserved than those we have at present, relevant to this numeral.

We may suppose that the first may have been used for counting such things as people, animals or things, while the second may have been used to express different measures in a system of measurement units (volumes and areas, for example).

	SIGNS	X	Y	VALUES
A		$\frac{1}{36,000}$ M	$\frac{1}{2}$ B	$\frac{1}{120}$
B		$\frac{1}{18,000}$ M	B	$\frac{1}{60}$
C		$\frac{1}{9,000}$ M	2 B	$\frac{1}{30}$
D		$\frac{1}{3,000}$ M	6 B	$\frac{1}{10}$
E		$\frac{1}{1,500}$ M	12 B	$\frac{1}{5}$
F		$\frac{1}{300}$ M	60 B	1
G		$\frac{1}{30}$ M	600 B	10
H		$\frac{1}{3}$ M	6,000 B	100
M		M	18,000 B = 300 × 60 B	300
N		10 M	180,000 B = 300 × 600 B	?
P		60 M	1,800,000 B = 300 × 6,000 B	?

FIG. 11.33. *The mathematical structure of the first proto-Elamite number-system*

These are of course only hypotheses, but the above results lend confirmation to the existence of cultural and economic relations between Elam and Sumer, at any rate from the end of the fourth millennium BCE, and to the influence exerted by the Sumerians upon Elamite civilisation.

CHAPTER 12

HOW THE SUMERIANS
DID THEIR SUMS

The arithmetical problems which the Sumerians had to deal with were quite complicated, as is shown by the many monetary documents which they have bequeathed to us. The question which we shall now address is to find out what methods they used in order to carry out additions, multiplications, and divisions. First of all, however, let us have a look at one very interesting document.

A FOUR-THOUSAND-YEAR-OLD DIVISION SUM

The tablet shown in Fig. 12.1 is from the Iraqi site of Fara (Šuruppak), and it dates from around 2650 BCE.

We shall present its complete decipherment according to A. Deimel's *Sumerisches Lexikon* (1947). This document provides us with the most valuable information on Sumerian mathematics in the pre-Sargonic era (the first half of the third millennium BCE). It shows the high intellectual level attained by the arithmeticians of Sumer, probably since the most archaic era.

The tablet is divided into two columns, each subdivided into several boxes.

From top to bottom, in the first box of the left-hand column is a narrow notch, followed by a cuneiform group (*še-gur₇*), which signifies "granary of barley".

In the box beneath is a representation of the number 7, preceded by a sign which is to be read *sìla*.

In the third box, the numeral 1 is followed by the sign for "man" (*lû*); below this is a group which is to be read *šu-ba-ti* (the word *šu* means "hand") and which might be translated as "given in the hand".

Finally, at the very bottom of the left-hand column is the sign for "man" again, above which is the character *bi* which is simply the indicative "these".

The literal translation of this column therefore is: "1 granary of barley; 7 *sìla*; each man, given in the hand; these men."

In the first box of the right-hand column, we can recognise the representation of 164,571 in the archaic numerals (see Fig. 8.20 above), and in the box below a succession of signs which represent the phrase "granary of barley, there remains: 3".

TRANSCRIPTION

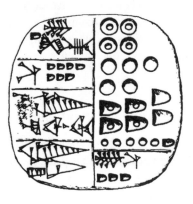

	(36,000) (36,000)
1 sĕ-gur₇	(36,000) (36,000)
sìla 7	(3,600) (3,600) (3,600)
	(3,600) (3,600)
1 lú	(600 (600) (60)
šu-ba-ti	(600) (600) (60)
	(10) (10) (10) (10) (10) (1)
lú-	
-bi	še sìla
	šu-kid
	3

LITERAL TRANSLATION

Left-hand register	Right-hand register
1 "granary of barley"	
7 *sìla* (of barley)	164,571
Each man in his hand receives	
Men these	*sìla* of barley remain 3

FIG. 12.1. *Sumerian tablet from Šuruppak (Fara). Date: c. 2650 BCE. Istanbul Museum. Ref. Jestin (1937), plate XXI, diagram 50 FS*

This tablet, which no doubt describes a distribution of grain, shows all the formal elements of arithmetical *division*: we have a *dividend*, a *divisor*, a *quotient*, and even, to an astonishing precision for the time, a *remainder*.

The *sìla* and the *še-gur₇* ("granary of barley") are units of measurement of volume. At that time, the former contained the equivalent of 0.842 of our litre, while the latter came to about 969,984 litres, namely 1,152,000 *sìla* [see M. A. Powell (1972)]:

$$1 \text{ } \textit{še-gur}_7 \text{ (1 granary of barley)} = 1,152,000 \text{ } \textit{sìla}$$

Thus this distribution involved the division of 1,152,000 *sìla* of barley between a certain number of people, each of whom is to receive 7 *sìla*.

Now let us do the calculation. 1,152,000 divided by 7 is 164,571, exactly the number in the first box of the right-hand column; and the remainder is 3, exactly the information given at the bottom of this column.

There is no doubt about it: you have before your very eyes the written testimony of the oldest known division sum in history – quite a complex one; and as old as Noah.

OFFICIAL DOCUMENT OR LEARNER'S EXERCISE?

One may suppose that this tablet was probably an official document in the archives of the ancient Sumerian city of Šuruppak, unless it happened to be an exercise for apprentice calculators.

On the first supposition, then its translation into plain language is as follows:

> We have divided 1 granary of barley between a certain number of people, giving 7 *sila* to each one. These men were 164,571 in number, and at the end of the distribution there were 3 *sila* remaining.

On the other hand, if it was really an exercise for learners, then the appropriate translation would be:

STATEMENT OF THE PROBLEM	SOLUTION OF THE PROBLEM
Given that a granary of barley has been divided between several men so that each man received 7 *sila*, find the number of men.	The number of men was 164,571 and 3 *sila* were left over after the distribution.

For convenience of exposition, we shall adopt the latter interpretation in what follows.

There is no indication whatever in the document as to the method of calculation to be used to obtain the result. Nor do we yet know of any formal description. One thing however is certain, and that is that the calculation was not carried out by means of Sumerian numerals, which do not encapsulate an operational capability in the way that our own numerals do.

Nonetheless the results of the previous chapter give us some basis for supposition as to what the means of calculation may have been. The Sumerians most probably made use of the *calculi* (the very ones shown in Fig. 10.4), as much prior to the emergence of their numerical notation as subsequently, since we find these tokens in various archaeological sites of the third millennium BCE, that is to say, at a time when *bullae* had almost entirely been displaced by clay tablets (see Fig. 10.2 above).

We shall now put forward a speculative but entirely plausible reconstruction of the technique of calculation which was most probably used.

CALCULATION WITH PELLETS, CONES, AND SPHERES

Let us imagine we are in the year 2650 BCE, in the Sumerian city of Šuruppak. We are in the school where scribes and accountants learn their skills, and the teacher has given a lesson on how to do a division. Now he begins the practical class, and sets the problem of dividing one granary of barley according to the conditions given.

The problem is therefore to divide 1,152,000 *sila* of barley between a certain number of persons (to be determined) so that each one gets 7 *sila* of barley, which comes down to dividing the first number by 7.

At this time, additions, multiplications, and divisions are carried out by means of the *calculi*, those good old *imnu* of former times which, in their several shapes and sizes, symbolise the different orders of magnitude of the units in the Sumerian number-system. Although their use has long disappeared from accounting practice, they are still the means that everyone uses for calculation. This has never worried any of the generations of scribes since the day when one of them thought of making replicas of the various *calculi* on clay tablets, to serve as numerical notations – a narrow notch for the small cone, a small hole for the pellet, a wide notch for the large cone, and so on (see Fig. 10.4 above).

Generally, the procedure for performing a division brings in successively: pierced spheres (= 36,000), plain spheres (= 3,600), large pierced cones (= 600), large plain cones (= 60), and so on. At each stage, the pieces are converted into their equivalents as multiples of smaller units whenever they are fewer than the size of the divisor.

Practically speaking, therefore, the above example proceeds as follows. In Sumerian, the dividend 1,152,000 is expressed in words (see Fig. 8.5 above) as

šàrgal-iá šàr-u-min

which corresponds to the decomposition

$$216,000 \times 5 + 36,000 \times 2 = 5 \times 60^3 + 2 \times (10 \times 60^2)$$

The largest unit of the written numerals at this time, however, is only 36,000 (see Fig. 10.4 above), which is also the value of the largest of the *calculi*. Therefore the dividend must first be expressed in multiples of this smaller unit, therefore by 32 pierced spheres each of which stands for 36,000 units:

$$1,152,000 = 32 \times 36,000$$

But we are to divide this by 7, so we arrange these as best we can in groups of 7:

36,000

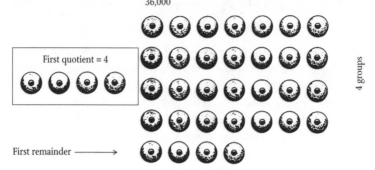

First quotient = 4

First remainder ———→

4 groups

FIG. 12.2A.

The number of groups, each with 7 pierced spheres, in this first arrangement is 4, which is the quotient from this first partial division. This, in the context of the problem, is also equal to the number (4 × 36,000) of the first group of people who will receive 7 *sìla* of barley each. In order not to lose track of this partial result, we shall put 4 pierced spheres on one side to represent it.

After this, we have 4 pierced spheres left over. We therefore must divide these 4 × 36,000 *sìla*. But when it is expressed in pierced spheres, worth 36,000 each, we find that 4 cannot be divided by 7. At this point, therefore, we convert each one of these into its equivalent number of the next lower order of magnitude.

Each pierced sphere (36,000) is equivalent to 10 plain spheres, each worth 3,600. The 4 pierced spheres therefore become 4 × 10 = 40 plain spheres, which we once again arrange in groups of 7:

3,600

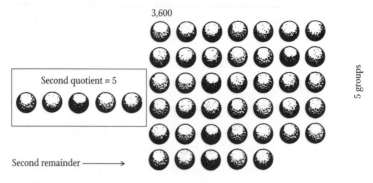

Second quotient = 5

Second remainder ———→

5 groups

FIG. 12.2B.

Now we find that there are 5 complete groups of 7 plain spheres, so we put on one side 5 plain spheres (corresponding to the second group, 5 × 3,600, of people who will each receive 7 *sìla* of barley).

But we find that there are 5 plain spheres left over at the end of this second division, and 5 is not divisible by 7, so we now replace each plain sphere by its equivalent number of pieces of the next lower order of magnitude.

Each "3,600" sphere is equivalent to 6 large pierced cones worth 600 each, so we convert the 5 pierced spheres left over into $5 \times 6 = 30$ large pierced cones which we again arrange in groups of 7:

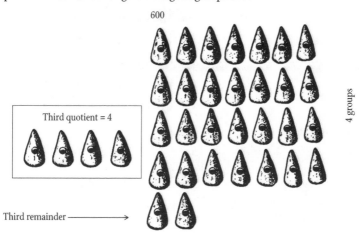

FIG. 12.2C.

Since we have 4 full groups of 7 pierced cones each, we therefore put aside 4 large pierced cones, corresponding to the third part of the men who will receive 7 *sila* of barley each (4×600).

However, we now have 2 large pierced cones left over, so we still have to divide 2×600 *sila* of barley.

Each "600" cone is equivalent to 10 large plain cones worth 60 each, so we convert the two large pierced cones left over into $2 \times 10 = 20$ large plain cones and we arrange these in groups of 7.

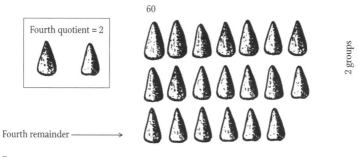

FIG. 12.2D.

We can form 2 complete groups of 7, with 6 large plain cones left over. As before, we put aside 2 cones to note the number of complete groups, corresponding to the 2×60 men who will each get 7 *sila* of barley at this fourth stage of the distribution.

Now we convert the 6 large plain cones left over, worth 60 each, into their equivalent in pellets worth 10 each, therefore into $6 \times 6 = 36$ pellets, and we arrange these into groups of 7, with 1 left over:

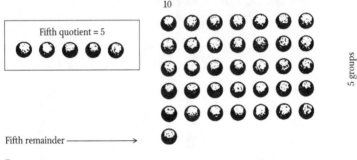

Fifth quotient = 5

10

5 groups

Fifth remainder ——————→

FIG. 12.2E.

Once again, we put aside 5 pellets corresponding to the 5×10 men who will each get 7 *sila* of barley at this fifth stage of the distribution.

The single pellet left over, worth 10, is now converted into 10 small cones each worth 1. This makes one complete group of 7, with 3 left over.

Sixth quotient = 1

1

1 group

Sixth remainder ——————→

FIG. 12.2F.

To note the one complete row, we put aside 1 small cone, and this corresponds to the number (10) of men who will each get 7 *sila* of barley at this sixth stage of the distribution. Since the number corresponding to the leftover cones is 3, and this is less than the divisor, we can proceed no further in the division of the original number into whole units, and we have finished.

The final quotient can now be easily obtained by totalling the values of the pieces which we successively set aside in the course of the division, as follows:

4 pierced spheres	(quotient from the first division)
5 plain spheres	(quotient from the second division)
4 large pierced cones	(quotient from the third division)
2 large plain cones	(quotient from the fourth division)
5 pellets	(quotient from the fifth division)
and	
1 small cone	(quotient from the sixth division)

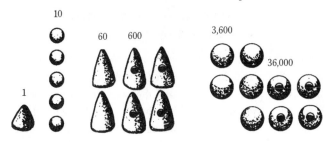

Fɪɢ. 12.2ɢ. *Result of the division*

In other words, the total number of people to whom the barley will be distributed is

$$4 \times 36{,}000 + 5 \times 3{,}600 + 4 \times 600 + 2 \times 60 + 5 \times 10 + 1 = 164{,}571$$

Back at the school of arithmetic, one student raises his hand and gives his answer, in Sumerian words pronounced in the following order:

šàr-u-limmu	=	$(3{,}600 \times 10) \times 4$	= 4 pierced spheres
šàr-iá	=	$3{,}600 \times 5$	= 5 spheres
geš-u-limmu	=	$(60 \times 10) \times 4$	= 4 large pierced cones
geš-min	=	60×2	= 2 large cones
ninnû	=	50	= 5 pellets
geš	=	1	= 1 small cone

Not forgetting to add, of course

še sìla šu-kid eš ("and there are 3 *sìla* of barley left over")

Another of the students, however, shows up his work to the teacher as he has traced it onto his clay tablet, which he has divided into boxes and filled up with Sumerian script. In the top right-hand box, in archaic numerals, he has written the answer (164,571) exactly as shown in Fig. 8.20 above:

- 4 large circular impressions with small circular impressions within (a direct representation of 4 pierced spheres, each worth 36,000);
- 5 large plain circular impressions (a direct representation of 5 spheres, each worth 3,600);
- 4 wide notches with small circular impressions within (recalling the 4 large pierced cones, each worth 600);
- 2 plain large notches (for the 2 plain large cones, each worth 60);
- 5 small circular imprints (for the 5 pellets each worth 10); and
- 1 narrow notch (for the small cone representing 1).

And, since the spoken word vanishes into thin air, while what is written remains, it is thanks to the latter that the division sum from Šuruppak has survived for the thousands of years since the students who solved it vanished from the face of the earth.

THE DISAPPEARANCE OF THE *CALCULI*
IN MESOPOTAMIA

We can infer that Sumerian arithmetic was done in this kind of way from the most archaic times down to the pre-Sargonic era. The tablet shown in Fig. 12.1 is one piece of evidence, and the *calculi* from this epoch found in those regions provide another; but the most solid proof is the reconstruction of the method which we have shown, for it can be easily demonstrated that the same principles may be applied equally well to multiplication, addition, and subtraction.

Nonetheless, the historical problems of Mesopotamian arithmetic have not been completely solved.

At the time at which the tablet we have been examining was made (around 2650 BCE), the *calculi* were still in use throughout the region, and in appearance they remained close to the archaic, or *curviform*, numerals which had then come into use. These numerals, however, while still present at the time of Sargon I (around 2350 BCE), gradually disappeared during the second half of this millennium. Finally, at the time of the dynasty of Ur III (around 2000 BCE) they had been replaced by the cuneiform numerals. Correspondingly, the *calculi* themselves are no longer found in the majority of the archaeological sites of Mesopotamia dating from this period or later (Fig. 10.2 above).

While undergoing this transformation from archaic curviform to cuneiform aspect, the written numerals lost all resemblance to the *calculi* which were their concrete ancestors. The Sumerian written number-system, moreover, was essentially a *static* tool with respect to arithmetic, since it was not adapted to manipulation for calculations: the numerals, whether curviform or cuneiform, instead of having inherent potential to take part in arithmetical processes, were graphical objects conceived for the purpose of expressing in writing, and solely for the sake of recall, the results of calculations which had already been done by other means.

Therefore the calculators of Sumer, at a certain point in time, faced the necessity of replacing their old methods with new in order to continue to function. They therefore substituted for the old system of the *calculi* a new "instrument" which I shall shortly describe. Meanwhile, we make a detour to prepare the ground.

FROM PEBBLES TO ABACUS

Only a few generations ago, natives in Madagascar had a very practical way of counting men, things, or animals. A soldier, for instance, would make his men pass in single file though a narrow passage. As each one emerged he would drop a pebble into a furrow cut into the earth. After the tenth had passed, the 10 pebbles would be taken out, and 1 pebble added to a parallel furrow reserved for tens. Further pebbles were then placed into the first furrow until the twentieth man had passed, then these 10 would be taken out and another added to the second furrow. When the second furrow had accumulated 10 pebbles, these in turn were taken out and 1 pebble was added to a third furrow, reserved for hundreds. And so on until the last man had emerged. So a troop of 456 men would leave 6 pebbles in the first furrow, 5 in the second, and 4 in the third.

Each furrow therefore corresponded to a power of 10: the ones, the tens, the hundreds, and so on. The Malagasies had unwittingly invented the abacus.

This was not unique to them, however. Very similar means have been devised since the dawn of time by peoples in every part of the earth, and the form of the instrument has also varied.

Some African societies used sticks onto which they slid pierced stones, each stick corresponding to an order of magnitude.

Amongst other peoples (the Apache, Maidu, Miwok, Walapai, or Havasupai tribes of North America, or the people of Hawaii and many Pacific islands) the practice was to thread pearls or shells onto threads of different colours.

Others, like the Incas of South America, placed pebbles or beans or grains of maize into compartments on a kind of tray made of stone, terracotta, or wood, or even constructed on the ground.

The Greeks, the Etruscans, and the Romans placed little counters of bone, ivory, or metal onto tables or boards, made of wood or marble, on which divisions had been ruled.

Other civilisations produced better implementations of the idea, by using parallel grooves or rods, with buttons or pierced pellets which could be slid along these. This is how the famous *suan pan* or Chinese Abacus came about, a most practical and formidable instrument which is still in common use throughout the Far East.

But before they used their abacus, the Chinese had for centuries used little ivory or bamboo sticks, called *chóu* (literally, "calculating sticks") which they arranged on the squares of a tiled floor, or on a table made like a chessboard.

The abacus did not evolve solely in form and construction. Far greater changes took place in the manner of its use.

The Madagascar natives, who did not profit fully from their great discovery, no doubt never understood that this way of representing numbers would give them the means to carry out complex calculations. So in order to add 456 persons to 328 persons, they would wait out the passage of the 456, and then of the 328 others, in order to finally observe the pebbles which gave the result.

Their use of the abacus was, therefore, purely for counting. Many other peoples were no doubt in the same state in the beginning. But, in seeking a practical approach to making calculations which were becoming ever more complex, they were able to develop procedures for the device by conceiving of a subtle game in which the pebbles were added or removed, or moved from one row to another.

To add one number to a number already represented on a decimal device, all they had to do was to represent the new number also on the abacus, as before, and then – after performing the relevant reductions – to read off the result. If there were more than 10 pebbles in a column, then 10 of these would be removed and 1 added to the next, starting with the lowest-order column. Subtraction can be done in a similar way, but by taking out pebbles rather than putting them in. Multiplication can be carried out by adding the results of several partial products.

The "heap of pebbles" approach to arithmetic, indeed the manipulation of various kinds of object for this purpose, thus once again is central in the history of arithmetic. These methods are at the very origin of the calculating devices which people have used throughout history, at times when the numerals did not lend themselves to the processes of calculation, and when the written arithmetic which we can achieve with the aid of "Arabic" numerals did not yet exist.

THE SUMERIAN ABACUS RECONSTRUCTED

It is logical, therefore, to suppose that Sumerian calculators themselves made use of some sort of abacus, at any rate once their *calculi* had disappeared from use.

Archaeological investigation in the land of Sumer has failed so far to yield anything of this kind, nor has any text been discovered which precisely describes it as well as its principles and its structure. Nonetheless, we can with the greatest plausibility reconstruct it precisely.

We may in the first place suppose that the instrument was based on a large board of wood or clay. It may equally well have been on bricks or on the floor.

The abacus consists of a table of columns, traced out beforehand, corresponding to the different orders of magnitude of the sexagesimal system.

We may likewise suppose that the tokens which were used in the device were small clay pellets or little sticks of wood or of reed, which each had a simple unit value (unlike the archaic system of the *calculi*, whose pieces stood variously for the different orders of magnitude of the same number-system).

We may determine the mathematical principles of the Sumerian abacus by appealing to their number-system itself.

Their number-system, as we have seen, used base 60. This theoretically requires memorisation of 60 different words or symbols, but the spacing between successive unit magnitudes was so great that in practice an intermediate unit was introduced to lighten the load on the memory. In this way, the unit of tens was introduced as a stepping stone between the sexagesimal orders of magnitude. The system was therefore based on a kind of compromise, alternating between 10 and 6, themselves factors of 60. In other words, the successive orders of magnitude of the sexagesimal system were arranged as follows:

first order	first unit	1	=	1	=		1
of magnitude	second unit	10	=	10	=		10
second order	first unit	60	=	60	=		10.6
of magnitude	second unit	600	=	10.60	=		10.6.10
third order	first unit	3,600	=	60^2	=		10.6.10.6
of magnitude	second unit	36,000	=	10.60^2	=		10.6.10.6.10
fourth order	first unit	216,000	=	60^3	=		10.6.10.6.10.6
of magnitude	second unit	2,160,000	=	10.60^3	=	10.6.10.6.10.6.10

On this basis, therefore, we can lay out the names of the numbers in a tableau as in Fig. 12.3. There are nine different units, five different tens, nine different sixties, and so on. From this table, therefore, we can clearly see that ten units of the first order are equivalent to one unit of the second, that six of the second are equivalent to one of the third, that ten of the third are equivalent to one of the fourth, and so on, alternating between bases of 10 and 6.

If, therefore, we accept that the Sumerians had an abacus, it must have been laid out as in Fig. 12.4.

Each column of the abacus therefore corresponded to one of the two sub-units of a sexagesimal order of magnitude. Since, moreover, the cuneiform notation of the numerals was written from left to right, in decreasing order of magnitude starting from the greatest, we may therefore reconstruct this subdivision in the following manner.

Proceeding from right to left, the first column is for the ones, the second for the tens, the third for the sixties, the fourth for the multiples of 600, the fifth for the multiples of 3,600, and so on (Fig. 12.4). To represent a given number on this abacus, therefore, one simply places in each column the number of counters (clay pellets, sticks, etc.) equal to the number of units of the corresponding order of magnitude.

SECOND SEXAGESIMAL ORDER   FIRST SEXAGESIMAL ORDER

Sub-order of the multiples of 600	Sub-order of the multiples of 60	Sub-order of the tens	Sub-order of the units
from 1×600 to 5×600	from 1×60 to 9×60	from 1×10 to 5×10	from 1 to 9

6   10   6   10

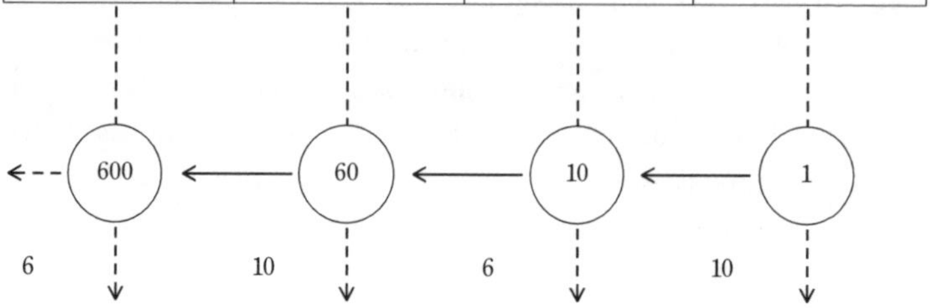

600 *geš-u* (= 3×600)	60 *geš* (= 1×60)	10 *u*	1 *geš*
1,200 *geš-u-min* (= 2×600)	120 *geš-min* (= 2×60)	20 *niš*	2 *min*
1,800 *geš-u-eš* (= 1×600)	180 *geš-eš* (= 3×60)	30 *ušu*	3 *eš*
2,400 *geš-u-limmu* (= 4×600)	240 *geš-limmu* (= 4×60)	40 *nimin*	4 *limmu*
3,000 *geš-u-iá* (= 5×600)	300 *geš-iá* (= 5×60)	50 *ninnû*	5 *iá*
	360 *geš-àš* (= 6×60)		6 *àš*
	420 *geš-imin* (= 7×60)		7 *imin*
	480 *geš-ussu* (= 8×60)		8 *ussu*
	540 *geš-ilimmu* (= 9×60)		9 *ilimmu*

FIG. 12.3. *Structure of the Sumerian number-system (see also Fig. 8.6 and 10.4)*

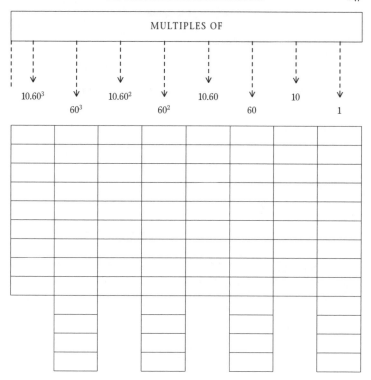

Fig. 12.4. *Form and structure of the divisions of the Sumerian abacus*

CALCULATION ON THE SUMERIAN ABACUS

Suppose one number is already laid out on the abacus, and we wish to add another number to it. To do this, lay out the second number on the abacus as well. Then, if there are 10 or more counters in the first column, replace each 10 by a single counter added to the second. Then replace each 6 in the second column by 1 added to the third, then each 10 in the third by 1 added to the fourth, and so on, alternating between 10 and 6. When the left-hand column has been reached, the result of the addition can be read off. Subtractions proceed in an analogous way, and multiplication and division are done by repeated additions or subtractions.

Let us return to the problem in the tablet shown in Fig. 12.1, and try to solve it on the abacus. We want to divide 1,152,000 by 7. We shall proceed by means of a series of partial divisions, each one on a single order of magnitude and beginning with the greatest.

First stage

In Sumerian terms, we are to divide by 7 the number whose expression, in number-names, is *šàrgal-iá šàr-u-min,*

which breaks down mathematically to:

$$5 \times 60^3 + 2 \times (10.60^2) = 5 \times 216{,}000 + 2 \times 36{,}000.$$

In the dividend there are therefore 5 units of order 216,000, and 2 of order 36,000. But, since the highest is present only five-fold and 5 is not divisible by 7, these units will be converted into multiples of the next lower order of magnitude, replacing the 5 counters in the highest order by the corresponding number of counters on the next.

One unit of order 216,000 is equal to 6 units of order 36,000, so we take $5 \times 6 = 30$ counters and add these to the 2 already there. There are, therefore, 32 counters on the board.

Now, 32 divided by 7 is 4, with remainder 4. I therefore place 4 counters (for the remainder) above the next column down (the 3,600 column) so as not to forget this remainder. Then the 4 counters (for the quotient) are placed in the 36,000 column. Then I remove the remaining counters.

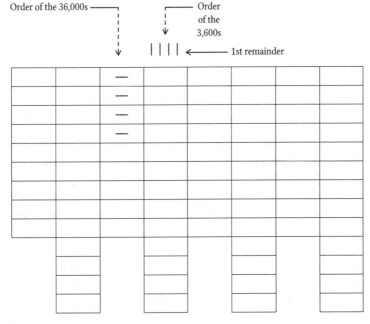

FIG. 12.5A.

Second stage

Now I convert the 4 counters of the preceding remainder into units of order 36,000.

One unit of 36,000 is 10 units of 3, 600, so I take $10 \times 4 = 40$ counters.

But 40 divided by 7 is 5, with remainder 5. Therefore I now place 5 counters (for this remainder) above the next column down (600) so as not to forget it.

Then I place the 5 counters (for the quotient) in the 3,600 column, and remove the remaining counters.

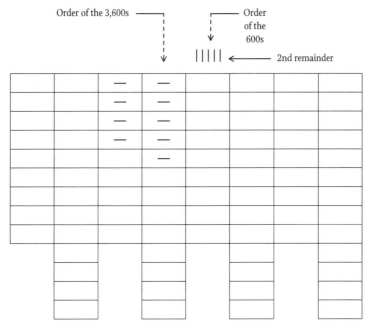

FIG. 12.5B.

Third stage

Now I convert the 5 counters for the preceding remainder into units of order 600. One unit of 3,600 is 6 units of 600, so I take 5 × 6 = 30 counters.

But 30 divided by 7 is 4, with remainder 2. I place 2 counters (for the remainder) above the next column down (60), as before.

Then I place 4 counters (for the preceding quotient) in the 600 column, and finally I remove the remaining counters (Fig. 12.5C,overleaf).

Fourth stage

Now I convert the 2 counters for the preceding remainder into units of order 60. One unit of 600 is 10 units of 60, so I take 2 × 10 = 20 counters.

Now 20 divided by 7 is 2, with remainder 6. So I place 6 counters (for the remainder) above the next column down (10). Then I place 2 counters (for the preceding quotient) in the 60 column. Then I remove the remaining counters (Fig. 12.5D,overleaf).

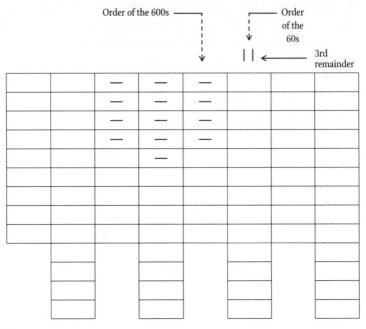

F<small>IG</small>. 12.5C.

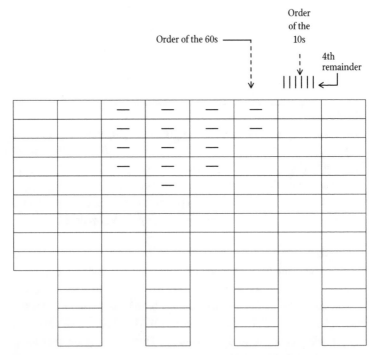

F<small>IG</small>. 12.5D.

Fifth stage

I convert the 6 counters for the preceding remainder into units of order 10. One unit of 60 is 6 units of 10, so I take 6 × 6 = 36 counters.

But 36 divided by 7 is 5 with remainder 1. So I place 1 counter (for the remainder) above the next column (units) and then 5 counters (for the preceding quotient) into the tens column. Then I remove the remaining counters.

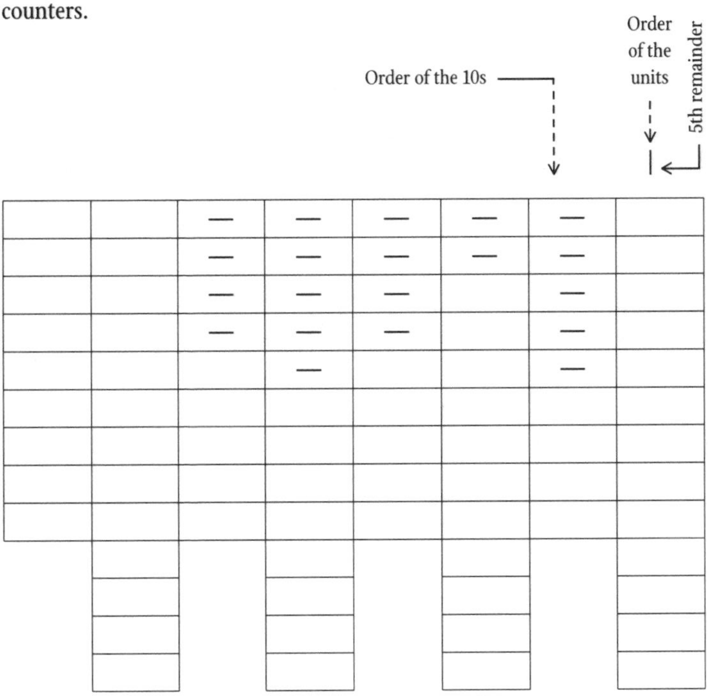

FIG. 12.5E.

Sixth and final stage

Now I convert the single counter for the preceding remainder into simple units. One unit of 10 is 10 simple units, so I take 10 counters.

But 10 divided by 7 is 1, with remainder 3. So I place 3 counters (for the remainder) to the right of the units column. Then I place 1 counter (for the preceding quotient) into the units column, and I remove the remaining counters.

Since I have now arrived at the final column, of simple units, the procedure is finished. To obtain the final result, I simply read off from the abacus to obtain the quotient (Fig. 12.5 F):

$$4 \times 36{,}000 + 5 \times 3{,}600 + 4 \times 600 + 2 \times 60 + 5 \times 10 + 1$$

and the 3 counters which I placed at the right of the last column give me the remainder.

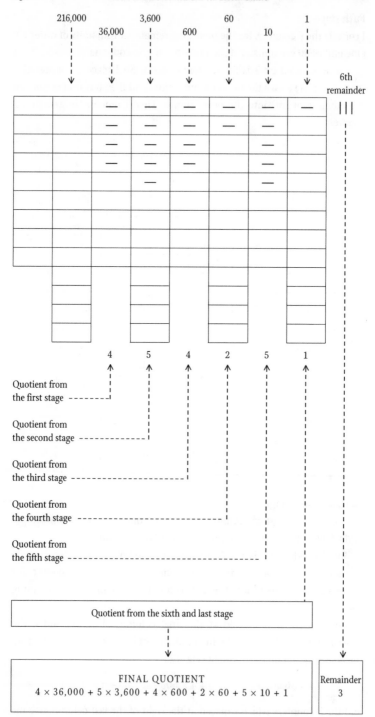

FIG. 12.5F.

On the abacus, therefore, the procedures for calculation were much simpler than for the much more ancient methods using the *calculi* of old. Undoubtedly both methods were in use together for a certain time, the more traditionally-minded tending to stay with the methods of their predecessors. These too were probably the same who continued to use the curviform notation of times past up until the end of the third millennium by which time the use of cuneiform notation had spread throughout Mesopotamia. We may therefore imagine the disputes between "calculists" and "abacists", the former standing to the defence of calculating by means of objects of different sizes and shapes, the latter attempting to demonstrate the many advantages of the new method.

What I have just said about the quarrel between the specialists is plausible, but it is merely a figment of my imagination. The rest of what I have been saying, though, is much more than merely probable.

CONFIRMATION OF THE SUMERIAN ABACUS AND ABACISTS

The reconstructions described above have in fact received confirmation as a result of recent discoveries.

I am referring to Sumero-Akkadian texts on cuneiform tablets dating from the beginning of the second millennium BCE, from various Sumerian archaeological sites (including Nippur), which have been meticulously collated, translated and interpreted by Liebermann (in *AJA*). These texts are all reports, and detailed analyses, in two languages (Sumerian and Ancient Babylonian) of various professions exercised at the time in Lower Mesopotamia. They are, in a way, "yearbooks" for these professions, and were made in several copies. The reports refer to each profession by giving a description of its representative, and a brief title of the kind "man of ...", but at the same time in each case they clearly specify the nature of any tools or devices used in each profession.*

Among all the many sorts of information in these texts, we find precisely the professions which are of prime interest for us. The lists give with great precision not only their official designation but also their tools and

---

* The bilingual texts from which have been taken the names given in Fig. 12.6 A–L occur mainly on tablets with the following museum references: – 3 NT 297, 3 NT 301 (cf. *Field Numbers of Tablets excavated at Nippur*); – IM 58433, IM 58496 (cf. *Tablets in the Collections of the Iraqi Museum of Baghdad*); – NBC 9830 (cf. *Tablets in the Babylonian Collection of the Yale University Library*, New Haven, Conn.); – MLC 653 and 1856 (cf. *Tablets in the Collection of the J. P. Morgan Library*, currently housed in the Babylonian Collection of the Yale University Library, New Haven). The article by S. J. Liebermann, of which the principal results are summarised here in a more accessible form and with some supplementary detail, provides the expert with all necessary philological information and correspondences, and all necessary bibliographical references, including those referring to the important publication by B. Landsberger (cf. *Materialen zum Sumerischen Lexikon*, Rome, 1937).

instruments, down to the very detail of their shape and material, and even which component goes with which instrument.

This is therefore a sufficiently significant discovery to justify a detailed philological explanation. The results will be displayed in successive diagrams each with three columns. On the left we shall place the Sumerian name (in capital letters); in the centre we shall place the Ancient Babylonian name (in italics); and, on the right, the equivalent English translation.

First we encounter a word which expresses the verb "to count":

ŠID	*ma-nû*	to count

Fig. 12.6a.

Remarkably, the Sumerian graphical etymology of this verb displays in itself evidence of the existence of the abacus. Originally, this verb was represented by the following pictogram (Fig. 12.6 B). Here we see a hand, or at any rate an extreme idealisation of one, doing something with a board in the shape of a frame or a tray and divided into rows and columns. Somewhat later on, the same verb was represented by a cuneiform ideogram, where we seem to see a frame divided into several columns and intersected by a vertical wedge resembling the figure for unity:

MOST ANCIENT FORM OF THE SIGN *(archaic Sumerian of the Uruk period)*	ARCHAIC CUNEIFORM SIGN *(Sumerian of the epoch of Jemdet Nasr)*	MORE RECENT SIGN *(classical Sumerian)*

Fig. 12.6b. *Sumerian notations of the verb "to count"* (šid). *Deimel (1947) no. 314*

Considering how ancient this sign for unity is (3000–2850 BCE), we are led to believe that the Sumerian abacus goes back to an even more ancient period than we had previously supposed. To come back to the "professional yearbooks", we find here also a clear reference to the system of *calculi*, which are referred to by a word which strictly means "small clay object":

IMNA (IMNA$_4$NA or NA$_4$IM)	*abnû*	*calculus, calculi* (small clay object)

Fig. 12.6c.

The "accounting" itself is denoted by a combination of the verb ŠID ("to count") and the word NIG̃ (total, sum):

NIG̃₂-ŠID	nik-kàs-si	accounting (making the total)

FIG. 12.6D.

Here again, the Sumerian etymology traces this to a suggestive origin, since the signs for the word NIGI (or NIGIN), meaning "total", "sum", "to collect together", clearly suggest the successive sections of the abacus:

MOST ANCIENT FORM OF THE SIGN (*Archaic Sumerian of the Uruk period*)	ARCHAIC CUNEIFORM SIGN (*Sumerian of the epoch of Jemdet Nasr*)	MORE RECENT SIGN (*Classical Sumerian*)

FIG. 12.6E.

Next we find the word for the expert in weights and measures, in a way the metrologist of that time and place, the "man of the stones":

LÚ NA₄NA	ša abnĕ e	man of the stones

FIG. 12.6F.

This designation obviously would not be confused with that of the calculator using the method of the *calculi*; this man is distinctly denoted in these texts by the terms:*

LÚ IMNA₄NA LÚ NA₄IM NA	ša ...(?)	man of the *calculi* (the man with the small clay objects)

FIG. 12.6G.

* The texts have been damaged by time at this point, and we cannot make out the corresponding Babylonian term. We see only its beginning, *ša*, which tells us little since *ša* is simply the Sumerian translation of the word for "man". But, following the work of A. L. Oppenheim (1959), we know that the Sumerian word for *calculi* was *abnü* (plural: *abnäti* or *abnĕ*, meaning literally "stone", "stone object", "kernel" or "hailstone". So we may suppose that the complete term was *ša abnäti-i*, where the scribe would use the second form of the plural in order to avoid confusion with *ša abnĕ-e*, unless he simply used the Sumerian word IMNA (*calculi*) in order to coin a term similar to *ša imnaki* (or *ša imnake*).

What follows next is even more thrilling, since it reveals not only the name but also the material used for the counter employed by the abacus-users of the period (the word "G̃EŠ" means "wood"):

G̃EŠ-ŠID-MA G̃EŠ-NIG$_2$-ŠID	iṣ-ṣi mi-nu-ti iṣ-ṣi nik-kàs-si	counting stick; stick for accounting

FIG. 12.6H.

Moreover, it yields not only what the counter is made of; we also learn its shape, since the profession which made use of it comes under the heading of "the men of the small wooden sticks".

As we have supposed above, they indeed made use of rods to perform their operations on the abacus (Fig. 12.5).

As to the abacus itself, we find that the texts refer to it clearly, using a figurative expression. To understand what this means, let us first note that, in Sumerian, "tablet" is said as DAB$_4$ and, in the absence of any further details, this word is always understood to mean "clay tablet", the dominant medium of the region for writing on. But in this case the material of the tablet is specified by the word G̃EŠ, meaning "wood". The word G̃EŠDAB$_4$ therefore means "wooden tablet" and in this context is therefore quite other than the "paper of Mesopotamia".

Another word which enters into the makeup of the Sumerian term for the abacus is DÍM. As a verb, this means "to fashion", "to form", "to model in clay", "to construct" and so on, and hence, by association of ideas, "to elaborate", "to perfect", "to create", "to invent". As a noun, it means "fashion", "form", "construction", and, by extension, "perfecting", "formation", "elaboration", "creation", or "invention" [A. Deimel (1947), no. 440].

Now we understand that the word DÍM frequently, by association of ideas, referred to the activities associated with Mesopotamian accounting, not only modelling and moulding clay (to make the *calculi* and the tablets) but also, and above all, to perfect, to elaborate results, and consequently to create and to invent something which nature did not provide in the raw state. Moreover, calculation is essential for shaping and fashioning objects, and also to architects for whom it is a vital necessity in their constructions.

Bringing all these terms together into a logical compound, by composing the expression G̃EŠDAB$_4$-DÍM to designate the instrument in question, the scribes must have had several simultaneous meanings present in their minds:

1. "wooden tablet" meaning perfecting;
2. "wooden tablet" meaning elaboration;
3. "wooden tablet" meaning creation;
4. "wooden tablet" meaning invention;

5. "wooden tablet" endowed with form (namely the tablet);
6. "wooden tablet" endowed with forms (the columns);
7. "wooden tablet" meaning the accounts;
8. "wooden tablet" meaning accounting; and so on.

Here we have therefore the characteristics and the many purposes of the abacus. The word G̃EŠDAB₄-DÍM can have but one translation.

G̃EŠDAB₄-DÍM	*g̃ešdab₄-dím mu*	abacus

Fig. 12.6I.

Even more significant is this other designation of the instrument of calculation:

G̃EŠŠU-ME-GE	*šu-me-ek-ku-ú*	abacus

Fig. 12.6J.

The word ŠU which is one component of this expression literally means "hand". In certain contexts, however, it also means "total", "totality", alluding to the hand which assembles and totalises [A. Deimel (1947), no. 354].

The word ME, for its part, means "rite", "prescription"; in other words "the determination of that which must be done according to the rules", or "an action which is performed according to a precise order as well as a prescribed order" [A. Deimel (1947), no. 532].

It would not in fact be at all surprising if the practice of calculating on the abacus corresponded to a genuine ritual, since the knowledge of abstract numbers and, even more, skill in calculation were not within the grasp of everyone as they are today. Those who knew how to calculate were rare indeed.

With all the peoples of the earth, calculation did not merely evoke admiration for those skilled in the art: they were feared, and regarded as magicians endowed with supernatural powers. This naturally gave rise to a certain element of sacred ritual in their activities, not to mention the numerous privileges which kings and princes often granted them.

In any case, in a context such as the present one, the word ME must be understood as "the determination of that which must be done in the precise order prescribed by the rules of calculation". This is something like what computer scientists of modern times would mean by "algorithm".

The term GE (or GI) is the word for a reed and stands for all the names of objects which can be made from this material [A. Deimel (1947), no. 85].

When we put them all together, these terms give the expression G̃EŠŠU-ME-GE, which corresponds to one or other of the following literal translations:

1. A hand (ŠU), a reed (GE), the rules (of arithmetic) (ME) and wood (G̃EŠ) ("of the tablet" understood);

2. The wood (of the tablet), a reed (GE), the rules (of arithmetic) (ME) and a total (i.e. provided by the hand) (ŠU).

In plain language, the expression G̃EŠŠU-ME-GE clearly refers to the abacus.

Lastly, for the "professional calculator", the texts use one or other of the following expressions:

LÚ G̃EŠ DAB₄-DÍM LÚ G̃EŠDAB₄	*ša da-ab-di-mi*	abacist

FIG. 12.6K.

The first of these means literally "the man (LÚ) of the wooden tablets for accounting (G̃EŠ DAB₄ DIM)", and the second simply means "the man (LÚ) of the wooden tablets (G̃EŠDAB₄)", there being no confusion possible about the material support.

We also find one or other of the two names

LÚ G̃EŠ ŠUMUN-GE LÚ ŠUMUN-GI₄	*ša šu-ma-ki-i*	abacist

FIG. 12.6L.

The first of these means literally a man (LÚ) who manipulates the rules (MUN) with a reed (GE) on the wood (G̃EŠ) ("of the tablet" understood), while the second corresponds to a symbolic variant of the first which could be translated as a man (LÚ) who finds the total (ŠU) with a reed (GI) according to the rules (MUN).

There is at this time no doubt that the abacus indeed existed in Mesopotamia, and even coexisted with the archaic system of *calculi*, most probably throughout the third millennium BCE.

This abacus consisted of a tablet of wood, on which were traced beforehand lines of division which exactly corresponded to the Sumerian sexagesimal system (Fig. 12.5) and therefore delimited, column by column, each of the order of unity of this numerical system (1, 10, 10.6, 10.6.10, 10.6.10.6, 10.6.10.6.10, and so on).

The counting tokens themselves were thin rods of wood or of reed, given the value of a simple unit, such that their subtle arrangement over the columns of the abacus allowed all the operations of arithmetic to be carried out. (No doubt it is as a result of their perishable nature that archaeologists have never brought any of these to light. Another reason may be that, as we may well suppose, whenever one of these experts did not have a "calculating board" to hand, he could simply draw the "tablet" on the loose soil.)

Lastly, as with writing but perhaps more so, the use of the abacus gave rise to a guild, perhaps even with the privileges of a special caste, so much would its complex rules and practices have been inaccessible to ordinary mortals: this was the caste of the professional abacists, who no doubt jealously preserved the secrets of their art.

CHAPTER 13

MESOPOTAMIAN NUMBERING
AFTER THE ECLIPSE OF SUMER

THE SURVIVAL OF SUMERIAN NUMBERS
IN BABYLONIAN MESOPOTAMIA

For some time after the decline of Sumerian civilisation, the sexagesimal system remained in use in Mesopotamia. Just as many French people still use "old francs" in their everyday reckonings even though "new francs" replaced them officially as long ago as 1960, so the inhabitants of Mesopotamia continued to use the "old counting" based on multiples and powers of 60.

The following examples come from an accounting tablet excavated at Larsa (near Uruk) and probably dating from the reign of Rîm Sîn (1822–1763 BCE). They are characteristic examples of the everyday reckonings that constitute the city archives, and give an account of sheep with the following numerical values:

61 (ewes)	60 1	96 (ewes)	60 30 6
84 (rams)	60 20 4	105 (rams)	60 40 5
145 (sheep)	120 20 5	201 (sheep)	180 20 1

FIG. 13.1. *Birot, tablet 42, p. 85, plate XXIV*

The numerals used are indeed those of the old Sumerian cuneiform system, with its characteristic difficulty in the representation of numbers such as 61, the vertical wedge signifying 60 being almost indistinguishable from the wedge used for 1 – and this is certainly the reason why the scribe leaves a large space between the two symbols, so as to avoid confusion with the number 2.

There is nothing at all surprising in the persistence of the old system in Lower Mesopotamia, since that is where the system first arose, in the lands of Sumer. What is less obvious is why the sexagesimal system survived for so long in the lands to the north, that is to say in Akkadian areas. However, the evidence is indisputable. This is an example from a tablet written in Ancient Babylonian, dating from the thirty-first year of the reign of Ammiditana of Babylon (1683–1647 BCE). It provides an inventory of calves and cows in the following manner:

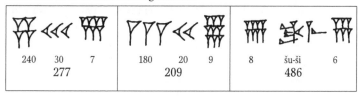

240	30	7	180	20	9	8	šu-ši	6
	277			209			486	

FIG. 13.2. *Finkelstein, tablet 348, plate CXIV, ll. 8–10*

The number 486 (the sum of 277 and 209) is represented not just by 8 large wedges each standing for 60 and 6 small wedges each standing for 1. The scribe has chosen to provide an additional phonetic confirmation of the number by putting the word *shu-shi* (the name of the number 60 in Akkadian) after the larger expression, rather like the way in which we write out cheques with a numerical and a literal expression of the sum involved.

All the same, these are just about the last traces of the unmodified system in use in Mesopotamia. Sumerian numbering was abandoned for good around the time that the first Babylonian Dynasty disappeared, in the fifteenth century BCE. By then, of course, modern Mesopotamian numerals, of Semitic origin, had been current for some time already.

WHO WERE THE SEMITES?

The term "Semite" derives from the passage in the Old Testament (Genesis 10) where the tribes of Eber (the Hebrews), Elam, Asshur, Aram, Arphaxad, and Lud are said to be the descendants of *Shem*, one of Noah's three sons, the brother of Ham and Japheth. However, though it may have represented a real political situation in the first millennium BCE, the biblical map of the nations of the Middle East makes the Elamites, who spoke an Asianic language, cousins to the Hebrews, Assyrians, and Aramaeans, whose languages belong to the Semitic group.

"Asianic" is the term used for the earlier inhabitants of the Asian mainland whose languages, mostly of the agglutinative kind, were neither Indo-European nor Semitic. It is generally believed that Mesopotamia was originally inhabited by Asianic peoples, prior to the arrival of Sumerians. It is thought that Semitic-speaking populations came in a second wave, and that Akkadian civilisation constitutes the earliest Semitic nation in the area. However, significant Semitic elements are to be found in the cultures of Mari and Kiš at the beginning of the third millennium BCE, and it is even possible that the people of El Obeid were of Semitic origin themselves, though absorbed and assimilated by the Sumerians. The discovery of the Ebla tablets revealed the existence of a state speaking a language of the Semitic family in the mid-third millennium BCE, and so it becomes ever less certain that the "cradle" of the Semitic languages was the Arabian peninsula, as was long held to be true. Nonetheless, Arabic is probably the closest to the proto-Semitic stem-language, which began to differentiate into numerous branches (Ancient Egyptian, some aspects of the Hamitic languages of eastern Africa, and possibly even Berber, spoken in Algeria and Morocco) as early as the Mesolithic era, that is to say (for the Middle East) in the tenth to eighth millennia BCE. That is too far back in time for it to be possible to say exactly where Semitic languages first arose or who the people were who brought them to different civilisations in the Middle East. Like the term "Indo-European", "Semitic" does not designate any ethnic or cultural entity, but serves only to define a broad family of languages.

There was no single "Semitic civilisation", just as there was never such a thing as an Aryan or an Indo-European culture. Each of the main Semitic-speaking civilisations of antiquity developed its own specific culture, even if there are some features common to several or all of them. It is therefore important to distinguish amongst the Semitic cultures those of the *Akkadians*, the *Babylonians*, the *Assyrians*, the *Phoenicians*, the *Hebrews*, the *Nabataeans*, the *Aramaeans*, the various peoples of *Arabia*, *Ethiopia*, and so on. (See Guy Rachet, *Dictionnaire de l'archéologie*, for further details.)

A BRIEF HISTORY OF BABYLON

At the beginning of the third millennium BCE Sumerians dominated the southern Mesopotamian basin, both numerically and culturally. To the north of them, between the Euphrates and the Tigris and on the northern and eastern edges of the Syro-Arabian desert, lived tribes of semi-nomadic pastoralists who spoke a Semitic language, called *Akkadian*. The Akkadian king, Sargon I The Elder, founded the first Semitic state when he defeated the Sumerians in c. 2350 BCE. His empire stretched over the whole of Mesopotamia and parts of Syria and Asia Minor. Its capital was at Agade (or *Akkad*), and, for one hundred and fifty years, it was the centre of the entire Middle East. As a result, Akkadian became the language of Mesopotamia and gradually pushed aside the unrelated language of Sumer. Assyrian and Babylonian are both descended from Akkadian and are thus Semitic, not Asianic, languages.

The Akkadian empire collapsed around 2150 BCE and for a relatively brief time thereafter Sumerians reasserted their control of the area. But that was the final period of Sumerian domination, for around 2000 BCE, the third empire of Ur collapsed under the simultaneous onslaughts of the

Elamites (from the east) and the Amorites (from the west). Sumerian civilisation disappeared with it for ever, and in its place arose a new culture, that of the *Assyro-Babylonians*.

The Amorites, a Semitic people from the west, settled in Lower Mesopotamia and founded the city of Babylon, which would become and remain for many centuries the capital of the country known as *Sumer and Akkad*. The famous king and law-maker Hammurabi (1792–1750 BCE) was one of the outstanding figures of the first Babylonian dynasty established by the Semites, who became masters of the region. Hammurabi extended Babylonian territory by conquest over the whole of Mesopotamia and as far as the eastern parts of Syria.

This huge and powerful kingdom was nonetheless seriously weakened, from the seventeenth century BCE, by the Kassites, Iranian highlanders who made frequent raids, and it finally surrendered in 1594 BCE to the Hittites, who came from Anatolia.

Babylon then remained under foreign domination until the twelfth century BCE, when another Semitic people, the Assyrians, from the hilly slopes between the left bank of the Tigris and the Zagros mountain range, entered the concert of nations. The Assyrians were bearers of a version of Sumerian culture, which they developed most fully in military conquest, establishing an empire which stretched out in all directions and which was one of the most fearsome and feared military powers in the ancient world, until in 612 BCE, Nineveh, the Assyrians' capital, was destroyed in its turn.

The Babylonians, although dominated by the Assyrians from the ninth to the seventh centuries BCE, nonetheless retained their own distinctive culture throughout this period. However, the fall of Nineveh (and with it of the whole Assyrian Empire) in 612 BCE allowed a great flowering of Babylonian culture, which was the prime force in the Middle East for over a century, most especially under the reign of Nebuchadnezzar II (604–562 BCE). But that was Babylon's last glory: it was conquered in 539 BCE by Cyrus of Persia, then in 331 BCE by Alexander the Great, and finally expired completely shortly before the beginning of the Common Era.

THE AKKADIANS, INHERITORS OF SUMERIAN CIVILISATION

In the Akkadian period (second half of the third millennium BCE) the Semites, who were now the masters of Mesopotamia, emerged as the preponderant cultural influence in the region. They naturally sought to impose their own language, and also to give it a written form. To do this they borrowed the cuneiform system of their predecessors, and adapted it progressively to their language and traditions.

By the time Sumerian cuneiform was adopted by the Akkadians, the writing system was already several centuries old. The ideas originally signified by the ideograms were mostly forgotten, and the signs were now purely symbolic. What the Akkadians found was a basically ideographic writing system with an already-established drift towards a phonetic system – a drift which the Akkadians accelerated, whilst retaining the ideographic meaning of some of the signs. They did so partly because their own language was less well suited to ideograms than Sumerian, and also because the signs which represented words for Sumerians represented only sounds to Akkadian ears.

The adaptation of cuneiform writing was however not a smooth or easy process. For one thing, Akkadian had sounds not present in Sumerian, and vice versa. The two ethnic groups of Akkadians (Babylonians and Assyrians) proceeded independently in this development, despite the numerous contacts between them. But by adopting the Sumerian cultural heritage, the Akkadians gave it its greatest flowering, leading it away from its origins in mnemotechnics and ultimately towards the creation of a true literary tradition.

THE NUMBERING TRADITIONS OF
SEMITIC PEOPLES

The spoken numbering system of the Semites was very different from the way Sumerians expressed numbers orally – not just linguistically, but also mathematically, since Semitic numbering was, and remains, strictly decimal.

However, Semitic numbering has one small grammatical oddity, in terms of the decimal numbering systems to which we are now accustomed.

Hebrew and Arabic numbering (see Fig. 13.3 below) provide characteristic examples.

In Hebrew as in Arabic, spoken numerals have feminine and masculine forms, according to the grammatical gender of the noun to which they are attached. For instance, the name of the number 1, treated as if it were an adjective, has one form if the noun it qualifies is masculine, and a different form if the noun is feminine. Similarly, the name of the number 2 agrees in gender with its noun. However, what is unusual is that for all numbers from 3, the number-adjective is feminine if the noun is masculine, and masculine if the qualified noun is feminine. In Hebrew, for example, where "men" is *anoshim* and "three" is *shalosh* (masculine) or *shloshah* (feminine), the expression "three men" is translated by *shloshah anoshim*, not, as you might expect from Latin or French grammar, by *shalosh anoshim*.

	HEBREW		ARABIC	
	Feminine	Masculine	Feminine	Masculine
1	'eḥad	'aḥat	'aḥadun	'iḥda
2	shnaim	shtei	'iṯnān	'iṯnatāni
3	shloshah	shalosh	ṯalāṯun	ṯalāṯatun
4	'arba 'ah	'arba'	'arba'un	'arba'atun
5	ḥamishah	ḥamesh	khamsun	khamsatun
6	shishah	shesh	situn	sitatun
7	shib'ah	sheba'	sab'un	sab'atun
8	shmonah	shmoneh	ṯamāny	ṯamānyatun
9	tishah	tesha'	tis'un	tis'atun
10	'asarah	'eser	'ashrun	'asharatun

FIG. 13.3.

Numbers from 11 to 19 are formed by the name of the unit followed by the word for 10, each having masculine and feminine forms, used according to the previous rule:

	HEBREW		ARABIC	
	Feminine	Masculine	Feminine	Masculine
11	'aḥad 'asar	'aḥat 'esreh	'aḥad 'ashara	'iḥda 'ashrata
12	shnaim 'asar	shtei 'esreh	'iṯnā 'ashara	'itnāta 'ashrata
13	shloshah 'asar	shlosh 'esreh	ṯalāṯut 'ashara	ṯalāṯa 'ashrata
14	'arba'ah 'asar	'arba 'esreh	'arba'ata 'ashara	'arba'a 'ashrata
15	ḥamishah 'asar	ḥamesh 'esreh	khamsata 'ashara	khamsa 'ashrata
16	shishah 'asar	shesh 'esreh	sitata 'ashara	sita 'ashrata

FIG. 13.4.

Apart from the number 20, which is derived from the dual form of the word for 10, the tens are derived from the name of the corresponding unit, with an ending that is derived from the customary mark of the plural:

	HEBREW	ARABIC	
20	'esrim	'isrūn	derived from dual of 10
30	shloshim	ṯalāṯūna	plural of name of 3
40	'arba'im	'arba'ūna	plural of name of 4
50	ḥamishim	khamsūna	plural of name of 5
60	shishim	sitūna	plural of name of 6
70	shibim	sib'ūna	plural of name of 7
80	shmonim	ṯamānūna	plural of name of 8
90	tishim	tis'ūna	plural of name of 9

FIG. 13.5.

The system has special names for 100 and 1,000, and proceeds thereafter by multiplication for multiples of each of these powers of the base:

HEBREW		ARABIC	
100	*me'ah*	*mi'ātun*	
200	*ma'taim*	*mi'atāny*	dual of 100
300	*shlosh me'ōt*	*ṭalāṭu mi'ātin*	(3 × 100)
1,000	*'elef*	*'alfun*	
2,000	*'alpaim*	*'alfāny*	dual of 1,000
3,000	*shloshet 'alafim*	*ṭalāṭu 'alāf*	(3 × 1,000)
10,000	*'aseret 'alafim*	*'asharat 'alāf*	(10 × 1,000)
20,000	*'esrim 'elef*	*'ishrūnat 'alāf*	(20 × 1,000)
30,000	*shloshim 'elef*	*ṭalāṭūnat 'alāf*	(30 × 1,000)

Note: Classical Hebrew also has the word *ribŏ* ("multitude") to designate 10,000, together with its multiples: *shtei ribot* for 20,000, *shlosh ribôt* for 30,000, etc. Similar words exist in other ancient Semitic languages: *ribab* (Elamite), *ribbatum* (Mari), *r(b)bt* (Ugaritic).

FIG. 13.6.

NUMBER-NAMES IN ASSYRO-BABYLONIAN

1	*ishtên*	10	*eshru, esheret*	100	*me'atu, me'at*	$= 10^2$
2	*sita, sinâ*	20	*eshrâ*	200	*sita metin*	$= 2 \times 100$
3	*shalâshu*	30	*shalâshâ*	300	*shalâsh me'at*	$= 3 \times 100$
4	*erbettu*	40	*arbâ*			
5	*khamshu*	50	*khamshâ*	1,000	*lim*	$= 10^3$
6	*sheshshu*	60	*shushshu, shushi*	2,000	*sinâ lim*	$= 2 \times 1,000$
7	*sîbu*	70	*	3,000	*shalâshat limi*	$= 3 \times 1,000$
8	*shamânu*	80	*			
9	*têshu*	90	*	10,000	*esheret lim*	$= 10 \times 1,000$
				20,000	*eshrâ lim*	$= 20 \times 1,000$

* The pronunciation of these numbers is not known

100,000	*me'at lim*	$= 100 \times 1,000$
200,000	*sita metin lim*	$= 2 \times 100 \times 1,000$

FIG. 13.7.

For intermediate numbers, addition and multiplication are used in conjunction. In Arabic, it should be noted, the units are always put before the tens: 57, for example, is *sab'un wa khamsūna* ("seven and fifty"), as in German (*siebenundfünfzig*).

The same order of expression is found in Ugaritic texts (Ugarit was a Semitic culture that flourished at Ras Shamra, in northern Syria, around the fourteenth century BCE) and in biblical Hebrew, most frequently in the Pentateuch and the Book of Esther. According to Meyer Lambert, this order of numbers is the archaic form.

However, the inverse order (hundreds followed by tens followed by units) is also found in the Hebrew Bible, and this is the commonest form in the first Books of the Prophets, and in most of the books written after the Exile (Haggai, Zechariah, Daniel, Ezra, Nehemiah, Chronicles). Modern Hebrew (Ivrit) also uses this order (except for numbers between 11 and 19), which is also the most frequent structure in Semitic languages as a whole (Assyro-Babylonian, Phoenician, Aramaic, Ethiopian, etc.).

All these numbering systems therefore demonstrate that they have a common origin, which gives all Semitic numbering its characteristic mark. It will now be easier to grasp how the Mesopotamian Semites radically transformed the cuneiform numerals of the Sumerians, and to understand the method that the western Semites (Phoenicians, Aramaeans, Nabataeans, Palmyreneans, Syriacs, the people of Khatra, etc.) invented to put their numbers in writing other than by spelling them out. (See Chapter 18 below, pp.227–32

THE SUMERO-AKKADIAN SYNTHESIS

When the Akkadians took over cuneiform sexagesimal numbering, they were naturally hampered by a written system whose organisation differed entirely from the strictly decimal base of their own long-standing oral number-name system. The cuneiform numerals had a sign for 1 (the vertical wedge) and for 10 (the chevron) – but, since there was no sign for 100 or for 1,000, it occurred to them to write out the names of these numbers phonetically. "Hundred" and "thousand" were respectively *me'at*

"6,657" IN ANCIENT & MODERN SEMITIC LANGUAGES					
ARABIC	*sitatunat 'aláf* six thousand 6 × 1,000 +	*sitatu mi'átin* six hundred 6 × 100 +	*sab'un* seven 7	*wa* & +	*khamsûna* fifty 50
UGARITIC	*ṭiṭ 'alpin* six thousand 6 × 1,000 +	*ṭiṭ mat* six hundred 6 × 100 +	*sab'a* seven 7	*l* & +	*khamishuma* fifty 50
CLASSICAL HEBREW	*sheshet 'alafim* six thousand 6 × 1,000 +	*sesh me'ôt* six hundred 6 × 100 +	*shib'ah* seven 7	*we* & +	*khamishim* fifty 50
CLASSICAL & MODERN HEBREW	*sheshet 'alafim* six thousand 6 × 1,000 +	*sesh me'ôt* six hundred 6 × 100 +	*khamishim* fifty 50	*we* & +	*shib'ah* seven 7
ASSYRO- BABYLONIAN	*sheshshu limi* six thousand 6 × 1,000 +	*seshshu me'at* six hundred 6 × 100 +	*khamsha* fifty 50	 +	*sibu* seven 7
ETHIOPIAN	*sassá ma'át* sixty hundred 60 × 100 +	*sadastú ma'át* six hundred 6 × 100 +	*khamsá* fifty 50	*wa* & +	*sab'atú* seven 7

FIG. 13.8.

and *lim* in Akkadian, so they represented these numbers as words, using the Sumerian cuneiform signs for ME and AT, on the one hand, and for LI and IM on the other – rather as if we made puzzle-pictures of "Hun" and "Dread" to represent the sound and thus the number "hundred":

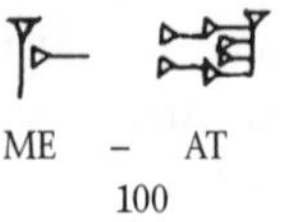

		Cuneiform "literals" for the Akkadian words for 100 (me'at) and 1,000 (lim)		
ME – AT			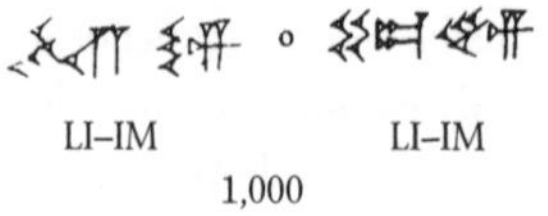 LI–IM	LI–IM
100				1,000

FIG. 13.9A.   FIG. 13.9B.

However, they did not stop at the "writing out" stage, they also created genuine numerals, even if these were derived from the phonetic notation of the number-names. The symbols chosen were of course no more than sound-signs from their point of view, since they had lost the meanings that they had had in Sumerian. The symbol for 100 was soon shortened to its first syllable, ME, and for 1,000 the Akkadians used the chevron (= 10) followed by the sign for ME (1,000 = 10 ME = 10 × 100). And since this was the sign for the *word* meaning "thousand", pronounced *lim*, the cuneiform chevron followed by ME came to have the phonetic value of the sound LIM and to be used in all Akkadian words containing the sound LIM.

	Akkadian cuneiform numerals for 100 and 1,000 as used from the second millennium BCE in everyday accounting documents	
ME		LIM
100		1,000

FIG. 13.10A.   FIG. 13.10B.

Because of the standard Semitic custom of counting orally in hundreds and thousands, the Akkadians therefore introduced *strictly decimal notations* into the sexagesimal numerals that they had adopted from the Sumerians. The result was a thoroughly mixed Akkadian number-writing system containing special signs for decimal and sexagesimal units, in the following manner:

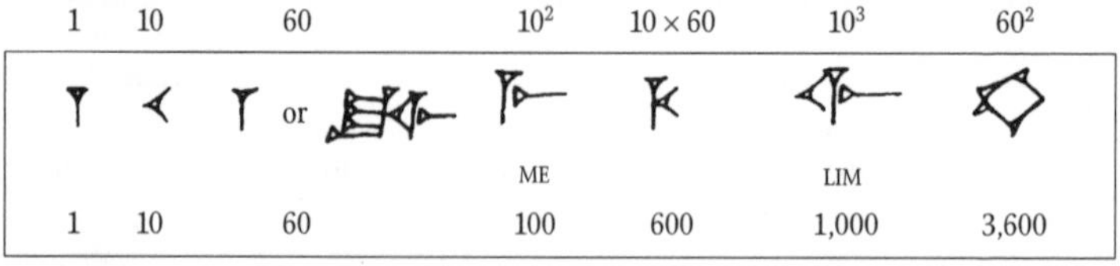

1	10	60	10^2	10×60	10^3	60^2
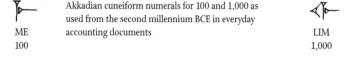						
			ME		LIM	
1	10	60	100	600	1,000	3,600

FIG. 13.11.

Let us look at a few characteristic examples. Those shown in Fig. 13.12 (M. J. E. Gautier, 1908, plates XVII, XLII and XLIII) come from clay tablets found at Dilbat, a small town in Babylonian territory that flourished in the nineteenth century BCE. Most of the tablets refer to the main events in the lives of members of a single family, and constitute as it were the family record.

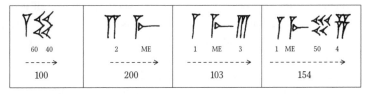

60 40	2 ME	1 ME 3	1 ME 50 4
-----→	--------→	----------→	------------→
100	200	103	154

FIG. 13.12. *See Gautier, plates XVII, XLII and XLIII*

The next figure is a transcription of a tally of cattle found in northern Babylon (M. Birot, 1970, tab. 33, plate XVIII) dating from the seventeenth year of the reign of Ami-Shaduqa of Babylon (1646–1626 BCE):

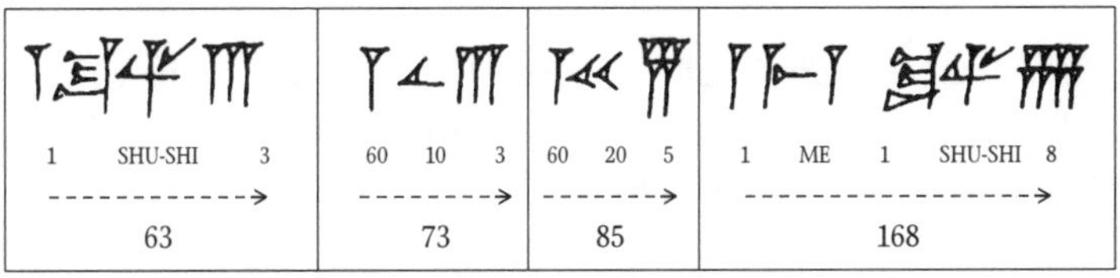

1 SHU-SHI 3	60 10 3	60 20 5	1 ME 1 SHU-SHI 8
-------------→	--------→	-------→	----------------→
63	73	85	168

FIG. 13.13. *See Birot, tablet 33, plate XVIII*

These examples show how in this period the Akkadians did not seek to overturn the sexagesimal system that was deeply rooted in local tradition.

However, for the numbers 60 and 61, and in many cases for multiples of 60, the Semites coped with the corresponding difficulties of the notation system rather better than had the Sumerians. It occurred to them to represent the number 60 by the sound-group *shu-shi*, which was how they pronounced the number in Akkadian (see Fig. 13.7 above) or, in abbreviated form, as *shu* (see Fig. 13.2, 13.13 and 13.14).

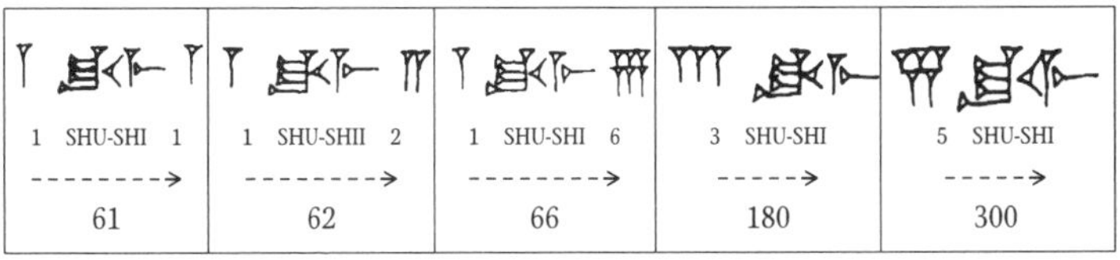

1 SHU-SHI 1	1 SHU-SHI 2	1 SHU-SHI 6	3 SHU-SHI	5 SHU-SHI
-------→	-------→	-------→	----→	----→
61	62	66	180	300

FIG. 13.14.

In short, up to the middle of the second millennium BCE, Mesopotamian scribes of public, private, economic, juridical, and administrative tablets had recourse *either* to sexagesimal Sumerian numbering, *or* to decimal Semitic numbering, *or* finally to a system constituted by a kind of *interference between the two bases*.

MESOPOTAMIAN DECIMALS

When Akkadian speech and writing finally supplanted their Sumerian counterparts in Mesopotamia, strictly decimal numbering became the norm in daily use. The ancient signs for 60, 600, 3,600, 36,000, and 216,000 progressively disappeared, and only the symbols ME (= 100) and LIM (= 1,000) remained, to provide the bases for the entire system of numerals.

As in classical Sumerian, units were represented by vertical wedges, repeated once for each unit, but whereas the Sumerians had grouped the wedges on a dyadic principle, the Akkadians put them in three groups:

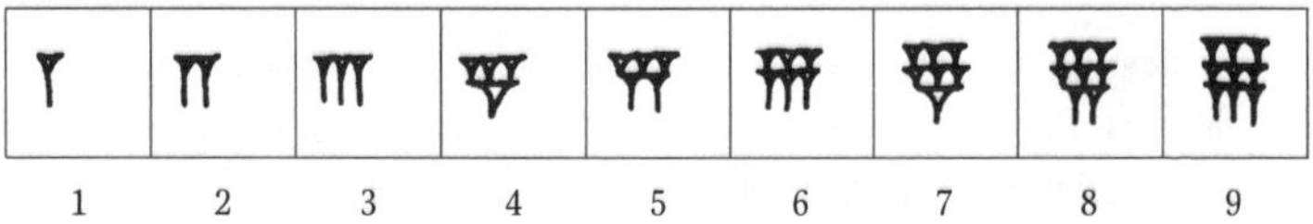

1	2	3	4	5	6	7	8	9

FIG. 13.15.

The tens were also usually represented by repetition of the chevron (= 10), but here again the layout or grouping of the repeated symbols was quite distinct from older Sumerian patterns:

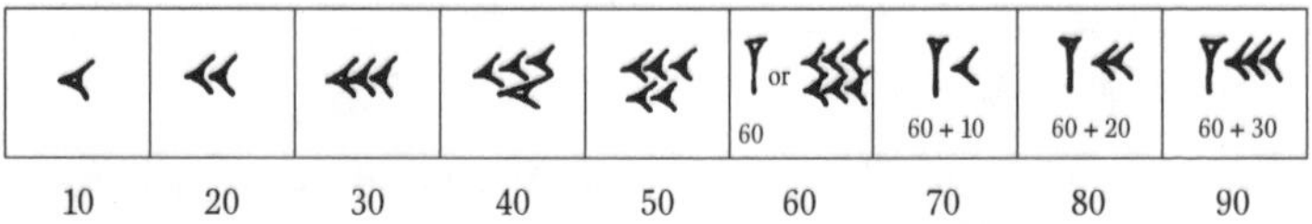

10	20	30	40	50	60	70	80	90
						60 + 10	60 + 20	60 + 30

FIG. 13.16.

As for the hundreds and thousands, they were symbolised by notations based on multiplication, that is to say in accordance with the analytical combinations that existed in the spoken language of the Akkadians:

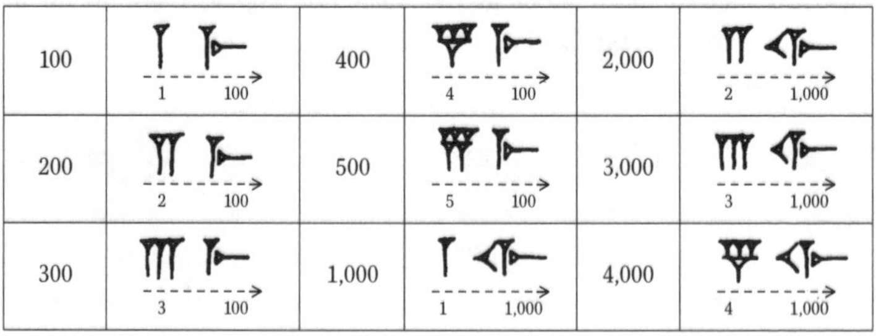

100	1 → 100	400	4 → 100	2,000	2 → 1,000
200	2 → 100	500	5 → 100	3,000	3 → 1,000
300	3 → 100	1,000	1 → 1,000	4,000	4 → 1,000

FIG. 13.17.

The following examples show just how radical the transformation of Sumerian cuneiform numerals was. The numbers shown relate to the booty taken during Sargon II's eighth campaign against Urartu (Armenia) in 714 BCE:

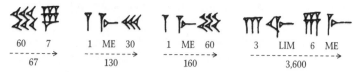

FIG. 13.18. *See Thureau-Dangin, lines. 380, 366 and 369*

As can be seen, 60 is now represented by six chevrons instead of the vertical wedge that formerly had this numerical value, and numbers such as 130, 160 and 3,600 are given strictly decimal representations.

We can also see that by grounding their written numerals on their spoken number-names, the Assyrians and Babylonians extended the arithmetical scope of their numeral system whilst restricting its basic figures to 100 and 1,000. All they needed to do was to combine these symbols with the multiplication principle, to produce expressions of the type $10,000 = 10 \times 1,000$, $40,000 = 40 \times 1,000$, $400,000 = 400 \times 1,000$, and so on. So Sargon II's scribe wrote out the number 305,412 in the following manner:

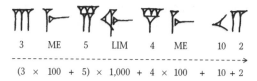

$(3 \times 100 + 5) \times 1,000 + 4 \times 100 + 10 + 2$

FIG. 13.19. *See Thureau-Dangin, l. 394*

RECONSTRUCTING THE DECIMAL ABACUS

The Akkadians must surely have possessed a calculating device, for they could not otherwise have performed their complex arithmetical operations save by the archaic device of *calculi*, of which barely a handful have been found in archaeological levels of the second millennium BCE. Indeed, as we also saw in Chapter 12, the Sumerians themselves must have had a kind of abacus, which we reconstructed in its most probable form along with the rules and procedures for its use. Furthermore, the Akkadians, at least in the Babylonian period, had specific terms for referring not only to the instrument and the tokens which went with it, but also to the operator of the abacus.

In Ancient Babylonian (see Fig. 12.6H above), the arithmetical "token", which must have been a stick of wood or a swatch of reed stems, was called either

- *iṣ-ṣi mi-nu-ti* ("wood-for-counting"); or
- *iṣ-ṣi nik-kàs-si* ("wood-for-accounts").

As for the abacus itself, it was referred to by one of the two following loan-words borrowed from the corresponding Sumerian terms (see Fig. 12.6I and 12.6J above):

- *g̃ešdab-dim mu* ("wooden-tablet-for-accounts");
- *šu-me-ek-ku-ú* (literally, from the corresponding Sumerian word G̃EŠŠUMEGE, "wood (i.e. of the tablet), hand, rule, reed" or alternatively "wood, sum, rule, reed".

The abacus operator or abacist had two official names (see Fig. 12.6K and 12. 6L above):

- *ša da-ab-di-mi* ("the man for the tablet for accounts");
- *ša šu-ma-ki-i* ("the man for the abacus").

Our knowledge of these terms comes from various bilingual tablets dating from the beginning of the second millennium BCE, which provide a kind of "Yellow Pages" in both Sumerian and Ancient Babylonian, each entry consisting of a brief description of a representative of a profession ("the man for ..."), followed by the name of the tools associated with the profession. (See Chapter 12 above, and for references to original sources, see S. J. Liebermann, in *AJA* 84.)

In view of all this, we have to suppose that the Akkadians first used the sexagesimal Sumerian abacus for as long as their arithmetic was dependent on Sumerian notation, but had to construct sexagesimal-decimal conversion tables for the requirements of their own decimal arithmetic during the long "transitional period" that lasted until the end of the first Babylonian dynasty, around the middle of the second millennium BCE. However, when Akkadian culture itself came to hold sway in Mesopotamia, the situation changed completely. The mathematical structure of the abacus had to be radically altered to adapt it to the modified cuneiform notation that was then used for strictly decimal arithmetic.

Indeed, the Assyro-Babylonian numeral system used base 10 and allowed all numbers up to one million to be represented by combinations of just these four signs:

| 1 | 10 | 100 (= ME) | 1,000 (= LIM) |

FIG. 13.20.

For numbers above 1,000, the system used analytical combinations of the given signs, that is to say it used the principle of multiplication to designate 10,000, 100,000, and 1,000,000, as follows:

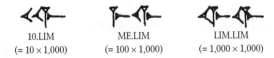

| 10.LIM | ME.LIM | LIM.LIM |
| (= 10 × 1,000) | (= 100 × 1,000) | (= 1,000 × 1,000) |

FIG. 13.21.

As we showed for the Sumerian abacus in Chapter 12 above, we can here show quite easily the most probable form of the Assyro-Babylonian abacus as it was used by "ordinary" counters (there are good reasons for thinking that there were two types of arithmeticians – the "ordinaries", whose arithmetic was exclusively decimal, and the "learned", who continued to use the sexagesimal system for mathematical and astronomical purposes). As for the way the abacus was used, it must have been very similar to the rules for the sexagesimal system, simply adapted to base 10:

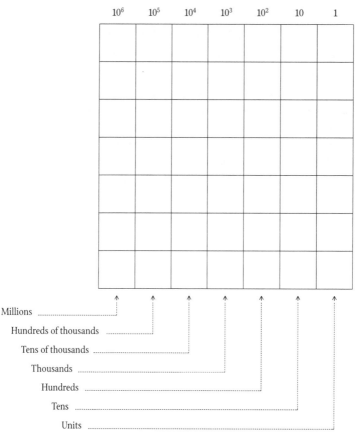

FIG. 13.22. *Reconstructed Assyro-Babylonian decimal abacus*

It should be noted that a brick marked with rows and columns as in Fig. 13.22 was discovered in the 1970s by the French Archaeological Delegation to Iran (DAFI) during the dig at the Acropolis of Susa, and that a few similar pieces were found in the same area during the Second World War. Up to now these objects have been taken as game-boards. *We suggest that they should rather be seen as arithmetical abaci.* Let us hope that further archaeological discoveries will provide suffficient evidence to confirm this interpretation. What we can be sure of, all the same, is that Susan accountants (and the Elamites in general) also used arithmetical tools, of which the first were of course the *calculi*. And there are very good reasons for thinking that the tools they used were similar to those of the Mesopotamians, for their operations were presumably just as complex as those being carried out a few hundred miles away by their Sumerian and Assyro-Babylonian counterparts.

THE LAST TRACES OF SUMERIAN ARITHMETIC IN THE ASSYRO-BABYLONIAN DECIMAL SYSTEM

In the hands of the Semites, cuneiform numerals and Mesopotamian arithmetic were gradually adapted and finally transformed into a system with a different base working on quite different principles. All the same, base 60 did not disappear entirely, and even continued to play a major role as "big unit" in "ordinary" Mesopotamian accounting. Although it was often represented (at least, from the start of the first millennium BCE) by the decimal expression ﷼﷼ Assyrians and Babylonians alike continued to represent the number 60 also by "spelling it out" inside numerical expressions, using either the sign for the sound *shu-shi* (which is how "sixty" was said in Semitic languages)

1   shu-shi

or in abbreviated form as *shu* (the first syllable of the word for "sixty")

1   shu

Above all, they went on figuring the numbers 70, 80, and 90 in the "old manner", that is to say in a way that carries the trace of the obsolete 60-based arithmetic of the Sumerians (just as, nowadays, the French words for 80 and 90 (*quatre-vingts, quatre-vingt-dix*) carry the trace of a vanished vigesimal arithmetic):

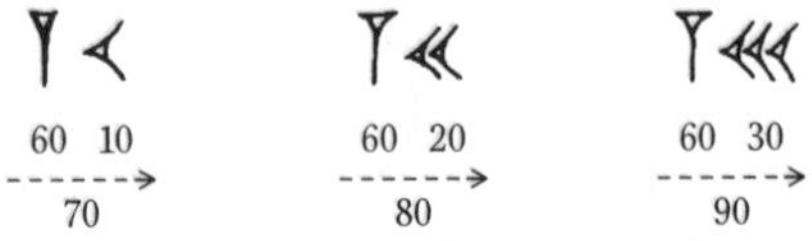

𒐕 𒌋	𒐕 𒌑	𒐕 𒌍
60 10	60 20	60 30
- - - - - →	- - - - - →	- - - - - →
70	80	90

FIG. 13.23.

The signs for the old base units of 600 and 3,600 never disappeared entirely either. They continue to crop up in contracts and financial statements, in auguries and in historical and commemorative texts. In these later usages, the sign for 3,600 underwent a graphic development in line with the evolution of Mesopotamian cuneiform writing:

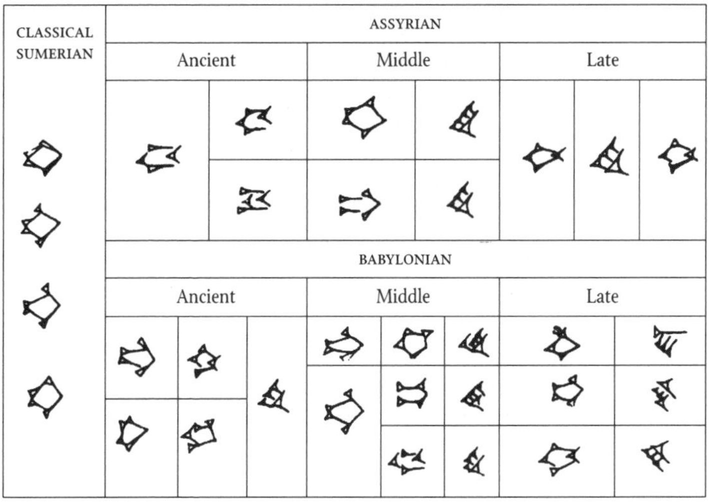
CLASSICAL SUMERIAN
ASSYRIAN
Ancient
Middle
Late
BABYLONIAN
Ancient
Middle
Late

FIG. 13.24. *Evolution and stability of the Sumerian sign* shar *(= 3,600)*

For example, when Sargon II of Assyria inscribed the dimensions of the walls of his fortress at Khorsabad – 16,280 cubits* – he had the figure written not in what had by then become the standard notation:

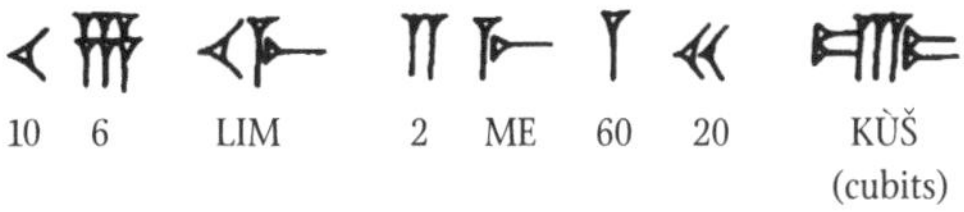
10
6
LIM
2
ME
60
20
KÙŠ (cubits)

FIG. 13.25.

but in this arithmetically more archaic manner:

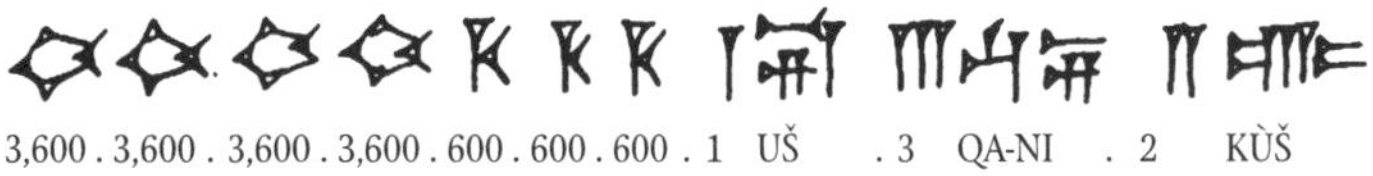
3,600 . 3,600 . 3,600 . 3,600 . 600 . 600 . 600 . 1 UŠ . 3 QA-NI . 2 KÙŠ

14,400 cubits	1,800 cubits	c. 60 cubits	3 × 6 cubits	2 cubits

FIG. 13.26. *See Lyon, p. 10, l. 65*

But such traces of the old system were mere relics, and had no influence at all on the strictly decimal arithmetic that the Assyro-Babylonians used throughout their history for everyday reckoning.

* The cubit (*kùš*) is a measure of length of approx. 50 cm. Six cubits make a *qânum*, and sixty cubits make an *uš*.

RECAPITULATION: FROM SUMERIAN TO ASSYRO-BABYLONIAN NUMBERING

There were, in brief, three main stages in Mesopotamian culture after the establishment of the Akkadian Empire:

- in the first, the Semites assimilated the cultural heritage of their Sumerian predecessors in the region;
- the second is an intermediate period;
- the third is the period of Semitic predominance in Mesopotamian culture.

	SUMERIAN SYSTEM (base 60 with 10 and 6 as auxiliary bases)	SUMERIAN-AKKADIAN SYNTHESIS (compromise between base 10 and base 60)	ORDINARY ASSYRO-BABYLONIAN SYSTEM (Strictly decimal base)
1	𒁹	𒁹	𒁹
10	𒌋	𒌋	𒌋
60	𒁹	𒁹 ... or 𒁹 ... 1 ŠU-ŠI 1 ŠU	
70	𒁹𒌋 60 10	𒁹𒌋	𒁹𒌋
80	𒁹𒌍 60 20	𒁹𒌍	𒁹𒌍
90	𒁹𒌎 60 30	𒁹𒌎	𒁹𒌎
100	𒁹... 60 40	𒁹... or 𒁹 ... 1 ME	𒁹 ... 1 ME
120	𒈫 60 60	𒈫 ... or 𒁹 ...𒌍 2 ŠU-ŠI 1 ME 20	𒁹 ...𒌍 1 ME 20
600	...	... ... 6 ME	... 6 ME
1,000	... ... 600 360 40	... ... or ... 1 LI-MI 1 LIM	... 1 LIM
3,600	◇	... ... 3 LIM 6 ME	... ... 3 LIM 6 ME

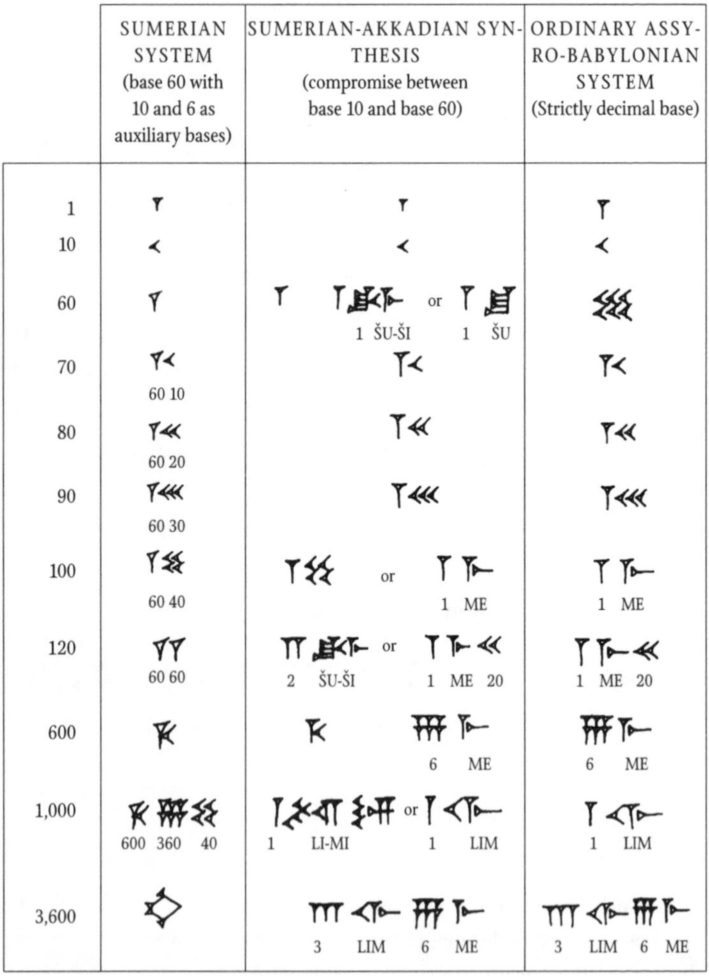

FIG. 13.27. *Evolution of popular Mesopotamian numerals before and after the eclipse of Sumerian civilisation (see also Fig. 18.9 below)*

As far as numbers and arithmetic are concerned, these periods correspond respectively to: pure and simple borrowing of Sumerian sexagesimal numbering; the emergence of a mixed system using a combination of decimal and sexagesimal signs; and the development of a strictly decimal system. This profound transformation of cuneiform numbers occurred under the pressure of oral number-names, whose strictly decimal structure is a common feature of all Semitic languages (see Fig. 13.7 and 13.19 above). But this is not where the development came to a full stop: as we shall see, the scribes of the city of Mari evolved their own unique version of a decimal numeral system.

THE ANCIENT SYRO-MESOPOTAMIAN CITY OF MARI

Various texts refer to the Sumero-Semitic city of Mari as an important place in the Mesopotamian world, but it was not until 1933 that André Parrot, led on by the suggestions of W. F. Albright and by the chance discovery of a statue, began to excavate at Tel-Hariri, on the border of Syria and Iraq. Over the following forty years, Parrot's team conducted a score of excavations and laid bare a whole civilisation.

The earliest traces of habitation at Mari date from the fourth millennium BCE, and by the first half of the third millennium it was already highly urbanised, with a ziggurat and a number of temples decorated with statuary and painted walls. The art and culture of Mari in this period resemble those of Sumer, but the facial types represented, as well as the names and the gods mentioned, are Semitic.

Mari became part of the Akkadian Empire, but regained some independence around the twenty-second century BCE. From the twentieth to the eighteenth century BCE Mari flourished as an independent and expanding city-state, but it was defeated and destroyed by Hammurabi around 1755 BCE. Though it continued to exist as a town, Mari never again regained any power or influence.

It was in the early eighteenth century BCE, under Zimri-Lim, that Mari built its most remarkable structures, including a 300-room palace occupying a ground area of 200 m × 120 m and in part of which were stored more than 20,000 cuneiform tablets, giving us a unique insight into the political, administrative, diplomatic, economic, and juridical affairs of a Mesopotamian state. The tablets include long lists of the palace's requirements (food, drink, etc.), and many letters written by women, which suggests that they played an important role in the life of the city.

WHAT IS THE RULE OF POSITION?

Just as an alphabet allows all the words of a language to be written by different arrangements of a very limited set of signs, so our current numerals allow us to represent all the integers by different arrangements of a set of only ten different signs. From an intellectual point of view, this system is therefore far superior to most numerical systems of the ancient world. However, that superiority does not derive from the use of base 10, since bases such as 2, 8, 12, 20, or 60 can produce the same advantages and be used in exactly the same way as our current decimal positional system. As we have already seen, moreover, 10 is by far the most widespread numerical base in virtue not of any mathematical properties, but of a particularity of human physiology.

What makes our written numeral system ingenious and superior to others is the principle that *the value of a sign depends on the position it occupies in a string of signs*. Any given numeral is associated with units, tens, hundreds, or thousands depending on whether it occupies the first, second, third, or fourth place in a numerical expression (counting the places from right to left).

These reminders allow us to understand fully the numbering system of Mari and of the learned men of Babylon ...

THE MARI SYSTEM

It has recently come to light that the scribes of Mari used, alongside "classical" Mesopotamian number-notation, a system of numerals quite different to all that had preceded it.

As in previous systems, the first nine units were represented by an equivalent number of *vertical wedges*:

Y	TT	TTT	Y	YY	TTT	YYY	YYY	YYY
1	2	3	4	5	6	7	8	9

FIG. 13.28.

Similarly, the representation of the tens was in line with previous traditions, since it was based on the use of an equivalent number of *chevrons*. However, unlike the Assyro-Babylonians, the scribes of Mari did not use the old sexagesimal character for 60, but carried on multiplying chevrons for the numbers 60, 70, 80, and 90:

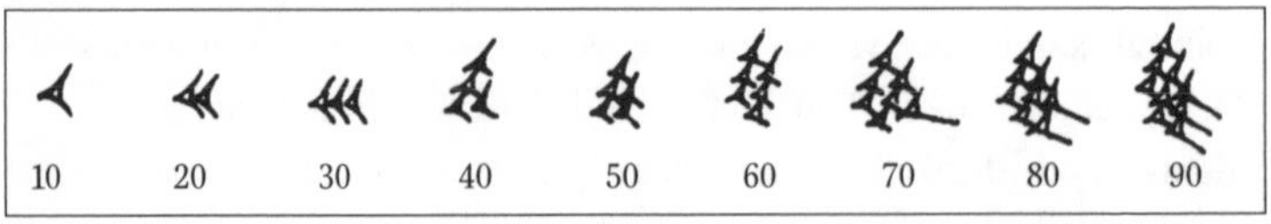

◄	◄◄	◄◄◄						
10	20	30	40	50	60	70	80	90

FIG. 13.29.

For 100, they did not use the old system of a wedge plus the sign for the word for 100 (ME), with the meaning 1 × "hundred": what they used was just the single vertical wedge. The number 200 was figured by two vertical wedges, 300 by three, and so on.

NOTATIONS OF THE HUNDREDS BY THE SCRIBES OF MARI

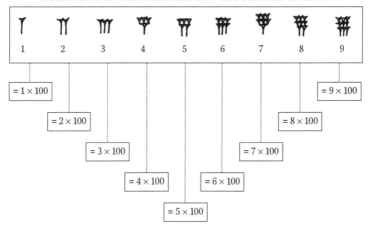

Fig. 13.30.

So a wedge represented either a unit or a hundred depending on where it came in the numerical expression.

For instance, to write "120", "130," and so on, the scribes of Mari put down 1 vertical wedge followed by 2, 3, etc. chevrons. And to represent a number such as 698, all that was needed was a representation of 6 followed by a representation of 98 (9 chevrons and 8 wedges):

[1; 10]	[1; 20]	[1; 30]	[6; 98]
= 1 × 100 + 10 = 110	= 1 × 100 + 20 = 120	= 1 × 100 + 30 = 130	= 6 × 100 + 98 = 698

Fig. 13.31.

It is clear that the scribes of Mari knew both the classical Mesopotamian decimal notation and also the positional sexagesimal system of the scholars (see below). When they drew up their tablets in Akkadian (a language which they handled with ease), they used the former for "current business" such as economic and legal documents, and the latter for "scientific" matters (tables, mathematical problems, and so on). In fact, the system we are now considering never was the official numeral-system of the city: for it is only found in quite particular places on the tablets (on the edges, on the reverse side, and in the margins) and, in most cases, the numbers worked out in the new system were written out again in one or the other of the two standard systems.

In other words, the new system seems to have served only as an aide-mémoire and checking device, to make doubly sure that the results written out in the traditional way were in fact correct. What we see is a kind of mathematical bilingualism, in which matching results reached by two separate notations resolve doubts about the correctness of the sums. And it is of course only because of the role that the system played, and because of the position of the new-style numerals on the tablets, that modern scholars have been able to read and interpret them.

The following examples come from the Royal Archives of Mari, as quoted, translated and decoded by D. Soubeyran. The first gives the last column of a tally of people, showing the totals for rows identified by the words in brackets, which refer to the categories of people counted:

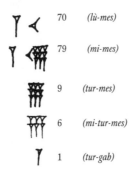

	70	(lù-mes)
	79	(mi-mes)
	9	(tur-mes)
	6	(mi-tur-mes)
	1	(tur-gab)

FIG. 13.32.

These numbers are written in classical Akkadian manner, so they represent: 70 + 79 + 9 + 6 + 1 = 165. However, after a space and before the title of the tablet comes the following expression:

FIG. 13.33A.   1   65

If this were a non-positional expression, its value would either be 1 + 65 = 66, or (allowing the vertical wedge to mean 60, as it often did in Akkadian arithmetic) 60 + 65 = 125. In neither case could it be a running total for the column which it follows after a space. However, if the wedge is given the value 100, then we do indeed get the running total of 165, in the following manner:

FIG. 13.33B.   [1;   65]   = 1 × 100 + 65 = 165

The second example is also a list of people, perhaps of nobles. Each entry is accompanied by a number, which perhaps indicates the number of servants owned. There is a running total of 183 brought forward, which is written in classical decimal form as

1 ME-AT 83 ("1 hundred 83")

A second subtotal gives the figure of 26 servants, and, as you might expect, the grand total comes to 209, which is expressed in the same way as:

2 ME-TIM 9 (2 × 100 + 9 = 209)

However, the side-edge of the tablet has the following expression:

FIG. 13.34A.   1   85

If this were taken in the classical (non-positional) way, then it would mean either 1 + 85 (= 86) or 60 + 85 (= 145), and the totals would not match at all.

However, if the figures on the edge are taken as a centesimal positional expression, then the sum is 185, which is roughly the same as the first count of servants, 183.

FIG. 13.34B.   [1;   85]   = 1 × 100 + 85 = 185

The last of the three tablets details a sequence of deliveries of copper scythes, with a running total, written in the standard way, of 471 scythes. But between the markings for the month and the year, there is this:

FIG. 13.35A.   4   76

Once again, this expression would not have much meaning if it were read in the classical manner, but, taking it as an expression in the positional system of the scribes of Mari, it would give 476, a good approximation of the previous running total (471):

FIG. 13.35B.   [4;   76]   = 4 × 100 + 76 = 476

According to Soubeyran, the minor discrepancies between these figures and the totals, as well as their position on the tablets and the rough and ready way they are written, shows that they are rough drafts or workings-out, intended to check figures before they were inscribed on the tablets in a formal way. That makes it all the more interesting to see the scribes of Mari *thinking* in a positional, centesimal-decimal system, before converting their results into sexagesimal notation.

For numbers between 100 and 1,000, the Mari system used the *rule of position*, and its base was not 10, but 100: the first "large" unit was the hundred, with the ten playing the role of auxiliary base. On the other hand, the system did not have a zero. If it had had such a thing, then it would have served to mark the absence of units in a given order. In other words, if there had been a zero in the Mari system, then the multiples of the base would have been written in the same way as we write multiples of our base (20, 30, 40, etc.), with a zero indicating the absence of units of the first order.

All the same, the scribes of Mari were perfectly aware that the value of the numerals they wrote down depended on their position in a specific numerical expression. This is all the more noteworthy because very few civilisations have ever reached such a degree of simplification in written numerals, and by the same token discovered the rule of position. This development took place very early on: the tablets that bear the trace of the rule of position are not later than the eighteenth century BCE.

However, the system was not strictly or consistently positional. Had that been the case, then 1,000 (= 10 × 100, or ten units of the second centesimal order) would have been represented by a chevron, 2,000 by two chevrons, and so on. As for 10,000, the square of the base of the second centesimal order, it would have been represented by a vertical wedge (had there been a zero, it would have been figured in the form [1; 0; 0], the first zero signifying the absence of any units of the first order (numbers between 1 and 99), the second the absence of units of the second order (multiples of 100 by a number between 1 and 99)). And since 200 = 2 × 100, represented by two vertical wedges, so 20,000 = 2 × 10,000 would similarly have been represented by two vertical wedges.

But it was not so: the Mari system had special signs for 1,000 and for 10,000. However, the "Mari thousand" was rather different from the classical numeral, and it was combined with a multiplier to make numbers like 2,000, 3,000, etc.:

	LI-IM	2 LI-IM
FIG. 13.36.	1,000	2,000

This adds up to a mixed system, using simultaneously all the basic rules, of addition (for the total), of multiplication (for the thousands), and of position (for numbers less than 1,000).

The Mari scribes used a figure derived from the thousand overlaid with a chevron (= 10) to represent 10,000 (which was then combined with units for multiples of 10,000):

FIG. 13.37.   10,000   =   1,000   ×   10

This is the only example amongst the decimal numerations of the whole Mesopotamian region where 10,000 is not written as an analytical combination of the numerals 10 and 1,000, and it is yet another way in which the Mari system is quite unique.

The Marian cuneiform sign for 10,000 (found not just in economic tablets, but in fields as diverse as tallies of bricks, of land areas, and of livestock) is related to the Sumerian ideogram GAL, which meant "large", and was pronounced *ribbatum* in the language of Mari, with the literal meaning of "multitude", whence "large number". So that was the *name* of the number 10,000, and it is clearly the same name as the one found at Ebla (*ri-bab*) in the twenty-fourth century BCE, at Ugarit (*r(b)bt*) in the fifteenth century BCE, and then in Syria (*ri-ib-ba-at*), and in Hebrew (*ribō*, pl. *ribōt*).

The following two examples from tablets found at Mari give a fuller view of how the system worked:

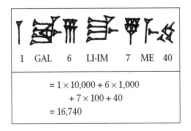

| 1 | GAL | 6 | LI-IM | 7 | ME | 40 |

= 1 × 10,000 + 6 × 1,000
+ 7 × 100 + 40
= 16,740

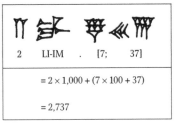

| 2 | LI-IM | . | [7; | 37] |

= 2 × 1,000 + (7 × 100 + 37)

= 2,737

FIG. 13.38. *See Durand (1987)*   FIG. 13.39. *See Soubeyran (1984)*

In etymological and graphological terms, 10,000 was the "biggest number" in the system of Mari. (The scribes could of course represent far larger numbers by using the multiplication principle, even if no really large numbers have yet been found in the tablets.) It was a quite unique centesimal numeral system, found exclusively in this one city on the common border of Syria and Mesopotamia, at the time of the patriarch Abraham. It might have developed into a fully positional system had the Babylonian king Hammurabi not razed the city to the ground in 1755 BCE, and buried with it a very large part of Mari's culture. Ironically, it was the Babylonians themselves who actually devised the world's first true positional system – but it was neither a variant of Akkadian decimal arithmetic, nor a centesimal system like that of Mari. Used for mathematical and astronomical reckonings right down to the dawn of the Common Era, the "learned" numerals of Babylon were a direct inheritance of Sumer, whose memory they have perpetuated, directly and indirectly, right down to the present day.

THE POSITIONAL SEXAGESIMAL SYSTEM OF THE LEARNED MEN OF MESOPOTAMIA

Although we cannot be sure about the exact date, the first real idea of a positional numeral system arose amongst the mathematicians and astronomers of Babylon in or around the nineteenth century BCE.

The Mesopotamian scholars' abstract numerals were derived from the ancient Sumerians' sexagesimal figures, but constituted a system far superior to anything else in the ancient world, anticipating modern notation in all respects save for the different base and the actual shapes used for the numerals.

Unlike the "ordinary" Assyro-Babylonian notation used for everyday business needs, the learned system used base 60 and was strictly positional. Thus a group of figures such as

$$[3; 1; 2]$$

which in modern decimal positional notation would express:

$$3 \times 10^2 + 1 \times 10 + 2$$

signified to Babylonian mathematicians and astronomers:

$$3 \times 60^2 + 1 \times 60 + 2$$

Similarly, the sequence [1; 1; 1; 1] which in our system would mean $1 \times 10^3 + 1 \times 10^2 + 1 \times 10 + 1$ (or $1,000 + 100 + 10 + 1$) signified in the Babylonian system $1 \times 60^3 + 1 \times 60^2 + 1 \times 60 + 1$ (or $216,000 + 3,600 + 60 + 1$).

Instances of this system of numerals have been known since the very dawn of Assyriology, in the mid-nineteenth century, and, thanks to excavations made throughout Mesopotamia and Iraq at that time, many examples have come to rest in the great European museums (Louvre, British Museum, Berlin) and in the university collections at Yale, Columbia, Pennsylvania, etc. The types of document in which the learned system is used (and which come from Elam and Mari, as well as from Nineveh, Larsa, and other Mesopotamian cities) are for the most part as follows: tables intended to assist numerical calculation (e.g. multiplication tables, division tables, reciprocals, squares, square roots, cubes, cube roots, etc.); astronomical tables; collections of practical arithmetical and elementary geometrical exercises; lists of more or less complex mathematical problems.

The system is sexagesimal, which is to say that 60 units of one order of magnitude constitute one unit of the next (higher) order of magnitude. The numbers 1 to 59 constitute the units of the first order, multiples of 60 constitute the second order, multiples of 3,600 (sixty sixties) constitute the third order, multiples of 216,000 (the cube of 60) constitute the fourth order, and so on.

In fact, there were really only two signs in the system: a vertical wedge representing a unit, and a chevron representing 10:

1   10

Numbers from 1 to 59 inclusive were built on the principle of addition, by an appropriate number of repetitions of the two signs. Thus the numbers 19 and 58 were written

(1 chevron + 9 wedges)   or   (5 chevrons + 8 wedges)

So far the system is exactly the same as its predecessors. However, beyond 60, the learned system became strictly positional. The number 69, for instance, was not written

60   9   but   [1;   9]

For example, this is how Asarhaddon, king of Assyria from 680 to 669 BCE, justified his decision to rebuild Babylon (wrecked by his father Sennacherib in 689 BCE) rather sooner than the holy writ prescribed:

> After inscribing the number 70 for the years of Babylon's desertion on the Tablet of Fate, the God Marduk, in his pity, changed his mind. He turned the figures round and thus resolved that the city would be reoccupied after only eleven years. [From *The Black Stone*, trans. J. Nougayrol]

The anecdote takes on its full meaning only in the light of Babylonian sexagesimal numbering. To begin with, Marduk, chief amongst the gods in the Babylonian pantheon, decides that the city will remain uninhabited for 70 years, and, to give full force to his decision, inscribes on the Tablet of Fate the signs:

Fig. 13.40A.   [1;   10]   ([1; 10] = 1 × 60 + 10)

Thereafter, feeling compassion for the Babylonians, Marduk inverts the order of the signs in the expression, thus:

FIG. 13.40B.   10 . 1   (= 10 + 1)

Since the new expression represents the number 11, Marduk decreed that the city would remain uninhabited only for that length of time, and could be rebuilt thereafter. The anecdote shows that the Mesopotamian public in general was at least aware of the rule of position as applied to base 60.

In the Babylonian system, therefore, the value of a sign varied according to its position in a numerical expression. The figure for 1 could for instance express

- a unit in first position from the right,
- a sixty in the second position,
- sixty sixties or 60^2 in third position,

and so on.

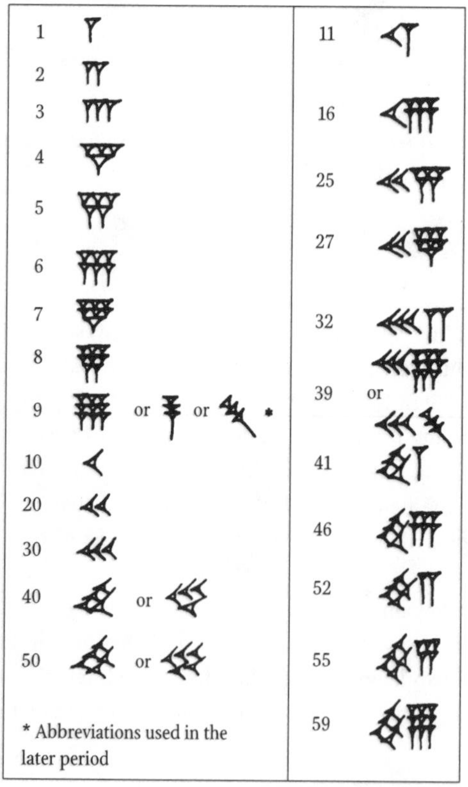

FIG. 13.41. *Representations of the fifty-nine significant units of the learned Mesopotamian numeral system*

For instance, to write the number 75 (one sixty and fifteen units) you put a "15" in first position and a "1" in second position, thus:

𒁹 𒌋𒌋𒌋   (= 1 × 60 + 15 = 75)

[1;   15]

FIG. 13.42.

And to write 1,000 (16 sixties and 40 units) you put a "40" in first position and a "16" in second position, thus:

𒌋𒌋𒌋 𒐏   (= 16 × 60 + 40 = 1,000)

[16   ;   40]

FIG. 13.43.

Conversely, an expression such as

𒐏𒌋𒌋𒌋 𒌋𒌋 𒐏𒁹

[48   ;   20   ;   12]

FIG. 13.44.

expresses the number:

$$48 \times 60^2 + 20 \times 60 + 12 = 48 \times 3,600 + 20 \times 60 + 12 = 174,012$$

in exactly the same way as we would express "174,012 seconds" as:

48 h 20m 12s

Similarly, an expression such as

𒁹 𒐏𒌋𒌋𒌋 𒌋𒌋𒌋𒌋 𒌋𒌋𒌋

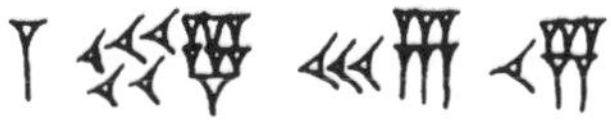

[1; 50 + 7 ; 30 + 6 ; 10 + 5] or [1; 57; 36; 15]

FIG. 13.45.

symbolises, in the minds of the Babylonian scholars, the number:

$$1 \times 60^3 + 57 \times 60^2 + 36 \times 60 + 15 \ (= 423,375)$$

The next examples come from one of the most ancient Babylonian mathematical tablets known (British Museum, BM 13901, dating from the period of the first kings of the Babylonian Dynasty), a collection of problems relating the solution of the equation of the second degree:

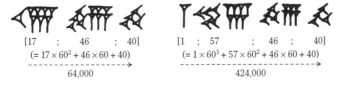

[17   ;   46   ;   40]
$(= 17 \times 60^2 + 46 \times 60 + 40)$
----------------→
64,000

FIG. 13.46.

[1   ;   57   ;   46   ;   40]
$(= 1 \times 60^3 + 57 \times 60^2 + 46 \times 60 + 40)$
----------------→
424,000

FIG. 13.47.

The difference between Sumerian numbers and the Babylonian "learned" system was simply this: the Sumerians relied on addition, the Babylonians on the rule of position. This can easily be seen by comparing the Sumerian and Babylonian expressions for the two numbers 1,859 and 4,818:

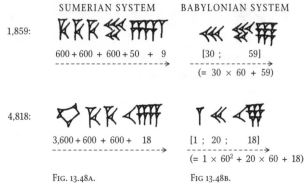

1,859:

$600 + 600 + 600 + 50 + 9$

$[30 ; \quad 59]$

$(= 30 \times 60 + 59)$

4,818:

$3,600 + 600 + 600 + 18$

$[1 ; \quad 20 ; \quad 18]$

$(= 1 \times 60^2 + 20 \times 60 + 18)$

FIG. 13.48A.   FIG 13.48B.

THE TRANSITION FROM SUMERIAN TO LEARNED BABYLONIAN NUMERALS

One of the reasons for the "invention" of the learned Babylonian system is easy to understand – it was the "accident" which gave 1 and 60 the same written sign in Sumerian, and which originally constituted the main difficulty of using Sumerian numerals for arithmetical operations.

Moreover, the path to the discovery of positionality had been laid out in the very earliest traces of Sumerian civilisation. The two basic units were represented, first of all, by the same name, *geš* (see Fig. 8.5A and 8.5B above); then, in the second half of the fourth millennium BCE, they were represented by objects of the same shape (the small and large cone) (see Fig. 10.4 above); then, from 3200–3100 BCE to the end of the third millennium, by two figures of the same general shape, the narrow notch and the thick notch (see Fig. 8.9 above); then, from around the twenty-seventh century BCE, by cuneiform marks of the same type, distinguished only by their respective sizes; and, finally, from the third dynasty of Ur onwards (twenty-second to twentieth century BCE), especially in the writings of Akkadian scribes, by the same vertical wedge.

In other words, as we can see from Asarhaddon's story in *The Black Stone*, and in the Assyro-Babylonian representations of the numbers 70, 80 and 90 (see Fig. 13. 23 above), the large wedge meaning 60 had evolved in line with the general evolution of cuneiform writing so as to be indistinguishable from the small wedge meaning 1.

In everyday usage, that evolution was seen as a problem, which was got round by "spelling out" 60 as *shu-shi* in numbers such as 61, 62, 63, where the confusion was potentially greatest (see Fig. 13.14 above), and eventually by replacing the sexagesimal unit with a multiple of a decimal one (Fig. 13.18 above).

But in the usage of the learned men of Mesopotamia, the graphical equivalence of the signs for 1 and 60 gave rise (at least for numbers with two orders of magnitude) to a true rule of position. As the following notations show:

SUMERIAN SYSTEM	SUMERIAN-AKKADIAN SYNTHESIS	LEARNED BABYLONIAN SYSTEM
 $60 + 50 + 7$	 $60 + 50 + 7$	 $[1 ; (50 + 7)]$
 $60 + 60 + 40 + 1$ $60 + 60$	 $60 + 60 + 40 + 1$ $60 + 60$	 $[4 ; (40 + 1)]$

FIG. 13.49.

Babylonian scholars realised therefore that the rule or principle could be generalised to represent all integers, provided that the old Sumerian signs for the multiples and powers of 60 were abandoned. The first to go was the 600 (= 60 × 10), for which was substituted as many chevrons (= 10) as there were 60s in the number represented. Then the sign for 3,600 (the square of 60) was dropped, and, since this number was a unit of the third sexagesimal order, it was henceforth represented by a single vertical wedge. Subsequently the sign for 36,000 was eliminated, and replaced by the sign for 10 in the position reserved for the third sexagesimal order, and so on.

For instance, instead of representing the number 1,859 by three signs for 600 followed by the notation of the number 59 (1,859 = 3 × 600 + 59), Babylonian scholars now used [30; 59] (= 30 × 60 + 59), as shown in Fig. 13.48 above, which also gives the example of the "old" and "new" representations of 4,818.

The vertical wedge thus came to represent not only the unit, but any and all powers of 60. In other words, 1 was henceforth figured by the same wedge that signified 60, 3,600, 216,000, and so on, and all 10-multiples of the base (600, 36,000, 2,160,000, etc.) by the chevron.

The discovery was extremely fruitful in itself, but, because of the very circumstances in which it arose, it gave rise to many difficulties.

THE DIFFICULTIES OF
THE BABYLONIAN SYSTEM

Despite their strictly positional nature and their sexagesimal base, learned Babylonian numerals remained decimal and additive within each order of magnitude. This naturally created many ambiguous expressions and was thus the source of many errors. For example, in a mathematical text from Susa, a number [10; 15] (that is to say, $10 \times 60 + 15$, or 615) is written thus:

[10 ; 15]

FIG. 13.50A.

However, this expression could also just as easily be read as

[25]   or   [10 ; 10 ; 5] $(= 10 \times 60^2 + 10 \times 60 + 5)$

FIG. 13.50B.

It is rather as if the Romans had adopted the rule of position and base 60, and had then represented expressions such as "10° 3′ 1‴" $(= 36{,}181″)$ by the Roman numerals X III I, which they could easily have confused with XI II I (11° 2′ 1″), X I III (10° 1′ 3″) , and so on. Scribes in Babylon and Susa were well aware of the problem and tried to avoid it by leaving a clear space between one sexagesimal order and the next. So in the same text as the one from which Fig. 13.50 is transcribed, we find the number [10; 10] $(= 10 \times 60 + 10)$, represented as:

[10   ;   10]

FIG. 13.51.

The clear separation of the two chevrons eliminates any ambiguity with the representation of the number 20.

In another tablet from Susa the number [1; 1; 12] $(= 1 \times 60^2 + 1 \times 60 + 12)$ is written

[1   ;   1   :   12]

FIG. 13.52A.

in which the clear separation of the leftmost wedge serves to distinguish the expression from

$$[2 \; ; \; 12] \qquad (= 2 \times 60 + 12)$$

FIG. 13.52B.

In some instances scribes used special signs to mark the separation of the orders of magnitude. We find double oblique wedges, or twin chevrons one on top of the other, fulfilling this role of "order separator"*:

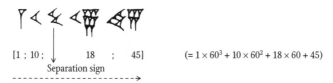

FIG. 13.53.

Here are some examples from a mathematical tablet excavated at Susa:

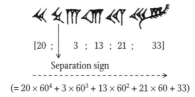

$$[1 \; ; \; 10 \; ; \downarrow \quad 18 \quad ; \quad 45] \qquad (= 1 \times 60^3 + 10 \times 60^2 + 18 \times 60 + 45)$$

Separation sign

FIG. 13.54A.

$$[20 \; ; \downarrow \quad 3 \; ; \; 13 \; ; \; 21 \; ; \quad 33]$$

Separation sign

$$(= 20 \times 60^4 + 3 \times 60^3 + 13 \times 60^2 + 21 \times 60 + 33)$$

FIG. 13.54B.

The sign of separation makes the first number above quite distinct from the representation of $[1; 10 + 18; 45]$ $(= 1 \times 60^2 + 28 \times 60 + 45)$; and for the same reason the second number above cannot be mistaken for $[20 + 3; 13; 21; 33]$ $(= 23 \times 60^3 + 13 \times 60^2 + 21 \times 60 + 33)$.

This difficulty actually masked a much more serious deficiency of the system – the *absence of a zero*. For more than fifteen centuries, Babylonian mathematicians and astronomers worked without a concept of or sign for zero, and that must have hampered them a great deal.

In any numeral system using the rule of position, there comes a point where a special sign is needed to represent units that are missing from the number to be represented. For instance, in order to write the number *ten* using (as we now do) a decimal positional notation, it is easy enough to place the sign for 1 in second position, so as to make it signify one unit of the higher (decimal) order – but how do we signify that this sign is indeed

---

* In commentaries on literary texts, the same sign was used to separate head-words from their explications; in multilingual texts, the sign was used to mark the switch from one language to another; and in lists of prophecies, the sign was used to separate formulae and to mark the start of an utterance.

in second position if we have nothing to write down to mean that there is nothing in the first position? *Twelve* is easy – you put "1" in second position, and "2" in first position, itself the guarantee that the "1" is indeed in second position. But if all you have for ten is a "1" and then nothing ... The

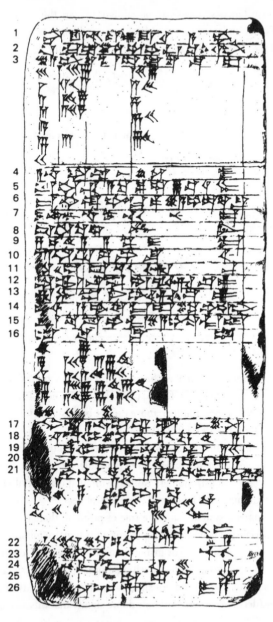

FIG. 13.55. *Important mathematical text from Larsa (Senkereh), dating from the period of the First Babylonian Dynasty (Louvre, AO 8862, side IV). See Neugebauer, tablet 38. Beneath line 16, note the representation of the number 18,144,220 as [1; 24; blank space; 3; 40].*

problem is obviously acute. Similarly, to write a number like "seven hundred and two" in a decimal positional system, you can easily put a "7" in third position and a "2" in first position, but it's not easy to tell that there's an arithmetical "nothing" between them if there is indeed *no thing* to put between them.

It became clear in the long run that such a *nothing* had to be represented by *something* if confusion in numerical calculation was to be avoided. The something that means nothing, or rather the sign that signifies the absence of units in a given order of magnitude, is, or would one day be represented by, zero.

The learned men of Babylon had no concept of zero around 1200 BCE. The proof can be seen on a tablet from Uruk (Louvre AO 17264) which gives the following solution:

"Calculate the square of 𝕋 ◀◀𝕎 *and you get*𝕋𝕋 𝕎   "

In decimal numbers using the rule of position, the first of these expressions ($2 \times 60 + 27$) is equal to 147, and the square of 147 is 21,609. This latter number can be expressed in sexagesimal arithmetic as $6 \times 3,600 + 0 \times 60 + 9$, and should therefore be written in learned Babylonian cuneiform numbers with a "9" in first position, a "6" in third position, and "nothing" in second position. If the scribe had had a concept of zero he would surely have avoided writing the square of [2; 27] as the expression [6; 9] which we see on the tablet – since the simplest resolution of [6; 9] is $6 \times 60 + 9 = 369$, which is not the square of 147 at all!

Another example of the same kind can be found on a Babylonian mathematical tablet from around 1700 BCE (Berlin Archaeological Museum, VAT 8528), where the numbers [2; 0; 20] ($= 2 \times 60^2 + 0 \times 60 + 20 = 7,220$) and [1; 0; 10] ($= 1 \times 60^2 + 0 \times 60 + 10 = 3,610$) are represented by

2 ; 20   1 ; 10

FIG. 13.56.

These notations are manifestly ambiguous, since they could represent, respectively, [2; 20] ($= 2 \times 60 + 20 = 140$) and [1; 10] ($= 1 \times 60 + 10 = 70$).

To overcome this difficulty, Babylonian scribes sometimes left a blank space in the position where there was no unit of a given order of magnitude. Here are some examples from tablets excavated at Susa (examples A, B, C) and from Fig. 13.58 below (example D, line 15). Our interpretations are not speculative, since the values given correspond to mathematical relations that are unambiguous in context:

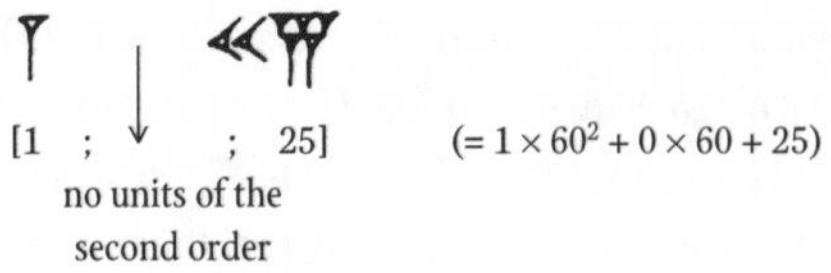

[1 ;   ↓   ;   25] $(= 1 \times 60^2 + 0 \times 60 + 25)$
            no units of the
            second order

FIG. 13.57A.

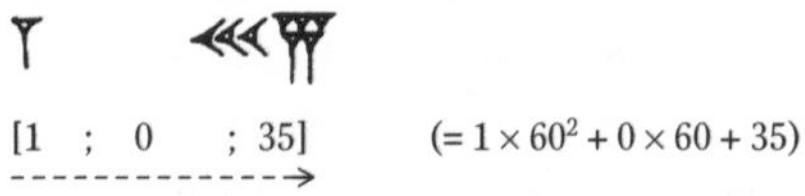

[1 ;   0   ; 35] $(= 1 \times 60^2 + 0 \times 60 + 35)$
------------→

FIG. 13.57B.

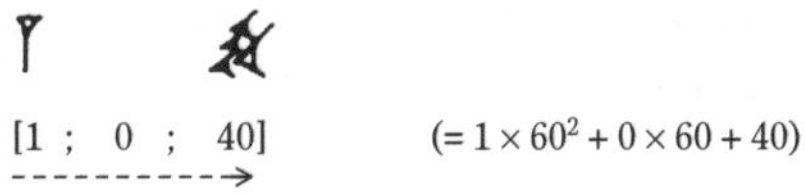

[1 ;   0   ;   40] $(= 1 \times 60^2 + 0 \times 60 + 40)$
----------→

FIG. 13.57C.

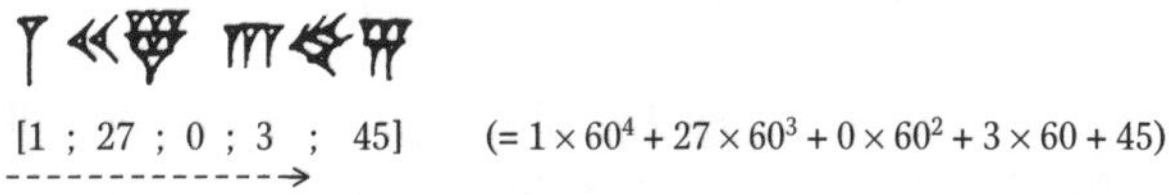

[1 ; 27 ; 0 ; 3 ;   45] $(= 1 \times 60^4 + 27 \times 60^3 + 0 \times 60^2 + 3 \times 60 + 45)$
------------→

FIG. 13.57D.

However, this did not solve the problem entirely. For a start, scribes often made mistakes or did not bother to leave the space. Secondly, the device did not allow a distinction to be made between the absence of units in one order of magnitude, and the absence of units in two or more orders of magnitude, since two spaces look much the same as one space. And finally, since the figure for 4, for instance, could mean 4×60, 4×60^2, 4×60^3, or 4×60^4, how could you know which order of magnitude was meant by a single expression?

These difficulties were compounded by fractions. Whereas their predecessors had given each fraction a specific sign (see Fig. 10.32 above for an example from Elam), the Babylonians used the rule of position for fractions whose denominator was a power of 60. In other words, positional sexagesimal notation was extended to what we would now call the negative powers of 60 ($60^{-1} = 1/60$, $60^{-2} = 1/60^2 = 1/3,600$, $60^{-3} = 1/60^3 = 1/216,000$, etc.). So the vertical wedge came to signify not just 1, 60, 60^2, etc., but also $1/60$, $1/3,600$, and so on. Two wedges could mean 2 or 120 or $1/30$ or $1/1,800$; the figure signifying 15 could also signify $1/4$ (= $15/60$), and the number 30 might just as easily mean $1/2$.

Numerals were written from right to left in ascending order of the powers of 60, and from left to right in ascending negative powers of 60, exactly as we now do with our decimal positional numbering – except that in Babylon there was nothing equivalent to the decimal point that we now use to separate the integer from the fraction.

LINE	TA-KI-IL-TI ŠI-LI-IP - TIM ŠA NA-AS-SÀ-HU-Ú-[M]A SAG-I ... -Ú	ÍB-SÁ SAG	ÍB-SÁ ŠI-LI-IP-TIM	MU-BI-IM	
1					
2					
3	[1; 59, 0,] 15	1; 59	2, 49	KI	1
4	[1; 56, 56,] 58, 14, 50, 6, 15	56, 7	3, 12; 1	KI	2
5	[1; 55,] 7, 41, 15, 33, 45	1, 16, 41	1, 50, 49	KI	3
6	[1; 5] 3, 10, 29, 32, 52, 16	3, 31, 49	5, 9, 1	KI	4
7	[1]; 48, 54, 1, 40	1, 5	1, 37	KI	5
8	[1]; 47, 6, 41, 40	5, 19	8, 1	KI	6
9	[1]; 43, 11, 56, 28, 26, 40	38, 11	59, 1	KI	7
10	[1]; 41, 33, 59, 3, 45	13, 19	20, 49	KI	8
11	[1]; 38, 33, 36, 36	9, 1	12, 49	KI	9
12	[1]; 35, 10, 2, 28, 27, 24, 26, 40	1, 22, 41	2, 16, 1	KI	10
13	[1]; 33, 45	45	1, 15	KI	11
14	[1]; 29, 21, 54, 2, 15	27, 59	48, 49	KI	12
15	[1]; 27, ★, 3, 45	7, 12; 1	4, 49	KI	13
16	[1]; 25, 48, 51, 35, 6, 40	29, 31	53, 49	KI	14
17	[1]; 23, 13, 46, 40	56	53	KI	

★ Blank space indicating the absence of units in a given order of magnitude

Fɪɢ. 13.58. *Mathematical tablet, 1800–1700 BCE, showing that Babylonian mathematicians were already aware of the properties of right-angled triangles (Pythagoras' theorem). If we take the numbers in the leftmost column A, the second column B, and the third column C, we find that the numbers obey the relationship*

$$A = \frac{a^2}{c^2};\ B = b;\ C = c,\ and\ a^2 = b^2 + c^2$$

This expresses the relationship by which in a right-angled triangle (with sides b and c and hypotenuse a) the square of the hypotenuse is equal to the sum of the squares on the other two sides. Columbia University, Plimpton 322. Author's own transcription

Naturally enough, this led to enormous difficulties, such as are suggested by the following three interpretations (out of many others possible) of a single expression:

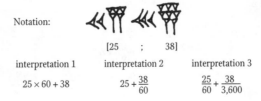

Notation:

$$[25 \quad ; \quad 38]$$

interpretation 1   interpretation 2   interpretation 3

$$25 \times 60 + 38 \qquad 25 + \frac{38}{60} \qquad \frac{25}{60} + \frac{38}{3,600}$$

FIG. 13.59.

All the same, Babylonian mathematicians and astronomers managed to perform quite sophisticated operations for over a thousand years despite the imperfections of their numeral system. Of course, they had the orders of magnitude present in their minds, and the ambiguities of the notation were resolved by the context (that is to say, the premises of the problem being tackled) or by the commentary of the teacher, who must presumably have indicated the magnitudes involved.

FIG. 13.60. *Mathematical tablet from Uruk, late third or early second century BCE, containing one of the earliest known instances of the Babylonian zero. Louvre, AO 6484 side B. See Thureau-Dangin, tablet 33, side B, plate LXII.*

THE BIRTH OF THE BABYLONIAN ZERO

At some point, probably prior to the arrival of the Seleucid Turks in 311 BCE, Babylonian astronomers and mathematicians devised a true zero, to indicate the absence of units of a given order of magnitude. They began to use, instead of a blank space, an actual sign wherever there was a missing order of the powers of 60, and the sign they used was a variant of the old "separator" sign discussed above (see Fig. 13.53):

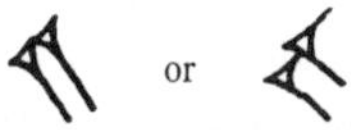

FIG. 13.61.

So, in an astronomical tablet from Uruk (now in the Louvre, AO 6456) from the Seleucid period, we can read:

$$[2; 0; 25; 38; 4]$$
$$(= 2 \times 60^4 + 0 \times 60^3 + 25 \times 60^2 + 38 \times 60 + 4)$$

written on the back of the tablet in the form:

[2 ; 0 ; 25 ; 38 ; 4]

FIG. 13.62.

The diagonal double wedge thus marks the absence of any sexagesimal units of the fourth order of magnitude.

On lines 10, 14 and 24 of the tablet reproduced in Fig. 13.60 above, we can read:

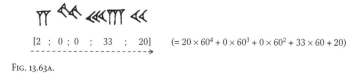

[2 ; 0 ; 0 ; 33 ; 20] $(= 20 \times 60^4 + 0 \times 60^3 + 0 \times 60^2 + 33 \times 60 + 20)$

FIG. 13.63A.

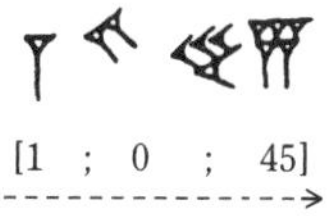

[1 ; 0 ; 45] $(= 1 \times 60^2 + 0 \times 60 + 45)$

FIG. 13.63B.

[1 ; 0 ; 0; 16 ; 40] $(= 1 \times 60^4 + 0 \times 60^3 + 0 \times 60^2 + 16 \times 60 + 40)$

FIG. 13.63C.

$$[1 \; ; \; 0 \; ; \; 7 \; ; \; 30] \qquad (= 1 \times 60^3 + 0 \times 60^2 + 7 \times 60 + 30)$$

FIG. 13.63D.

The Babylonian mathematical documents that have been published to date show the zero only in median positions. For that reason, some historians of science have inferred that Mesopotamian scholars only ever used their zero in intermediate positions and that it would therefore be unwise to treat their zero as identical with ours. They argue that although Mesopotamians wrote expressions such as [1; 0; 3] or [12; 0; 5; 0; 33], they would never have thought of expressions of the form [5; 0] or [17; 3; 0; 0]. More recently, however, O. Neugebauer has shown that Babylonian astronomers used the zero not only in median, but also in initial and terminal positions. For instance, in an astronomical tablet from Babylon (Seleucid period), we find 60 written thus:

$$(= 1 \times 60 + 0)$$

$$[1 \; ; \; 0]$$

FIG. 13.64A.

Here the double slant chevron is used not as a separator, but to mark the absence of units of the first order. On the back of the same tablet, we also find 180 represented in the same way:

$$[3 \; ; \; 0] \quad (= 3 \times 60 + 0)$$

FIG. 13.64B.

And in another astronomical tablet from Babylon of the same period (British Museum, BM 34581), the number

$$[2; 11; 46; 0] \; (= 2 \times 60^3 + 11 \times 60^2 + 46 \times 60 + 0)$$

is represented as:

$$[2 \; ; \; 11 \; ; \; 46 \; ; \; 0]$$

FIG. 13.65.

The final zero in this last example is written in a rather special way, like a "10" with an elongated tail. Has the upper chevron of the zero just been omitted? Is it a scribal decoration? Or just a sign of haste? The latter seems the most likely, since there are other examples in tablets from the same period and the same astronomical source. The following example comes from one such tablet on which the zero is also represented several times in the normal manner:

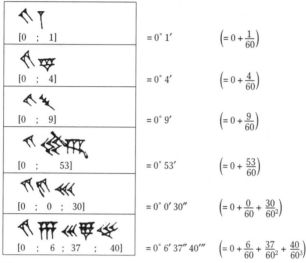

$$[3 \; ; \; 0 \; ; \; 18]$$
$$\dashrightarrow$$
$(= 3 \times 60^2 + 0 \times 60 + 18)$

FIG. 13.66.

The double oblique wedge or chevron in initial position also allowed Babylonian astronomers to represent sexagesimal fractions unambiguously. Here are some examples from the tablet previously quoted:

$[0 \; ; \; 1]$	$= 0° \, 1'$	$\left(= 0 + \frac{1}{60}\right)$
$[0 \; ; \; 4]$	$= 0° \, 4'$	$\left(= 0 + \frac{4}{60}\right)$
$[0 \; ; \; 9]$	$= 0° \, 9'$	$\left(= 0 + \frac{9}{60}\right)$
$[0 \; ; \; 53]$	$= 0° \, 53'$	$\left(= 0 + \frac{53}{60}\right)$
$[0 \; ; \; 0 \; ; \; 30]$	$= 0° \, 0' \, 30''$	$\left(= 0 + \frac{0}{60} + \frac{30}{60^2}\right)$
$[0 \; ; \; 6 \; ; \; 37 \; ; \; 40]$	$= 0° \, 6' \, 37'' \, 40'''$	$\left(= 0 + \frac{6}{60} + \frac{37}{60^2} + \frac{40}{60^3}\right)$

FIG. 13.67.

To summarise: the learned men of Mesopotamia perfected an abstract, strictly positional system of numerals at the latest around the middle of the second millennium BCE, a system far superior to any other in the Ancient World. At a much later date, they also invented zero, the oldest zero in history. Mathematicians seem only to have used it in median position; but the astronomers used it not only in the middle, but also in the final and initial positions of numerical expressions.

THE DATING OF THE EARLIEST ZERO IN HISTORY

As we have seen, there is no zero in scientific texts of the First Babylonian Dynasty, and the figure is hardly attested in any texts prior to the third century BCE. Does that mean that the Mesopotamians only invented the zero in the Seleucid period? That cannot be so easily said, for there are distinctions to be made between the presumed date of an invention, the period of its propagation, and the dates of its first occurrence in texts that have come down to us. It is perfectly possible for an invention to have been made several generations before its use became widespread, just as it is possible for the "oldest documents known to bear a trace" to be several centuries later than the invention itself – either because the earlier documents have perished, or because they have not yet been discovered.

It is therefore legitimate to believe that the Babylonian zero arose several centuries before the third century BCE. This supposition is all the more plausible because we now know that the literary tablets of the Seleucid period are actually copies of much earlier documents (see H. Hunger, 1976): mathematical tablets of the Seleucid period may therefore not all be contemporary texts.

But these are only suppositions. Only further archaeological discoveries can provide definite proof.

HOW WAS ZERO CONCEIVED?

The double wedge or double chevron had the meaning of "void", or rather of the "empty place" in the middle of a numerical expression, but it does not appear to have been imagined as "nothing", that is to say as the result of the operation 10 – 10.

In a mathematical tablet from Susa a scribe tried to explain the result of such an operation, thus:

20 minus 20 comes to ... you see?

Similarly, in another mathematical text from Susa, at the end of an operation (referring to the distribution of grain) where you would expect the sum of 0 to occur, the scribe writes simply that "the grain is finished".

The concepts of "void" and "nothing" both certainly existed. But they were not yet seen as synonyms.

HOW DID BABYLONIAN SCIENTISTS
DO THEIR SUMS?

There are no known accounts of the computational methods used by Babylonian mathematicians and astronomers. These methods can nonetheless be reconstructed from the numerous mathematical texts that have been found and deciphered.

Although the rule of position had been adopted, learned Babylonian numerals remained close to their Sumerian roots in the sense that they remained sexagesimal, with 10 serving as an auxiliary base within each order of magnitude. Now, given that we have proved the existence of a Sumerian abacus and shown what shape it must have had, we can assume fairly safely that the tool was handed down to Babylonian scholars as part of their Sumerian heritage, and used by them for the same purposes. That is very probably how things happened, at least at the beginning of this story.

But there are very good reasons for believing that the rules and the shape of the abacus changed very quickly, and that the method became simpler as the centuries passed.

The simplification of the abacus counting method must have required as its counterpart the memorisation of "tables" for the numbers between 1 and 60 – these tables constituting the necessary mental "baggage" to be able to use the abacus for arithmetical operations.

In fact, the Babylonians never bothered to learn such number-tables by heart: they wrote them out once and for all, and handed the tablets down from generation to generation. Consequently, the mathematical tablets that have been discovered include a great number of multiplication tables.

Fig. 13.68 below is a typical example. The transcription can easily be followed by looking at the face of a clock or watch, and imagining the units of the first order as minutes, and the units of the second order as hours. It can then be seen that the tablet on the left gives the numbers from 1 to 20 followed by 30, 40, and 50, and on the right gives the result of multiplying those numbers by 25. It is therefore a 25-times table, completely analogous to one we could construct using our current decimal system:

1	(times 25 equals)	25
2	(times 25 equals)	50
3	(times 25 equals)	75
4	(times 25 equals)	100
5	(times 25 equals)	125
6	(times 25 equals)	150
7	(times 25 equals)	175
etc.		

Generally speaking, the multiplication tables give the products of a number n (smaller than 60) of the first twenty integers, then of the numbers 30, 40, and 50. This clearly suffices to provide the product of n multiplied by any number between 1 and 60.

With such tables in support, multiplications could be done fairly easily on an abacus.

The rule of position must have led rather quickly to the realisation that wooden tablets of the sort shown in Fig. 12.4 above were no longer necessary, and that the divisions of the Sumerian abacus did not have to be reproduced. All that was now needed was to draw parallel lines to create vertical columns, one for each of the magnitudes of the sexagesimal system. Since clay is easier to work than wood, we can surmise that the columns

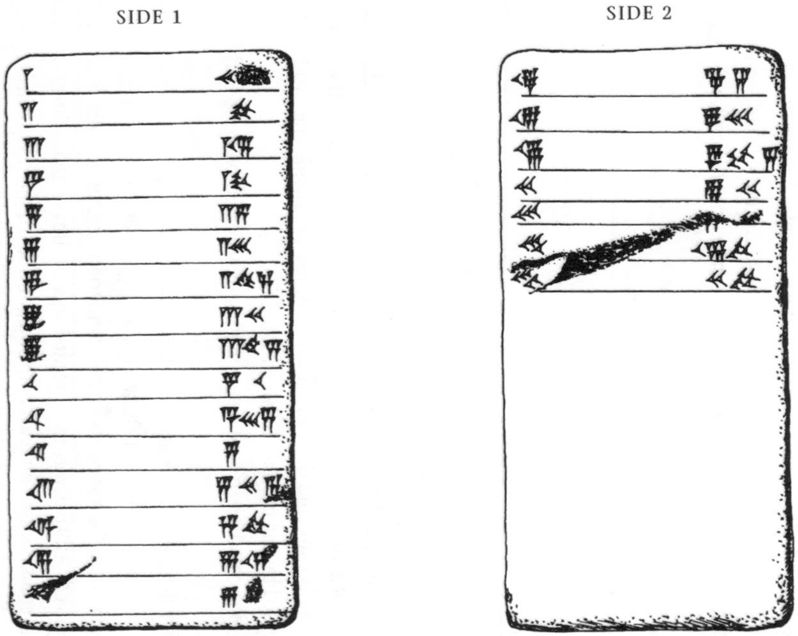

		TRANSCRIPTION	TRANSLATION (decimal positional system)	
SIDE 1	1	[25]	1	25
	2	[50]	2	50
	3	[1;15]	3	75
	4	[1;40]	4	100
	5	[2;05]	5	125
	6	[2;30]	6	150
	7	[2;55]	7	175
	8	[3;20]	8	200
	9	[3;45]	9	225
	10	[4;10]	10	250
	11	[4;35]	11	275
	12	[5;]	12	300
	13	[5;25]	13	325
	14	[5;50]	14	350
	15	[6;15]	15	375
	16	[6;40]	16	400
SIDE 2	17	[7;05]	17	425
	18	[7;30]	18	450
	19	[7;45] *	19	465*
	20	[8;20]	20	500
	30	[12;30]	30	750
	40	[16;40]	40	1,000
	50	[20;50]	50	1,250

*Scribal error

FIG. 13.68. *Twenty-five-times table, Susa, first half of second millennium BCE. Louvre. See MDP, XXXIV, text iv, tablet K*

were drawn onto the wet clay of a tablet made afresh for each calculation. Sticks and tokens would not have been needed any longer, since the numbers involved could be drawn straight onto the clay in the relevant columns and wiped or scored out as the calculation proceeded. This reconstruction of the Mesopotamian abacus is of course only a speculation, but it is in our view a highly plausible one.

Here is an example of how it might have worked, using the multiplication table shown in Fig. 13.68.

The task is to multiply 692 by 25, or, in Babylonian terms, to multiply [11; 32] (= 11 × 60 + 32) by 25.

Let us begin by scoring the first three columns onto the wet clay tablet, in which the result will be entered in the three orders of magnitude, starting from the right (numbers from 1 to 59 will be entered in the rightmost column, multiples of 60 from 1 to 59 in the middle column, and multiples of 3,600 from 1 to 59 in the leftmost column).

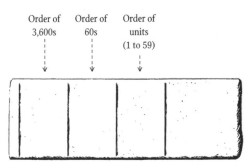

FIG. 13.69A.

To the right of the columns, let us inscribe the multiplicand [11; 32] (= 692) in cuneiform notation:

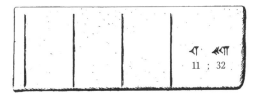

FIG. 13.69B.

Using the 25 × multiplication table, we look for the product of 2; finding 50, we enter that number in cuneiform notation in the units column of the abacus tablet:

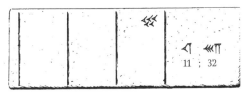

FIG. 13.69C.

We can now rub out the 2 from the multiplicand on the right of the tablet, and proceed to look up 30 in the 25 × multiplication table. The product supplied is [12; 30], so we enter 30 in the rightmost column of the units on our abacus, and 12 in the middle column, reserved for multiples of 60.

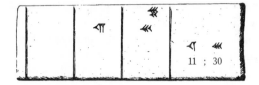

FIG. 13.69D.

So we rub out the 30 from the multiplicand on the right of the tablet, and proceed to look up 11 in the 25 × multiplication table. The product supplied is [4; 35], so we enter 35 in the middle column (since we have changed our order of magnitudes) and 4 in the leftmost column, the one reserved for multiples of 3,600.

So we can now rub out the 11 from the multiplicand, and find that there is nothing left on the right of the tablet. The first stage of the operation is complete.

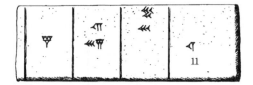

FIG. 13.69E.

The rightmost column now has 8 chevrons in it. Since this is more than the 6 chevrons which make a unit of the next order, we rub out 6 of them and "carry" them into a wedge which we enter in the middle column, leaving 2 chevrons in the units column.

FIG. 13.69F.

So we now have 4 chevrons and 8 wedges altogether in the 60s column. The sum of these being not greater than 60, we simply rub out the numerals in the column and replace them with the numeral signifying 48, the sum of 4 chevrons (4 × 10) and 8 wedges (= 8). And as there is only a 4 in the column of the third order, the result of the multiplication is now fully entered on the abacus:

$$[11; 32] \times 25 = [4; 48; 20]$$
$$(= 4 \times 3,600 + 48 \times 60 + 20 = 17,300)$$

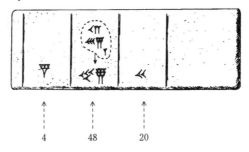

FIG. 13.69G. 4   48   20

The Babylonians also had tables of squares, square roots (Fig. 13.70), cube roots, reciprocals, exponentials, etc., for all numbers from 1 to 59, which enabled far more complex calculations to be performed. For instance, division was done by using the reciprocal table, i.e. to divide one number by another, you multiplied it by its reciprocal.

All this goes to show the great intellectual sophistication of the mathematicians and astronomers of Mesopotamia from the beginning of the second millennium BCE.

TRANSCRIPTION AND
RECONSTRUCTION

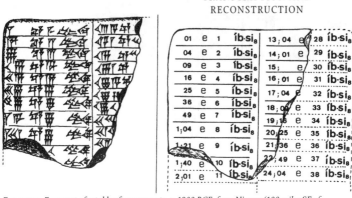

FIG. 13.70. *Fragment of a table of square roots, c. 1800 BCE, from Nippur (100 miles SE of Baghdad). University of Pennsylvania, Babylonian section, CBS 14233 side 2.*

THE BABYLONIAN LEGACY

The abstract system of the learned men of Babylon has had a powerful influence over the scientific world from antiquity down to the present.

From at least the second century BCE, Greek astronomers used the Babylonian system for expressing the negative powers of 60. However, instead of using cuneiform numerals, the Greeks used an adapted version of their own alphabetic numerals. For example, they wrote expressions like 0° 28′ 35″ and 0° 17′ 49″ in the following way:

$$\text{ꚍ KH ΛE} \qquad \left(= 0 + \frac{28}{60} + \frac{35}{60^2}\right)$$
$$[0 \ ; \ 28 \ ; \ 35]$$
$$\dashrightarrow$$

$$\overline{\text{o}} \text{ IZ MΘ} \qquad \left(= 0 + \frac{17}{60} + \frac{49}{60^2}\right)$$
$$[0 \ ; \ 17 \ ; \ 49]$$
$$\dashrightarrow$$

FIG. 13.71.

FIG. 13.72. *Greek astronomical papyrus, second century CE (after 109). Copied from Neugebauer, plate 2*

TRANSCRIPTION

IΘ	...	..	....	IB	.....		KE]
K	...	..	....	IΓ	.....		KϚ]
KA	...	..	....	IΔ	.....		KZ]
KB	...	A	[.·····]	IE	.....		[KH]
KΓ	B MΓ	B	KΔ ΛϚ	IϚ	NB	ME	KΘ
KΔ	Δ Λ	Γ	KϚ Θ	IZ	NΔ	IϚ	Λ
KE	Θ KϚ	Δ	KZ MA	IH		TAYPOY	ΔΙΔΥΜ
KϚ	Ϛ MZ	E	KΘ IΓ	IΘ	○ KΘ NϚ		○K[·]
KZ	H Θ	Ϛ	Λ ΛϚ	K	○ NΘ NB		○M[·]
KH	I NB	H	ΛB IH	KA	○ KB MH		A
KΘ	IB Γ	Θ	ΛΓ N	KB	B NΔ	ΛϚ	B
Λ	IΓ ΛE	I	ΛE KB	KΓ	Δ KΘ		Γ
	IΔ NϚ	IA	ΛE NE	KΔ	E NΘ		Δ

TRANSLATION

19	....	..	....	12	....		25]
20	....	..	....	13	....		26]
21	....	..	....	14	....		27]
22	....	1	....	15	....		[28]
23	2 43	2	24 36	16	52	45	29
24	4 30	3	26 9	17	54	16	30
25	9 26	4	27 41	18		BULLS	GEMINI
26	6 47	5	29 13	19	0 29 56		0 20[.]
27	8 9	6	30 36	20	0 59 52		0 40[.]
28	10 52	8	32 18	21	0 22 48		1
29	12 3	9	34 50	22	2 54 36		2
30	13 35	10	35 22	23	4 29		3
	14 56	11	35 55	24	5 59		4

FIG. 13.73. *Transcription and translation of a Greek astronomical table, from a third-century papyrus. University of Michigan Papyrus Collection, Inv. 924. See J. Garett Winter, pp. 118–20*

GREEK PAPYRI

ꝛ ꝛ ꝛ	♋︎	ο	ꝺ
first century	after + 109	2nd century	467 CE
Pap. Aberdeen	Pap. Lund	Pap. London	Pap. Michigan
No. 128	Inv. 35A	No. 1278	Inv. 1454

FIG. 13.74A. *The "sexagesimal" zero of Greek astronomers*

ARABO-PERSIAN MANUSCRIPTS

♋︎	ↄ	ꝫ	ʃ
+ 1082	+1436	+ 1680	+ 1788
Bodleian Library,	Leyden Univ. Lib,	Princeton, Firestone Library,	
Ms Or. 516	Cod. Or. 187 B	ELS 147	ELS 1203

FIG. 13.74B. *Scribal variants of the "sexagesimal zero" in Arabic and Jewish astronomical texts*

Arab and Jewish astronomers also followed the Greeks' borrowing of the Babylonian system, which they "translated" into their own alphabetic numerals, giving the following forms for the illustrative numbers shown in Fig. 13.71 above:

FIG. 13.75A.   FIG. 13.75B.

Thus the learned Babylonian system has come down to us and is perpetuated in the way we express measures of time in hours, minutes and seconds and in the way we count arcs and angles, despite the strictly decimal nature of the rest of our numerals and metric weights and measures. It is largely due to the Arabs that the system was transmitted to modern times.

FIG. 13.76. *Bilingual (Latin-Persian) astronomical table, transcribed by Thomas Hyde, 1665. British Library 757 cc 11 (1), pp. 6–7*

CODES AND CIPHERS IN CUNEIFORM NUMERALS

In some periods and in some fields, the scribes of Susa and Babylon were much given to playing cryptic games with numerals. Some of these games involved numerical transposition, that is to say the use of numerical expressions in lieu of words or ideograms, generally based on some coherent system of "coding", or on complex numerological symbolism.

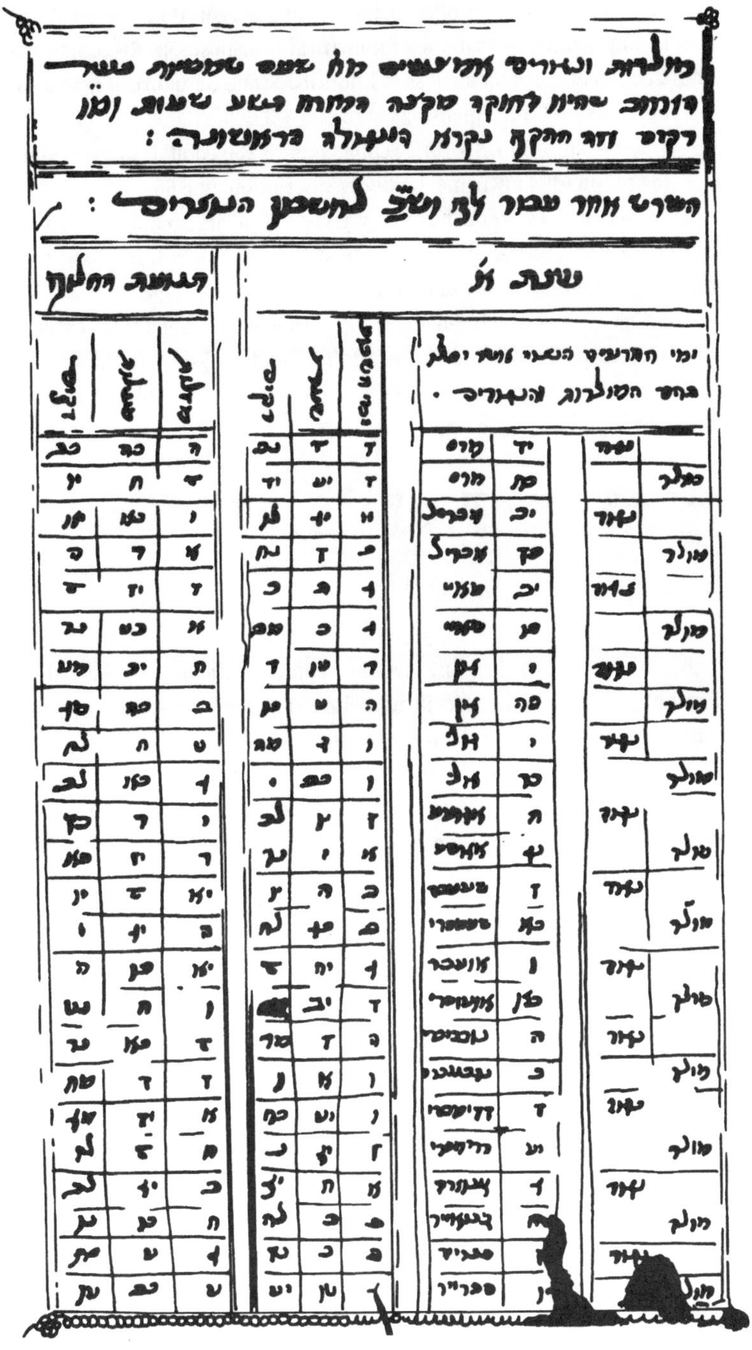

FIG. 13.77. *Astronomical table by Levi Ben Gerson, a French-Jewish savant, 1288–1344 CE. British Museum, Add. 26 921, folio 20b. Transcribed by B. R. Goldstein, table 36.1*

One of the inscriptions of the name of King Sargon II of Assyria (722–705 BCE) provides an example of numerical transposition. Recording the construction of the great fortress of Khorsabad (Dur Šarukin), Sargon says:

> I gave its wall the dimensions of (3,600 + 3,600 + 3,600 + 3,600 + 600 + 600 + 600 + 60 + 3 × 6 + 2) cubits [i.e. 16,280 cubits] corresponding to the sound of my name. [Cylinder-inscription, line 65]

However, this assertion has not yet yielded all its secrets: we cannot reconstitute the coding system by which the name was transposed into numerals from this single example alone.

Another type of number-name game is shown in a tablet from Uruk of the Seleucid period. At the end of the *Exaltation of Ishtar* (published by F. Thureau-Dangin in 1914) the scribe indicates that the tablet belongs to someone called

| 21 | 35 | 35 | 26 | 44 | son of | 21 | 11 | 20 | 42 |

Fig. 13.78.

But who is he? The last line gives his name and the name of his father, but both names are written in numerals. The scribe gives us a puzzle without giving us the key [Thureau-Dangin (1914)].

Numerical cryptograms were also widely used for *haruspicy*, the "secret science" of divination or fortune-telling. Seers and fortune-tellers used several different numerical combinations for mystifying the profane and for ensuring that their magical texts remained impenetrable to the uninitiated (Fig. 13.79). Commenting on the *Esagil* tablet, which gives the dimensions of the great temple of Marduk at Babylon and of the tower of Babel, G. Contenau wrote:

> This difficult text looks on first reading like a bland statement of the dimensions of yards and terraces – a mere sequence of numbers, as on a stock list, with all it has to say stated plainly. However, the scribe has peppered his account with the intercalated formula so often found in hieratic texts:
>
> > *May the initiated explain this to the initiated*
> > *And the uninitiated see it not!*

We should not forget the significant role played by the oral teaching of the pupil by the master which accompanied the lessons of the invariably summary texts themselves. Even texts which appear to be utterly ordinary hid esoteric meanings which we cannot imagine.

The scribes of Susa and Babylon also used cryptograms for word-games, or rather, writing-games, which are worth some attention. For instance, the

FIG. 13.79. *Astrological table with cryptograms in a code that remains to be deciphered. (Line 5, for instance, reads: 3; 5; 2; 1; 12; 4; 31). British Museum, 92685 side 1. Copy made by H. Hunger*

combination "3; 20" is often found used as an ideogram for the word meaning "king", which was pronounced *šàr* or *šarru* in Akkadian. An inscription on a brick from the reign of Šušinak-Šár-Ilâni, king of Susa (twelfth–eleventh century BCE) bears this formulation of the king's name:

ŠUŠINAK – **𝕿𝕴𝕶** – ILÂNI   **𝕿𝕴𝕶** SUSI

3 . 20   3 . 20

("Shushinak-Shar-Ilâni, king of Susa")

Now, why is the numerical combination "3; 20" a logogram for "king"? In Akkadian, the word for "king" was *šár*, pronounced more or less exactly the same way as *šàr*, the name of the "higher" sexagesimal unit of counting in the Sumero-Babylonian system, that is to say 3,600. Elamite scribes thus seem to have made a pun by replacing the word "king" by a numerical combination [3; 20] which represented 3,600 according to a specific rule of interpretation.

But what was the rule? It clearly has nothing to do with the positional sexagesimal system of the learned men of Mesopotamia, since in that way of reckoning [3; 20] = 3 × 60 + 20 = 200, which is not the right answer. On the other hand, we could decompose *šàr* into a kind of "literal" numerical expression that would be represented by "sixty sixties", or, in cuneiform:

60   SHU-SHI

Punning scribes could have written this out, as a game, in this alternative way:

20   20   20   SHU-SHI

or finally as

$$\mathrm{Ⲙ}\,\text{«}\quad \text{SHU-SHI}$$

3 × 20   SHU-SHI

So we can see that Susan scribes regarded the sequence [3; 20] as expressing the product of 3 × 20 (implicitly, × 60), that is to say 3,600, making a pun on *šár*, "the king".

Assyro-Babylonian scribes also used the combination [3; 20] to refer to the king, but they sometimes added a chevron, making [3; 30]. This latter variant cannot be accounted for by 3 × 60 + 30 = 210, nor by (3 × 30) × 60 = 5,400. However, if the addition of a chevron (= 10) to the expression [3; 20] is understood as the mark of a *multiplication* of [3; 20] by 10, then the symbol can be understood as:

$$[3; 30] = [3; 20] \times 10 = 3,600 \times 10 = 36,000$$

This gives the number called *šàr-u* in Sumerian, written in that older system as a 3,600 with an additional chevron in the middle:

3,600   3,600 × 10

The word *šàr-u* (which means "ten *šàr*" or "the great *šàr*") is thus what is meant by the numerical expression

$$\text{𝕸𝕸𝕸⫷⫷⫷} \quad = \quad \text{𝕸𝕸𝕸⫷⫷} \quad \text{⟨}$$

3 ; 30   3 ; 20   ×   10
                                        ŠAR-U

– because this Sumerian number-name has exactly the same sound as the Akkadian word *sharru*, meaning king. So when a scribe referred to the king by writing [3; 30], we can deduce that he meant to say "the great king".

There are many other Babylonian cryptograms which remain unsolved, however. For instance, we have no idea why the concepts of "right" and "left" came to be written by the cuneiform numerals [15] and [2; 30] respectively, nor why [1; 20] was used as an ideogram for "throne", nor, finally, why the vertical wedge, the sign for the unit in the numerical system, had the role of the determiner ("the man who ... ") in the names of the main male functions.

SIDE 1

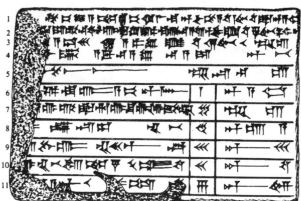

SIDE 2

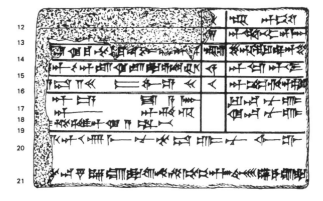

FIG. 13.80A. *Cuneiform tablet listing names of gods and their corresponding numbers. Seventh century BCE, from the "library" of Assurbanipal. British Museum K 170. Trans. J. Bottero*

CODED CRYPTOGRAMS AND
MYSTICAL NUMEROLOGY

Coded cryptograms were also used in theological speculation, for Mesopotamian scribes accorded great weight to the numerical transposition of the names of the gods. Indeed, the religions of the Assyro-Babylonians assumed that the celestial world was a "numerologically harmonious" one, in which the numerical value of a name was an essential attribute of the individual to which it belonged. For this reason, from the early part of the second millennium BCE and consistently throughout the first millennium, some of the Babylonian gods were represented by cuneiform numerals. Fig. 13.80 reproduces a tablet from the seventh century BCE which gives the names of the gods and for each one a number which could be used as that god's ideogram. These are the main points made on the tablet:

  • **ANU**, god of heaven, is attributed the number 60, the higher unit of the sexagesimal Sumerian and Babylonian system, and considered to be the number of perfection, because, the scribe says, "Anu is the first god and the father of all the other gods";

  • **ENLÍL**, god of the earth, is represented by 50;

  • **EA**, god of water, is represented by 40 (elsewhere, she is sometimes ascribed the number 60);

TRANSCRIPTION AND TRANSLATION OF THE TWO RIGHTMOST COLUMNS				
	LINE 6	𒁹		1 or 60	ᵈ A-num
	7			50	ᵈ En-líl
SIDE 1	8			40	ᵈ E-a
	9			30	ᵈ Sîn (name written as 30)
	10			20	ᵈ Shamash
	11			6	ᵈ Adad
	LINE 12			10	ᵈ Bêl ᵈ Marduk (the Lord Marduk)
	13			15	ᵈ Ishtar be-lit ilî (Ishtar queen of the heavens)
SIDE 2	14			En-líl	50 ᵈ Nin-urta, mâr 50 (50 Nin-urta, son of the god Enlíl) (written as 50)
	15			14	ᵈ U + gur, ᵈ Nergal
	16			10	ᵈ Gibil, ᵈ Nusku

FIG. 13.80B.

- Sîn, the lunar god, corresponds to 30, because, the scribe says (line 9, column 1, side 1), "he is the lord of the decision of the month", or, in other words, Sîn is the god who regulates the 30 days of the month
- Shamash, the sun-god, is worth 20;
- Adad here has the number 6 (more frequently, he has the number 10);
- Ishtar, daughter of Anu lord of the heavens and held to be the "queen of the heavens", has the number 15;
- Ninurta, son of Enlíl, has the same number as his father, 50;
- Nergal has the number 14;
- Gibil and Nusku are both represented by 10 because, according to the scribe (line 16, column 1, side B), "they are the companions of god 20 (= Shamash): $2 \times 10 = 20$".

The numerological values of the gods of Babylon had all sorts of consequences. For example, the Babylonian *Creation Epic* concludes with a list of the "names" of **Marduk**, a series of epithets defining his virtues and powers and intended to demonstrate that he is truly the supreme god and the most godly of all. First comes a list of ten names, because Marduk's "number" is 10, then a second group of forty names, because Marduk is the son of Eá, whose number is 40; which adds up to fifty names, because 50 is the number of Enlíl, and the main point of the epic is to show how Marduk replaced Enlíl at the head of the universe of gods and men.

CHAPTER 14

THE NUMBERS OF
ANCIENT EGYPT

Egypt in the time of the Pharaohs had writing and written numerals. They first arose around 3000 BCE, that is to say, at about the same time that words and numbers were first written down in Mesopotamia.

We now know that there were regular contacts between Egypt and Mesopotamia before the end of the third millennium BCE. However, that does not mean that the Egyptians derived their writing or their counting from Sumerian models. Egyptian hieroglyphs, Jacques Vercoutter explains, use signs derived exclusively from the flora and fauna of the Nile basin; moreover, the tools used for making written signs existed in Egypt from the fourth millennium BCE.

The pictograms of Egyptian hieroglyphic writing are very different from Sumerian ideograms, even when we compare signs intended to represent the same idea or object; the shapes of such signs also seem quite unrelated. The media of the two systems likewise have little in common. As we saw, the Sumerians only ever wrote words and numbers by scoring clay tablets with a stylus, or else by pressing a shaped instrument onto wet clay; whereas the Egyptians carved their numerals and hieroglyphs in stone, with hammer and chisel, or else used the bruised tip of a reed to paint them on shards of stone or earthenware, or onto sheets made by flattening the dried-out, fibrous, and fragile stems of the papyrus reed.

Egyptian numerals are also quite different from Sumerian ones from a mathematical point of view. As we have seen, Sumer used a sexagesimal base; whereas the system of Ancient Egypt was strictly decimal.

So if there was something borrowed by Egypt from Sumer, it could only have been the idea of writing down numbers in the first place, and not any part of the way it was done.

Peoples very distant from each other in time and in place but facing similar situations and needs have discovered quite independently some of the same paths to follow, and have arrived at similar, if not identical, results. The Indus civilisation, the Chinese, and the pre-Columbian populations of Central America (Zapotecs, Maya, etc.), were all faced with situations probably very similar to those of the Sumerians, and made much the same mathematical discoveries for themselves. So it seems most sensible to suppose that around the dawn of the third millennium BCE the social, psychological, and economic conditions of Ancient Egypt were such that the invention of writing and of written numerals arose there of its own accord.

In fact, Egyptian society was already advanced, urbanised, and expanding rapidly long before 3000 BCE. Administrative and commercial logic led to the slow realisation that human memory could no longer suffice to fill all the needs of the state without some material support; oral culture must have come up against its natural limits. We must then suppose that the Egyptians felt an increasing need to record and thus to retain thoughts and words, and to fix in a durable form the accounts and inventories of their commercial activities. And, since necessity is the mother of invention, the Egyptians overcame the limits of oral culture by devising a system for writing down words and numbers.

WHAT ARE EGYPTIAN HIEROGLYPHS?

Although they no longer knew how to read them, the Ancient Greeks recognised the signs carved on the many monuments of the Nile Valley (temples, obelisks, tombs, and funeral *stelae*) as "sacred signs" and thus called them *grammata hiera*, or, more precisely, *grammata hierogluphika* ("carved sacred signs"), whence our word, *hieroglyph*. It is from these carved signs, which the Ancient Egyptians considered to be "the expression of the words of the gods", that we have derived our knowledge of the spoken language of Ancient Egypt. The basic writing system for the representation of this language was designed for and was used for the most part only on stone monuments, and it is this writing system (rather than the language it represents) that we commonly call *hieroglyphs*.

HOW TO READ HIEROGLYPHS

Hieroglyphs are very detailed pictograms representing humans, in various positions, all kinds of animals, buildings, monuments, sacred and profane objects, stars, plants, and so on.

to adore	phallus	quail	bird in flight	fish	*uraeus*
woman	bull	screech-owl	beetle	snake	flowering reed
pregnant woman	hare	falcon	bee	hooded viper	lotus

FIG. 14.1. *Some hieroglyphs*

Hieroglyphs may be written in lines from left to right or right to left, or in columns from top to bottom or from bottom to top. The direction of reading is indicated by the orientation of the animate figures (humans or animals) – they are "turned" so as to face the start of the line. So they look left in a text written/read from left to right:

Fig. 14.2.

and they look right in a text written/read from right to left:

Fig. 14.3.

Hieroglyphic signs could be used and understood, first of all, as "integral picture-signs" or *pictograms*: pictures that "meant" what they showed. In the second place, they could also be used as *ideograms*: that is to say, signs meaning something more than, or something connected to, what they showed. For example, an image-sign of a human leg could mean, first, "leg", as a pictogram, but also, as an ideogram, the related ideas or actions of "walking", "running", "running away", etc. Similarly the image of the sun's orb could mean "day", "heat", "light", or else refer to the sun-god. The ideogrammatic interpretation of a sign did not supplant its pictogrammatic meaning, but coexisted alongside it. The interpretation of a hieroglyphic sign is therefore open to infinite subjective variation.

Pictograms and ideograms cannot easily cope with every nuance of language. How can such a system represent actions such as *wishing, desiring, seeking, deserving,* and so on? Or abstract notions like *thought, luck, fear,* or *love*? Moreover, pictograms cannot represent the articulations of spoken language, and are independent of any particular language spoken.

To overcome these limitations, the Ancient Egyptians used their signs in a third way, quite at variance with their pictogrammatic and ideogrammatic values. A sign could also represent the *sound* of the *name* of the thing represented pictographically, and then be used in combination with other *sounds* represented by the ideograms of other words, to make a kind of visual pun or *rebus*. For instance, let us suppose that in Britain today we had only a hieroglyphic writing system and in that system had no pictogram for the things we call "carpets"; but did have a conventional pictogram for "car": 🚗; and, given that we are a nation of dog-lovers, represented the general idea of "pet" by the ideogram: 🐕. Were we to proceed with the system as the Ancient Egyptians did, we would not invent a new pictogram for "carpet" but would create a compound picture-pun

CAR-PET

Such a system has a built-in propensity towards ambiguity. This is not just because ideograms by their very nature have variable interpretations, but also and most especially because a rebus may make a sense in more than one reading of the phonetic value of the ideograms. To take an equally fictitious example from hieroglyphs to be realised by speakers of English, where the pictogram ♠ has the full pictorial meaning of "fir" and the broader meaning "wood" when taken as an ideogram, and the ideogram 🏠 has the meanings of "house", "inn" or "home", the expression (read left to right)

could be realised phonetically as INN-FIR, with the punning meaning of the verb *infer*; or else, read right to left as HOME-WOOD, with the punning meaning of *homeward*. In order to reduce the number of total misapprehensions of that sort, Egyptian hieroglyphs therefore needed an additional sign in each compound expression, a kind of ideogrammatic hint or determiner that showed which way the sound-signs were to be taken. To continue our example, the determiner •🦶• when added to ♠ 🏠 would ensure that it was taken in the directional sense. So

would indeed be read as INFER, whereas

would be read as HOMEWARD.

That is roughly how Egyptian hieroglyphs evolved from pictorial evocations of things to phonetic representations of words. For example: the Ancient Egyptian for "quail chick" was pronounced Wa; the sign depicting a quail chick signified a quail chick, but also represented the sound Wa. Similarly, "seat" was pronounced Pe, and the drawing of a seat came to represent the sound Pe; "mouth" was eR, and a drawing of a mouth meant the syllable eR; a picture of a hare (WeN) stood for the sound WeN, a picture of a beetle (KhePeR) made the sound KhePeR, and so on.

| i | W | P | R | WN | Ḥʀᴘ | Fɪɢ. 14.4. |

Like Hebrew and other Semitic scripts, Egyptian hieroglyphs are consonantal, that is to say they represented only the consonants, leaving the vowels to be "understood" by convention and habit. (Where vowels are put in in modern transcriptions of the language, they are hypothetical and conventional: there is in fact no way of knowing how Ancient Egyptian was actually vocalised.) Since words in Ancient Egyptian contained either one, two, or three consonants, hieroglyphs used as sound-signs also belonged to one of three classes: uniliteral (representing a single consonantal sound), biliteral (representing two sounds), or triliteral (representing a group of three consonants). With their signs used simultaneously as pictograms, as ideograms, and with syllabic value, the Ancient Egyptians were thus able to represent all the words of their language.

An early example is Narmer's Palette (c. 3000–2850 BCE), which commemorates the victory of King Narmer over his enemies in Lower Egypt.

FIG. 14.5. *King Narmer's Palette, from Hieraconpolis, c. 3000–2850 BCE. Cairo Museum*

The king can be seen in the centre of the panel, wielding his club over a captive. The king's name, written in the cartouche above his regal headgear, is composed of the hieroglyphs "fish" and "scissor". The word meaning "fish" was pronounced N'R, and the word meaning "scissor" was pronounced M'R: the two together thus make N'RMR, or Narmer.

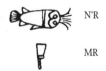

FIG. 14.6.

In similar fashion, the word for "woman", pronounced SeT, was represented by the image of a bolt (the word for "bolt" being a uniliteral with value S) and an image of a piece of bread ("piece of bread" being also a uniliteral, with value T). However, to ensure that S + T was read in the right way, a pictogram of a woman (unrealised in speech) was added as a determiner:

FIG. 14.7.

Likewise the vulture, NeReT in Ancient Egyptian, was represented by N ("stream of water"), R ("mouth") and T ("piece of bread"), plus the determiner, "bird", to ensure that the sound-signs were read as a word belonging to the class of birds.

FIG. 14.8.

Hieroglyphic writing did not use only these kinds of determiners, however. In many cases, biliteral and triliteral signs are disambiguated by a phonetic "complement" which gives a supplementary clue as to how to read the sign. For instance, the hieroglyph of "hare", a word pronounced WeN, would be "confirmed" as meaning the biliteral sound WeN by the addition of the sign for "stream of water", a uniliteral sound pronounced N, as follows:

FIG. 14.9.

It is as if in our imaginary English hieroglyphs we added ☕ to the sign ⬠ to ensure that

was recognised as a syllable containing the uniliteral consonant T (as in "cup of *tea*") and thus pronounced PET, and not seen as a pictogram meaning (for example) "Labrador".

In Ancient Egyptian the name of the god Amon was represented by the signs whose pronunciation was í ("reed in flower") and mn ("crenellation"), supplemented by a determiner (the ideogram signifying the class of gods) *plus* a phonetic complement, the sign for "stream of water", pronounced N, whose sole function was to confirm that the syllable was to be read in a way that made it include the sound n.

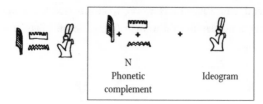

N
Phonetic   Ideogram
FIG. 14.10.   complement

HIEROGLYPHIC NUMERALS

Written Egyptian numerals from their first appearance were able to represent numbers up to and beyond one million, for the system contained specific hieroglyphs for the unit and for each of the following powers of 10: 10, 100 (= 10^2), 1,000 (= 10^3), 10,000 (= 10^4), 100,000 (= 10^5), and 1,000,000 (= 10^6).

The unit is represented by a small vertical line. Tens are signified by a sign shaped like a handle or a horseshoe or an upturned letter "U". The hundreds are symbolised by a more or less closed spiral, like a rolled-up piece of string. Thousands are represented by a lotus flower on its stem, and ten thousands by a slightly bent raised finger. The hundred thousand has the form of a frog, or a tadpole with a visible tail, and the million is depicted by a kneeling man raising his arms to the heavens.

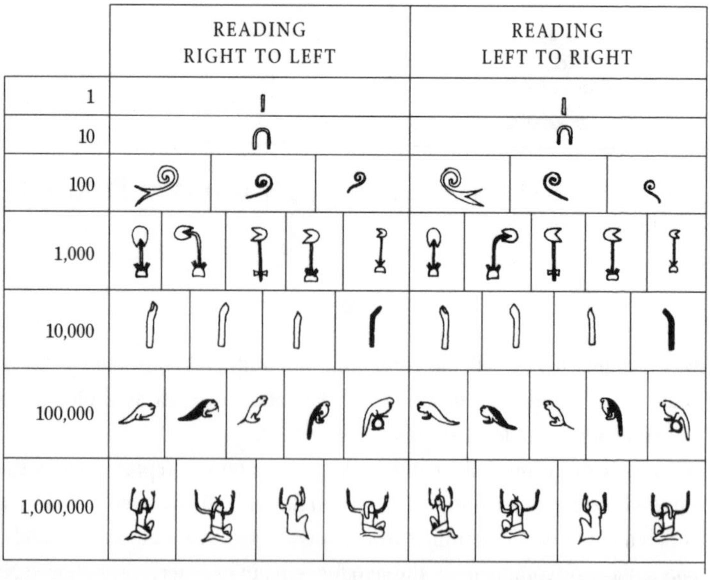

	READING RIGHT TO LEFT			READING LEFT TO RIGHT		
1						
10						
100						
1,000						
10,000						
100,000						
1,000,000						

FIG. 14.11. *The basic figures of hieroglyphic numerals with their main variants in stone inscriptions. Note that the signs change orientation depending on which way the line is to be read: the tadpole (100,000) and the kneeling man (1,000,000) must always face the start of the line.*

One of the oldest examples that we have of Egyptian writing and numerals is the inscription on the handle of the club of King Narmer, who united Upper and Lower Egypt around 3000–2900 BCE.

Fig. 14.12. *Tracing of the knob of King Narmer's club, early third millennium BCE*

Apart from King Narmer's name, written phonetically, the inscription on the club also provides a tally of the booty taken during the king's victorious expedition, consisting of so many head of cattle and so many prisoners brought back. The tally is represented as follows:

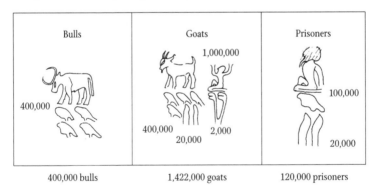

Fig. 14.13.

Are these real numbers, or are they purely imaginary figures whose sole aim is to glorify King Narmer? Scholars disagree. But we should note that the livestock tallies found on the *mastabas* of the Old Kingdom also often give very high numbers for individual owners, and that here we are dealing with the looting of an entire country.

Another example of high numbers can be found on a statue from Hieraconpolis, dating from c. 2800 BCE, where the number of enemies slain by a king called KhaSeKhem are shown as 47,209 by the following signs:

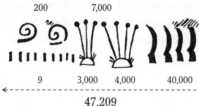

FIG. 14.14.   47,209

To represent a given number, then, the Egyptians simply repeated the numeral for a given order of decimal magnitude as many times as necessary, starting with the highest and proceeding along the line to the lowest order of magnitude (thousands before hundreds before tens, etc.).

Early examples show rather irregular outlines and groupings of the signs. In Fig. 14.13 above, for example, the number of goats (1,422,000) is written in a way that is contrary to the rules that were later laid down by Egyptian stone-cutters, since the figure for the million is placed to the right of the beast and on the same line, whilst the remainder of the number-signs are inscribed on the line below. The normal rule was for the signs to go from right to left in descending order of magnitude on the line below the sign for the object being counted, thus:

FIG. 14.15.

Similarly, Figure 14.14 shows rather primitive features in the representation of the finger (= 10,000), the grouping of the thousands (lotus flowers) into two distinct sets, and the relatively poor alignment of the unit signs. However, from the twenty-seventh century BCE, the execution of hieroglyphic numerals became more detailed and more regular. Also, to avoid making lines of numerals over-long, the custom emerged of grouping signs for the same order of magnitude onto two or three lines, which made them easier to add up by eye:

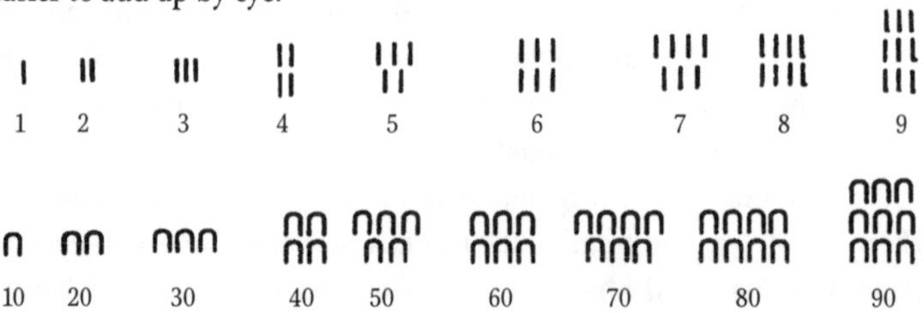

1	2	3	4	5	6	7	8	9

10	20	30	40	50	60	70	80	90

FIG. 14.16.

The evolution of Egyptian numerals can be traced as follows:

1: Old Kingdom period: funerary inscriptions of Sakhu-Rê, a Pharaoh of the Fifth Dynasty, who lived at the time of the building of the pyramids, around the twenty-fourth century BCE:

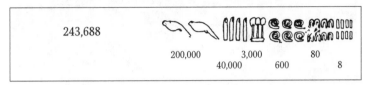

243,688				
200,000	3,000	600	80	8
	40,000			

FIG. 14.17.

A	B	C
10,000 3,000 40 20,000 400	200,000 3,000 20,000 400	30,000 400 10 2,000 3
123,440 + ?	**223,400**	**32,413**

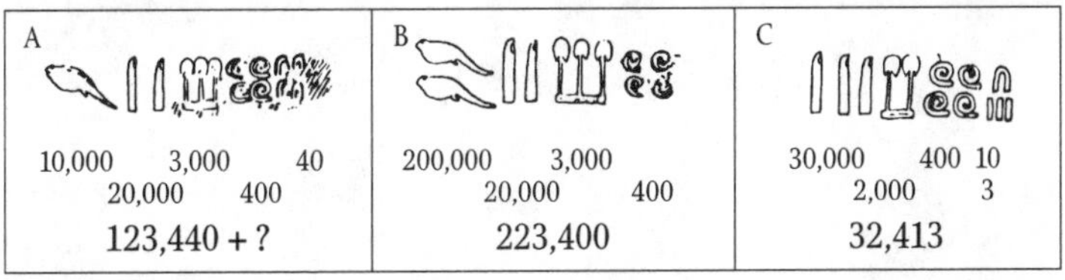

FIG. 14.18.

Although some parts of them have deteriorated somewhat from age, the hieroglyphic numerals are entirely recognisable. The tadpoles are all facing left, and thus these numerical expressions are read from left to right (see Fig. 14.11 above). In Fig. 14.17, the number 200,000 has been written along the line, unlike example B in figure 14.18, where the two tadpoles are put one above the other. The thousands are represented by lotus flowers connected at the base, a custom which disappeared by the end of the Old Kingdom period.

2: End of the First Intermediate period (end of third millennium BCE), from a tomb at Meir:

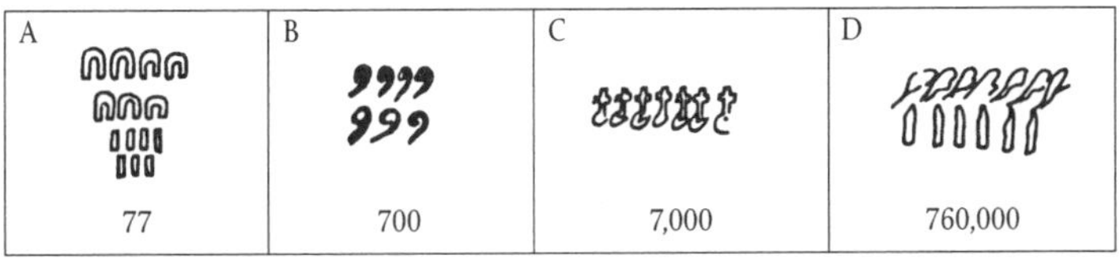

A	B	C	D
77	700	7,000	760,000

FIG. 14.19.

3: From the Annals of Thutmosis (1490–1436 BCE), a list of the plunder of the twenty-ninth year of the Pharaoh's reign (see Fig. 14.21):

The numerals can be transcribed as:

276   4,622   FIG. 14.20.

Fig. 14.21. *Stone bas-relief from Karnak. Louvre*

4: Numerical expression from the *stela* of Ptolemy V at Pithom, 282–246 BCE:

660,000

FIG. 14.22.

THE ORIGINS OF EGYPTIAN NUMERALS

The numerical notation of Ancient Egypt was in essence a written-down trace of a concrete enumeration method that was probably used in earlier periods. The method was to represent any given number by setting out in a line or piling up into a heap the corresponding number of standard objects or tokens (pebbles, shells, pellets, sticks, discs, rings, etc.), each of which was associated with a unit of a given order of magnitude.

	UNITS	TENS	HUNDREDS	THOUSANDS		TENS OF THOUSANDS	HUNDREDS OF THOUSANDS
1							
2							
3							
4							
5							
6							
7							
8							
9							

FIG. 14.23. *Hieroglyphic representations of the consecutive units in each decimal order*

Unlike Sumerian numerals, however, the hieroglyphs give no clue as to the nature of the tokens used in concrete reckoning prior to the invention of writing. It seems pretty unlikely that lotus flowers (1,000) or tadpoles (100,000), were ever practical counting tokens at any period of time. The spiral, the finger, and the kneeling man with upraised arms pose just as awkward and still unanswered questions.

It seems most likely to me that the origins of Egyptian numerals are much more complex than the origins of the written numbers of Sumer and Elam, and that their inventors used not one but several different principles at the same time. What follows are no more than plausible hypotheses about the origins of hieroglyphic numerals, unconfirmed by any hard evidence.

The origin of the numeral 1 could have been "natural" – the vertical line is just about the most elementary symbol that humans have ever invented for representing a single object. It was used by prehistoric peoples from over 30,000 years ago when they scored notches on bone, and as we have seen a whole multitude of different civilisations have given the line or notch the same unitary value over the ages.

In addition, the line (for 1) and the horseshoe (for 10) could well be the last traces in hieroglyphic numerals of the archaic system of concrete numeration. The line could have stood for the little sticks used with a value of 1, and the horseshoe might in fact have been at the start a drawing of the piece of string with which bundles of ten sticks were tied to make a unit of the next order.

As for the spiral and the lotus, they most probably arose through phonetic borrowing. We could imagine that the original Egyptian words for "hundred" and "thousand" were complete or partial homophones of the words for "lotus" and "spiral"; and that to represent the numbers, the Egyptians used the pictograms which represented words which had exactly or approximately the same sound, irrespective of their semantic meaning, as they did for many other words in their language and writing.

Parallels for such procedures exist in many other civilisations. In classical Chinese writing, for instance, the numeral 1,000 was written with the same character as the word "man", because "man" and "thousand" are reckoned to have had the same pronunciation in the archaic form of the language.

On the other hand, the Egyptian hieroglyph for 10,000, the slightly bent raised finger, seems to be a reminiscence of the old system of finger-counting which the Egyptians probably used. The system relies on various finger positions to make tallies up to 9,999.

The hieroglyphic sign for 100,000 may derive from a more strictly symbolic kind of thinking: the myriads of tadpoles in the waters of the Nile, the vast multiplication of frogspawn in the spring ...

The hieroglyphic numeral for 1,000,000 might more plausibly be ascribed a psychological origin. The Egyptologists who first interpreted this sign thought that it expressed the awe of a man confronted with such a large number. In fact, later research showed that the sign (which also means "a million years" and hence "eternity") represented in the eyes of the Ancient Egyptians a *genie holding up the vault of heaven*. The pictogram's distant origin lies perhaps in some priest or astronomer looking up to the night sky and taking stock of the vast multitude of its stars.

SPOKEN NUMBERS IN ANCIENT EGYPTIAN

The spoken numbers of Egyptian have been reconstructed from its modern descendant, Coptic, together with the phonetic transcriptions of numerical expressions found in hieroglyphic texts on the pyramids. Here are their syllabic transcriptions with their approximate phonetic realisations:

1	*w'*	[wa']	10	*mḏ*	[medj]
2	*snw*	[senu]	20	*dwty*	[dwetye]
3	*khmt*	[khemet]	30	*m'b'*	[m'aba']
4	*fdw*	[fedu]	40	*khm*	[khem]
5	*díw*	[diwu]	50	*díyw*	[diyu]
6	*srsw*	[sersu]	60	*sí*	
7	*sfkh*	[sefekh]	70	*sfkh*	[sefekh]
8	*khmn*	[khemen]	80	*khmn*	[khemen]
9	*psḏ*	[pesedj]	90	*psḏ*	[pesedj]

št [shet]	*kh'* [kha']	*ḏb'* [djebe']	*ḥfn* [hefen]	*ḥḥ* [heh]
100	1,000	10,000	100,000	1,000,000

Note that 7, 8, and 9 have the same consonantal structure as 70, 80, and 90 respectively. The Egyptians may well have pronounced them slightly differently in order to avoid confusion: for instance, *sefekh* for 7 and *sefakh* for 70, *khemen* for 8 and *kheman* for 80, etc.

The spoken numerals, as can be seen, were strictly decimal. Compound numbers were expressed along the lines of the following example:

4,326:

fdw	*kh*	*khmt*	*sht*	*dwty*	*srsw*
"four	thousand	three	hundred	twenty	six"

FRACTIONS AND THE DISMEMBERED GOD

Fractions were mostly expressed in Ancient Egyptian writing by placing the hieroglyph "mouth", pronounced eR and having in this context the specific sense of "part", over the numerical expression of the denominator, thus:

$$\frac{1}{3} \qquad \frac{1}{5} \qquad \frac{1}{6} \qquad \frac{1}{10} \qquad \frac{1}{100}$$

FIG. 14.24.

When the denominator was too large to go entirely beneath the eR sign, the remainder of it was placed to the right, thus:

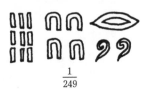

$$\frac{1}{249}$$

FIG. 14.25.

There were special signs for some fractions:

VALUE		MEANING
$\frac{1}{2}$	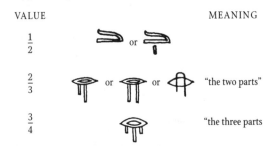	
$\frac{2}{3}$		"the two parts"
$\frac{3}{4}$		"the three parts

FIG. 14.26.

Save for the last two expressions in Fig. 14.26, the only numerator used in Egyptian fractions was the unit. So to express (for instance) the equivalent of what we write as $\frac{3}{5}$, they did not write $\frac{1}{5} + \frac{1}{5} + \frac{1}{5}$ but decomposed the number into a sum of fractions with numerator 1.

$$= \frac{1}{2} + \frac{1}{10} = \frac{3}{5} \qquad = \frac{1}{3} + \frac{1}{4} + \frac{1}{5} = \frac{47}{60}$$

FIG. 14.27.

Measures of volume (dry and liquid) had their own curious system of notation which gave fractions of the *ḥeqat*, generally reckoned to have been equivalent to 4.785 litres. These volumetric signs used "fractions" of the hieroglyph representing the painted eye of the falcon-god Horus:

FIG. 14.28.

The name of Horus's eye was *oudjat*, written phonetically in hieroglyphs as follows:

Wa   DJ   'a   T
- ->   FIG. 14.29.

The *oudjat* was simultaneously a human and a falcon's eye, and thus contained both parts of the cornea, the iris and the eyebrow of the human eye, as well as the two coloured flashes beneath the eye characteristic of the falcon. Since the most common fractions of the *ḥeqat* were the half, the quarter, the eighth, the sixteenth, the thirty-second and the sixty-fourth, the notation of volumetric fractions attributed to each of the parts or strokes in the *oudjat* sign the value of one of these fractions, as laid out in Fig. 14.30 below.

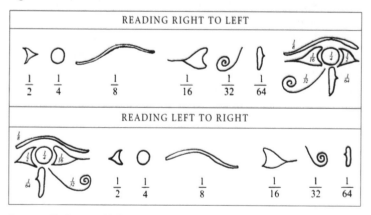

FIG. 14.30. *The fractions of the* ḥeqat

Horus was the son of Isis and Osiris, the god murdered and cut up into thirteen pieces by his brother Seth. When he grew up, Horus devoted himself to avenging his father, and his battles with his uncle Seth were long and bloody. In one of these combats, Seth ripped out Horus's eye, tore it into six pieces and dispersed the pieces around Egypt. Horus gave as good as he got, and castrated Seth. In the end, according to legend, the assembly of the gods intervened and put a stop to the fighting. Horus became king of Egypt and then the tutelary god of the Pharaohs, the guarantor of the legitimacy of the throne. Seth became the cursed god of the Barbarians and the Lord of Evil. The assembly of the gods instructed Thot, the god of learning and magic, to find and to reassemble Horus's eye and to make it healthy again. The *oudjat* thus became a talisman symbolising the wholeness of the body, physical health, clear vision, abundance and fertility; and so the scribes (whose tutelary god was Thot) used the *oudjat* to symbolise the fractions of the *ḥeqat*, specifically for measures of grain and of liquids.

An apprentice scribe one day observed to his master that the total of the fractions of the *oudjat* came to less than 1:

$$\frac{1}{2} + \frac{1}{4} + \frac{1}{8} + \frac{1}{16} + \frac{1}{32} + \frac{1}{64} = \frac{63}{64}$$

His master replied that the missing $\frac{1}{64}$ would be made up by Thot to any scribe who sought and accepted his protection.

HIERATIC SCRIPT AND CURSIVE NUMERALS IN ANCIENT EGYPT

With its minutely complex and decorative signs, the hieroglyphic system of writing words and numbers was only really suitable for memorial inscriptions, and was used mainly, if not quite exclusively, on stone monuments such as tombs, funeral *stelae*, obelisks, palace and temple walls, etc. When Ancient Egyptians needed to note down or record accounts, censuses, inventories, reports, or wills, for example, or when they penned administrative, legal, economic, literary, magical, mathematical, or astronomical works, they had far more frequent recourse to a script that was easier to handle at speed, namely hieratic script.

Hieratic script uses signs that are simplifications and schematisations of the corresponding hieroglyphs, with fewer details and with shapes reduced to skeleton forms. In some cases, the hieratic versions can be recognised as variants of the original sign; but most often the relationship between the "cursive" and the "monumental" form is impossible to guess and has to be learned sign by sign.

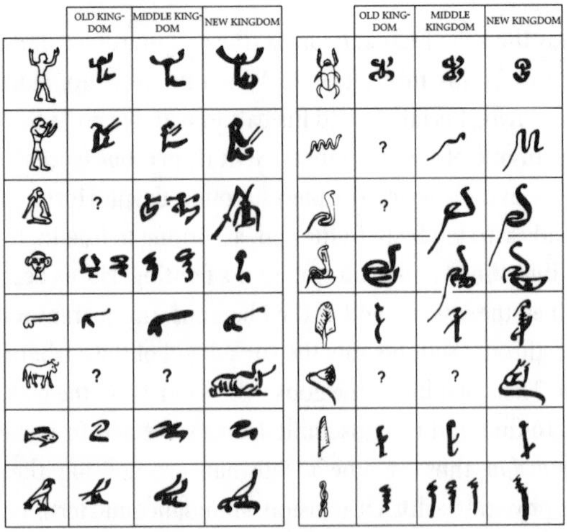

FIG. 14.31. *Some hieroglyphs and their hieratic equivalents*

There were also hieratic versions of the hieroglyphic numerals. These are the numerical signs found in the Harris Papyrus (British Museum), dating from the Twentieth Dynasty, which gives the possessions of the temples at the death of Ramses III (1192–1153 BCE):

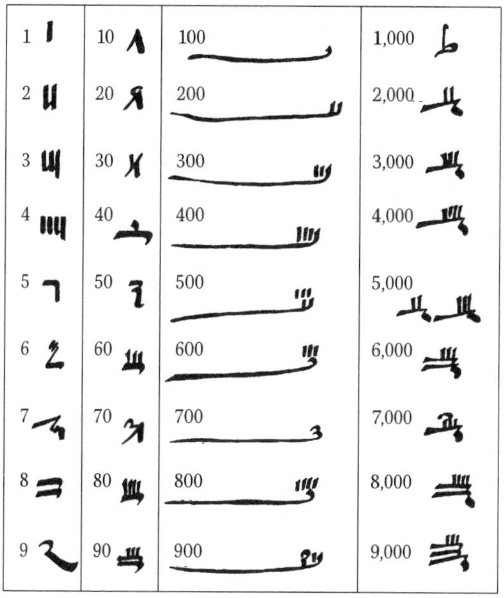

Fig. 14.32.

As can be seen, the hieratic numerals are for the most part visually quite unrelated to their equivalent hieroglyphs. Although the signs for the first four units are fairly self-explanatory ideograms, all the other numerals seem quite devoid of visually intuitive meaning.

So do hieratic numerals constitute a genuinely independent numbering system? Should we consider the numerals found in the Harris Papyrus as an arbitrary shorthand, invented by scribes for jotting down numbers intended to be written quite differently on stone monuments?

In fact, hieratic numerals, like the syllabic signs of this script, are developed from the corresponding hieroglyphs, and do not constitute an independent system. However, the changes in the shapes of the signs were very considerable, imposed in part by the characteristics of the reed-brushes used for hieratic characters (which, unlike hieroglyphs, were always written from right to left) and in part by a tendency to use ligatures, that is to say to run several signs together to produce single compounds. That is why the groups of five, six, seven, eight, and nine vertical lines became single signs devoid of any intuitive meaning:

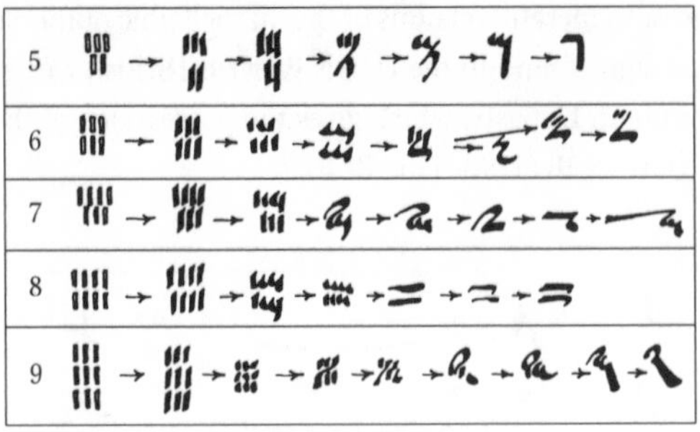

5	...						
6	...						
7	...						
8	...						
9	...						

FIG. 14.33.

The relationship between hieratic numerals and hieroglyphs is difficult to see, but it was probably no more difficult for an Ancient Egyptian than it is for us to see the equivalence between the following ways of writing our own letters:

A B C D E F K R S
A B C D E F K R S
a b c d e f k r s

Imagine how hard it would be for a speaker of Chinese or Arabic, for example, with no knowledge of the Latin alphabet, to work out that the signs on the second and third lines have exactly the same value as the signs in the corresponding position on the first line!

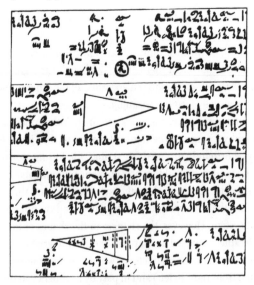

FIG. 14.34. *Detail from the Rhind Mathematical Papyrus (RMP), an important mathematical document written in hieratic script. From the Hyksos (Shepherd Kings) period (c. seventeenth century BCE), the RMP is a copy of an earlier document probably going back to the Twelfth Dynasty (1991–1786 BCE). The RMP is in the British Museum.*

Hieratic script was therefore not a form of "shorthand", in the sense that modern shorthand consists of purely arbitrary signs visually unrelated to the letters of the alphabet which they represent. Hieratic signs were indeed derived from hieroglyphs and represent the terminus of a long but specifically graphical evolution. Hieratic script never replaced the monumental script used for inscriptions on stone, and never had much impact on the shape of the hieroglyphs. The two systems were used in parallel for nearly 2,000 years, from the third to the first millennium BCE, and throughout this period hieratic script, despite its apparent difficulty, provided a perfectly serviceable tool for all administrative, legal, educational, magical, literary, scientific, and private purposes.

Hieratic script was gradually displaced from about the twelfth century BCE by a different cursive writing, called demotic. It survived in specific uses – notably in religious texts and in sacred funeral books – until the third century CE, which is why the Greeks called it *hieratikos*, meaning "sacred", whence our term "hieratic".

FROM HIEROGLYPHIC TO HIERATIC NUMERALS

Hieratic numerals of the third millennium BCE are still fairly close to their hieroglyphic models; but over the centuries, the use of ligatures and the introduction of diacritics turn them little by little into apparently quite different signs with no intuitive resemblance to the original hieroglyphs. The end result was a set of numerals with distinctive signs for each of the following numbers:

1	2	3	4	5	6	7	8	9
10	20	30	40	50	60	70	80	90
100	200	300	400	500	600	700	800	900
1,000	2,000	3,000	4,000	5,000	6,000	7,000	8,000	9,000

So though they began with a very basic additive numeration, the Egyptians developed a rapid notation system that was quite strikingly simple, requiring (for example) only four signs to represent the number 3,577, whereas in hieroglyphs it takes no fewer than 22 signs:

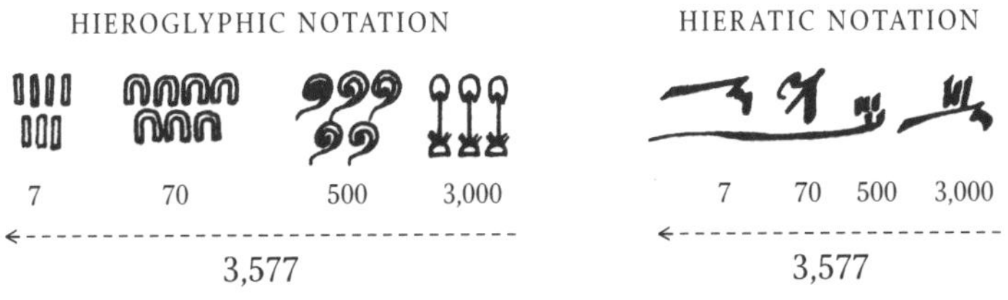

HIEROGLYPHIC NOTATION				HIERATIC NOTATION			
7	70	500	3,000	7	70	500	3,000

3,577   3,577

FIG. 14.35.

The main disadvantage of the hieratic system was of course that it required its users to memorise a very large number of distinct signs, and was thus quite impenetrable to all but the initiate. Here are the shapes that a hieratic mathematician had to know as well as we know 1 to 9:

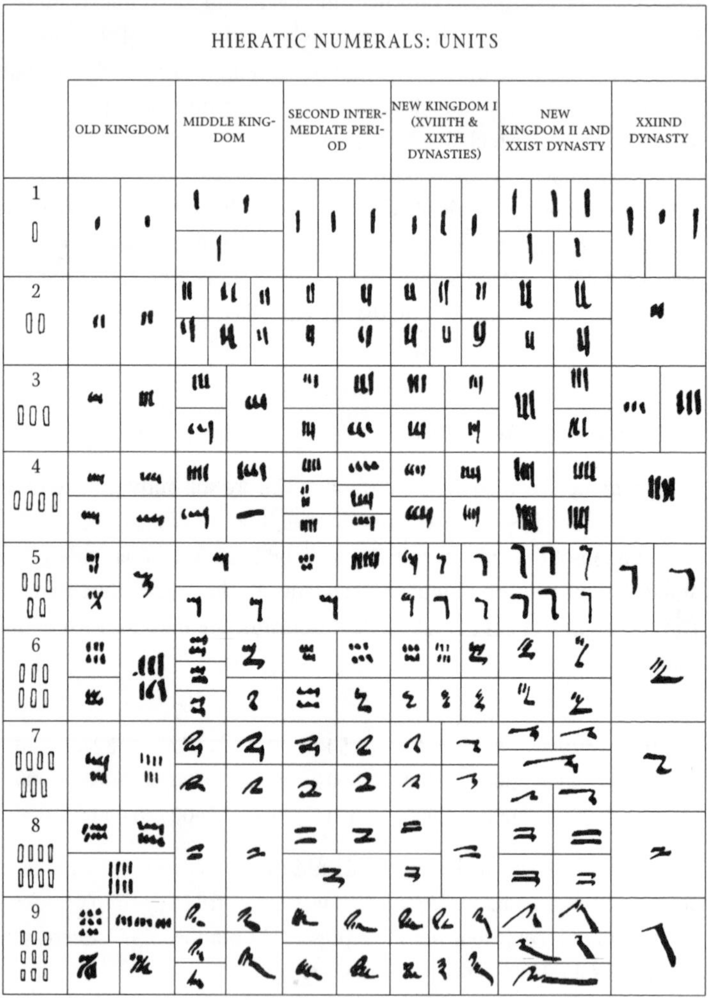

FIG. 14.36A.

HIERATIC NUMERALS: TENS

	OLD KINGDOM		MIDDLE KING-DOM		SECOND INTER-MEDIATE PERI-OD		NEW KINGDOM I (XVIIITH & XIXTH DYNASTIES)		NEW KINGDOM II AND XXIST DYNASTY		XXIIND DYNASTY	
10												
20												
30												
40												
50												
60												
70												
80												
90												

FIG. 14.36B.

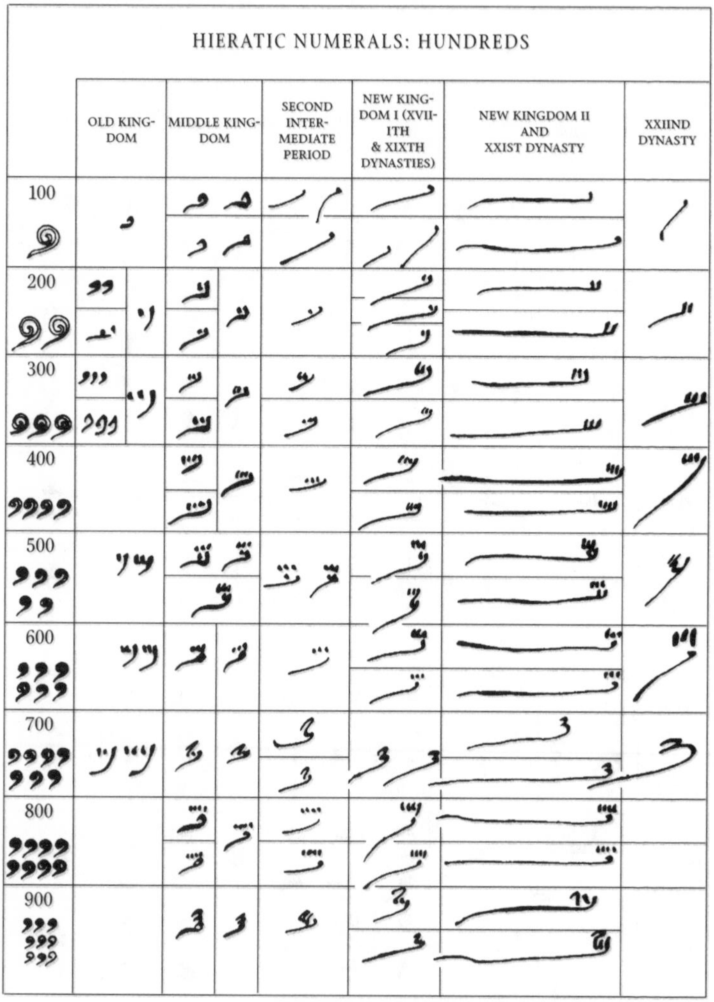

Fig. 14.36c.

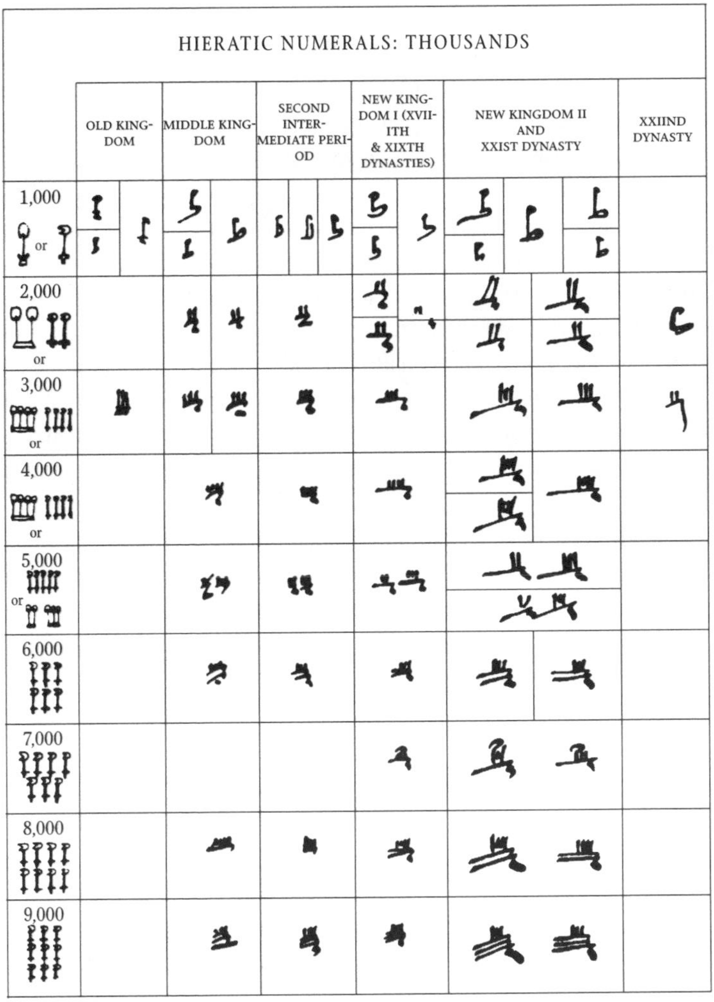

FIG. 14.36D.

DOING SUMS IN ANCIENT EGYPT

Let us imagine we're at a farm near Memphis, in the autumn of the year 2000 BCE. The harvest is in, and an inspector is here to make an assessment on which the annual tax will be calculated. So he orders some of the farm workers to measure the grain by the bushel and to put it into sacks of equal size.

This year's harvest includes white wheat, einkorn, and barley. So as to keep track of the different varieties of grain, the workers stack the white wheat in rows of 12 sacks, the einkorn in rows of 15 sacks, and the barley in rows of 19 sacks, and for each the total number of rows are respectively 128, 84, and 369.

When this is done, the inspector takes a piece of slate to use as a "notepad" and starts to do some sums on it in hieroglyphic numerals. For despite the primitive nature of their numerals, the Egyptians have known for centuries how to do arithmetic with them.

Adding and subtracting are quite straightforward. To add up, all you do is to place the numbers to be summed one above the other (or one alongside the other), then to make mental groups of the identical symbols and to replace each ten of one set of signs by one sign of the next higher decimal order.

For instance, to add 1,729 and 696, you first place (as in Fig. 14.37 below) 1,729 above 696. You then make mental groupings respectively of the vertical lines, the handles, the spirals, and the lotus flowers. By reducing them in packets of 10 to the sign of the next higher order, you get the correct result of the addition:

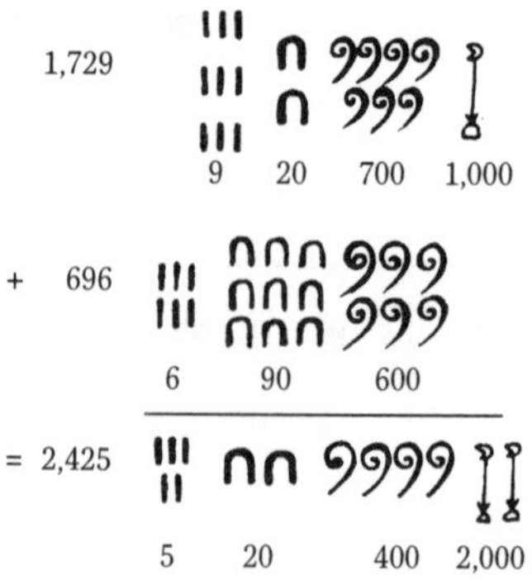

Fig. 14.37.

It is also quite easy to multiply and to divide by 10 in Egyptian hieroglyphics: to multiply, you replace each sign in the given number by the sign for the next higher order of decimal magnitude (or the next lower, for division by 10). So to multiply 1,464 by 10 you take:

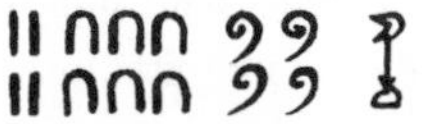

4	60	400	1,000

FIG. 14.38.

and by following the regular procedure it becomes:

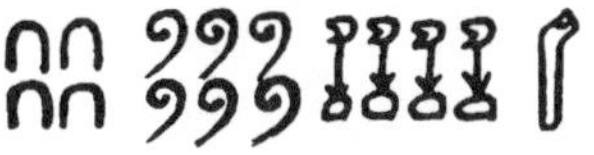

40	600	4,000	10,000

FIG. 14.39.

However, to multiply and to divide by any other factor, the Egyptians went about it quite differently. They knew only their two times table, and so they proceeded by a sequence of *duplications*.

To come back now to the tax-collector who needs to know the total number of sacks of white wheat in this year's harvest, and therefore needs to multiply 12 by 128. He goes about it like this:

1	12
2	24
4	48
8	96
16	192
32	384
64	768
128	1,536

That is to say, he writes the multiplier 12 in the right-hand column of his slate, and opposite it, in the left-hand column, he writes the number 1. He then doubles each of the two numbers in successive rows until the multiplier 128 appears on the left. As the number 1,536 appears on the right in the row where the left column shows 128, this is the result of the operation: $12 \times 128 = 1,536$.

To discover how many sacks of einkorn there are, he now has to multiply 84 by 15. His "doubling table" would look like this:

1	15
2	30
4	60
8	120
16	240
32	480
64	960

As the next doubling would take the multiplier beyond the required figure of 84, he stops there, and looks down the left-hand column to see which of the multipliers entered would sum to 84. He finds that he can reach 84 with just three of the multiplications already computed, and he checks the left-hand column numbers by making a little mark next to them, and putting an oblique stroke beside their right-hand column products, thus:

1	15
2	30
–4	60/
8	120
–16	240/
32	480
–64	960/

He can then add up the numbers with the oblique check-mark and arrive at the result:

$$84 \times 15 = 960 + 240 + 60 = 1{,}260$$

To compute the number of sacks of barley, the inspector now has to multiply 369 by 19. He goes about it in exactly the same way, putting 1 in the left-hand column of his slate and 19 in the right-hand column, and then doubling the two terms successively as he goes down the rows. He stops when the left-hand column reaches 256, since the next step would give a multiplier of 512, which is higher than the required figure of 369:

–1	19/
2	38
4	76
8	152
–16	304/
–32	608/
–64	1,216/
128	2,432
–256	4,864/

Then he looks down the left-hand column to find those numbers whose sum is 369, finds that they are 256, 64, 32, 16, and 1, and thus adds up the corresponding right-hand figures to arrive at his total:

$$369 \times 19 = 4{,}864 + 1{,}216 + 608 + 304 + 19 = 7{,}011$$

So the harvest adds up to 1,536 sacks of white wheat, 1,260 sacks of einkorn, and 7,011 sacks of barley. And since the Pharaoh's share of that is one tenth, the inspector can easily calculate the tax payable as 153 sacks of white wheat, 126 sacks of einkorn, and 701 sacks of barley.

So multiplication in the Egyptian manner is really quite simple and can be done without any multiplication tables other than the table of 2. Division is done similarly by successive duplication, but in reverse, as we shall see.

Let us suppose that in the time of Ramses II (1290–1224 BCE) robbers have just stripped the tomb of one of the Pharaohs of the preceding dynasty. They have stolen diadems, ear-rings, daggers, breast-plates, pendants – a whole mass of precious jewellery decorated with gold leaf and glass beads. Altogether there are 1,476 items in the robbers' haul, and the leader of the gang proposes to divide them equally amongst his eleven men and himself. So he has to divide 1,476 by 12. He goes about it just as if he were doing a multiplication, putting 12 in the right-hand column, and stopping when the right-hand figure reaches 768 since the next step would take the sequence beyond the total number of items to be shared:

/1	12–
/2	24–
4	48
/8	96–
/16	192–
/32	384–
/64	768–

He now has to find which of the numbers in the *right*-hand column total 1,476 and after various attempts to make the total he finds that 768, 384, 192, 96, 24, and 12 come out exactly right. So he makes a little mark against these figures in the right-hand column and puts an oblique against their corresponding numbers in the left-hand column. So he can now add up the checked numbers on the left to come out with the exact answer to the question: how many twelves go into 1,476?

$$1,476/12 = 64 + 32 + 16 + 8 + 2 + 1 = 123$$

So each of the robbers takes 123 pieces from the haul, and off they go with their fair shares.

This method of division only works when there is no remainder; where the dividend is not a multiple of the divisor, the Egyptians had a much more complicated method involving the use of fractions, which will not be explained here.*

The arithmetical methods of Pharaonic Egypt did not therefore require any great powers of memorisation, since, to multiply and to divide, all that you needed to know by heart was your two times table. Compared

*The method is explained in Richard J. Gillings, *Mathematics in the Time of the Pharaohs* (Cambridge, MA: MIT Press, 1972).

to modern arithmetic, however, Egyptian procedures were slow and very cumbersome.

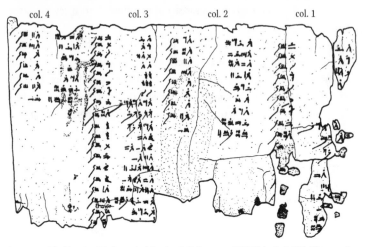

FIG. 14.40. *The Egyptian Mathematical Leather Roll (known as EMLR) in the British Museum. It contains, in hieratic notation, and in duplicate, twenty-six additions done in unit fractions and was probably used as a conversion table. [See Gillings (1972), pp. 89–103]*

ANCIENT EGYPTIAN NUMBER-PUZZLES

Egyptian carvers, especially in the later periods, indulged in all sorts of puns and learned word-games, most notably in the inscriptions on the temples of Edfu and Dendara. Some of these word-games involve the names of the numbers, and the following tables (based on the work of P. Barguet, H. W. Fairman, J. C. Goyon, and C. de Wit) give a small sample of the innumerable curious scribal inventions for the representation of the numbers in hieroglyphs. The references are to Chassinat's transcription of the inscriptions on the walls of the temples of Edfu ("E") and Dendara ("D").

VALUE	SIGN & MEANING	EXPLANATION	REFERENCE
1	harpoon	Homophony: "one" and "harpoon" are both pronounced *wa'*	E.VII, 18, 10
1	sun	Because there is only one sun	E.IV, 6, 4
1	moon	Because there is only one moon	E.IV, 6, 4
1	fraction 1/30	Only used in the expression "one day" or "the first day": 1/30 of a month is 1 day	E.IV, 8, 4; E.IV, 7, 1

VALUE	SIGN & MEANING	EXPLANATION	REFERENCE
2		Two × harpoon = 2 × 1	E.IV, 14, 4
2		Sun + moon = 1 + 1	E.VI, 7, 5
3		Three × harpoon = 3 × 1	E.VII, 248, 10
4	*jubilaeum*	No known explanation	E.IV, 6, 5; E.IV, 6, 6; E.VII, 15, 1
5	5-pointed star	Self-evident	E.IV, 6, 3; E.IV, 6, 5; E.VII, 6, 4
6		Standard sign for 1 + star = 1 + 5	E.IV, 5, 4
7	human head	The head has seven orifices: two eyes, two nostrils, two ears, mouth	E.IV, 4, 4; E.V, 305, 1
7		Standard sign for 2 + star = 2 + 5	E.IV, 6, 5
7	$\frac{1}{5} + \frac{1}{30}$	Only in the expression "seven days": 1/5 of a month = 6 days + 1/30 = 1 day	E.IV, 8, 4; E.IV, 7, 1
8	ibis	The sacred ibis was the incarnation of the god Thot, the principal divinity of the city of Hermopolis, formerly *Khmnw* or *Khemenu*, meaning "the city of eight"	E.III, 77, 17; E.VII, 13, 4; E. VII, 14, 2
8		A curious "re-formation" in hieroglyphics of the hieratic numeral 8	E.VI, 92, 13
8		Standard notation of 3 + star = 3 + 5 = 8	E.IV, 5, 2
8		Moon + head = 1 + 7 = 8	E.IV, 6, 4
8		Standard notation of 1 + head = 1 + 7 = 8	E.IV, 9, 3

VALUE	SIGN & MEANING	EXPLANATION	REFERENCE
9	to shine	Homophony: "nine" and "shine" are both pronounced *psd*	E.IV, 8, 2; E.VII, 8, 8
9	scythe	Based on the fact that in hieratic the numeral 9 and the sign for scythe were identical	E.VII, 15, 3; E.VII, 15, 9; E.VII, 17, 3
9		Standard notation of 4 + star = 4 + 5 = 9	E.IV, 6, 1
9		Standard notation of 2 + head = 2 + 7 = 9	D.II, 47, 3
10	falcon	The falcon-god Horus was the first to be added to the original nine divinities of Heliopolis, and thus represents 10	E.V, 6, 5
14		falcon + *jubilaeum* = 10 + 4 = 14	E.V, 6, 5
15	fraction $\frac{1}{2}$	Only in the expression "15 days" or "fortnight": 1/2 of a month = 15 days	E.VII, 7, 6
17		Standard notation of 10 + head = 10 + 7 = 17	E.VII, 248, 9
18	$\frac{1}{2} + \frac{1}{10}$	Only in the expressions "18 days" or "the 18th day": 1/2 month + 1/10 month = 15 + 3 = 18	E.IV, 9, 1; E.VII, 7, 6; E.VII, 6, 1
19		Standard notation of 10 + scythe = 10 + 9 = 19	E.VII, 248, 4
20		Two falcons = 2 × 10 = 20	E.VII, 11, 8
107		Standard notation of 100 + head = 100 + 7 = 107	E.VII, 248, 11

CHAPTER 15

COUNTING IN THE TIMES OF THE CRETAN AND HITTITE KINGS

THE NUMBERS OF CRETE

Between 2200 and 1400 BCE, the island of Crete was the centre of a very advanced culture: Minoan civilisation, as it is called, from the name of the legendary priest-king Minos who, according to Greek mythology, was one of the first rulers of Knossos, the ancient Cretan capital near the modern port of Heraklion (Candia).

The very existence of Minoan civilisation was almost completely unknown until the end of the last century, and it is only relatively recently that archaeologists have uncovered a brilliant and original culture which was, in many respects, the precursor of Greek civilisation.

When Minoan civilisation fell, around 1400 BCE, probably as a result of some natural disaster or of the invasion of the island by the Mycenaeans (of Greek origin), it disappeared almost without trace save for what was preserved in the fables and legends of the Ancient Greeks.

We owe the most spectacular discoveries – such as the famous Palace of Knossos – to the indefatigable enthusiasm and energy of the British archaeologist Sir Arthur Evans (1851–1941). He was the first to show that the Greek legends had a historical basis, and constituted a living trace of one of the oldest known European civilisations.

Since the end of the last century, archaeological investigations carried out mainly on the sites of Knossos and Mallia have brought to light a large number of documents whose analysis has revealed the existence of a "hieroglyphic" script between 2000 and 1660 BCE.

Cretan hieroglyphics have still not been deciphered, and these documents remain enigmatic. Nevertheless they show evidence of an accounting system adapted to a "bureaucracy" no doubt born within the earliest palaces of Minoan civilisation. In proof of this we find clay blocks and tablets covered with figures and hieroglyphic signs, which are more or less schematic drawings of all kinds of objects. These appear to be accounts giving details of inventories, supplies, deliveries, or exchanges. The purpose of the symbols was to note down the quantities of the different kinds of goods.

The numerical notation of Crete was strictly decimal, and was based on the additive principle. Unity was represented by a short slightly oblique stroke, or by a small circular arc which could be oriented anyhow. Cretan

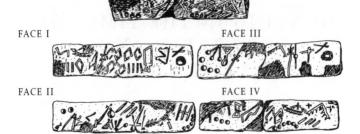

FACE I

FACE III

FACE II

FACE IV

FIG. 15.1. *Inscriptions on bars of clay, showing Cretan hieroglyphic signs and numerals. Palace of Knossos, 2000–1500 BCE. [Evans (1909), Doc. P 100]*

hieroglyphic writing went from left to right and from right to left, in *boustrophedon* (as a ploughman ploughs a field from side to side). 10 was represented by a circle (or, on clay, by a small circular imprint as would be made by the pressure of a round-tipped stylus held perpendicularly to the surface of the clay). 100 was represented by a large oblique bar (distinctly different from the small stroke of unity), and 1,000 was represented by a kind of lozenge.

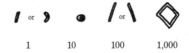

1   10   100   1,000

FIG. 15.2. *Cretan hieroglyphic figures*

With these as starting points, the Cretans represented other numbers by repeating each one as many times as required. The hieroglyphic figures were not, however, the only forms used. Other excavations have revealed a second script, no doubt derived from the hieroglyphic, in which the picture symbols give way to schematic drawings which, often, we cannot identify now. Analysis of these documents led Evans to distinguish two variants of this kind of writing, which he called "Linear A" and "Linear B".

The system known as "Linear A" is the older. It was in use from the start of the second millennium BCE up to around 1400 BCE, that is to say at about the same time as the hieroglyphic script.

The sites which have yielded documents in Linear A are several, notably Haghia Triada, Mallia, Phaestos, and Knossos. From Haghia Triada we have a large collection of accounting tablets, unfortunately in a somewhat sloppy script [Fig. 15.4]. These are, therefore, inventories, with ideograms and numbers; the tablets are in the format of small pages. But Linear A can be found as well on a wide variety of objects: vases (with inscriptions cut, painted, or written in ink), seals, stamps, and labels of clay; ritual objects (libation tables); large copper ingots; and so on. This writing may therefore be very widely found, not only in administrative environments but also in holy places and probably also in people's homes. [O. Masson (1963)]

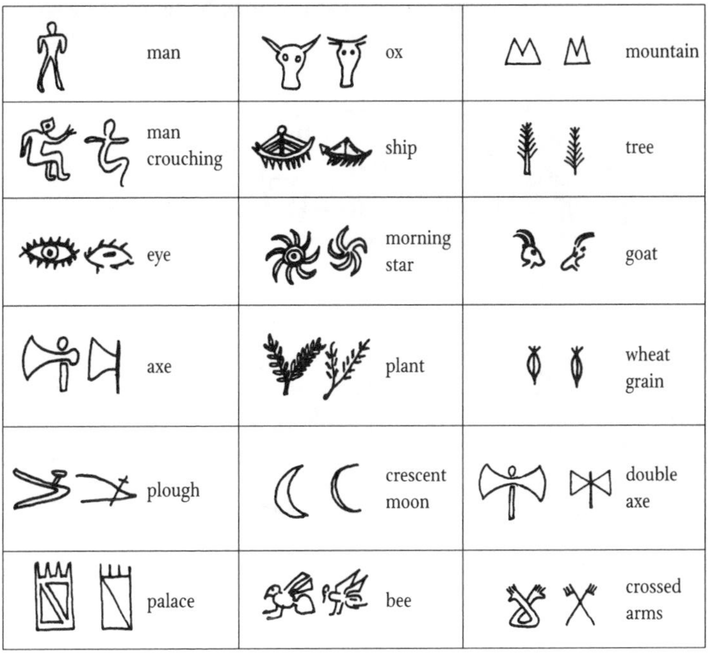

	man		ox		mountain
	man crouching		ship		tree
	eye		morning star		goat
	axe		plant		wheat grain
	plough		crescent moon		double axe
	palace		bee		crossed arms

FIG. 15.3. *A selection of Cretan hieroglyphics [after Evans]*

FIG. 15.4. *Cretan tablet with signs and numerals from the "Linear A" script. Haghia Triada, sixteenth century BCE. [GORILA (1976), HT 13, p. 26]*

The script known as "Linear B" is the more recent, and the best known, of the Cretan scripts. It is usually dated to the period between 1350 BCE and 1200 BCE. At this time, the Mycenaeans had conquered Crete, and ancient Minoan civilisation had spread onto the Greek mainland, especially in the region of Mycenae and Tyrinth.

The signs of this script were engraved on clay tablets, which were first unearthed in 1900. Since then, 5,000 tablets have been found in Crete (at Knossos only, but in large numbers) and on mainland Greece (mainly at Pylos and Mycenae). Linear B, therefore, may be found outside Crete. We may also note that this script, apparently derived by modification of Linear A, was used to record an archaic Greek dialect, as demonstrated by Michael Ventris, the English scholar who first deciphered it. It is the only Creto-Minoan script to have been deciphered to date (Linear A and the hieroglyphic script correspond to a language which still remains largely unknown).

A

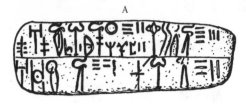

B

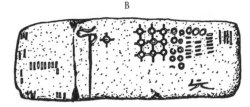

FIG. 15.5. *Cretan tablets with signs and numerals from the "Linear B" script, fourteenth or thirteenth century BCE. [Evans and Myres (1952)]*

Both Linear A and Linear B used practically the same number-signs (Fig. 15.6). These were:

• a vertical stroke for unity;

• a horizontal stroke (or, solely in Linear A, sometimes a small circular imprint) for 10;

• a circle for 100;

• a circular figure with excrescences for 1,000;

• the same, with a small horizontal stroke inside, for 10,000 (found only in Linear B inscriptions: Fig. 15.6, last line).

		1	10	100	1,000	10,000
	Hieroglyphic system c. 2000 to c. 1500 BCE	*various marks*	●	/ \	◇	?
	"Linear A" system c. 1900 to c. 1400 BCE	❘	● —	○	⊕	?
FIG. 15.6. *Cretan numerals*	"Linear B" system c. 1350 to c. 1200 BCE	❙	—	○	⊕	⊖

Examples from hieroglyphic documents from the Palace at Knossos, first half of the second millennium BCE			Examples from tablets inscribed in Linear A from the archives of Haghia Triada, around 1600–1450 BCE		
42		SM I P.103 C	86		GORILA HT 107
160		SM I P.101 C	95		GORILA HT 104
170		SM I P.101 C	161		GORILA HT 21
407		SM I P.109 D	684		GORILA HT 15
1,640		SM I P.103 C	976	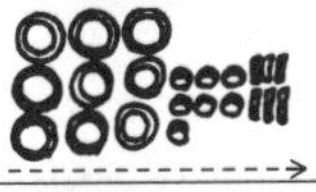	GORILA HT 102
2,660		SM I P.100 D	3,000		GORILA HT 31

Fig. 15.7. *The principle of the Cretan numerals*

To represent a given number, it was enough to repeat each of the above as many times as needed (Fig. 15.7).

The number-systems used in Crete in the second millennium BCE (hieroglyphic, Linear A, and Linear B) had, therefore, exactly the same intellectual basis as the Egyptian hieroglyphic notation and, for the whole time they were in use, underwent no modification of principle. (Similarly, the drawing of signs and numbers on clay did not give way to a cuneiform system, as happened in Mesopotamia). As in the monumental Egyptian system, these number-systems were founded on base 10 and used the principle of juxtaposition to represent addition. Moreover, the only numbers to which each system gave a special sign were unity, and the successive powers of 10.

The number 10,000 (found only in Linear B inscriptions) is derived from the number 1,000 by adding a horizontal bar in the interior of the latter. By all appearances, therefore, a multiplicative principle has been used (10,000 = 1,000 × 10), since the horizontal bar is simply the symbol for 10 in this system (Fig. 15.6).

THE HITTITE HIEROGLYPHIC NUMBER-SYSTEM

From the beginning of the second millennium BCE the Hittites (a people of Indo-European origin) settled progressively in Asia Minor, no doubt by a process of slow immigration. Between the eighteenth and the sixteenth centuries BCE, they there established a great imperial power of which there were two principal phases: the Ancient Empire (pre-1600 to around 1450 BCE) and the New Empire (1450–1200 BCE).

In the course of the imperial era, the Hittites, with many successes and failures, pursued a policy of conquest in central Anatolia and northern Syria. But at the start of the thirteenth century BCE, no doubt under attack from the "Peoples of the Sea", this powerful empire abruptly collapsed. A renaissance, however, ensued from the ninth century BCE in the north of Syria where several small Hittite states maintained elements of the imperial tradition in the midst of mixed populations. This was the beginning of what is called the "neo-Hittite" phase of the civilisation. Finally, however, in the seventh century BCE, all these small states were absorbed by the Assyrian Empire.

The Hittites had two writing systems. One was a hieroglyphic system which seems to have been of their own creation, of which the earliest known evidence is from the fifteenth century BCE. The other was a cuneiform system borrowed from Assyro-Babylonian civilisation whose introduction dates from around the seventeenth century BCE.*

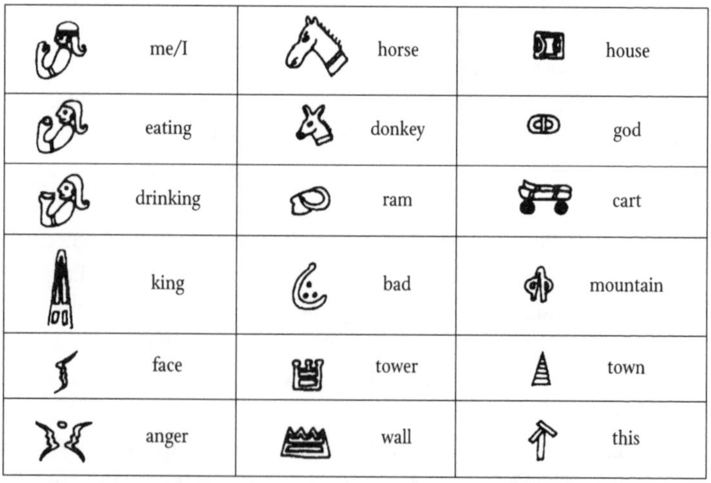

	me/I		horse		house
	eating		donkey		god
	drinking		ram		cart
	king		bad		mountain
	face		tower		town
	anger		wall		this

FIG. 15.8. *The meanings of some of the Hittite hieroglyphics [after Laroche (1960)]*

Thus, for at least three centuries (1500–1200 BCE) the hieroglyphic lived alongside the cuneiform in Anatolia, and they constituted the dual medium of expression of the Hittite state. For a people to practise

* The cuneiform system, of Assyro-Babylonian origin, was adapted into at least three Hittite dialects: Nesitic, spoken in the capital of the empire; Louvitic, employed in southern Anatolia, and Palaitic in the north. Cuneiform characters were used for the numerous tablets making up the royal archives of the town of Hattuša, capital of the Hittite Empire, at the place which is now Boğazköy in Turkey, about 150 km east of Ankara; thanks to these documents, the history and language of the Hittites have been partially reconstructed.

two writing systems at the same time is not a frequent phenomenon. We are now able to perceive the reasons which induced the Hittites into this paradoxical situation. The scribes of Hattuša, who were the keepers of the Babylonian tradition, were a small and privileged group who had sole access to their literature and to the documents on clay. The establishment of a library answered a need, and the use of the cuneiform ensured that the kingdom could maintain communication with its representatives abroad. But the tablet was, in effect, a banned document: it made no public proclamation of the sublimity of the god, nor of the grandeur of the king. Without doubt the Hittites felt that these imprinted cuneiform characters, mechanical and lacking expression, should take second place to a different writing more visual, more monumental, more apt for writing of divine effigies and royal profiles.... The hieroglyphs are made to be gazed upon, and contemplated upon walls of rock: they give life to a name just as a relief brings the whole person to life. [E. Laroche (1960)]

All the same, hieroglyphic writing survived the cuneiform after the destruction of the Hittite Empire around 1200 BCE. It served not only for religious and dedicatory purposes, but also, and perhaps above all, for lay purposes in business documents.

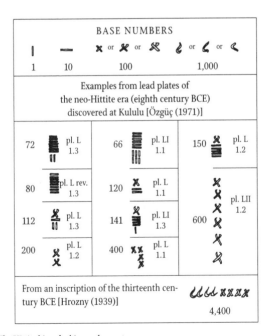

FIG. 15.9. *The Hittite hieroglyphic number-system*

In the Hittite hieroglyphic number-system, a vertical stroke represented unity. For the successive integers, small groups of two, three, four or five strokes were used to allow the eye to grasp the total sum of the units. The number 10 was represented by a horizontal stroke, a 100 by a kind of Saint Andrew's cross, and 1,000 by a sign which looked like a fish-hook (Figure 15.9). On this basis, the representation of intermediate numbers presented no difficulty, since it was sufficient to repeat each sign as many times as required.

The Hittite hieroglyphic number-system was, after the fashion of the Egyptian, strictly decimal and additive, since the only numbers to have specific signs were unity and the successive powers of 10.

CHAPTER 16

GREEK AND ROMAN NUMERALS

THE GREEK ACROPHONIC NUMBER-SYSTEM

Let us now visit the world of the Ancient Greeks, and look at the number-systems used in the monumental inscriptions of the first millennium BCE. The Attic system, which was used by the Athenians, assigns a specific sign to each of the numbers

$$1 \quad 5 \quad 10 \quad 50 \quad 100 \quad 500 \quad 1,000 \quad 5,000 \quad 10,000 \quad 50,000$$

and is based above all on the additive principle (Fig. 16.1).

1	I	100	H	10,000	M
2	II	200	HH	20,000	MM
3	III	300	HHH	30,000	MMM
4	IIII	400	HHHH	40,000	MMMM
5	Γ	500	ᴿ	50,000	ᴾ
6	ΓI	600	ᴿH	60,000	ᴾM
7	ΓII	700	ᴿHH	70,000	ᴾMM
8	ΓIII	800	ᴿHHH	80,000	ᴾMMM
9	ΓIIII	900	ᴿHHHH	90,000	ᴾMMMM
10	Δ	1,000	X		
20	ΔΔ	2,000	XX		
30	ΔΔΔ	3,000	XXX		
40	ΔΔΔΔ	4,000	XXXX		
50	ᴾ	5,000	ᴿ		
60	ᴾΔ	6,000	ᴿX		
70	ᴾΔΔ	7,000	ᴿXX		
80	ᴾΔΔΔ	8,000	ᴿXXX		
90	ᴾΔΔΔΔ	9,000	ᴿXXXX		

FIG. 16.1. *System of numerical annotation found in Attic inscriptions from around the fifth century BCE until the start of the Common Era. [Franz (1840); Guarducci (1967); Guitel; Gundermann (1899); Larfeld (1902–7); Reinach (1885); Tod]*

The Attic system has an interesting feature: with the exception of the vertical bar representing 1, the figures are simply the initial letters of the Greek names of the corresponding number, or are combinations of these: this is what is meant by an *acrophonic* number-system.

To show this:

THE SIGN	WHICH IS THE SAME AS THE LETTER	WHOSE VALUE IS	IS THE FIRST LETTER OF THE WORD	WHICH IS THE GREEK NAME OF THE NUMBER
Γ	PI (the archaic form of the letter Π)	5	Πεντε (Pente)	Five
Δ	DELTA	10	Δεκα (Deka)	Ten
H	ETA	100	Ηεκατον (Hekaton)	Hundred
X	KHI	1,000	Χιλιοι (Khilioi)	Thousand
M	MU	10,000	Μυριοι (Murioi)	Ten thousand

FIG. 16.2.

The signs for the numbers 50, 500, 5,000, and 50,000 are, as can be seen, made up by combining the preceding signs according to the multiplicative principle:

50	⌐Δ⌐ = Γ.Δ	5 ×	10
500	⌐H⌐ = Γ.H	5 ×	100
5,000	⌐X⌐ = Γ.X	5 ×	1,000
50,000	⌐M⌐ = Γ.M	5 ×	10,000

FIG. 16.3.

In other words, in the Attic system, in order to multiply the value of one of the alphabetic numerals Δ, H, X and M by 5, it is placed inside the letter Γ = 5.

This system, which in fact only recorded cardinal numbers, was used in metrology (to record weights, measure, etc.) and for sums of money. We shall later see it used for the Greek abacus.

Originally, ordinal numbers were spelled out in full, but from the fourth century BCE (probably, indeed, from the fifth) a different system was used to write these numbers, which we shall study later.

To write down a sum expressed in *drachmas*, the Athenians made use of these figures, repeating each one as often as required to add up to the quantity; each occurrence of the vertical bar for "1" was replaced by the symbol ⊢ which stood for "drachma":

XXX ⌐H⌐ H ΔΔΔ ⊢⊢⊢
3,000   500 100   30   3
- →
3,633 drachmas   FIG. 16.4.

For multiples of the *talent*, which was worth 6,000 *drachmas*, they used the same number of signs but with T (the first letter of TALANTON) instead of ⊢:

⌐H⌐ ⌐M⌐ΔΔΔΔ Γ T T T
500 50   40   5   3
- →
598 talents   FIG. 16.5.

For divisions of the *drachma* (the *obol*, the half- and the quarter-*obol*, and the *chalkos*) special signs were used:

1 CHALKOS (or 1/8 of an obol)	X	X: initial letter of ΧΑΛΚΟΥΣ
1 QUARTER-OBOL	Ɔ or T	T: initial letter of TETAPTHMOPION
1 HALF-OBOL	C	
1 OBOL (1/6 of a drachma)	I or O	O: initial letter of OBOΛION

FIG. 16.6.

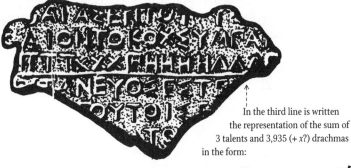

In the third line is written
the representation of the sum of
3 talents and 3,935 (+ *x*?) drachmas
in the form:

3	3,000	500	400	30	5
TALENTS		DRACHMAS			

FIG. 16.7. *Greek inscription (fragment) from Athens dating from the fifth century BCE. (Museum of Epigraphy, Athens. Inv. Em12 355)*

By the use of these signs, the Athenians were able to write easily those sums of money which were of relatively frequent occurrence. The following examples give the idea. (A quite similar system was also used for weights and measures such as the *drachma, mina,* and *stater.*)

ΔΔ ΗΗΗ	III C T	23 drachmas and (3 + 1/2 + 1/4) obols
20 3	3 ½ ¼	
drachmas	obols	
ΗΔΔΔ	IIII	*read:* 40 drachmas and 4 obols
40	4	
XX ᴨ HΔΔΔII		*read:* 2,630 drachmas and 2 obols
2,000 500 100 30 2		

XXXΗΗᴨ ΔTTT	XX ᴨΔΔΔΔΗΗΗ	IIIII
3,000 200 50 10 3	2,000 500 40 4	5
talents	drachmas	obols
3,263 talents	2,544 drachmas and 5 obols	

FIG. 16.8.

In the other states of the Ancient Greek world, the citizens also used similar acrophonic symbols in their various monumental inscriptions during the latter half of the first millennium BCE (Fig. 16.9 and 16.10). The Attic system itself, which is the oldest known of the Greek acrophonic systems, became more widespread at the time of Pericles, when the city of Athens was the capital of a number of Greek republics.

However it would be wrong to think that these different number-systems were all strictly identical to the Athenian one. Each had features which distinguished it from the others. We should not forget that each Greek state had its own system of weights and its own system of coinage (by this period the use of money was widespread throughout the Mediterranean). Furthermore the very notion of a unified metric system, on the lines of an international monetary system, was foreign to the Greek spirit.[*]

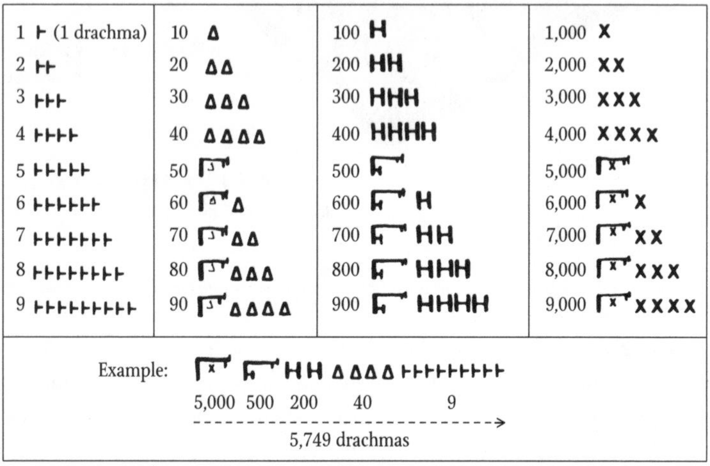

1 Ⱶ (1 drachma)	10 Δ	100 H	1,000 X
2 ⱵⱵ	20 ΔΔ	200 HH	2,000 XX
3 ⱵⱵⱵ	30 ΔΔΔ	300 HHH	3,000 XXX
4 ⱵⱵⱵⱵ	40 ΔΔΔΔ	400 HHHH	4,000 XXXX
5 ⱵⱵⱵⱵⱵ	50 Ⲅᴰ	500 Ⲅ	5,000 Ⲅˣ
6 ⱵⱵⱵⱵⱵⱵ	60 ⲄᴰΔ	600 Ⲅ H	6,000 Ⲅˣ X
7 ⱵⱵⱵⱵⱵⱵⱵ	70 ⲄᴰΔΔ	700 Ⲅ HH	7,000 Ⲅˣ XX
8 ⱵⱵⱵⱵⱵⱵⱵⱵ	80 ⲄᴰΔΔΔ	800 Ⲅ HHH	8,000 Ⲅˣ XXX
9 ⱵⱵⱵⱵⱵⱵⱵⱵⱵ	90 ⲄᴰΔΔΔΔ	900 Ⲅ HHHH	9,000 Ⲅˣ XXXX

Example: Ⲅˣ   Ⲅ HH ΔΔΔΔ ⱵⱵⱵⱵⱵⱵⱵⱵⱵ

     5,000   500   200   40   9

------------------------------→

5,749 drachmas

FIG. 16.9. *Numerical notation in Greek inscriptions from the island of Cos (third century BCE).* [*Tod*]

1 drachma	Ⱶ[*] or I[**]	
5	Ⲅ[*]	Π: first letter of Πεντε, "five"
10	ᗡ[**] or Δ[*]	Δ: first letter of Δεκα, "ten"
50	ⲅE or Ⲅᴱ[*]	ΠΕ: abbreviation of Πεντεδεκα, "fifty"
100	ⱵE	HE: abbreviation of Ηεκατον, "hundred"
300	ꞂE[*]	T.HE: abbreviation of Τριακοσιοι, "three hundred"
500	ⲤⱵE or ⲤⱵE	Π.HE: abbreviation of Πεντακοσιοι, "five hundred"
1,000	Ψ	Ancient Boeotian form of the letter X: first letter of Χιλιοι, "thousand"
5,000	Ⲅ	Π.X: abbreviation of Πενταχιλιοι, "five thousand"
10,000	M	Letter M: first letter of Μυριοι, "ten thousand"

[*] Found only at THESPIAE
[**] Found only at ORCHOMENOS

FIG. 16.10. *Numerical notation in Greek inscriptions from Orchomenos and from Thespiae (third century BCE)* [*Tod*]

[*] As P. Devambez (1966) explains: "Money was in the first place defined in terms of weight. Each state chose from its system of weights one unit to be the standard, and the others were multiples or sub-multiples of this. For instance, at Aegina in the Peloponnese, the standard unit of weight for commerce was the *mina* which weighed 628 gm. The unit of money was chosen to be one hundredth of this, the *drachma*, which therefore weighed 6.28 gm. The *didrachma* or *stater* was about twice this (12.57 gm). The sub-unit, the *obol*, weighing 1.04 gm, was a sixth of the *drachma*. At Euboea and in Attica, where the *mina* weighed 436 gm, the *drachma* was 4.36 gm; its multiples, the *didrachma* and *tetradrachma*, weighed twice and four times this, or 8.73 gm and 17.46 gm respectively; the *obol*, a sixth of the unit, weighed 0.73 gm."

Drachmas	Signs	Identifications
1	• I ⊢ P (1 2 3 4 5	1 Epidaurus, Argos, Nemea 2 Karystos, Orchomenos 3 Attica, Cos, Naxos, Nesos, Imbros, Thespiae 4 Corcyra (Corfu), Hermione (Kastri) 5 Troezen, Chersonesus Taurica (Korsun), Chalcidice
5	Γ ᴦ Ⲅ Γ Π 6 7 8 9 10	6 Epidaurus 7 Thera 8 Troezen 9 Attica, Corcyra, Naxos, Karystos, Nesos, Thebes, Thespiae, Chersonesus Taurica 10 Chalcidice, Imbros
10	O ⊙ — ⋛ 11 12 13 14 ↑ Λ Δ ▷ 15 15 16 17	11 Argos 12 Nemea 13 Epidaurus, Karystos 14 Troezen 15 Corcyra, Hermione 16 Attica, Cos, Naxos, Nesos, Mytilene, Imbros, Chersonesus Taurica, Chalcidice, Thespiae 17 Orchomenos, Hermione
50	Γ Ρ ᴘ ᴘ ᴀ 18 19 20 21 ᴘ Ψ ΓΕ Γᵋ 22 23 24 24	18 Argos 19 Epidaurus, Troezen, Cos, Naxos, Karystos 20 Nemea, Cos, Nesos, Attica, Thebes 21 Imbros 22 Troezen 23 Chersonesus Taurica 24 Thespiae, Orchomenos
100	目 H ⱶE 25 26 27 ⌐ ⊢ E 28 29 30	25 Epidaurus, Argos, Nemea, Troezen 26 Attica, Thebes, Cos, Epidaurus, Corcyra, Naxos, Chalcidice, Imbros 27 Thespiae, Orchomenos 28 Karystos 29 Chersonesus Taurica 30 Chersonesus Taurica, Chios, Nesos, Mytilene
500	ℾ ᴦ⸺ ᴦ ᴦ⁀ 31 32 33 34 ⱶ ᴦ ᴦ ᴦE 35 36 37 38	31 Troezen 32 Epidaurus 33 Karystos 34 Cos 35 Naxos 36 Epidaurus 37 Epidaurus, Troezen, Imbros, Thebes, Attica 38 Thespiae, Orchomenos

1,000	X Ψ
	 39 40
		39 Attica, Thebes, Epidaurus, Argos, Cos, Naxos, Troezen,
		     Karystos, Nesos, Mytilene, Imbros, Chalcidice,
		     Chersonesus Taurica
		 40 Thespiae, Orchomenos

5,000	ᴦˣ ᴦ⌐ 41 42	41 Attica, Cos, Thebes, Epidaurus, Troezen, Chalcidice, Imbros 42 Thespiae, Orchomenos
10,000	M Λ Χ 43 43 44	43 Attica, Epidaurus, Chalcidice, Imbros, Thespiae, Orchomenos 44 Attica
50,000	ᴦᴹ ᴘ 45 46	45 Attica 46 Imbros

Fig. 16.11. *Table of the numerical signs found in various Greek inscriptions of the period 1500–1000 BCE, used to express sums of money (in general, the numbers shown here refer to amounts in drachmas). When they are collected together as here we can see the common origin of all of the Greek acrophonic numerals which were in use at this time.* [Tod]

Bringing together all the different systems, we can observe their common origin (Fig. 16.11A and B).

FIG. 16.12. *The Ancient Greek world*

Looking now at Fig. 16.14, 16.15, and 16.16, we can see that the original number-systems were quite similar to the Egyptian hieroglyphic system and to the Cretan and Hittite systems.

The inconvenient feature of this kind of notation, in that it required multiple repetitions of identical symbols, led the Greeks to seek a simplification by assigning a specific sign to each of the numbers:

1	5	10	50	100	500	1,000	5,000	10,000

| | auxiliary base | | 5×10 | 10^2 | 5×10^2 | 10^3 | 5×10^3 | 10^4 |

FIG. 16.13.

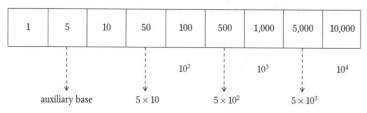

• 1 drachma	− 10 drachmas	⊟ 100 drachmas	X 1,000 drachmas
2 **:**	20 **=**	200 **⊟⊟**	2,000 **XX**
3 **∴**	30 **=−**	300 **⊟⊟⊟**	3,000 **XXX**
4 **::**	40 **==**	400 **⊟⊟⊟⊟**	4,000 **XXXX**

FIG. 16.14A.

:: • 5 drachmas	==− 50 drachmas	⊟⊟⊟⊟⊟ 500 drachmas	xxxxx 5,000 drachmas
6 :::	60 ===	600 ⊟⊟⊟⊟⊟⊟	6,000 xxxxxx
7 :::•	70 ===−	700 ⊟⊟⊟⊟⊟⊟⊟	7,000 xxxxxxx
8 ::::	80 ====	800 ⊟⊟⊟⊟⊟⊟⊟⊟	8,000 xxxxxxxx
9 ::::•	90 ====−	900 ⊟⊟⊟⊟⊟⊟⊟⊟⊟	9,000 xxxxxxxxx

⊟ : ancient form of the letter H; first letter of Ηεκατον, "hundred"
x : first letter of Χιλιοι, "thousand"

Example: x ⊟⊟⊟⊟⊟⊟⊟⊟⊟ ===− :::•
        1,000   900   70   9
        - →
                      1,979 DRACHMAS

FIG. 16.14B. *System of numerical notation in Ancient Greek inscriptions from Epidaurus (beginning of the fourth century BCE). This system, which is based on exactly the same principle as the Cretan number-systems, and is acrophonic for the numbers 100 and 1,000 only, has no symbols for 5, 50, 500, or 5,000. [Tod]*

1 •	10 ⊙	100 ⊟
2 :	20 ⊙⊙	200 ⊟⊟
3 :•	30 ⊙⊙⊙	300 ⊟⊟⊟
4 ::	40 ⊙⊙⊙⊙	400 ⊟⊟⊟⊟
5 ::•	50 Γ⊓	500 ⊟⊟⊟⊟⊟
6 :::	60 Γ⊓⊙	600 ⊟⊟⊟⊟⊟ ⊟
7 :::•	70 Γ⊓⊙⊙	700 ⊟⊟⊟⊟⊟ ⊟⊟
8 ::::	80 Γ⊓⊙⊙⊙	800 ⊟⊟⊟⊟⊟ ⊟⊟⊟
9 ::::•	90 Γ⊓⊙⊙⊙⊙	900 ⊟⊟⊟⊟⊟ ⊟⊟⊟⊟

Γ⊓ : sign Π.Δ. Abbreviation of Πεντε Δεκα, "fifty"

⊟ ⊟ ⊟ ⊟ Γ⊓⊙ ⊙ ⊙ ⊙ ::::
400   50   40   8
- →
            498 DRACHMAS

FIG. 16.15. *System of numerical notation in Greek inscriptions from Nemaea (fourth century BCE): a decimal system with a supplementary sign for 50 only [Tod]*

• 1 drachma	− 10 drachmas	⊟ H 100 drachmas	X 1,000 drachmas	M 10,000 drachmas
2 ••	20 =	200 HH	2,000 XX	20,000 MM
3 •••	30 =−	300 ⊟⊟⊟	3,000 XXX	30,000 MMM
4 ••••	40 ==	400 HHHH	4,000 XXXX	40,000 MMMM

FIG. 16.16A.

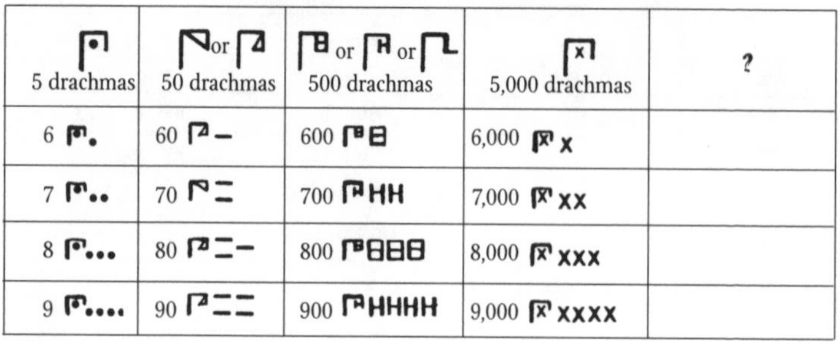

$\begin{array}{c}\fbox{$\cdot$}\\ \text{5 drachmas}\end{array}$	$\begin{array}{c}\text{or}\\ \text{50 drachmas}\end{array}$	$\begin{array}{c}\text{or}\,\,\text{or}\\ \text{500 drachmas}\end{array}$	$\begin{array}{c}\fbox{x}\\ \text{5,000 drachmas}\end{array}$	?
6 ⌐⌐.	60 ⌐⌐ –	600 ⌐⌐B	6,000 ⌐x⌐ x	
7 ⌐⌐..	70 ⌐⌐ ⁼	700 ⌐⌐HH	7,000 ⌐x⌐ xx	
8 ⌐⌐...	80 ⌐⌐ ⁼–	800 ⌐⌐BBB	8,000 ⌐x⌐ xxx	
9 ⌐⌐....	90 ⌐⌐ ⁼⁼	900 ⌐⌐HHHH	9,000 ⌐x⌐ xxxx	

FIG. 16.16B. *Numerals in late inscriptions from Epidaurus (end of fourth to middle of third centuries BCE) [Tod]*

They therefore arrived at a mathematical system equivalent to the one used by the South Arabs and the Romans.

Thereafter, no more than fifteen different signs were required in order to represent the number 7,699 for example, instead of the thirty-one that were needed in the Cretan and the archaic Greek system.

$$\underset{5,000}{\fbox{x}} \quad \underset{2,000}{\text{X X}} \quad \underset{500}{\text{⌐}} \quad \underset{100}{\text{H}} \quad \underset{50}{\fbox{⌐}} \quad \underset{40}{\Delta\Delta\Delta} \quad \underset{5}{\Gamma} \quad \underset{4}{\text{IIII}}$$

FIG. 16.17.

Nevertheless, this advance in notation was a step backwards in the evolution of arithmetic itself. In the beginning, the Greeks had assigned specific symbols only to unity and to each power of the base, and they were able to do written arithmetic after the fashion of the Egyptians. But once they had introduced supplementary figures into their initial set, the Greeks deprived it of all operational capability. As result, the Greek calculators thenceforth had to resort to "counting tables".

THE NUMBERS OF THE KINGDOM OF SHEBA

We now consider the numerical notation used by the ancient people of South Arabia, especially the Minaeans and the Shebans who shared what is now Yemen during the first millennium BCE. [M. Cohen (1958); J. C. Février (1959); M. Höfner (1943)]

The inscriptions which have come down to us from these peoples concern the most varied subjects: buildings constructed on several floors, irrigation systems retained by large dikes, offerings to the astral gods, animal sacrifices, tales of conquest, inventories of booty, and so forth. The writing, in which were written the neighbouring Semitic Arab languages, was no doubt derived (with some major changes) from the ancient Phoenician writing and had twenty-nine consonants represented by characters of geometric form, almost all of the same size. [M. Cohen (1958); J. G. Février (1959); M. Rodinson (1963)]

The system used by these people was based on the additive principle. A distinct symbol was assigned to each power of 10, and also to the number five and to the number fifty (Fig. 16.19).

Like the Greek systems which we have just analysed, this system was acrophonic in nature. Except for the signs for 1 and 50, all the others are letters of the alphabet, and are in fact the initial letters of the Semitic names of the numbers 5, 10, 100, and 1,000. (Quite possibly the South Arabs were influenced in this respect by the Greeks. This is conjectural, though we do in fact know from other studies that there were contacts between the Greeks, the Shebans, and the Minaeans.)

1	**I**			Simple vertical bar
5	**ᑌ** (a)	or	**ᑌ** (b)	Letter HA: first letter of HAMSAT, Southern Arabic word for "five"
10	**o**			Letter 'AYIN: first letter of the word 'ASARAT, "ten"
50	**ᑭ** (a)	or	**◀** (b)	Half of the sign for 100
100	**ᗷ** (a)	or	**◀** (b)	Letter MIM: first letter of the word MI'AT, "hundred"
1,000	**ᔓ** (a)	or	**ᔓ** (b)	Letter 'ALIF, first letter of the word 'ALF, "thousand"
(a) reading from left to right			(b) reading from right to left	

FIG. 16.18.

In the Minaean and Sheban inscriptions, numerals are usually enclosed between a pair of signs **ᒐ** and **ᒐ** in order to avoid confusion between letters representing numbers and letters standing for themselves (Fig. 16.22 and 16.24). It often happens, also, that the figures change orientation within the same inscription, since the South Arab writing was in *boustrophedon* (alternately from right to left and from left to right).

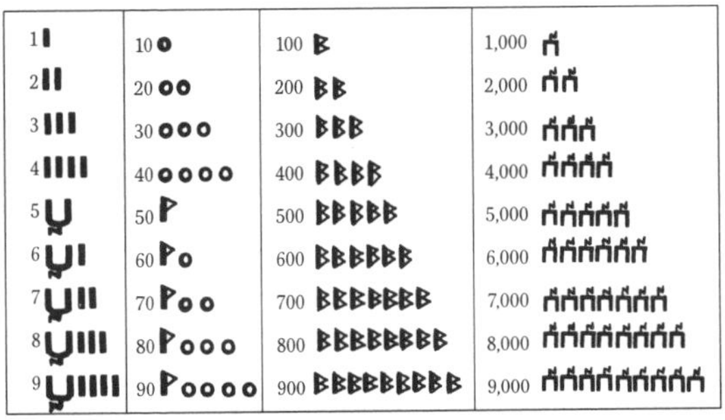

1 **I**	10 **o**	100 **ᗷ**	1,000 **ᔓ**
2 **II**	20 **oo**	200 **ᗷᗷ**	2,000 **ᔓᔓ**
3 **III**	30 **ooo**	300 **ᗷᗷᗷ**	3,000 **ᔓᔓᔓ**
4 **IIII**	40 **oooo**	400 **ᗷᗷᗷᗷ**	4,000 **ᔓᔓᔓᔓ**
5 **ᑌ**	50 **ᑭ**	500 **ᗷᗷᗷᗷᗷ**	5,000 **ᔓᔓᔓᔓᔓ**
6 **ᑌI**	60 **ᑭo**	600 **ᗷᗷᗷᗷᗷᗷ**	6,000 **ᔓᔓᔓᔓᔓᔓ**
7 **ᑌII**	70 **ᑭoo**	700 **ᗷᗷᗷᗷᗷᗷᗷ**	7,000 **ᔓᔓᔓᔓᔓᔓᔓ**
8 **ᑌIII**	80 **ᑭooo**	800 **ᗷᗷᗷᗷᗷᗷᗷᗷ**	8,000 **ᔓᔓᔓᔓᔓᔓᔓᔓ**
9 **ᑌIIII**	90 **ᑭoooo**	900 **ᗷᗷᗷᗷᗷᗷᗷᗷᗷ**	9,000 **ᔓᔓᔓᔓᔓᔓᔓᔓᔓ**

FIG. 16.19. *The symbols, and the principle, of the Southern Arabian number-system. This system is known only from the period from the fifth to the second or first centuries BCE. On inscriptions dating from after the beginning of the Common Era, it seems, numbers are spelled out in full.*

There is one interesting and important difference between the number-system of the South Arabs – at any rate those of Sheba – and the otherwise similar Greek system, in that the Arab system incorporated a rudimentary principle of position.

In fact, when one of the figures

o	◀ or ▶		◀ or ▷	
10	50		100	FIG. 16.20.

is placed to the right of the sign for 1,000 (when reading from right to left), then this figure is (mentally) multiplied by 1,000. In the following, for example:

2,000	30	50	200

← - - - - - - - - - - - - - - - - - -   FIG. 16.21.

we would at first be inclined to read the value

$$200 + 50 + 30 + 2,000 = 2,280$$

according to the traditional usage of the additive principle, whereas in fact it represented, in Sheba, the value

$$(200 + 50 + 30) \times 1,000 + 2,000 = 282,000$$

CIS IV: inscr. 924		5	Notice the irregularity
RES: inscr. 2740, 1.7		50	
RES: inscr. 2868, 1.4	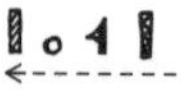	60	
RES: inscr. 2743, 1.10	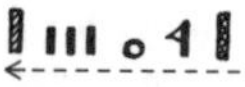	63	
RES: inscr. 2774, 1.4		47	
RES: inscr. 2965, 1.4	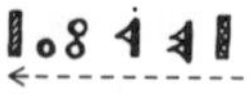	180	Note the unusual manner of writing the number 30: ○ / ○○ instead of ○○○

FIG. 16.22. *Examples taken from Minaean inscriptions (third to first century BCE). The numbers shown above refer to the volume capacity of certain recipients offered to the astral gods of ancient Southern Arabia, or to lists of offerings to these gods, or to animals which have been sacrificed. [C. Robin (personal communication)].*

Similarly, when reading from left to right, the same effect is produced by placing the figure to the left of the sign for 1,000. Thus

200	50	30	2,000

- - - - - - - - - - - - - - - - - - →   FIG. 16.23.

gives 282,000 (and not 2,280!).

RES: inscr. 3945, 1.13		500
RES: inscr. 3945, 1.13		3,000
RES: inscr. 3945, 1.19		12,000
CIS IV: inscr. 413, 1.2	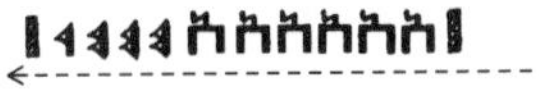	6,350
RES: inscr. 3945, 1.4	 10 6,000	16,000
RES: inscr. 3943, 1.2	 30 1,000	31,000
RES: inscr. 3943, 1.3	 40 5,000	45,000

Fig. 16.24. *Examples from inscriptions from the ancient kingdom of Sheba (fifth century BCE). These inscriptions, principally from the site of Sirwah, tell of military conquests and give various inventories: numbers of soldiers, material resources, booty, prisoners, and so on. [C. Robin, personal communication]*

However, this practice must have surely given rise to confusion among the readers, and the Sheban stone-cutters therefore took the precaution of also writing out in words the number represented by the figures.

A lucky precaution, for it has enabled us today to arrive at an unambiguous interpretation of this number-system!

ROMAN NUMERALS

Like the preceding systems, the Roman numerals allowed arithmetical calculation only with the greatest difficulty.

To be convinced of this, let us try to do an addition in these figures. Without translating into our own system, it is very difficult, if not impossible, to succeed.

The example which is most often cited is the following:

	CCXXXII		232
+	CCCCXIII	+	413
+	MCCXXXI	+	1,231
+	MDCCCLII	+	1,852
=	MMMDCCXXVIII		3,728

Roman numerals, in fact, were not signs which supported arithmetic operations, but simply abbreviations for writing down and recording numbers. This is why Roman accountants, and the calculators of the Middle Ages after them, always used the abacus with counters for arithmetical work.

As with the majority of the systems of antiquity, Roman numerals were primarily governed by the principle of addition. The figures (I= 1, V= 5, X= 10, L = 50, C = 100, D = 500 and M = 1,000) being independent of each other, placing them side by side implied, generally, addition of their values:

$$CLXXXVII = 100 + 50 + 10 + 10 + 10 + 5 + 1 + 1 = 187$$
$$MDCXXVI = 1,000 + 500 + 100 + 10 + 10 + 5 + 1 = 1,626$$

The Romans proceeded to complicate their system by introducing a rule according to which every numerical sign placed to the left of a sign of higher value is to be subtracted from the latter.

Thus the numbers 4, 9, 19, 40, 90, 400, and 900, for example, were often written in the forms

IV (= 5 – 1) instead of IIII   XC (= 100 – 10) instead of LXXXX
IX (= 10 – 1) instead of VIIII   CD (= 500 – 100) instead of CCCC
XIX (= 10 + 10 –1) instead of XVIIII   CM (= 1,000 – 100) instead of DCCCC
XL (= 50 – 10) instead of XXXX

It is remarkable that a people who, in the course of a few centuries, attained a very high technical level, should have preserved throughout that time a system which was needlessly complicated, unusable, and downright obsolete in concept.

In fact, the writing of the Roman numerals as well as its simultaneous use of the contradictory principles of addition and subtraction, are the vestiges of a distant past before logical thought was fully developed.

Roman numerals as we know them today seem at first sight to have been modelled on the letters of the Latin alphabet:

I	V	X	L	C	D	M
1	5	10	50	100	500	1,000

Fig. 16.25.

However, as T. Mommsen (1840) and E. Hübner (1885) have remarked, these graphic signs are not the first forms of the figures in this system. They were in fact preceded by much older forms which had nothing to do with letters of an alphabet. They are late modifications of much older forms.*

---

* The oldest known instances of the use of the letters L, D and M as numerals do not go back earlier than the first century BCE. As far as we know, the earliest Roman inscription which uses the letter L for 50 dates only from 44 BCE (CIL, I, inscr. 594). The earliest known use of the numerals M and D is in a Latin inscription which dates from 89 BCE, in which the number 1,500 is written as MD (CIL, IV, inscr. 590).

Originally, 1 was represented by a vertical line, the number 5 by a drawing of an acute angle, 10 by a cross, 50 by an acute angle with an additional vertical line, 500 by a semi-circle at an angle, and 1,000 by a circle with a superimposed cross (of which the *denarii* figure for 500 is geometrically one half):

ı	v	x	ѵ	✗	⟁	⊗
FIG. 16.26. 1	5	10	50	100	500	1,000

In an obvious way, the original figures for 1, 5 and 10 were assimilated to the letters I, V and X.

The original figure for 50 (which can still be found as late as the reign of Augustus, 27 BCE – 14 CE*) evolved progressively as shown below, finally merging with the letter L around the first century BCE:

$$\psi \rightarrow \downarrow \rightarrow \bot \rightarrow \bot \rightarrow \Bbb{L} \rightarrow L$$

FIG. 16.27.   50

The original figure for 100 initially evolved in a similar way towards a more rounded form:)(and then, for the sake of abbreviation, was split into one or other of the forms) or (. By similarity of shape, and under the influence of the initial letter of the Latin word *centum* ("one hundred"), it was finally assimilated to the letter C.

The original figure for 500 first of all underwent an anticlockwise rotation of 45°. It then evolved towards the sign Ð (these signs can still be found on texts from the Imperial period[†]) and finally was assimilated to the letter D:

$$Ⱥ \rightarrow Ᵽ \rightarrow ᵬ \rightarrow Ð \rightarrow D$$

FIG. 16.28.   500

The figure for 1,000 first of all evolved towards the form Φ. This gave rise to the many variant forms shown below for which, progressively, the letter M came to be substituted, from the first century BCE, under the influence of the first letter of the Latin word *mille*:

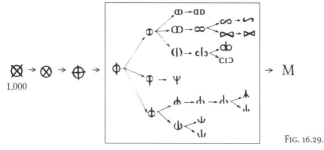

FIG. 16.29.

* CIL, IV, inscr. 9934
† CIL, VIII, inscr. 2557

1	I	CIL I 638, 1449		837	ƉCCCXXXVII	CIL I 638
2	II	CIL I 638, 744			⊕	CIL I 1533, 1578, 1853 and 2172
3	III	CIL I 1471			or ⋃	CIL X 39
4	IIII	CIL I 638, 587, 594			or ∞	CIL I 594 and 1853
5	Λ or V	CIL I 1449 / CIL I 590, 809, 1449, 1479, 1853		1,000	or ⋈	CIL X 1019
6	VI	CIL I 618			or ⋈	CIL VI 1251a
7	VII	CIL I 638			or M	CIL I 593
8	VIII	CIL I 698, 1471			M	CIL I 590
9	VIIII	CIL I 594, 590		1,200	∞ CC	CIL I 594
10	X	CIL I 638, 594, 809, 1449		1,500	MD	CIL I 590
14	XIIII	CIL I 594		2,000	∞∞	CIL I 594
15	XV	CIL I 1479		2,320	∞∞ CCCXX	CIL I 1853
19	XVIIII	CIL I 809		3,700	⊕⊕⊕ ƉCC	CIL I 25
20	XX	CIL I 638			Ɗ	CIL X 817
24	XXIIII	CIL I 1319		5,000	or Ɩ)	CIL I 1853 and 1533
40	XXXX	CIL I 594			or ☍	CIL I 2172
	↓	CIL I 214, 411 and 450		5,000	IƆƆ	CIL I 590, 594
	or ↓	CIL I 1471, 638, 1996		7,000	☍ ⊕⊕	CIL I 2172
50	or ⊥	CIL I 617, 1853		8,670	Ɩ) ⊕⊕⊕ DCLXX	CIL I 1853
	or ⌐	CIL I 744, 1853			(⊕)	CIL I 1252, 198
	or L	CIL I 594, 1479 and 1492			or ⍟	CIL I 583
51	↓	CIL I 638		10,000	or CCIƆƆ	CIL I 1474
74	↓ XXIIII	CIL I 638			or ⍟	CIL I 744
95	LXXXXV	CIL I 1479			or ⋃	CIL I 1724
100	C	CIL I 638, 594, 25, 1853		12,000	⍟ ⊕⊕	CIL I 1578
100	Ɔ	CIL VIII 21, 701		21,072	⋀⋀⊕ .LXXII	CIL I 744
300	CCC	CIL I 1853		30,000	CCIƆƆ CCI ƆƆ CCIƆƆ	CIL I 1474
400	CCCC	CIL I 638		30,000	⋃⋃⋃⋃	CIL I 1724
	Ð	CIL I 638, 1533 and 1853		50,000	IƆƆƆ	CIL I 593
500	or D	CIL I 590			((⍟))	CIL I 801
				100,000	or ((⍟))	CIL I 801
					or CCCIƆƆƆ	CIL I 594

FIG. 16.30. *Written numbers from monumental Latin inscriptions, dating from the Republican and early Imperial periods*

The various forms associated with the number 1,000 in Fig. 16.29 were mainly used during the period of the Republic, but they can also be found in some texts of the Imperial period.* A few of them even survived long after the fall of the Roman civilisation, since they can be found in quite a few printed works from the seventeenth century (Fig. 16.69 and 16.70).

28	XXIIX	CIL I 1319	140	CXL	CIL I 1492
45	XⱢV	CIL I 1996	268	CⱢXIIX	CIL I 617
69	LXIX	CIL I 594	286	CCXXCVI	CIL I 618
74	LXXIV	CIL I 594	340	CCCXⱢ	CIL I 1529
78	LXXIIX	CIL I 594	345	CCCXⱢV	CIL I 1853
79	LXXIX	CIL I 594	1,290	∞ CCXC	CIL I 1853

FIG. 16.31. *Latin inscriptions from the Republican era showing the use of the principle of subtraction. Use of this principle (which undoubtedly reflects the influence of the popular system on the monumental system) was nevertheless unusual on well-styled inscriptions.*

FIG. 16.32A. *Milestone engraving found at the Forum Popilii in Lucania (southern Italy), and made by C. Popilius Laenas, Consul in 172 BCE and 158 BCE. Now in the Museo della Civiltà Romana, Rome. [CIL, I, 638]*

line 4	ⱢI	51	line 7	CCXXXI	231
line 4	XXCIIII	84	line 7	CCXXXVII	237
line 5	Ⱡ XXIIII	74	line 8	CCCXXI	321
line 5	CXXIII	123	line 12	ꓷCCCCXVII	917
line 6	C Ⱡ XXX	180			

FIG. 16.32B. *Written numbers on the inscription shown in Fig. 16.32 A*

* CIL, IV, inscr. 1251; CIL, X, inscr. 39 and 1019; CIL, IL, inscr. 4397: etc.

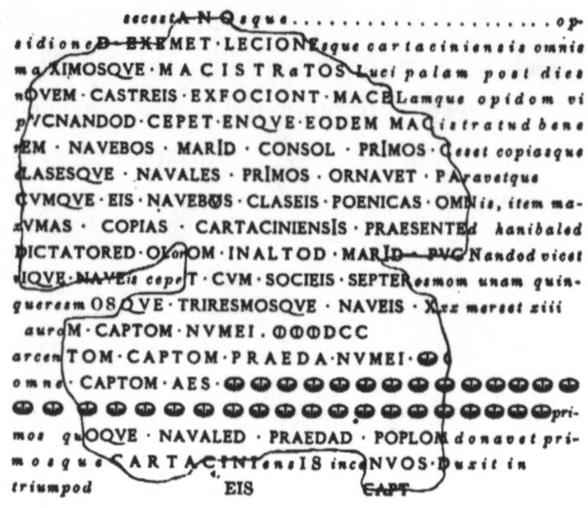

FIG. 16.33. Elogium *of Duilius, who conquered the Carthaginians at the battle of Mylae, 260 BCE. The inscription was re-cut at the start of the Imperial period, during the reign of Claudius (41– 54 CE), in the style of the third century BCE. Found in the Roman Forum at the place of the rostra* (columna rostrata)*, and now in the Palazzo dei Conservatori in Rome*
[CIL, I, 195].

In lines 15 and 16, the figure for 100,000 is repeated at least 23 times (and at most 33, according to the restoration by the Corpus). On line 13, the number 3,700 is written in the form:

ⅭⅠↃⅭⅠↃⅭⅠↃ DCC

Note: The capital letters (in upright characters in the figure) correspond to that part of the inscription which remains intact. The italic letters correspond to the restoration (by the Corpus) of the part which has been damaged.

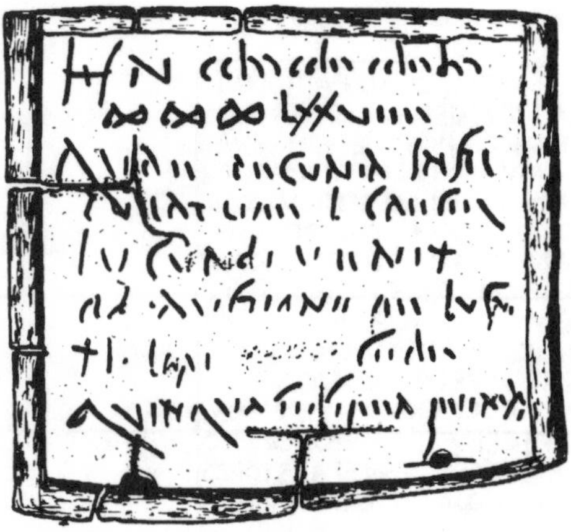

TRANSCRIPTION

HS n. ccIↄↄ ccIↄↄ ccIↄↄ Iↄↄ
∞ ∞ ∞ LXXVIII*

quæ pecunia in stipulatum
L. Caecili Iucundi venit ob
auctione (m) M. Lucreti Leri
[mer] cede quinquagesima
minu [s]

* See Fig. 16.29 and 16.30

FIG. 16.34. *Second panel of a triptych found at Pompeii, therefore prior to 79 CE (the year the city was destroyed)*

ETRUSCAN NUMERALS

Roman numerals reached their standardised form, identical to letters of the Latin alphabet, late in the history of Rome; but in reality they began life many hundreds of years, maybe even thousands, before Roman civilisation, and they were invented by others.

The Etruscans, a people whose origins and language both remain largely unknown, dominated the Italian peninsula from the seventh to the fourth century BCE, from the plain of the Po in the north to the Campania region, near Naples, in the south. They vanished as a distinct people at the time of the Roman Empire, becoming assimilated into the population of their conquerors.

Several centuries before Julius Caesar the Etruscans, and the other Italic peoples (the Oscans, the Aequians, the Umbrians, etc.), had in fact invented numerals with form and structure identical to those of the archaic Roman numerals.

1	**I**	CIE 5710	38	**X̅I̅XXX**	CIE 5707	
2	**II**	CIE 5708 TLE 26	42	**IIX XXX**	CIE 5710	
3	**III**	CIE 5741	44	**IIIIXXXX**	CIE 5748	
4	**IIII**	CIE 5748	50	**↑** or **Λ̅** or **↑**	CIE 5708, 5695, 5705, 5706, 5677 and 5763 / Buonamici, p. 245	
5	**Λ** or **∩**	CIE 5705, 5706, 5683, 5677 and 5741 / ACII, Table IV 114	52	**II↑**	CIE 5708	
6	**IΛ**	CIE 5700	55	**Λ↑**	CIE 5705 and 5706	
7	**IIΛ**	CIE 5635	60	**X↑**	CIE 5695	
8	**IIII∩**	ACII, Table IV 114	75	**ΛX·X↑**	CIE 5677	
9	**IIIIΛ**	CIE 5673	82	**II+++↑**	TLE 26	
10	**X** or **X̅** or **+**	CIE 5683, 5741, 5710 5748, 5695, 5763, 5797 5707, 5711 and 5834 / CIE 5689 and 5677 / TLE 126	86	**IIIIIXXX↑**	CIE 5763	
19	**XIX**	CIE 5797	100	**✳** or **✱** or **⫫C**	ACII, Table IV 114 / Buonamici, p. 473	
36	**IΛXXX**	CIE 5683	106	**IΛ✳**	SE, XXIII, series II (1965), p. 473	
38	**IIIΛXXX**	CIE 5741				

FIG. 16.35. *Written numbers from Etruscan inscriptions*

FIG. 16.36. *Etruscan coins dating from the fifth century BCE bearing the numbers*

Λ and X
5 10

Collection of the Landes museum, Darmstadt [Menninger (1957) vol. II, p. 48]

For many centuries, they used these figures according to the principles of addition and subtraction simultaneously. This is evidenced by several Etruscan inscriptions of the sixth century BCE, where the numbers 19 and 38 are written on the subtractive principle as 10 + (10 − 1) and 10 + 10 + 10 + (10 − 2) (Fig. 16.35).

FIG. 16.37. *Fragments of an Etruscan inscription bearing the numbers:*

160   208   15

[ACII table IV 114]

A QUESTIONABLE HYPOTHESIS

A hypothesis commonly accepted nowadays asserts that all of these numerals derived from Etruscan numerals, themselves of Greek origin.

We should recall that Latin writing derived from Etruscan writing, and that this comes directly from Greek writing. The Greek alphabets fall into two groups: the Western type which (like the Chalcidean alphabet, for example) assigned the sound "kh" to the letter **Y** or **↓** or **V**; the Eastern type which (like the alphabet of Miletus or Corinth, for example) assigned to this symbol the sound "ps", while the sound "kh" is represented by the letter + or x. Etruscan writing, for several reasons, is associated with the Western type.

Therefore it has come to be supposed that the Etruscan alphabet "was borrowed from a Greek alphabet of Western type on the land of Italy itself, since the oldest of the Greek colonies which had such an alphabet, that of Kumi, dates from 750 BCE, and its establishment precedes the birth of the Tuscan civilisation by half a century." [R. Bloch (1963)]

On this basis, having compared the forms of the letters, many specialists in the Roman numbering system have therefore inferred that the ancient Latin signs for the numbers 50, 100 and 1,000 come respectively from the following letters, which belong to the Chalcidean alphabet (a Greek alphabet of Western type used, as it happens, in the Greek colonies in Sicily). These letters represented sounds which did not occur in Etruscan or in Latin, and later became assimilated to the Latin forms which we know.

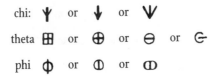

According to this hypothesis, the Greek letter *theta* Θ (originally ⊞ or ⊕) gradually turned into C, under the influence of the initial letter of the Latin word *centum*.

This explanation (which many Hellenists, epigraphers, and historians of science now hold as dogma) is seductive, but it cannot be accepted.

Why, in fact, should three particular foreign characters be introduced into the Roman number-system, and three only? And why should they be letters of the alphabet? No doubt, one may reply, because the Greeks themselves had often used letters of their alphabet as number-signs.

In antiquity, it is true, the Hellenes used two different systems of written numerals whose figures were in fact the letters of their alphabet. One of these used the initial letters of the names of the numbers. The other made use of all the letters of the alphabet (see Fig. 17.27 below):

A	Alpha	1	I	Iota	10	P	Rho	100
B	Beta	2	K	Kappa	20	Σ	Sigma	200
Γ	Gamma	3	Λ	Lambda	30	T	Tau	300
Δ	Delta	4	M	Mu	40	Υ	Upsilon	400
E	Epsilon	5	N	Nu	50	Φ	Phi	500
...	...	...	Ξ	Xi	60	X	Chi	600
H	Eta	8	O	Omicron	70	Ψ	Psi	700
Θ	Theta	9	...	...	...	Ω	Omega	800

Now the letter *chi*, which was supposed to be borrowed for the number 50 in Latin, has value 1,000 in the first of these systems, and 600 in the second; the letter *theta*, "borrowed" for the number 100 in Latin, has value 9 in the second Greek version; and the letter *phi*, supposed to have been borrowed for the Roman numeral for 1,000, is worth 500 in the second system. Why the differences?

If the Romans had borrowed the following Greek signs for the numbers 50 and 100:

chi: Ψ or Ѵ or ↓

theta ⊕ or ⊖ or Ͼ

then the same would probably have been borrowed by the Etruscans as well. How then can we explain that, for the same values, the Etruscans in fact used quite different figures, namely (see Fig. 16.35):

$$\text{\Large A} \quad \text{or} \quad \text{\Large ↑} \quad \text{for } 50$$

$$\text{and} \quad \text{\Large ✱} \quad \text{or} \quad \text{\Large ✸} \quad \text{for } 100$$

One can see that the hypothesis is not very sound. The error is due to the fact that specialists have believed through many generations that Roman numerals are the children of Etruscan numerals, whereas in fact they are cousins.

THE ORIGIN OF ROMAN NUMERALS

Though long obscure, the question is no longer in doubt. The signs I, V and X are by far the oldest in the series. Older than any kind of writing, older therefore than any alphabet, these figures, and their corresponding values, come naturally to the human mind under certain conditions. In other words, the Roman and Etruscan numerals are real prehistoric fossils: they are descended directly from the principle of the notched stick for counting, a primitive arithmetic, performed by cutting notches on a fragment of bone or on a wooden stick, which anyone can use in order to establish a one-to-one correspondence between the objects to be counted and the objects used to count them.

Let us imagine a herdsman who is in the habit of noting the number of his beasts using this simple prehistoric method.

Up to now, he has always counted as his forebears did, cutting in a completely regular manner as many notches as there are beasts in his herd. This is not very useful, however, because whenever he wants to know how many beasts he has, he has to count every notch on his stick, all over again.

The human eye is not a particularly good measuring instrument. Its capacity to perceive a number directly does not go beyond the number 4. Just like everyone else, our herdsman can easily recognise at a glance, without counting, one, two, three, or even four parallel cuts. But his intuitive perception of number stops there for, beyond four, the separate notches will be muddled in his mind, and he will have to resort to a procedure of abstract counting in order to learn the exact number.

Our herdsman, who has perceived the problem, is beginning to look for a way round it. One day, he has an idea.

As always, he makes his beasts pass by one by one. As each one passes, he makes a fresh notch on his tally stick. But this time, once he has made four marks he cuts the next one, the fifth, differently, so that it can be recognised at a glance. So at the number 5, therefore, he creates a new unit of counting which of course is quite familiar to him since it is the number of fingers on one hand.

For any individual, cutting into wood or bone presents the same problems, and will lead to the same solutions, whether in Africa or Asia, in Oceania, in Europe or in America.

Our herdsman only has a limited number of options. To distinguish the fifth notch from the first four, the first idea he has is simply to change the direction of cut. He therefore sets this one very oblique to the other four, and thereby obtains a representation all the more intuitive in that it reflects the angle that the thumb makes with the other four fingers.

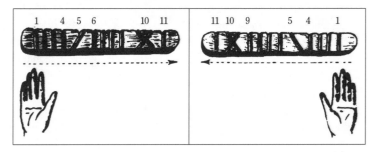

FIG. 16.38.

Another idea is to augment the fifth notch by adding a small supplementary notch (oblique or horizontal), so that the result is a distinctive sign in the form of a "t", a "Y" or a "V", variously oriented:

V ∧ < > Y ⅄ ⊣ ⊢ ⅄ ⋏

He resumes cutting notches in the same way as the first four, counting his beasts up to the ninth. But, at the tenth, he finds he must once again modify the notch so that it can be recognised at a glance. Since this is the total number of fingers on the two hands together, he therefore thinks of a mark which shall be some kind of double of the first. And so, as in all the numeral systems, he comes to make a mark in the form of an "X" or a cross:

✗ ⤬ ⤬ ✛

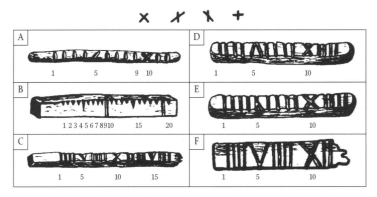

FIG. 16.39. *Anyone who counts by cutting notches on sticks will come to represent the numbers 1, 5, 10, 15, and so on in one of the above ways.*

So he has now created another numerical unit, the ten, and counting on the tally stick henceforth agrees with basic finger-counting.

Reverting to his simple notches, the herdsman continues to count beasts until the fourteenth and then, to help the eye to distinguish the fifteenth from the preceding ones, he again gives it a different form. But this time he does not create a new symbol. He simply gives it the same form as the "figure" 5, since it is like "one hand after the two hands together".

He carries on as before up to 19, and then he makes the twentieth the same as the tenth. Then again up to 24 with the ordinary notches, and the twenty-fifth is marked with the figure 5. And so on up to $9 + 4 \times 10 = 49$.

This time, however, he must once more imagine a new sign to mark the number 50, because he is not able to visually recognise more than four signs representing 10.

This is naturally done by adding a third cut to his notch, so he naturally chooses one of the following which can be made by adding one notch to one of the representations of the number 5:

$$\text{V} \quad \text{∧} \quad \text{Ν} \quad \text{Ν} \quad \text{Κ} \quad \text{Я} \quad \text{Λ} \quad \text{Υ} \quad \text{Λ} \quad \text{У} \quad \text{Ⴖ} \quad \text{Ⴉ}$$

Having done this, he can now proceed in the same way until he has gone through all the numbers from 50 to $50 + 49 = 99$.

At the hundredth, our herdsman once again faces the problem of making a distinct new mark. So equally naturally he will choose one of the following which can be made either by adding a further notch to one of the representations of 10, or by making a double of one of the representations of 50:

$$\text{✶} \quad \text{Ɱ} \quad \text{Ⴗ} \quad \text{Ⴗ} \quad \text{Ⴟ} \quad \text{Η} \quad \text{Ⱨ} \quad \text{K}$$

Again as before, he continues counting up to $100 + 49 = 149$. For the next number, he re-uses the sign for 50 and then continues in the same way up to $150 + 49 = 199$.

At 200, he re-uses the figure for 100 and continues up to $200 + 49 = 249$. And so on until he reaches $99 + 4 \times 100 = 499$.

Now he creates a new sign for 500 and continues as before until $500 + 499 = 999$. Then another new sign for 1,000 which will allow him to continue the numbers up to 4,999 ($= 999 + 4 \times 1,000$), and so on.

And so, despite not being able to perceive visually a series of more than four similar signs, our herdsman, thanks to some well-thought-out notch-cutting, can now nonetheless perceive numbers such as 50, 100, 500, or 1,000, without having to count all the notches one by one. And if he runs out of space on his tally stick and cannot reach one of these numbers, then all he needs to do is to make as many more tally sticks as he needs.

When the notches are cut in a structured way like this, it is possible to go up to quite large numbers, as large as are likely to be needed in practice, without ever having to take account of any series of more than four signs of the same kind. Such a technique is therefore like a lever, the mechanical instrument which allows someone to raise loads whose weight far exceeds his raw physical strength.

The procedure also defines a written number-system which gives a distinct figure to each of the terms of the series

$$
\begin{aligned}
1 & \\
5 & \\
10 & = 5 \times 2 \\
50 & = 5 \times 2 \times 5 \\
100 & = 5 \times 2 \times 5 \times 2 \\
500 & = 5 \times 2 \times 5 \times 2 \times 5 \\
1{,}000 & = 5 \times 2 \times 5 \times 2 \times 5 \times 2 \\
5{,}000 & = 5 \times 2 \times 5 \times 2 \times 5 \times 2 \times 5
\end{aligned}
$$

Our herdsman's approach to cutting notches on sticks therefore gives rise to a decimal system in which the number 5 is an auxiliary base (and the numbers 2 and 5 are alternating bases), and its successive orders of magnitude are exactly the same as in the Roman system; furthermore, it will naturally give rise to graphical forms for the figures which are closely comparable with those in the archaic Roman and Etruscan systems.

Again, the use at the same time of both the additive and the subtractive principles in the Etruscan and Roman systems is yet another relic of this ancient procedure.

To return to our herdsman. Now that he has counted his various beasts under various categories, he wants to transcribe the results of this breakdown onto a wooden board. In total 144, his beasts are distributed as:

26 dairy cows
35 sterile cows
39 steers
44 bulls

In order to write down one of these numbers, say the steers, the first idea which occurs to him is to mark these by simply copying the marks of the tally stick onto the board:

IIII	V	IIII	X	IIII	V	IIII	X	IIII	V	IIII	X	IIII	V	IIII
1	5		10		15		20		25		30		35	39

Fig. 16.40.

But he soon becomes aware that such a *cardinal notation* is very tedious, because it brings in all of the successive marks made on the stick. To get around this difficulty, he therefore thinks of an ordinal kind of representation, much more abridged and convenient than the preceding one. For the numbers from 1 to 4, he at first adopts a cardinal notation writing them successively as

I II III IIII

He can hardly do otherwise for, to indicate that one of the lines is the third in the series, he must mark two others before it, in order that it shall be clear that it is indeed the third.

He does not do the same for the number 5, however, since this already has its own sign ("V", say), which distinguishes it from the preceding four. Therefore this "V" is sufficient in itself and dispenses with the need to transcribe the four notches that precede it on the tally stick. Instead of transcribing this number as IIIIV, all he needs to do is to write V.

Starting from this point, the number 6 (the next notch after the V) can be written simply VI, and not IIIIVI; the number 7 can be written as VII, and so on up to VIIII (= 9).

In turn, the sign in the shape of an "X" can represent the tenth mark in the series all on its own, and renders the nine preceding signs superfluous. On the same principle, the numbers 11, 12, 13, and 14 can be written as XI, XII, XIII, and XIIII (and not IIIIVIIIIXI, etc.). Now the number 15 can be written simply XV (and not IIIIVIIIIXIIIIV nor XIIIIV): each X can erase the nine preceding marks, and the last V the four preceding marks. The numbers from 16 to 19 can be written XVI, XVII, XVIII, XVIIII. Then, for the number 20, which corresponds to the second "X" in the series, we can write XX. And so on.

When he has counted his animals by means of the notches on his sticks, our herdsman can now transcribe the breakdown onto his wooden board:

XXVI	(= 26)	for the dairy cows
XXXV	(= 35)	for the sterile cows
XXXVIIII	(= 39)	for the steers
XXXXIIII	(= 44)	for the bulls

However, looking for ways of shortening the work, our herdsman comes up with another idea. Instead of writing the number 4 using four lines (IIII), he writes it as IV, which is a way of marking the "I" as the fourth in the series on the stick, since this is the one that comes before the "V":

IIII ➤ (III)IV ➤ IV

In this way he cuts down on the number of symbols to write, saving 2. In the same way, instead of writing the number 9 as VIIII, he writes it as IX since this likewise marks the "I" as the ninth mark in the series on the notched stick:

IIIIVIIII ➤ (IIIIVIII)IX ➤ IX

He again cuts down on the number of symbols, saving 3. He does likewise for the numbers 14, 19, 24, and so on.

This is how one can explain why the Roman and Etruscan number-systems use forms such as IV, IX, XIV, XIX, etc., as well as IIII, VIIII, XIIII, XVIIII.

We can now conceive that all of the peoples who, for long ages, had been using the principle of notches on sticks for the purpose of counting should, in the course of time, with exactly the same motives as our herdsman and quite independently of any influence from the Romans or Etruscans themselves, be led to invent number-systems which are graphically and mathematically equivalent to the Roman and Etruscan systems.

This hypothesis seems so obvious that it could be accepted even if there were no concrete evidence for it. But such evidence exists, and in plenty.

A REVEALING ETYMOLOGY

It is hardly an accident that the Latin for "counting" should refer to the practicalities of this primitive method of doing it.

In Latin, "to count" is *rationem putare*.* As M. Yon (cited by L. Gerschel) points out, the term *ratio* not only refers to counting,[†] but also has a meaning of "relationship" or "proportion between things".[‡]

Surely this is because, for the Romans, this word referred originally to the practice of notching, since counting, in a notch-based system, is a matter of establishing a correspondence, or one-to-one relationship, between a set of things and a series of notches. Gerschel has demonstrated this with a large number of examples.

As for the word *putare*:

> This strictly means to remove, to cut out from something what is superfluous, what is not indispensable, or what is damaging or foreign to that thing, leaving only what appears to be useful and without flaw. In everyday life it was employed above all to refer to cutting back a tree, to pruning. [L. Gerschel (1960)]

To sum up:

> In the method of counting described by the expression *rationem putare*, if the term *ratio* means representing each thing counted by a corresponding mark, then the action denoted by *putare* consists of cutting into a stick with a knife in order to create this mark: as many as there

---

* To one of his contemporaries, Plautus wrote: *Postquam comedit rem, post rationem putat.* "It is now that he has consumed his resources that he counts the cost!" (*Trin.* 417).

† We find an example of this use in Cicero (*Flacc.* XXVII, 69): *Auri ratio constat; aurum in aerario est.* "The count of the gold is correct; the gold is in the public treasury."

‡ Cato (in *Agr.* I, 5) uses the expression *pro ratione* to mean "in proportion", giving *ratio* an arithmetical sense; Vitruvius (III, 3, 7) also uses the expression to mean "architectural proportion".

are things to count, so many are the notches on the stick, made by cutting out from the wood a small superfluous portion, as in the definition of *putare*. In a way, *ratio* is the mind which sees each object in relation to a mark; *putare* is the hand which cuts the mark in the wood. [L. Gerschel (1960)]

FURTHER CONFIRMATION

A different confirmation from the preceding is given by F. Škarpa (1934) in a detailed study of the different kinds of notches used since time immemorial by herdsmen of Dalmatia (in the former Yugoslavia). In one of these, the number 1 is represented by a small line, the number 5 by a slightly longer one, and the number 10 by a line which is much longer than the others (which is very reminiscent of the measuring scales on rulers and on thermometers).

Another type of marking used by the Dalmatian herdsmen represented the number 1 by a vertical stroke, the number 5 by an oblique stroke, and the number 10 by a cross. In a third type, for the numbers 1, 5 and 10 we find:

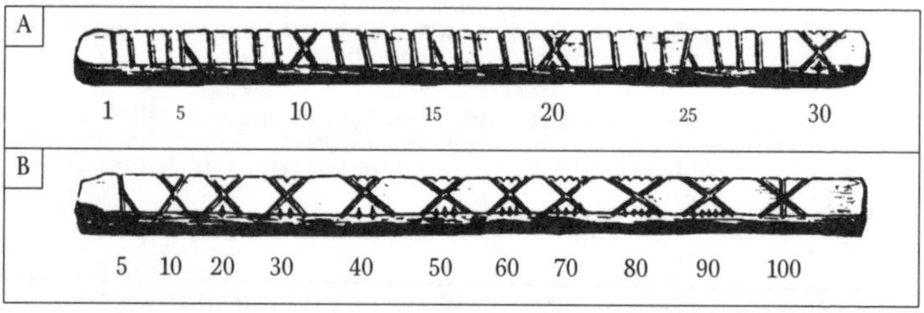

FIG. 16.41. *Herdsmen's tally sticks from Dalmatia [Škarpa (1934), Table II]*

Does this not rather closely resemble the Roman and Etruscan numerals? This is all the more striking in that this same type of tally (Fig. 16.41) shows that the sign for 100, as below, is identical to the Etruscan figure for 100:

FIG. 16.42.

We may well ask why these people used such a figure for 100, but not the "half-figure" for 50 as the Etruscans did (Fig. 16.35). But close examination of the tally stick in question tells us why. We find that the marks for the tens, from 20 to 90, differ from the others by having very small notches on the edge of the stick, above and below, and the number of these small notches gives the corresponding number of tens.

This method of marking can dispense with a separate sign for 50. Suppose that a herdsman wants to keep a record of the fact that he has 83 dairy cows and 77 sterile cows, having already counted them as above. All he needs to do is write the results as shown below on a separate piece of wood which he will keep by him:

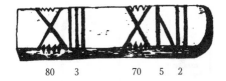

FIG. 16.43.   80   3   70   5   2

So this herdsman has no need for a special symbol for 50, since he has executed the idea described above as shown in Fig. 16.44:

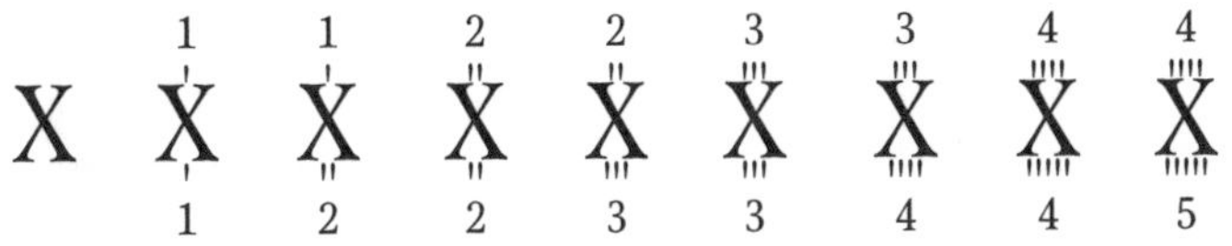

10	20	30	40	50	60	70	80	90

FIG. 16.44.

A final type of tally known from Dalmatia gives the following figures:

I	V	X	N	●
1	5	10	50	100

FIG. 16.45.

The presence of the sign N for 50 seems fairly natural since it can be made by adding a vertical bar to the figure for 5, just as the sign for 100 is made by adding a vertical bar to the sign X (Fig. 16.41).

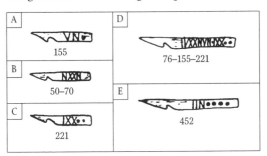

FIG. 16.46. *Herdsmen's tally sticks from Dalmatia [Škarpa(1934), Table IV]*

Very similar series are to be found in the Tyrol and in the Swiss Alps. They are found at Saanen on peasant double-entry tallies, at Ulrichen on milk-measuring sticks, as well as at Visperterminen on the famous *tallies of capital,* where sums of money lent by the commune or by religious foundations to the townsfolk were noted using the figures:

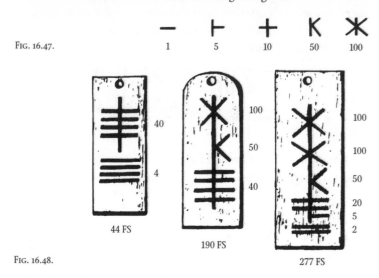

Fig. 16.47.

Fig. 16.48.

Further evidence can be found in the *calendrical ciphers,* strange numerical signs on the calendrical boards and sticks which were in use from the end of the Middle Ages up to the seventeenth century in the Anglo-Saxon and West Germanic world, from Austria to Scandinavia (Fig. 16.52 to 16.54).

1 2 3 4	5	6 7 8 9	10	11 12 13 14	15	16 17 18 19

Fig. 16.49.

These wooden almanacs give the Golden Number of the nineteen-year Metonic cycle in the graphical variants of the numerals shown on Fig. 16.52 to 16.54. See E. Schnippel (1926).

The figures used in the English clog almanacs of the Renaissance (see Fig. 16.54) are the following:

1 2 3 4	5	6 7 8 9	10	11 12 13 14	15	16 17 18 19	20	21 22

Fig. 16.50.

and those used in Scandinavian runic almanacs have the prototype:

·	:	:·	::	>	>>>> · :· :: ::·	+	++++ · :· :: ::·	+ >	++++ >>>> · :· :: ::·	+ +	++ ++ · ·
1	2	3	4	5	6 7 8 9	10	11 12 13 14	15	16 17 18 19	20	21 22

FIG. 16.51.

In all of these notations, which appear dissimilar at first sight but which on close examination prove to originate with the tally-stick principle, the signs given to the numbers 1, 5 and 10 are unmistakably similar to the Roman numerals I, V and X and to the Etruscan numerals I, Λ, and + or X.

FIG. 16.52. *"Page" from a wooden almanac (Figdorschen Collection, Vienna, no. 799) [Riegl (1888), Table I]*

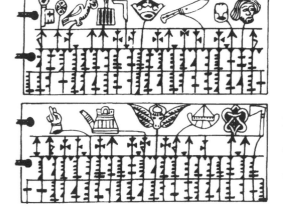

FIG. 16.53. *Two "pages" from a wooden almanac from the Tyrol (15th century) (Figdorschen Collection, Vienna, no. 800) [Riegl (1888), Table V]*

Fig. 16.54. *English clog almanac from the Renaissance (Ashmolean Museum, Oxford, Clog C)* *[Schnippel (1926), Table IIIa]*

Even better, in the nineteenth century the Zuñis (Pueblo Indians of North America living in New Mexico, at the Arizona frontier, whose traditions go back 2,000 years) still used "irrigation sticks" [F. H. Cushing (1920)] inscribed with numerals that were just as close to Roman figures:

- a simple notch for the number 1;
- a deeper notch, or an oblique one, for 5;
- a sign in the shape of an X for 10.

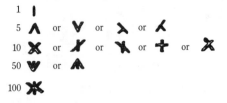

Fig. 16.55. *Zuñi irrigation stick (New Mexico). The tally marked towards the right of the stick totals 24, which is marked as a number at the left-hand end in the form XXI\,which is reminiscent of the Roman representation XXIV where the principle of subtraction has been applied. [Cushing (1920)]*

There can now be no possible doubt: Roman and Etruscan numerals derive directly from counting on tally sticks.

We are now in a position to put forward the following explanation of the genesis of such numerals.

Pastoral peoples, who lived in Italy long before the Etruscans and the Romans, since earliest antiquity (and possibly even in prehistoric times) counted by the method of tally sticks, and the Dalmatian herdsmen or the Zuñis, for example, independently discovered the same for their own use. In a quite natural way, all came to make use of the following signs:

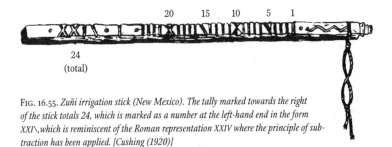

Fig. 16.56.

Inheriting this ancient tradition, the Etruscans and the Romans who came after them retained from these only the following:

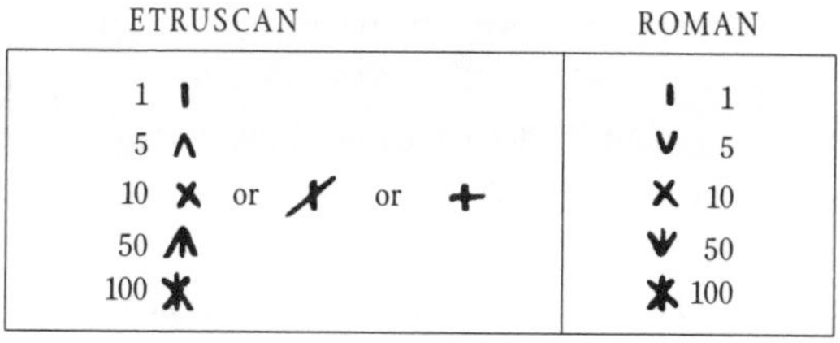

ETRUSCAN					ROMAN	
1	**I**				**I**	1
5	**Λ**				**V**	5
10	**X**	or	**✗**	or	**X**	10
50	**Λ**				**V**	50
100	**✗**				**✗**	100

FIG. 16.57A.

The Romans then completed the series by adding a sign for 500, and another for 1,000 (the former was the right-hand half of the latter, itself generated by drawing a circle on top of the figure for 10 (see Fig. 16.29, 16.30 and 16.35):

FIG. 16.57B.   Figures for 1,000 ⊗ or ⊕

In their hands, these signs changed form over the centuries until they were replaced by the alphabetic numerals which we know.

This, therefore, is the most plausible explanation of the origins of the Roman and Etruscan numerals. The following does not gainsay it. A. P. Ninni (1899) reports that the Tuscan peasants and herdsmen were still using, in the last century, in preference to Arabic numerals, the following signs which they call *cifre chioggiotti*:

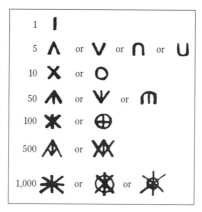

FIG. 16.58.

G. Buonamici (1932) saw these as descending from Etruscan or Roman numerals; may we not with more reason see them as a survival of the ancient practice of counting by cutting notches, a practice older than any writing and one which is to be found in every rural community on earth?

ROMAN NUMERALS FOR LARGE NUMBERS

The largest number for which Roman numerals as we know them (and still sometimes use them) had a separate symbol is 1,000. The simple application of the additive principle to the seven basic figures of this system would only take us up to 5,000. Therefore, when we come to make use of these numerals, we find it effectively impossible to write large numbers. How do we represent, say, 87,000, except by writing down 87 copies of the letter M?

The ancients had some trouble getting round the problem, and adopted a variety of conventions for writing large numbers. The difficulties which they encountered, as did their successors in the European Middle Ages, deserve special consideration.

In the Republican period, the Romans had a simple graphical procedure by which they could assign a special notation to the numbers 5,000, 10,000, 50,000, and 100,000. The principal ones (found sporadically as late as the Renaissance) are the following:

5,000

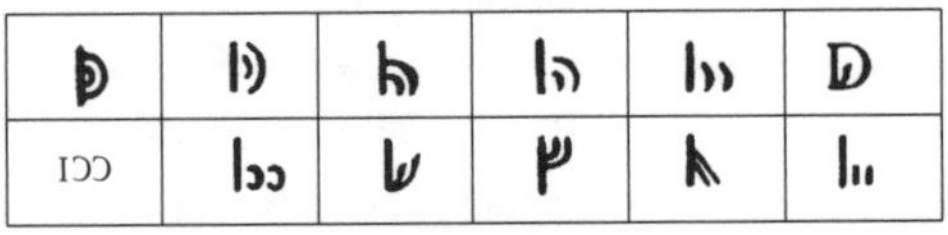

FIG. 16.59A.

10,000

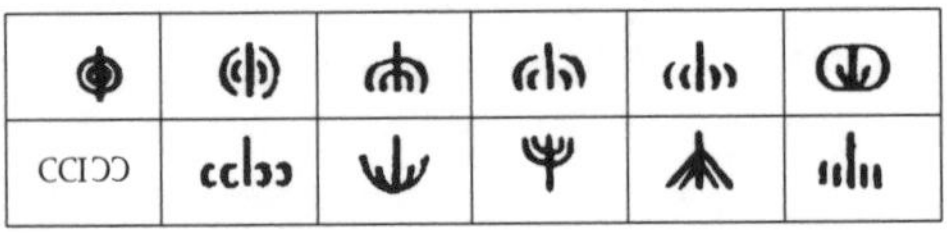

FIG. 16.59B.

50,000

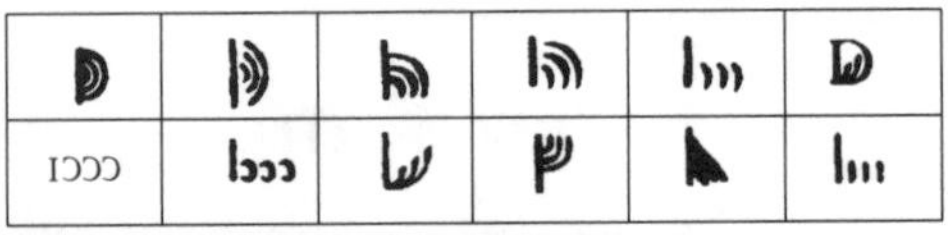

FIG. 16.59C.

100,000

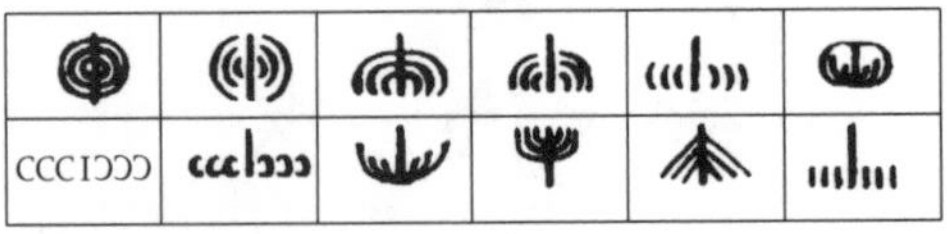

FIG. 16.59D.

Comparing these with each other, and with the various ancient forms of the symbol for 1,000 (Fig. 16.30), we realise that they have a common origin. In fact, they are simply stylisations (more or less recognisable) of the original five signs.

The idea governing the formation of four of these consists of an extremely simple geometrical procedure. Taking as a starting point the primitive Roman sign for 1,000 (originally a circle divided in two by a vertical line), the signs for 10,000 and 100,000 were made by drawing one or two circles, respectively, around it; and the signs for 5,000 and for 50,000 were made by using the right-hand halves of these (Fig. 16.62):

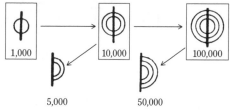

FIG. 16.59E.   5,000   50,000

Following the same principle, the Romans were able to write the numbers 500,000, 1,000,000, 5,000,000, etc., in the following forms:

FIG. 16.60.   500,000: ⫸ or IƆƆƆƆ   1,000,000: 🎯 or CCCCIƆƆƆƆ, etc.

But this kind of graphical representation is complicated, and it is difficult to recognise numbers above 100,000 at a glance; the Romans do not seem to have taken it any further. An additional possible reason is that there is no special word in Latin for numbers greater than 100,000: for example, Pliny (*Natural History*, XXXIII, 133) notes that in his time, the Romans were unable to name the powers of 10 above 100,000. For a million, for example, they said *decies centena milia*, "ten hundred thousand".

Nevertheless such a representation may be found, for numbers up to one million, in a work published in 1582 by a Swiss writer called Freigius (Fig. 16.61, 16.62 and 16.70):

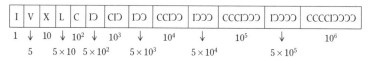

I	V	X	L	C	IƆ	CIƆ	IƆƆ	CCIƆƆ	IƆƆƆ	CCCIƆƆƆ	IƆƆƆƆ	CCCCIƆƆƆƆ

$1 \downarrow 10 \downarrow 10^2 \downarrow 10^3 \downarrow 10^4 \downarrow 10^5 \downarrow 10^6$

$5 \quad 5 \times 10 \quad 5 \times 10^2 \quad 5 \times 10^3 \quad 5 \times 10^4 \quad 5 \times 10^5$

FIG. 16.61.

Other conventions were frequently used by the Romans, and may be found in use in the Middle Ages, which simplified the notation of numbers above 1,000 and allowed considerably larger numbers to be reached.

In one of these, a horizontal bar placed above the representation of a number meant that that number was to be multiplied by 1,000. In this way all numbers from 1,000 to 5,000,000 could be easily written.

It should however be noted that this convention could sometimes cause confusion with another older convention, in which, in order to distinguish between letters used to denote numbers from those used to write words, the Romans were in the habit of putting a line above the letters being used as numerals, as can be found in certain Latin abbreviations such as

$$\overline{\text{II}}\text{VIR} = \text{duumvir}; \quad \overline{\text{III}}\text{VIR} = \text{triumvir}$$

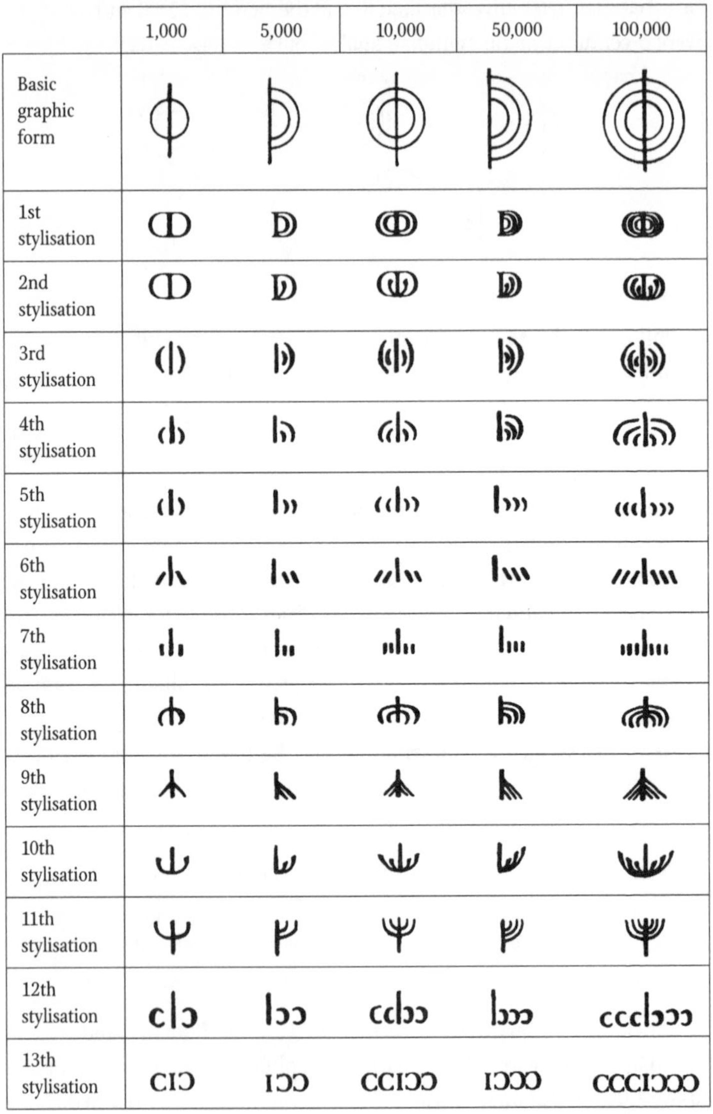

	1,000	5,000	10,000	50,000	100,000
Basic graphic form	φ	⊅	⊕	⊅	◎
1st stylisation	⊂⊃	Ð	⊂⊂⊃	Ð	⊂⊂⊃
2nd stylisation	⊂⊃	Ð	⊂⊂⊃	Ð	⊂⊂⊃
3rd stylisation	(I)	I)	(I)	I)	(I)
4th stylisation	⊂I⊃	I⊃	⊂I⊃	I⊃	⊂I⊃
5th stylisation	(I)	I»	((I))	I»»	((I))»
6th stylisation	⋏	I⋏	⋰I⋏	I⋏⋏	⋰⋰I⋏⋏
7th stylisation	ılı	Iıı	ıılıı	Iııı	ıılııı
8th stylisation	⊄⊅	⋔	⊄⊅	⋔	⊄⊅
9th stylisation	木	乄	木	乄	木
10th stylisation	⍦	⍦	⍦	⍦	⍦
11th stylisation	Ψ	Ψ	Ψ	Ψ	Ψ
12th stylisation	c|ɔ	Iɔɔ	ccIɔƆ	Iɔɔɔ	cccIɔɔɔ
13th stylisation	CIƆ	IƆƆ	CCIƆƆ	IƆƆƆ	CCCIƆƆƆ

Fig. 16.62.

$\bar{\mathrm{V}}$	=	5,000 =	5 × 1,000	CIL, VIII, 1577
$\bar{\mathrm{X}}$	=	10,000 =	10 × 1,000	CIL, VIII, 98
$\overline{\mathrm{LXXXIII}}$	=	83,000 =	83 × 1,000	CIL, I, 1757

Examples from Latin inscriptions of which the oldest date from the end of the Republic

Fig. 16.63.

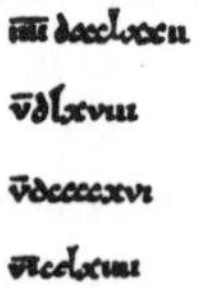

ꟾꟾꟾ dccLxxɪɪ	$\overline{\mathrm{IIII}}$DCCCLXXII	4,872
v̄δ̄Lxvɪɪɪ	$\overline{\mathrm{VD}}$LXVIII	5,568
v̄δccccxvɪ	$\overline{\mathrm{VD}}$CCCCXVI	5,916
v̄ɪcc̄Lxɪɪɪɪ	$\overline{\mathrm{VI}}$CCLXIIII	6,264

Examples from a Latin astronomical manuscript of the 11th or 12th centuries CE (Bibliothèque nationale, Paris, Ms. lat. 14069, folio 19)

Fig. 16.64.

Probably it is for this reason that at the time of the Emperor Hadrian (second century CE) the multiplication by 1,000 was indicated by placing a vertical bar at either side, as well as the horizontal line on top.

Reconstructed examples:

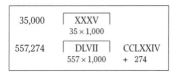

35,000	XXXV	
	35 × 1,000	
557,274	DLVII	CCLXXIV
	557 × 1,000	+ 274

FIG. 16.65.

However, this notation was generally reserved for a quite different purpose.

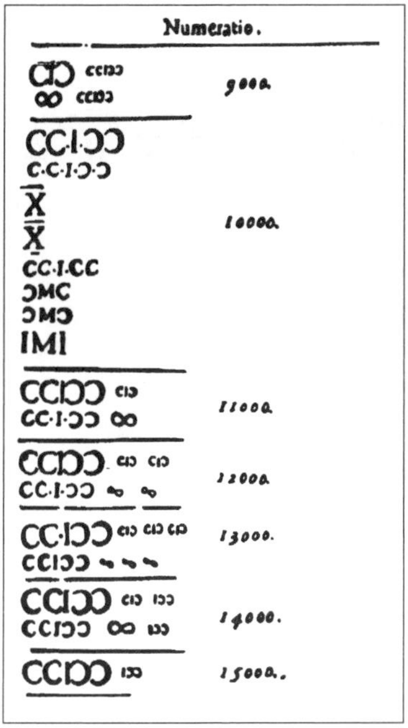

Numeratio.	
⊂Ɔ ᴄᴄᴉᴐ ᴏᴏ ᴄᴄᴉᴐ	*9000.*
ᴄᴄ·ɪ·ᴐᴐ ᴄ·ᴄ·ɪ·ᴐ·ᴐ	
X̄ X̱ ᴄᴄ·ɪ·ᴄᴄ ᴐᴍᴄ ᴐᴍᴐ ɪᴍɪ	*10000.*
ᴄᴄᴉᴐᴐ ᴄɪᴐ ᴄᴄ·ɪ·ᴐᴐ ᴏᴏ	*11000.*
ᴄᴄᴉᴐᴐ ᴄɪᴐ ᴄɪᴐ ᴄᴄ·ɪ·ᴐᴐ ∾ ∾	*12000.*
ᴄᴄ·ɪ·ᴐᴐ ᴄɪᴐ ᴄɪᴐ ᴄɪᴐ ᴄᴄɪᴐᴐ ∾ ∾ ∾	*13000.*
ᴄᴄᴐᴐ ᴄɪᴐ ɪᴐᴐ ᴄᴄɪᴐᴐ ᴏᴏ ɪᴐᴐ	*14000.*
ᴄᴄᴉᴐᴐ ɪᴐᴐ	*15000.*

FIG. 16.66. *The archaic Roman numerals being used in a work by Petrus Bungus on the mystical significance of numbers (*Mysticae numerorum significationes opus ...) *published at Bergamo in 1584–1585. (Bibliothèque nationale, Paris [R. 7489])*

Every Roman numeral enclosed in a kind of incomplete rectangle was, in fact, usually supposed to be multiplied by 100,000, which allowed the representation of all numbers between 1,000 and 500,000,000.

Examples from Latin inscriptions from the Imperial period in Rome:

XII	a	1,200,000	12 × 100,000
XIII	b	1,300,000	13 × 100,000
∞ ∞	c	200,000,000	2,000 × 100,000

a. Cf. CIL, I, 1409
b. Cf. CIL, VIII, 1641
c. Inscription from Ephesus, 103 CE [Cagnat (1899)]

FIG. 16.67.

According to some authors, the logical continuation of the convention of placing a line above the number was to place a double line to represent multiplication by 1,000,000, thus allowing the representation of numbers up to 5,000,000,000:

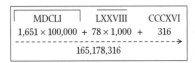

$$1,000,000,000 \quad \overline{\overline{M}} \qquad = 1,000 \times 1,000,000$$

$$2,300,000,000 \quad \overline{\overline{MMCCC}} = 2,300 \times 1,000,000$$

However no evidence of this in currently known Roman inscriptions has been found.

MDCLI	LXXVIII	CCCXVI
$1,651 \times 100,000$ +	$78 \times 1,000$ +	316

$$\dashrightarrow$$

165,178,316

FIG. 16.68.

FIG. 16.69.
Frontispieces from works by Descartes (published 1637) and Spinoza (published 1677). The dates are written in the archaic Roman numerals.

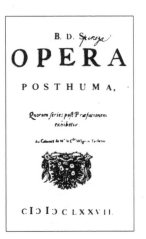

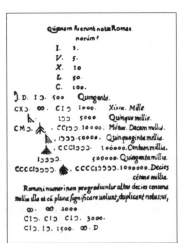

FIG. 16.70. *Archaic Roman numerals in a work by Freigius, published in 1582 [Smith, D.E (1958)]*

FIG. 16.71. *Detail of a page of a Portuguese manuscript of 1200 CE, referring to the Venerable Bede's method of calculation (Lisbon Public Library, MS Alcobaça 394 (426), folio 252) [Burnham in PIB, plate XV]*

Figures on $\overline{\text{XL}}$ XXX = 40,030
the drawings: $\overline{\text{L}}$ XC = 50,090
$\overline{\text{LX}}$ = 60,000

But these kinds of notations could only cause confusion and errors of interpretation – as a future Roman Emperor learnt to his cost, according to Seneca (*Galba*, 5).

On succeeding to his mother Livia, Emperor Tiberius had to pay large sums of money to her legatees. Tiberius's mother had written the amount of her legacy to young Galba in the form: $\overline{\text{CCCCC}}$.

But Galba had not taken the precaution of checking that the amount was written out in words. So when he presented himself to Tiberius, Galba thought that the five Cs had been enclosed in vertical lines, and that therefore the sum due to him was

$$500 \times 100,000 = 50,000,000 \text{ sesterces.}$$

But Tiberius took advantage of the fact that the two sides-bars were very short, and claimed that this representation was a simple line above the five Cs. "My mother should have written them as $\lceil\overline{\text{CCCCC}}\rceil$ if you were to be right," he said. Since the simple line only represented multiplication by 1,000, Galba only received from Tiberius the sum of

$$500 \times 1,000 = 500,000 \text{ sesterces}$$

Which goes to show that an unstable notation system can turn a large fortune into a mere pittance!

The Romans also devised other conventions. Instead of repeating the letters C and M for successive multiples of 100 or 1,000, they first wrote the number of hundreds or thousands they wanted, and then placed the letter C or M either as a coefficient or as a superscript index:

| 200: | II.C | or | II^C | 2,000: | II.M | or | II^M |
| 300: | III.C | or | III^C | 3,000: | III.M | or | III^M |

However, instead of simplifying the system, these various conventions only complicated it, since the principle of addition was completely subverted by the search for economy of symbols.

We therefore see the complexity and the inadequacy of the Roman number-system. *Ad hoc* conventions based on principles of quite different kinds made it incoherent and inoperable. There is no doubt that Roman numerals constituted a long step backwards in the history of number-systems.

THE GREEK AND ROMAN ABACUSES

Given such a poor system of numerals, the Greeks, Etruscans and Romans did not use written numbers when they needed to do sums: they used abacuses.

The Greek historian Polybius (c. 210–128 BCE) was no doubt referring to one of these when he put the following words into the mouth of Solon (late seventh century to early sixth century BCE).

> Those who live in the courts of the kings are exactly like counters on the counting table. It is the will of the calculator which gives them their value, either a *chalkos* or a *talent*. (*History,* V, 26)

We can all the better understand the allusion when we know that the *talent* and the *chalkos* were respectively the greatest and the least valuable of the ancient Greek coins, and they were represented by the leftmost and rightmost columns of the abacus.

Fig. 16.72. *Detail of the Darius Vase from Canossa, c. 350 BCE (Museo Archeologico Nazionale, Naples)*

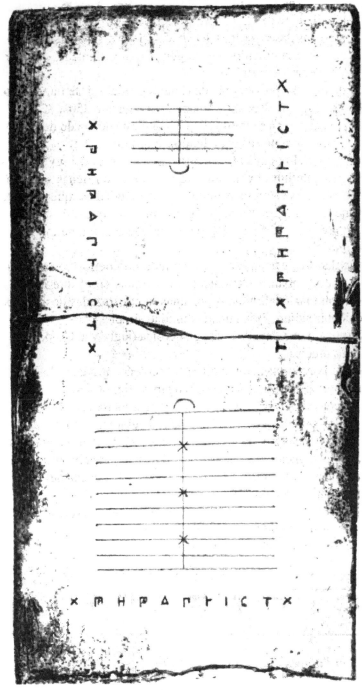

FIG. 16.73. *The Table of Salamis, originally considered to be a gaming table, which is in fact a calculating apparatus. Date uncertain (fifth or fourth century BCE). (National Museum of Epigraphy, Athens)*

The writings of many other Greek authors from Herodotus to Lysias also bear witness to the existence and use of the abacus.

Descriptions of the Greek abacus are not only to be found in literary text, but also in images. The "Darius Vase" is the most famous example (Fig. 16.72). It is a painted vase from Canossa in southern Italy (formerly a Greek colony) and dates from around 350 BCE. The various scenes painted on it are supposed to describe the activities of Darius during his military expeditions.

In one detail of the vase, we can see the King of Persia's treasurer using counters on an abacus to calculate the tribute to be levied from a conquered city. In front of him, a personage hands him the tribute, while another begs the treasurer to allow a reduction of taxes which are too heavy for the city he represents.

The Greek calculators stood by one of the sides of the horizontal table and placed pebbles or counters on it, within a certain number of columns marked by ruled lines. The counters or pebbles each had the value of 1.

A document from the Heroic Age (fifth century BCE) gives us a more detailed idea. It is a large slab of white marble, found on the island of Salamis by Rhangabes, in 1846 (Fig. 16.73).

It consists of a rectangular table 149 cm long, 75 cm wide and 4.5 cm thick, on which are traced, 25 cm from one of the sides, five parallel lines; and, 50 cm from the last of these lines, eleven other lines, also parallel, and divided into two by a line perpendicular to them: the third, sixth and ninth of these lines are marked with a cross at the point of intersection.

Furthermore, three almost identical series of Greek letters or signs are arranged in the same order along three of the sides of the table. The most complete of the series has the following thirteen symbols in it:

$$\text{T} \quad \text{Γ} \quad \text{X} \quad \text{Γ} \quad \text{H} \quad \text{Γ} \quad \Delta \quad \text{Γ} \quad \text{Ⱶ} \quad \text{I} \quad \text{C} \quad \text{T} \quad \text{X}$$

FIG. 16.74.

As we saw at the beginning of this chapter, these in fact correspond to the numerical symbols of the acrophonic number system (Fig. 16.1), and they serve here to represent monetary sums expressed in *talents, drachmas, obols,* and *chalkoi,* that is to say in multiples and sub-multiples of the *drachma.*

These symbols represented, from left to right in the order shown, 1 *talent* or 6,000 *drachmas*, then 5,000, 1,000, 500, 100, 50, 10, 5 and 1 *drachmas*, then 1 *obol* or one sixth of a *drachma*, 1 *demi-obol* or one twelfth of a *drachma*, 1 *quarter-obol* or one twenty-fourth of a *drachma*, and finally 1 *chalkos* (one eighth of an *obol* or one forty-eighth of a *drachma*). (Fig. 16.75)

T	1 talent	First letter of TALANTON, "talent"
⌐	5,000 drachmas	
X	1,000 drachmas	First letter of CHILIOI, "thousand" (drachmas)
⌐	500 drachmas	
H	100 drachmas	First letter of HEKATON, "hundred" (drachmas)
⌐	50 drachmas	
Δ	10 drachmas	First letter of DEKA, "ten" (drachmas)
Γ	5 drachmas	First letter of PENTE, "five" (drachmas)
⊢	1 drachma	
I	1 obol	Unit mark for counting obols
C	1/2 obol	Half of the letter O, first letter of OBOLION
T	1/4 obol	First letter of TETARTHMORION
X	1 chalkos	First letter of CHALKOUS
	1 talent = 6,000 drachmas	
	1 drachma = 6 obols	
	1 obol = 8 chalkos	

FIG. 16.75.

In the abacus of Salamis, each column was associated with a numerical order of magnitude.

The pebbles or counters disposed on the abacus changed value according to the position they occupied (see Fig. 16.76).

The four columns at the extreme right were reserved for fractions of a *drachma*, the one on the extreme right being for the *chalkos*, the next for the *quarter-obol*, the third for the *demi-obol*, and the last for the *obol*.

The next five columns (to the right of the central cross on Fig. 16.75) were associated with multiples of the *drachma*, the first on the right being for the units, the next for the tens, the third for the hundreds, and so on. In the bottom half of each column, one counter represented one unit of the value of the column. In the upper half, one counter represented five units of the value of the column.

The last five columns (to the left of the central cross in Fig. 16.76) were associated with *talents*, tens of *talents*, hundreds, and so on. One *talent* being worth 6,000 *drachmas*, the calculator would replace counters corresponding to 6,000 by one counter in the *talents* columns (sixth from the right).

As a result of this method of dividing up the table, additions, subtractions and multiplications could be done (Fig. 16.77 and 16.78).

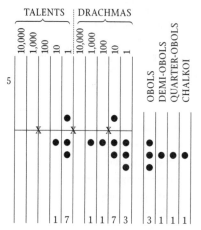

FIG. 16.76. *The principle of the Greek abacus from Salamis, showing the representation of the sum "17 talents, 1,173 drachmas, 3 obols, 1 demi-obol, 1 quarter-obol, and 1* chalkos". *(Ch a l k o i is the plural of c h a l k o s.)*

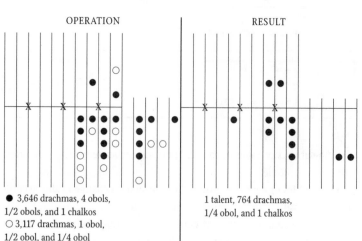

● 3,646 drachmas, 4 obols, 1/2 obols, and 1 chalkos
○ 3,117 drachmas, 1 obol, 1/2 obol, and 1/4 obol

1 talent, 764 drachmas, 1/4 obol, and 1 chalkos

FIG. 16.77. *The method of addition on the Salamis abacus, showing the addition of "3,646 drachmas, 4 obols, 1/2 obol and 1 chalkos" (shown in black) and "3,117 drachmas, 1 obol, 1/2 obol and 1/45 obol" (shown in white). By reducing the counters according to the rules, the result is obtained as "1 talent, 764 drachmas, 1/4 obol, and 1 chalkos".*

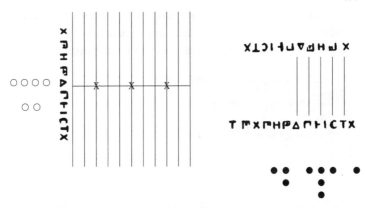

FIG. 16.78. *To multiply "121 drachmas, 3 obols, 1/2 obol, and 1 chalkos" by 42, for example, we start by placing the multiplier 42 on the abacus, by laying out the corresponding counters under the appropriate number-signs on the left of the table. Then the multiplicand, the sum of money, is laid out under the number-signs of one of the two series on the right (black circles). Then by manoeuvring the counters the result is obtained (see a similar method in Fig. 16.84).*

The Etruscans and their Roman successors also employed abacuses with counters. In Fig. 16.79 we reproduce an Etruscan medallion, a carved stone which shows a man calculating by means of counters on an abacus, noting his results on a wooden tablet on which Etruscan numerals can be seen (Fig. 16.35).

Many Roman texts mention it:

> *Coponem laniumque balneumque, tonsorem tabulamque calculosque et paucos ... haec praesta mihi, Rufe ...*

An innkeeper, a butcher, baths, a barber, a calculating table (= *tabulamque calculosque*) with its counters ... fetch me all that, Rufus ... *

> *Computat, et cevet. Ponatur calculus, adsint cum tabula pueri; numeras sestertia quinque omnibus in rebus; numerentur deinde labores.*

He calculates, and he wriggles his rear. Let the counter (= *calculus*) be placed, let the slaves bring the (calculating) table: you find five thousand sesterces in all; now make the total of my works.†

FIG. 16.79. *The medallion with the Etruscan calculator (date uncertain). (Coin Room, Bibliothèque nationale, Paris. Intaille 1898)*

At Rome, the abacus with counters was a table, on which parallel lines

---

* Martial, *Epigrams*, Vol. 1, book 2, 48
† Juvenal, *Satires*, IX, 40–43

separated the different numerical orders of magnitude of the Roman number-system. The Latin word *abacus* denotes a number of devices with a flat surface which serve for various games, or for arithmetic (Fig. 16.80).

FIG. 16.80. *A Roman abacus with* calculi *(reconstruction)*

Each column generally symbolised a power of 10. From right to left, the first was associated with the number 1, the next with the tens, the third with the hundreds, the fourth with the thousands, and so on. To represent a number, as many pebbles or counters were placed as required. The Greeks called these counters *psephoi*, ("pebble" or "number") and the Romans called them *calculi* (singular: *calculus*). Certain authors (notably Cicero, *Philosophica Fragmenta*, V, 59) called them *aera* ("bronze"), alluding to the material they were often made of after the Imperial epoch (Fig. 16.81).

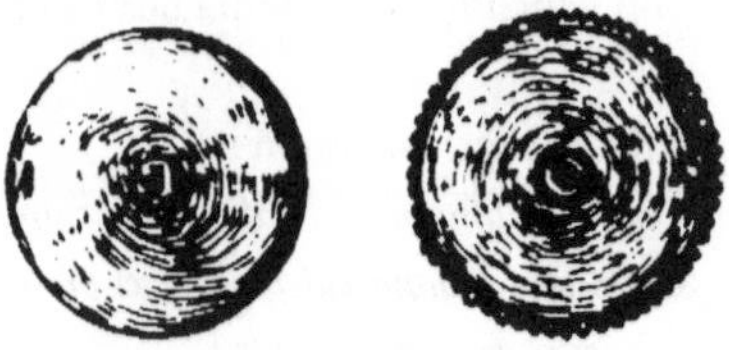

FIG. 16.81. *Roman calculating counters. After the originals in the Städtisches Museum, Wels, Germany*

To represent the number 6,021 on the columns of the abacus we therefore place one counter in the first column, two in the second, none in the third, and six in the fourth.

For 5,673 we place three in the first, seven in the second, six in the third, and five in the fourth (Fig. 16.82).

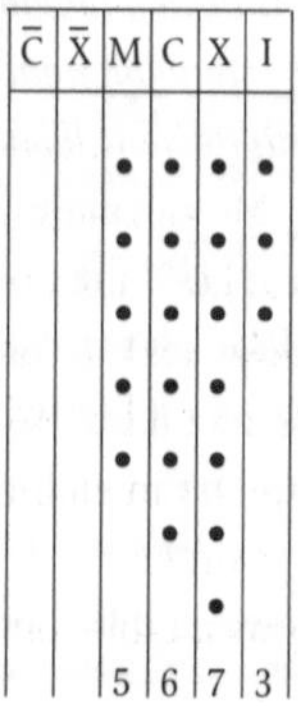

FIG. 16.82. *The principle of the Roman abacus with* calculi

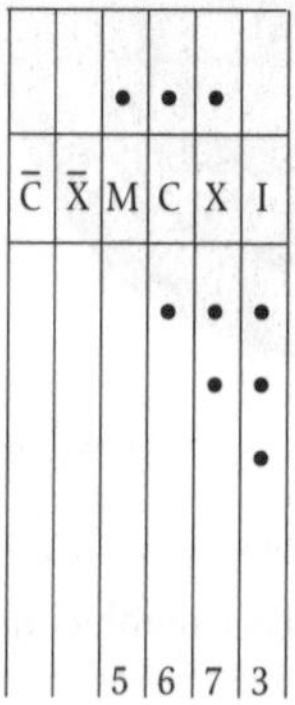

FIG. 16.83. *Simplification of the principle of the Roman abacus with* calculi

To simplify calculation, each column is divided into an upper and a lower part. A counter in the lower half represents one unit of the value of the column, and a counter in the upper half represents half of one unit of the value of the next column (or five times the value of the column it is in). For the upper halves we therefore have five for the first column, fifty for the second, 500 for the third, and so on (Fig. 16.83).

By cleverly moving the counters between these divisions (adding to and taking away from the counters in each division) it is possible to calculate.

To add a number to a number which has already been set up on the abacus, it is set up in turn, and then the result is read off after the various manipulations have been performed. In a given column, if ten or more counters are present at any time then ten of these are removed and one is placed in the next higher column (to the left) (Fig. 16.82). On the simplified abacus, this procedure is somewhat modified. If there are five or more in the lower half, then five are removed and one is placed in the upper half; while if two or more are present in the upper half then two are removed, and one is placed in the lower half of the next column, to the left (Fig. 16.83). Subtraction is carried out in a similar way, and multiplication is done by addition of partial products.

For example, to multiply 720 by 62, we start setting up the numbers 720 and 62 as shown in Fig. 16.84A. Then the 7 of 720 (worth 700) and the 6 of 62 (worth 60) are multiplied, to give 42 (worth 42,000). Therefore two counters are placed in the fourth column and four counters are placed in the fifth.

First partial product: $6 \times 7 = 42$

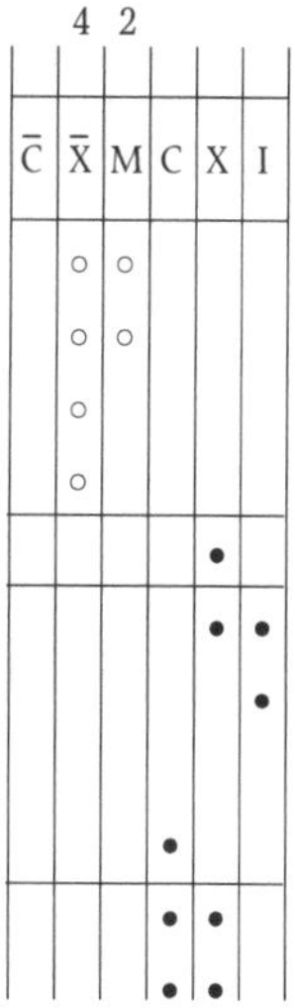

62 Multiplier

720 Multiplicand

FIG. 16.84A.

Then the 7 of 720 (worth 700) and the 2 of 62 (worth 2) are multiplied to give 14 (worth 1,400), and four counters are placed into the third column and one is placed into the fourth.

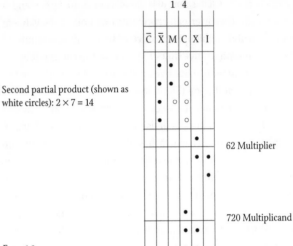

Second partial product (shown as white circles): $2 \times 7 = 14$

62 Multiplier

720 Multiplicand

FIG. 16.84B.

Now the 7 of 720 has done its work, and can be removed. Next we multiply to 2 of 720 (worth 20) by the 6 of 62 (worth 60) to get 12 (worth 1,200), and so two counters are placed in the third column and one is placed in the fourth.

Finally, the 2 of 720 (worth 20) and the 2 of 62 (worth 2) are multiplied to get 40. Therefore four counters are placed in the second column.

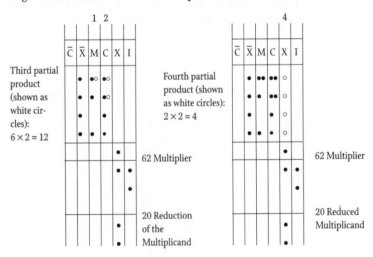

Third partial product (shown as white circles): $6 \times 2 = 12$

62 Multiplier

20 Reduction of the Multiplicand

FIG. 16.84C.

Fourth partial product (shown as white circles): $2 \times 2 = 4$

62 Multiplier

20 Reduced Multiplicand

FIG. 16.84D.

Now the various counters on the table are reduced as explained above to give the required result of the multiplication:

$$720 \times 62 = 44{,}640$$

4 4 6 4 0

C̄	X̄	M	C	X	I
	o				
o	o	o	o		
o	o		o		
o	o		o		
o	o		o		

Result

FIG. 16.84E.

Calculating on the abacus with counters was therefore a protracted and difficult procedure, and its practitioners required long and laborious training. It is obvious why it remained the preserve of a privileged caste of specialists.

But traditions live on, and for centuries these methods of calculation remained extant in the West, deeply attached to Roman numerals and their attendant arithmetic. They even enjoyed considerable favour in Christian countries from the Middle Ages up to relatively recent times.

All the administrations, all the traders and all the bankers, the lords and the princes, all had their calculating tables* and struck their counters from base metal, from silver or from gold, according to their importance, their wealth, or their social standing. "I am brass, not silver!" was said at the time to express that one was neither rich nor noble. The clerks of the British Treasury, until the end of the eighteenth century, used these methods to calculate taxes, employing exchequers, or *checkerboards* (because of the way they were divided up). This is why the British Minister of Finance is still called "Chancellor of the Exchequer".

FIG. 16.85A. *The use of abacuses with counters continued in Europe until the Renaissance (and in some places until the French Revolution). Here we see an expert calculator in a German illustration from the start of the sixteenth century. [Treatise on Arithmetic by Köbel, published at Augsburg in 1514]*

* The existence of large numbers of treatises on practical arithmetic which mention these procedures throughout Europe in the sixteenth, seventeenth, and eighteenth centuries gives an idea of how widespread these practices were before the French Revolution.

FIG.16.85B."*Madame Arithmetic" teaching young noblemen the art of calculation on the abacus (sixteenth-century French tapestry). (Cluny Museum)*

FIG. 16.86. *Calculating counter bearing the arms of Montaigne (and surrounded with the chain of the order of Saint Michel de Montaigne). This counter was found earlier, in the ruins of the Château de Montaigne, though its original diestamp was found in the nineteenth century.* [Brieux (1957)]

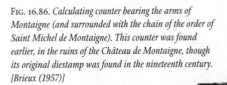

FIG. 16.87. *Fifteenth-century calculating table: one of the rare known abacuses from this period. (Historical Museum of Dinkelsbühl, Germany)*

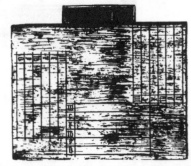

FIG. 16.88. *Calculating table with three divisions, sixteenth–seventeenth century, as formerly used in Switzerland and Germany to calculate rates and taxes. The letters to be seen on it are (from the top): d for the* deniers (denarius); *s for the* sols *or shillings* (solidus); *lb or lib for the pounds* (libras); *then X, C and M for 10, 100 and 1,000 pounds. (Historical Museum of Basel. Inv. 1892.209. Neg. 1500)*

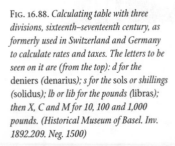

At the time of the Renaissance, many writers make reference to this. Thus Montaigne (1533–1592):

> We judge him, not according to his worth, but from the style of his counters, according to the prerogatives of his rank. (*Essays*, Book III, Bordeaux edition, 192, I, 17)

Likewise Georges de Brébeuf (1618–1661), adapting the formula of Polybius:

> Courtesans are counters;
> Their value depends on their place;
> If in favour, why then it's millions,
> But zero if they're in disgrace.

Again Fénelon (1651 – 1715), who makes Solon say:

> The people of the Court are like the counters used for reckoning: they are worth more or less depending on the whim of the Prince.

And Boursault (1638–1701):

> Never forget, if I may have your grace,
> Whatever more power either of us might have had
> We are still but counters stamped with value by the King.

Finally, Madame de Sévigné, who sent these words to her daughter in 1671:

> We have found, thanks to these excellent counters, that I would have had five hundred and thirty thousand pounds if I counted all my little successions.

The abacus of this period also consisted of a table marked out into divisions corresponding to the different orders of magnitude (Fig. 16.87 and 16.88). Numbers were set up on the table with counters (made of the most diverse materials), whose values depended on where they were placed. Placed on successive lines, from bottom to top, a counter would be worth 1, 10, 100, 1,000, and so on. Between successive lines, a counter was worth five of the value of the line below it (Fig. 16.89 and 16.90).

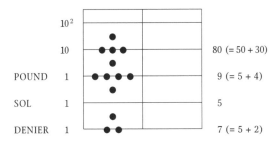

Fig. 16.89. *The layout of the sum "89 pounds, 5 sols and 7 deniers" on the French calculating table (sixteenth–eighteenth century).*

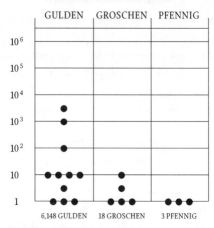

FIG. 16.90. *Representation of the sum of 6,148* gulden, *18* groschen *and 3* pfennigs *on the German calculating table (sixteenth–eighteenth century).*

The counting tables facilitated addition or subtraction, but lent themselves with difficulty to multiplication or division and even less well to more complex operations.

Arithmetical operations practised by this means had little in common with the operations of modern arithmetic with the same names. Multiplication, for example, was reduced to a sum of partial products or to a series of duplications. Division was reduced to a succession of separation into equal parts.

Such difficulties were at the origin of the fierce polemic which, from the beginning of the sixteenth century, ranged the *abacists* on one side, clinging to their counters and to archaic number-systems like the Greek and the Roman, against the *algorists* on the other, who vigorously defended calculation with pen and paper, the ancestor of modern methods.

Here, for example, is what Simon Jacob (who died in Frankfurt in 1564) had to say about the abacus:

It is true that it seems to have some use in domestic calculations, where it is often necessary to total, subtract, or add, but in serious calculations, which are more complicated, it is often an embarrassment. I do not say that it is impossible to do these on the lines of the abacus, but every advantage that a man walking free and unladen has over he who stumbles under a heavy load, the figures have over the lines.

Pen and paper soon gained the day amongst mathematicians and astronomers. The abacus was in any case used almost exclusively in finance and in commerce. Only with the French Revolution would the use of the abacus finally be banished from schools and government offices.

ABACUS IN WAX AND ABACUS IN SAND

The Latin word *abacus* derives from the Greek *abax* or *abakon* signifying "tray", "table" or "tablet", which possibly in turn derives from the Semitic word *abq*, "sand", "dust".

It is true that the "abacus in sand" is part of these oriental traditions, but it is mentioned also in the Graeco-Roman West, along with the abacus with counters, especially by Plutarch and by Apuleus. It consisted of a table with a raised border which was filled with fine sand on which the sections were marked off by tracing the dividing lines with the fingers or with a point. (Fig. 16.91).

FIG. 16.91. *Mosaic showing Archimedes (287?–212 BCE) calculating on an abacus with numerals (sand or wax), at the moment when a Roman soldier was about to assassinate him (eighteenth century). (Städtische Galerie/Liebieghaus, Frankfurt)*

Another type of calculating instrument used in Rome was the abacus in wax. It was a true portable calculator which was carried hanging from the shoulder, and it consisted of a small board of wood or of bone coated with a thin layer of black wax; the columns were marked by tracing in the wax with a pointed iron stylus (whose other end, being flat, was used to erase marks by pressing on the surface of the wax).

A specimen from Rome, dating from the sixth century, has been described by D. E. Smith. It is in the collections of the John Rylands Library in Manchester. It is made of bone, and consists of two rectangular iron plates joined by an iron hinge, with three iron styluses.

Horace (65–8 BCE) was perhaps alluding to this instrument in this passage from the first book of *Satires*:*

> ... *causa fuit pater his, qui macro pauper agello noluit in Flavi ludum me mittere, magni quo pueri magnis e centurionibus orti laevo suspensi loculos tabulanque lacerte ibant octonos referentes Idibus aeris* ...

> I owe this to my father who, poor and with meagre possessions, did not wish to send me to the school of Flavius, where the noble sons of noble centurions went, their box and their board (*tabulanque*) hanging from their left shoulder, paying at the Ides their eight bronze coins....

The Europeans of the Middle Ages probably also used one or other of these, as well as the abacus with counters.

In his *Vocabularium* (1053), Papias (who may be considered one of the authorities on the knowledge of his time) also talks of the abacus as "a table covered with green sand", which is exactly what can be found in Rémy d'Auxerre in his commentary on the *Arithmetic* of Martianus Capella (c. 420–490 CE) where he describes it as "a table sprinkled with a blue or green sand, where the *figures* [the numbers] are drawn with a rod".

As for the abacus in wax, Adelard of Bath (c. 1095–c. 1160) alludes to it as follows [B. Boncompagni (1857)]:

> *Vocatur (Abacus) etiam radius geometricus, quia cum ad multa pertineat, maxime per hoc geometricae subtilitates nobilis illuminantur.*

> (The abacus) is also called the "geometrical radius" since it permits so many operations. In particular, thanks to it the subtleties of geometry become perfectly clear and comprehensible.

Finally, it is perfectly possible that Radulph de Laon (c. 1125) was thinking of one or other of these in writing [D. E. Smith & L. C. Karpinski (1911)]:

> ... *ad arithmaticae speculationis investigandas rationes, et ad eos qui musices modulationibus deserviunt numeros, necnon et ad ea quae astrologorum sollerti industria de variis errantium siderum cursibus* ... *Abacus valde necessarius inveniatur.*

> For the examination of the rules of mathematical thought and of the numbers which are at the base of musical modulations, and for the calculations which, thanks to the skilful industry of the astrologers, explain the various trajectories of the moving stars, the abacus shows itself absolutely indispensable.

---

* *Satires*, I, VI, 70–75

These authors do not however say what kinds of numeral were used with the abacuses of these two types, though especially at the time of Papias, Adelard and Radulph the Arab numerals were used and were already well known in Europe. But the Greek numerals were used also (from $\alpha = 1$ to $\theta = 9$) which had been much better known before this time, as well as the Roman numerals which were in a way the "official" numerals of mediaeval Europe.

In any case, which figures were used is not of great importance with instruments of this type for, by reason of its structure (which assigns variable values to the symbols according to their positions), the columns of the abacus in sand or the abacus in wax can render even the most primitive figures operational. Of which the proof follows, for the Roman numerals.

Let us again take up the multiplication of 720 by 62, and try to do it with Roman numerals on a tablet covered with sand or with wax.

The technique works for any decimal number-system whatever, provided the figures greater than or equal to 10 are not used. We start by writing the 720 and the 62 in the bottom lines. (Fig. 16.92A)

$\overline{C}$	$\overline{X}$	$\overline{I}$	C	X	I	
				VI	II	←Multiplier (62)
			VII	II		←Multiplicand (720)

FIG. 16.92A.

Now we multiply the 7 (700) of 720 by the 6 (60) of 62 and get 42 (42,000). Therefore we write this result at the top, the 2 in the fourth column and the 4 in the fifth. (Fig. 16.92B)

Then we multiply the 7 (700) of 720 by the 2 of 62 and get 14 (1,400), and we write this result at the top below the last one, with a 4 in the third column and a 1 in the fourth. (Fig. 16.92C)

First partial
product:
$6 \times 7 = 42 \rightarrow$

$\bar{C}$	$\bar{\bar{X}}$	$\bar{I}$	C	X	I
	IV	II			
				VI	II
			VII	II	

FIG. 16.92B.

Second partial
product:
$2 \times 7 = 14 \rightarrow$

$\bar{C}$	$\bar{\bar{X}}$	$\bar{I}$	C	X	I
	IV	II			
		I	IV		
				VI	II
			VII	II	

FIG. 16.92C.

Now we can forget the 7 of 762, and multiply the 2 (20) of 720 by the 6 (60) of 62, and get 12 (1,200) which we again write at the top below the last result: 2 in the third column and 1 in the fourth. (Fig. 16.92D)

C̄	X̄	İ	C	X	I
	IV	II			
		I	IV		
		I	II		
				VI	II
				II	

Third partial product: $6 \times 2 = 12 \rightarrow$

FIG. 16.92D.

Finally we multiply the 2 (20) of 720 by the simple 2 of 62 and get 4 (40), so we write a 4 in the second column.

C̄	X̄	İ	C	X	I
	IV	II			
		I	IV		
		I	II		
				IV	
				VI	II
				II	

Fourth partial product: $2 \times 2 = 4 \rightarrow$

FIG. 16.92E.

We can now erase the 720 and the 62, and proceed to reduce the figures which remain. Here we can start with the second column.

Since this figure is less than 10, we pass immediately to the next column, the third. We add 4 and 2 and get 6, which is less than 10, we erase the two figures 4 and 2 and write in 6.

Then we pass to the fourth column, where we add 2, 1 and 1 to get 4, which is less than 10, so we erase the three figures and write a 4 in the fourth column.

The fifth column will remain unchanged since the single figure in it is less than 10.

It only remains to read the result directly off the columns:

$$720 \times 62 = 44,640$$

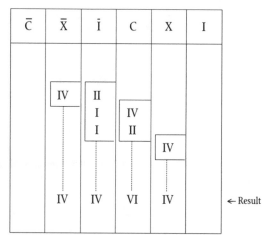

FIG. 16.92F.

THE FIRST POCKET CALCULATOR

As well as the "desk-top models for school and office", some of the Roman accountants used a real "pocket calculator" whose invention undoubtedly predates our era. The proof of this is a bas-relief on a Roman sarcophagus of the first century, which shows a young calculator* standing before his master, doing arithmetic with the aid of an instrument of this type (Fig. 16.96). This instrument consisted of a small metal plate, with a certain number of parallel slots (usually nine). Each slot was associated with an order of magnitude, and mobile beads could slide along them.†

Ignoring for the moment the two rightmost slots, the remaining seven are divided into two distinct segments, a lower and an upper. The lower one

---

* Among the Romans, the word calculator meant, on the one hand, a "master of calculation" whose principal task was to teach the art of calculation to young people using a portable abacus or an abacus with counters; and, on the other hand, the keeper of the accounts or the intendant in the important houses of the patricians, where he was also called *dispensator*. If these were slaves, they were called *calculones*; but if they were free men then they were called *calculatores* or *numerarii*.

† As well as the abacus shown in Fig. 16.94, we know of at least two other examples. One is in the British Museum in London, and the other in the Museum of the Thermae in Rome.

contains four sliding beads, and the upper one, which is the shorter, contains only one.

In the space between these two rows of slots a series of signs is inscribed, one for each slot. These are figures expressing the different powers of 10 according to the classical Roman number-system which the bankers and the publicans used to count by *as*, by *sestertii*, and by *denarii** (Fig. 16.62 and 16.67 above):

$\boxed{X}$	$(\!(\!\flat\!)\!)$	$(\!\flat\!)$	$\flat$	C	X	I
10^6	10^5	10^4	10^3	10^2	10	1

FIG. 16.93.

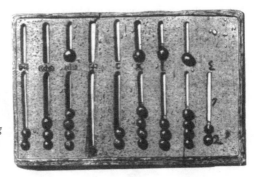

FIG. 16.94. *Roman "pocket abacus" (in bronze), beginning of Common Era. (Cabinet des médailles, Bibliothèque nationale, Paris.) (br. 1925)*

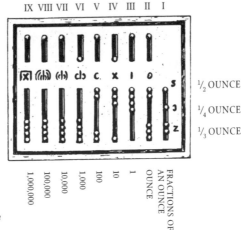

IX VIII VII VI V IV III II I

$\frac{1}{2}$ OUNCE

$\frac{1}{4}$ OUNCE

$\frac{1}{3}$ OUNCE

FIG. 16.95. *The principle of the portable Roman abacus. This specimen belonged to the German Jesuit Athanasius Kircher (1601–1680). (Museum of the Thermae, Rome)*

1,000,000
100,000
10,000
1,000
100
10
1
OUNCE
FRACTIONS OF
AN OUNCE

* The unit of the Roman monetary system was the *as* of bronze. Its weight continually diminished, from the origin of the monetary system around the fourth century BCE until the Empire. It successively weighed 273 gm, 109 gm, 27 gm, 9 gm, and finally 2.3 gm. Its multiples were the *sestertius* (first silver, later bronze, then brass), the *denarius* (silver), and, from the time of Caesar, the *aureus* (gold). In the third century BCE, 1 *denarius* was 2.5 *as*, 1 *sestertius* was 4 *denarii* or 10 *as*. From the second century BCE, after a general monetary reform, 1 *sestertius* was 4 *as*, 1 *denarius* was 4 *sestertii* or 16 *as*, and 1 *aureus* was 25 *denarii* or 400 *as*.

Each of these seven slots was therefore associated with a power of 10. From right to left, the third corresponded to the number 1, the fourth to the tens, the fifth to the hundreds and so on (Fig. 16.95).

If the number of units of a power of 10 did not exceed 4, it was indicated in the lower slot by pushing the same number of beads upwards. When it exceeded 4, the beads in the upper slot was pulled down towards the centre, and 5 units were removed from the number and this was represented in the lower slot.

If we are considering a calculation in *denarii*, the number represented on the abacus in Fig. 16.95 corresponds (leaving aside the first two slots on the right) to the sum of 5,284 *denarii*: 4 beads up in the lower slot III means 4 ones or 4 *denarii*; the upper bead down and 3 beads up in the lower slot IV means (5 + 3) tens or 80 *denarii*; two beads up in the lower slot V means two hundreds or 200 *denarii*; and finally the upper bead down in slot VI means five thousands or 5,000 *denarii*.

Fɪɢ. 16.96. *Bas-relief on a sarcophagus from a Roman tomb dating from the first century CE. (Capitoline Museum, Rome.)*

The first two slots on the right were used to note divisions of the *as*.* The second slot, marked with a single O, has an upper part with a single bead, and a lower which has not four, but five beads: it was used to represent multiples of the *uncia* (ounce) or twelfths of the *as*, each lower bead being worth one ounce and the upper bead being worth six ounces, which

---

* In Roman commercial arithmetic, fractions of a monetary unit were always expressed in terms of the *as*, the basic unit of money which was divided into twelve equal parts called *unciae* ("ounces") – which gave its name to the corresponding unit of value whatever its nature. Each multiple or sub-multiple of the *as* (or of the unit which the *as* represented) was then given a particular name. For example, for the sub-multiples we have 1/2: *as semis*; 1/3: *as triens*; 1/4: *as quadrans*; 1/5: *as quincunx*; 1/6: *as sextans*; 1/7: *as septunx*; 1/8: *as octans*; 1/9: *as dodrans*; 1/10: *as dextans*; 1/11: *as deunx*; 1/12: *as uncia*; 1/24: *as semuncia*; 1/48: *as sicilicus*; 1/72: *as sextula*.

allows counting up to 11/12 of an *as*. The first slot, divided into three and carrying four sliding beads, was used for the half ounce, the quarter ounce and the *duella*, or third part of the ounce. The upper bead was was worth 1/2 ounce or 1/24 *as* if it was placed at the level of the sign: **S** or **Ƨ** or **Ƨ** : the sign of *as semuncia*, 1/24 of an *as*.

The middle bead was worth 1/4 ounce or 1/48 *as* if it was placed at the level of the sign: **Ɔ** or **)** or **7**: the sign of *as sicilicus*, 1/48 of an *as*.

Finally, either of the two beads at the bottom of the slot was worth 1/3 ounce or 2/72 *as* if it was placed at the level of the sign: **Z** or **2** or **Ƨ** : the sign of *as duae sextulae*, 2/72 of an *as* or *duella*.

The four beads of the first slot probably had different colours (one for the half ounce, one for the quarter ounce, and one for the third of an ounce) in case the three should find themselves on top of each other (as in Fig. 16.95). In certain abacuses these three beads ran in three separate slots.

Therefore we have here a calculating instrument very much the same as the famous Chinese abacus which still occupies an important place in the Far East and in certain East European countries.

With a highly elaborate finger technique executed according to precise rules, this "pocket calculator" (one of the first in all history) allowed those who knew how to use it to rapidly and easily carry out many arithmetic calculations.

Why did Western Europeans of the Middle Ages – the direct heirs of Roman civilisation – carry on using ancient calculating tables in preference to this more refined, better conceived, and far more useful instrument? We still do not know. Perhaps the invention belonged to one particular school of arithmeticians, which disappeared along with its tools at the fall of the Roman Empire.

CHAPTER 17

LETTERS AND NUMBERS

THE INVENTION OF THE ALPHABET

The invention of the alphabet was a huge step in the history of human civilisation. It constituted a far better way of representing speech in any articulated language, for it allowed all the words of a given language to be fixed in written form with only a small set of phonetic signs called *letters*.

This fundamental development was made by northwestern Semites living near the Syrian-Palestinian coast around the fifteenth century BCE. The Phoenicians were bold sailors and intrepid traders: once they had broken with the complex writing systems of the Egyptians and Babylonians by inventing their simpler method for recording speech, they took it with them to the four corners of the Mediterranean world. In the Middle East, they brought the idea of an alphabet to their immediate neighbours, the Moabites, the Edomites, the Ammonites, the Hebrews, and the Aramaeans. These latter were nomads and traders too, and thus spread alphabetic writing to all the cultures of the Middle East, from Egypt to Syria and the Arabian peninsula, from Mesopotamia to the confines of the Indian subcontinent. From the ninth century BCE, alphabetic writing of the Phoenician type also began to spread around the Mediterranean shores, and was gradually adopted by speakers of Western languages, who adapted it to their particular needs by modifying or adding some characters.

The twenty-two letters of the Phoenician alphabet thus gave rise directly to Palaeo-Hebraic writing (in the era of the Kings of Israel and Judaea), whence came the modern alphabet of the Samaritans, who have maintained ancient Jewish traditions. Aramaic script developed a little later, whence came the "square" or black-letter Hebrew alphabet, as well as Palmyrenean, Nabataean, Syriac, Arabic, and Indian writing systems. At the same time, Phoenician letters gave birth to the Greek alphabet, the first one to include full and rigorous representation of the vowel sounds. From Greek came the Italic alphabets (Oscan, Umbrian and Etruscan as well as Latin), and at a later stage the alphabets used for Gothic, Armenian, Georgian, and Russian (the Cyrillic alphabet). In brief, almost all alphabets in use in the world today are descended directly or indirectly from what the Phoenicians first invented.

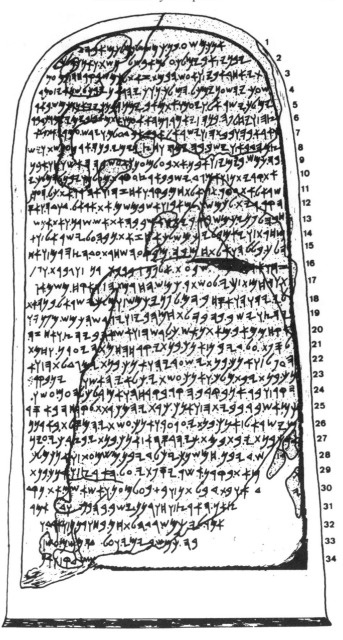

F<small>IG</small>. 17.1. Stela *of the Moabite king Mesha, a contemporary of the Jewish kings Ahab (874–853 BCE) and Joram (851–842 BCE), in the Louvre (M. Lidzbarski, vol. II, tab. 1). This is one of the oldest examples of palaeo-Hebrew script (used here to write in Moabitic, a dialect of Canaan close to Hebrew and Phoenician). This stela, put up in 842 BCE at Dibon-Gad, the Moabite capital, gives several clues to the relations that existed between Moab and Israel at that time; it is also the only document of the period found so far outside of Palestine in which the name of the God Yahweh is explicitly mentioned.*

LETTERS AND ALPHABETIC NUMBERING

It is a remarkable fact that the names and the order of the twenty-two letters of the original Phoenician alphabet have been maintained more or less intact by almost all derivative alphabets, from Hebrew to Aramaic, from Etruscan to ancient Arabic, from Greek to Syriac. According to J. G. Février (1959), we can be sure of the order of the Phoenician letters because there are alphabetic primers in Etruscan dating from 700 BCE the order of whose letters is the same as the one encoded in many acrostics in biblical Hebrew (the lines of Psalms 9, 10, 25, 34, 111, 112, etc. begin with each of the letters of the Hebrew alphabet, in alphabetic order). In fact, the same order of the letters is even found in Ugaritic primers, dating from the fourteenth century BCE. These primers contain thirty letters written in cuneiform: however, as M. Sznycer has shown, the eight "extra" Ugaritic signs, intercalated or appended to the original twenty-two, do not alter the fundamental Phoenician order of the letters.

It is because the order of the ABC ... is so ancient and so fixed that letters were able to play an important role in numbering systems.

	ARCHAIC PHOENICIAN		PALAEO-HEBREW SCRIPT				Aramaic cursive, Elephantine, 5th century BCE	HEBREW		
	Inscription of Akhiram, 11th century BCE	Inscription of Yehimilk, 10th century BCE	Stela of Mesha, 842 BCE	Samarian Ostraca, 8th century BCE	Arad Ostraca, 7th century BCE	Lakhish Ostraca, 6th century BCE		Dead Sea Scrolls	Rabbinical cursive	Black-letter Hebrew
aleph										
bet										
gimmel										
dalet										
he										
vov										
zayin										
het										
tet										
yod										
kof										
lamed										
mem										
nun										
samekh										
ayin										
pe										
tsade	[⊬]									
quf	[φ]									
resh										
shin										
tav										

FIG. 17.2. *Western Semitic alphabets*

PHOENICIAN 10th to 6th centuries BCE	HEBREW Early	HEBREW Modern	ANCIENT GREEK 5th century BCE	Oscan	Umbrian	Etruscan	
aleph			alpha				a
bet			beta				b
gimmel			gamma				g
dalet			delta				d
he			epsilon				e
vov			digamma				v
zayin			zeta				z
het			eta				h
tet			theta				th
yod			iota				i
kof			kappa				k
lamed			lambda				l
mem			mu				m
nun			nu				n
samekh			xi				s?
ayin			omicron				o
pe			pi				p
tsade			san				ṣ
quf			koppa				q
resh			rho				r
shin			sigma				s
tav			tau				t
			upsilon				u
			phi				f
			chi				kh
			psi				dh
			omega				c

FIG. 17.3. *Phoenician and Hebrew alphabets compared to Greek and Italic*

PHOENICIAN et ARAMAEAN		HEBREW		SYRIAC		ANCIENT ARABIC		GREEK	
'aleph	(')	aleph	(')	olap	(')	alif	(')	alpha	(a)
bet	(b)	bet	(b, v)	bet	(b)	ba	(b)	beta	(b)
gimmel	(g)	gimmel	(g)	gomal	(g)	jim	(j)	gamma	(g)
dalet	(d)	dalet	(d)	dolat	(d)	dal	(d)	delta	(d)
he	(h)	he	(h)	he	(h)	ha	(h)	epsilon	(e)
waw	(w)	vov	(v)	waw	(w)	wa	(w)	faw*	(f)
zayin	(z)	zayin	(z)	zayin	(z)	zay	(z)	zeta	(z)
het	(ḥ)	het	(h)	het	(ḥ)	ḥa	(ḥ)	eta	(h)
tet	(t)	tet	(t)	tet	(ṭ)	ṭa	(ṭ)	theta	(th)
yod	(y)	yod	(y)	yud	(y)	ya	(y)	iota	(i)
kaf	(k)	kof	(k, kh)	kop	(k)	kaf	(k)	kappa	(k)
lamed	(l)	lamed	(l)	lomad	(l)	lam	(l)	lambda	(l)
mem	(m)	mem	(m)	mim	(m)	mim	(m)	mu	(m)
nun	(n)	nun	(n)	nun	(n)	nun	(n)	nu	(n)
samekh	(s)	samekh	(s)	semkat	(s)	sin	(s)	ksi	(ks)
'ayin	(')	ayin	(')	'e	(')	'ayin	(')	omicron	(o)
pe	(p)	pe	(p, f)	pe	(p, f)	fa	(f)	pi	(p)
sade	(ṣ)	tsade	(ts)	ṣode	(ṣ)	ṣad	(ṣ)	san	(s)
qof	(d)	quf	(q)	quf	(q)	qaf	(q)	qoppa	(q)
resh	(r)	resh	(r)	rish	(r)	ra	(r)	rho	(r)
shin	(s, sh)	shin	(s, sh)	shin	(sh)	shin	(sh)	sigma	(s)
taw	(t)	tav	(t)	taw	(t)	ta	(t)	tau	(t)
						tha	(th)	upsilon	(u)
						kha	(kh)	phi	(ph)
						...		...	

* or digamma, subsequently dropped

FIG. 17.4. *The order of the twenty-two Phoenician letters has in most cases been preserved unaltered. The names here given to the Phoenician letters are only confirmed from the sixth century BCE, but their order and phonetic values go back much further, to at least the fourteenth century BCE.*

FIG. 17.5. *Ugaritic alphabet primer, fourteenth century BCE, found in 1948 at Ras Shamra. Damascus Museum. Transcription made by the author from a cast. See PRU II (1957), p. 199, document 184 A*

SILENT NUMBERS

North African shepherds used to count their flock by reciting a text that they knew by heart: "Praise be to Allah, the merciful, the kind ... ". Instead of using the fixed order of the number-names (*one, two, three* ...), they would use the fixed order of the words of the prayer as a "counting machine". When the last of the sheep was in the pen, the shepherd would simply retain the last word that he had said of the prayer as the name of the number of his flock.

This custom corresponds to an ancient superstition in this and many other cultures that counting aloud is, if not a sin, then a hostage to the forces of evil. In this view, numbers do not just express arithmetical quantities, but are endowed with ideas and forces that are sometimes benign and sometimes malign, flowing under the surface of mortal things like an underground river. People who hold such a belief may count things that are not close (such as people or possessions belonging to others), but must not count aloud their own loved ones or possessions, for *to name an entity is to limit it*. So you must never say how many brothers, wives or children you have, never name the number of your cattle, sheep or dwellings, or state your age or your total wealth. For the forces of evil could capture the hidden power of the number if it were stated aloud, and thus dispose of the people or things numbered.

The North African shepherd using the prayer as a counting device was therefore doing so not only to invoke the protection of Allah, but also to avoid using the actual names of numbers. In that sense his custom is similar to the use of counting-rhymes by children – fixed rhythmic sequences of words which when recited determine whose go it is at a game. In Britain, for instance, children chant as they point round at each other in a circle: *eeny meeny miny mo – catch a blackman by his toe – if he hollers let him go – eeny meeny miny mo!* The child whose "go" it is is the one to whom the finger is pointing when the reciter reaches *mo!*

The use of a fixed sequence like this is reminiscent of the archaic counting methods of pre-numerate peoples, for whom points of the body functioned much like a counting rhyme. Similarly, disturbed children (and sometimes quite normal ones) invent their own counting sequences: one boy I got to know counted *André, Jacques, Paul, Alain, Georges, Jean, François, Gérard, Robert,* (for 1, 2 ... 9) in virtue of the position of his dorm-mates' bunks with respect to his own; and G. Guitel (1975) reports the case of a girl who counted things as *January, February, March ...* etc.

The girl could of course have used instead the invariable order of the letters of the alphabet (A, B, C, D, E ...), for any sequence of symbols can be used as a counting model – provided that the order of its elements is immutable, as it is with the alphabet. And for that reason many civilisations have thought of representing numbers with the letters of their alphabet, still set in the order given them by the Phoenicians.

From the sixth century BCE, the Greeks developed a written numbering system from 1 to 24 by means of alphabetic letters, known as acrophonics:

A	1	I	9	P	17
B	2	K	10	Σ	18
Γ	3	Λ	11	T	19
Δ	4	M	12	Υ	20
E	5	N	13	Φ	21
Z	6	Ξ	14	X	22
H	7	O	15	Ψ	23
Θ	8	Π	16	Ω	24

FIG. 17.6.

The tablets of Heliastes, like the twenty-four songs of the *Iliad* and the *Odyssey*, used this kind of numbering, which is also found on funerary inscriptions of the Lower Period. However, what we have here is really only a simple substitution of letters for numbers, not a proper alphabetic number-system which, as we will now see, calls for a much more elaborate structure.

HEBREW NUMERALS

Jews still use a numbering system whose signs are the letters of the alphabet, for expressing the date by the Hebrew calendar, for chapters and verses of the Torah, and sometimes for the page numbers of books printed in Hebrew.

Hebrew characters, in common with most Semitic scripts, are written right to left and, a little like capital letters in the Latin alphabet, are clearly separated from each other. Most of them have the same shape wherever they come in a word: the five exceptions are the final *kof, mem, nun, pe,* and *tsade* (respectively, the Hebrew equivalents of our K, M, N, P, and a special letter for the sound TS):

	kof	*mem*	*nun*	*pe*	*tsade*
Regular form	כ	מ	נ	פ	צ
Final form	ך	ם	ן	ף	ץ

FIG. 17.7.

Black-letter ("square") Hebrew script is relatively simple and well-balanced, but care has to be taken with those letters that have quite similar graphical forms and which can mislead the unwary beginner:

פ כ ב	ך ר ד	ם ט	ו ז
b k p	d r k final	m t	v z
נ ג	ה ח ת	ס ם	ע צ
g n	h kh t	s m final	(guttural) ts

FIG. 17.8.

Hebrew numerals use the twenty-two letters of the alphabet, in the same order as those of the Phoenician alphabet from which they derive, to represent (from *aleph* to *tet*) the first nine units, then from *yod* to *tsade*, the nine "tens", and finally from *kof* to *tav*, the first four hundreds (see Fig. 17.10).

א aleph	ו vov	כ kof	ע ayin	ש shin
ב bet	ז zayin	ל lamed	פ pe	ת tav
ג gimmel	ח het	מ mem	צ tsade	
ד daleth	ט tet	נ nun	ק kuf	
ה he	י yod	ס samekh	ר resh	

FIG. 17.9. *The Modern Hebrew alphabet*

Letter	Name	Sound	Value		Letter	Name	Sound	Value
א	aleph	(h) a	1		ל	lamed	l	30
ב	bet	b	2		מ	mem	m	40
ג	gimmel	g	3		נ	nun	n	50
ד	daleth	d	4		ס	samekh	s	60
ה	he	h	5		ע	ayin	guttural	70
ו	vov	v	6		פ	pe	p	80
ז	zayin	z	7		צ	tsade	ts	90
ח	het	kh	8		ק	kuf	k	100
ט	tet	t	9		ר	resh	r	200
י	yod	y	10		ש	shin	sh	300
כ	kof	k	20		ת	tav	t	400

FIG. 17.10. *Hebrew numerals*

Compound numbers are written in this system, from right to left, by juxtaposing the letters corresponding to the orders of magnitude in descending order (i.e., starting with the highest). Numbers thus fit quite easily in Hebrew manuscripts and inscriptions. But when letters are used as numbers, how do you distinguish numbers from "ordinary" letters?

THIS IS THE MONUMENT OF ESTHER DAUGHTER OF ADAIO, WHO DIED IN THE MONTH OF SHIVRAT OF YEAR 3 (**ג**) OF THE "SHEMITA". YEAR THREE HUNDRED AND 46 (**ומ**) AFTER THE DESTRUCTION OF THE TEMPLE (OF JERUSALEM)*
PEACE! PEACE BE WITH HER!

* Year 346 of the Shemita +70 = 416 CE

FIG. 17.11. *Jewish gravestone (written in Aramaic), dated 416 CE. From the southwest shore of the Dead Sea. Amman Museum (Jordan). See IR, inscription 174*

HERE LIES AN
INTELLIGENT WOMAN
QUICK TO GRASP ALL
THE PRECEPTS OF FAITH
AND WHO FOUND THE
FACE OF GOD THE
MERCIFUL AT THE
TIME THAT COUNTS (?)
WHEN HANNA DEPARTED
SHE WAS 56 YEARS OLD

נ ו = נ׳ ו׳

6 50
←-----
  56

Fig. 17.12. Part of a bilingual (Hebrew-Latin) inscription carved on a soft limestone funeral stela found at Ora (southern Italy), seventh or eighth century CE. See CII, inscription 634 (vol. 1, p. 452).

Numbers that are represented by a single letter are usually distinguished by a small slanted stroke over the upper left-hand corner of the character, thus:

ש׳ פ׳ ל׳ ג׳ א׳

300   80   30   3   1

Fig. 17.13.

When the number is represented by two or more letters, the stroke is usually doubled and placed between the last two letters to the left of the expression (Fig. 17.14). But as these accent-strokes were also used as abbreviation signs, scribes and stone-cutters sometimes used other types of punctuation or "pointing" to distinguish numbers from letters (Fig. 17.15):

ש נ׳ב   ל״ה

2 + 50 + 300   5 + 30
←------------   ←------

Fig. 17.14.

150	קָנ ←-----	IHE, 183 date: 1389-90 CE	25	כ׳ה ←-----	IHE, 26 date: 1239 CE
175	קע׳ה ←-----	IHE, 100 date: 1415 CE	27	כ׳ז ←-----	IHE, 27 date: 1240 CE
196	קֻצו ←-----	IHE, 201 date: 1436-37 CE	28	כ׳ח ←-----	IHE, 45 date: 1349 CE
219	רי׳ט ←-----	BM Add. 27 106 date: 1459 CE	32	לב ←-----	IHE, 110 date: 1271–72 CE
312	שיב ←-----	BM Add. 27 146 date: 1552 CE	44	מ׳ד ←-----	IHE, 139 date: 1283–84 CE

FIG. 17.15. *Numerical expressions found in mediaeval Hebrew manuscripts and inscriptions.*
See Cantera and Millas

The highest Hebrew letter-numeral is only 400, so this is how higher numbers were expressed:

תתק	תת	תש	תר	תק
100 400 400	400 400	300 400	200 400	100 400
900	800	700	600	500

FIG. 17.16.

So for numbers from 500 to 900, the customary solution was to combine the letter *tav* (= 400) with the letters expressing the complement in hundreds. Compound numbers in this range were written as follows:

תתס׳ט 9 60 400 400 ←----------- 869	IHE, 102 date: 1108 CE
תתק׳ם 40 100 400 400 ←----------- 940	IHE, 107 date: 1180 CE
תתקמג 3 40 100 400 400 ←----------- 943	IHE, 108 date: 1183 CE

FIG. 17.17. *Expressions found on Jewish gravestones in Spain*

The numbers 500, 600, 700, 800, and 900 could also be represented by the final forms of the letters *kof, mem, nun, pe,* and *tsade* (see Fig. 17.7 above). However, this notation, which is found, for example, in the Oxford manuscript 1822 quoted by Gershon Scholem, was adopted only in Cabbalistic calculations. So in ordinary use, these final forms of the letters simply had the numerical value of the corresponding non-final forms of the letters.

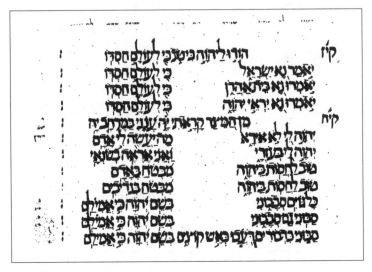

FIG. 17.18. *Page from a Hebrew codex, 1311 CE, giving Psalms 117 and 118. The numbers can be seen in the right-hand margin, in Hebrew letter-numbers. (Vatican Library, Cod. Vat. ebr. 12, fol. 58)*

To represent the thousands, the custom is to put two points over the corresponding unit, ten, or hundred character. In other words, when a character has two points over it, its numerical value is multiplied by 1,000.

א → אָ̈	ב → בָ̈	מ → מָ̈	צ → צָ̈
1 1,000	2 2,000	40 40,000	90 90,000

FIG. 17.19.

The Hebrew calendar in its present form was fixed in the fourth century CE. Since then, the months of the Jewish year begin at a theoretical, calculated date and not, as previously, at the sighting of the new moon. The foundation point for the calculation was the *neomenia* (new moon) of Monday, 24 September 344 CE, fixed as 1 Tishri in the Hebrew calendar, that is to say New Year's Day. As it was accepted that 216 Metonic cycles, in other words 4,400 years, sufficed at that point to contain the entire Jewish past, the chronologists calculated that the first *neomenia* of creation took place on Monday, 7 October 3761 BCE. As a result, the Jewish year 5739, for example, corresponds to the period from 2, October 1978 to 21 September 1979, and it is expressed on Jewish calendars (of the kind you can find in any kosher grocer's or corner shop) as:

9   30   300   400   5000

Fig. 17.20.

Jewish scribes and stone-carvers did not always follow this rule, but exploited an opportunity to simplify numerical expressions that was implicit in the system itself. Consider the following expression found on a gravestone in Barcelona: it gives the year 5060 of the Jewish calendar (1299–1300 CE) in this manner:

$$\dot{ס} \quad \ddot{ה} \qquad (= 5 \times 1,000 + 60)$$

60   5

Fig. 17.21.

Here, the points simply signify that the letters are to be read as numbers, not letters. But the expression appears to break one of the cardinal rules of Hebrew numerals – that the highest number always comes first, counting from right to left, which is the direction of writing in the Hebrew alphabet. So in any regular numerical expression, the letter to the right has a higher value than the one to its left. For that reason, the expression on the Barcelona gravestone is entirely unambiguous. Since the letter *he* can only have two values – 5 and 5,000, and the letter *samekh* counts for 60, the character to the right, despite not having its double point, must mean 5,000.

... שמואל בּרׄ חלאבו ...

... בשנת תתׄדׄ

804

"... SAMUEL SON OF KHALABU ...
... IN THE YEAR 804"

Fig. 17.22. *Fragment of a Jewish gravestone from Barcelona. The date is given as 804, for 4804 (4804 – 3760 = 1044 CE).*

Here are some other examples:

5,109	ṭ 9	ק 100	ה 5	Toledo, 1349 CE; IHE, no. 85

5,156	וֹ 6	נ 50	ק 100	ה 5	MS dated 1396; BM Add. 2806, fol. 11a

There is an even more interesting "irregularity" in the way some mediaeval Jewish scholars wrote down the total number of verses in the Torah [see G. H. F. Nesselmann (1842), p. 484]. The figure, 5,845, was written by using only the letters for the corresponding units, thus:

ה חם ׳ה

He	Mem	Het	He
5	40	8	5

←-------------------

Because of the rule that we laid out above, this expression is not ambiguous. The letter *het*, for example, whose normal value is 8, cannot have a lower value than *mem*, to its left, and whose value is 40; nor can it be 8,000, since it is itself to the left of *he*, whose value must be larger. For that reason the *het* can only mean 800.

It is not difficult to account for this particular variant of Hebrew numerals. In speech, the number 5,845 is expressed by:

KHAMISHAT	ALAFIM	SHMONEH	ME'OT	ARBA'IM	VE	KHAMISHA
"Five	thousand	eight	hundred	forty	&	five"

The names of the numbers thus make the arithmetical structure of the number apparent:

$$5 \times 1,000 + 8 \times 100 + 40 + 5$$

This could be transposed into English as "five thousand eight hundred forty (&) five", or in Hebrew as:

ה׳אלפים ח מאותמ׳ה

5	40	hundred	8	thousand	5

←--

FIG. 17.23.

"Mixed" formulations like these, combining words and numerals, are found on Hispano-Judaic tombstones (IHE, no. 61) and in some mediaeval manuscripts (for example, BM Add. 26 984, folio 143b). It is easy to see how such expressions can safely be abbreviated by leaving out the words for "hundred" and "thousand".

Another particularity arises in Hebrew numerals with the numbers 15 and 16. The regular forms would be:

5   10   6   10

←------   ←-----

FIG. 17.24.

However, the letter-values of these numbers spell out parts of the name of *Yahweh* – and it is forbidden, in Jewish tradition, to write the name of the Lord, even if its literal form of four letters (the "divine tetragrammaton", יהוה "yahve") is perfectly well-known. To avoid writing the tetragrammaton, various abbreviations were devised (יהו, הו, יו, יה) but these two were covered by the prohibition on writing the name of God. So the regular forms of the numbers 15 and 16 could not be used, and were replaced by the expressions 9 + 6 and 9 + 7 respectively:

ט ו   ט ז

6   9   7   9

←-----   ←-----

FIG. 17.25.

These are the main features of Hebrew numerals. It was by no means the only one to use the letters of the alphabet for expressing numbers. Let us now look at the Greek system of alphabetic numbering.

GREEK ALPHABETICAL NUMERALS

The Greek alphabet is absolutely fundamental for the history of writing and for Western civilisation as a whole. As C. Higounet (1969) explains, the Greek alphabet, quite apart from its having served to transmit one of the richest languages and cultures of the ancient world, forms the "bridge" between Semitic and Latin scripts. Historically, geographically, and also graphically, it was an intermediary between East and West; even more importantly, it was a structural intermediary too, in the sense that it first introduced regular and complete representations of the vowel sounds.

There is no question but that the Greeks borrowed their alphabet from the Phoenicians. Herodotus called the letters *phoinikeia grammatika*, "Phoenician writing"; and the early forms of almost all the Greek letters as well as their order in the alphabet and their names support this tradition. According to the Greeks themselves, Cadmos, the legendary founder of Thebes, brought in the first sixteen letters from Phoenicia; Palamedes was supposed to have added four more during the Trojan War; and four more were introduced later on by a poet, Simonides of Ceos.

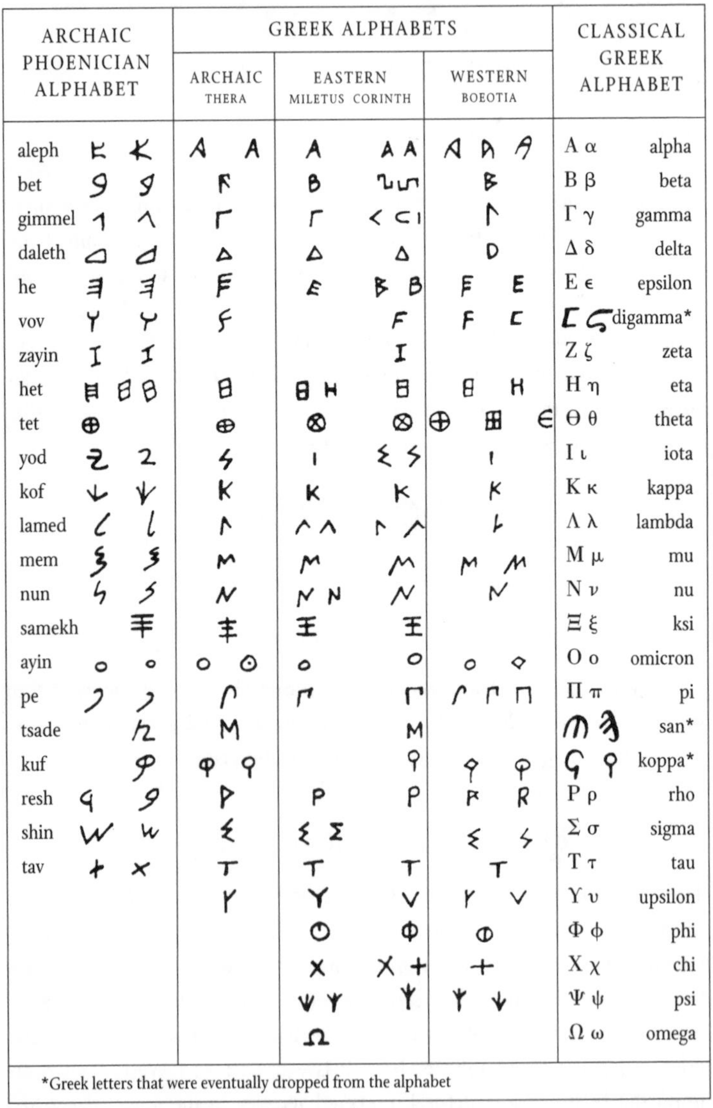

ARCHAIC PHOENICIAN ALPHABET		GREEK ALPHABETS			CLASSICAL GREEK ALPHABET	
		ARCHAIC THERA	EASTERN MILETUS CORINTH	WESTERN BOEOTIA		
aleph					A α	alpha
bet					B β	beta
gimmel					Γ γ	gamma
daleth					Δ δ	delta
he					E ε	epsilon
vov					Ϝ Ϛ	digamma*
zayin					Z ζ	zeta
het					H η	eta
tet					Θ θ	theta
yod					I ι	iota
kof					K κ	kappa
lamed					Λ λ	lambda
mem					M μ	mu
nun					N ν	nu
samekh					Ξ ξ	ksi
ayin					O o	omicron
pe					Π π	pi
tsade					san*	
kuf					Ϙ ϙ	koppa*
resh					P ρ	rho
shin					Σ σ	sigma
tav					T τ	tau
					Y υ	upsilon
					Φ φ	phi
					X χ	chi
					Ψ ψ	psi
					Ω ω	omega

*Greek letters that were eventually dropped from the alphabet

FIG. 17.26. *Greek alphabets compared to the archaic Phoenician script.*
See Février (1959) and Jensen (1969)

The oldest extant pieces of writing in Greek date from the seventh century BCE. Some scholars believe that the original borrowing from Phoenician occurred as early as 1500 BCE, others think it did not happen until the eighth century BCE: but it seems most reasonable to suppose that it happened around the end of the second millennium or at the start of the first. At any rate, the Greek alphabet did not arise in its final form at all quickly. There was a whole series of regional variations in the slow adaptation of Phoenician letters to the Greek language, and these non-standard forms are generally categorised under the following headings: archaic alphabets (as found at Thera and Melos), Eastern alphabets (Asia Minor and its coastal archipelagos, the Cyclades, Attica, Corinth, Argos, and the Ionian colonies in Sicily and southern Italy), and Western alphabets (Eubeus, the Greek mainland, and non-Ionian colonies). Unification and standardisation did not occur until the fourth century BCE, following the decision of Athens to replace its local script with the so-called Ionian writing of Miletus, itself an Eastern form of the alphabet.

Early Greek writing was done right to left, or else in alternating lines (*boustrophedon*), but it settled down to left-to-right around 500 BCE. Since letters are formed from the direction of writing, this change of orientation has to be taken into account when we compare Greek characters to their Semitic counterparts.

The names of the original Greek letters are:

alpha, beta, gamma, delta, epsilon, digamma,
zeta, eta, theta, iota, kappa, lambda, mu, nu,
ksi, omicron, pi, san, koppa, rho, sigma, tau.

Of these, the *digamma* was lost early on, and the *san* and *koppa* were also subsequently abandoned. However, a different form of the Semitic *vov* provided the *upsilon*, and three new signs, *phi, chi,* and *psi,* were added to represent sounds that do not occur in Semitic languages. Finally, *omega* was invented to distinguish the long *o* from the *omicron*. So the classical Greek alphabet, from the fourth century BCE, ended up having twenty-four letters, including vowels as well as consonants.

Semitic languages can be written down without representing the vowels because the position of a word in a sentence determines its meaning and also the vowel sounds in it, which change with different functions. In Greek, however, the inflections (word-endings) alone determine the function of a word in a sentence, and the vowel sounds cannot be guessed unless the endings are fully represented. The Phoenician alphabet had letters for guttural sounds that do not exist in Greek; Greek, for its part, had aspirated consonants with no equivalents in Semitic languages. So the

Greeks converted the Semitic guttural letters, for which they had no use, into vowels, which they needed. The "soft breathing sound" *aleph* became the Greek *alpha*, the sound of *a*; the Semitic letter *he* was changed into *epsilon* (*e*), and the *vov* first became *digamma* then *upsilon* (*u*); the Hebrew *yod* was converted into *iota* (*i*); and the "hard breathing sound" *ayin* became an *omicron* (*o*). For the aspirated consonants, the Greeks simply created new letters, the *phi*, *chi* and *psi*. In brief, the Greeks adapted the Semitic system to the particularities of their own language. But despite all that is clear and obvious about this process, the actual origin of the idea of representing the vowel sounds by letters remains obscure.

This survey of the development of the Greek alphabet allows us now to look at the principles of Greek numbering, often called a "learned" system, but which is in fact entirely parallel to Hebrew letter-numbers.

We can get a first insight into the system by looking at a papyrus (now in the Cairo Museum, Inv. 65 445) from the third quarter of the third century BCE (Fig. 17.31).

O. Guéraud and P. Jouguet (1938) explain that this papyrus is a "kind of exercise book or primer, allowing a child to practise reading and counting, and containing in addition various edifying ideas ... As he learned to read, the child also became familiar with numbers. The place that this primer gives to the sequence of the numbers is quite natural, coming as it does after the table of syllables, because the Greek letters also had numerical values. It was logical to give the child first the combination of letters into syllables, and then the combinations of letters into numbers."

The numeral system the papyrus gives uses the twenty-four letters of the classical Greek alphabet, plus the three obsolete letters, *digamma*, *koppa* and *san* (see Fig. 17.26 above). These twenty-seven signs are divided into three classes. The first, giving the units 1 to 9, uses the first eight letters of the classical alphabet, plus *digamma* (the old Semitic *vov*), inserted in the sequence to represent the number 6. The second contains the eight following letters, plus the obsolete *koppa* (the old *quf*), to give the sequence of the tens, from 10 to 90. And the third class gives the hundreds from 100 to 900, using the last eight letters of the classical alphabet plus the *san* (the Semitic *tsade*) (for the value of 900) (see Fig. 17.27).

Intermediate numbers are produced by additive combinations. For 11 to 19, for instance, you use *iota*, representing 10, with the appropriate letter to its right representing the unit to be added. To distinguish the letters used as numerals from "ordinary" letters, a small stroke is placed over them. (The modern printing convention of placing an accent mark to the top right of the letter is not used in most Greek manuscripts.)

UNITS				TENS				HUNDREDS		
A	α	alpha	1	I	ι	iota	10	P	ρ	rho	100
B	β	beta	2	K	κ	kappa	20	Σ	σ	sigma	200
Γ	γ	gamma	3	Λ	λ	lambda	30	T	τ	tau	300
Δ	δ	delta	4	M	μ	mu	40	Υ	υ	upsilon	400
E	ε	epsilon	5	N	ν	nu	50	Φ	φ	phi	500
ϛ	ϛ	digamma*	6	Ξ	ξ	ksi	60	X	χ	chi	600
Z	ζ	zeta	7	O	o	omicron	70	Ψ	ψ	psi	700
H	η	eta	8	Π	π	pi	80	Ω	ω	omega	800
Θ	ϑ	theta	9	ϟ	ϟ	koppa	90	ϡ	ϡ	san (sampi)	900

*In manuscripts from Byzantium, 6 is written στ (sigma+tau). In Modern Greek, where alphabetic numerals are still used for specific purposes (rather like Roman numerals in our culture), this sign is called a *stigma*.

FIG. 17.27. *Greek alphabetic numerals*

The beginning of the primer scroll has the remnants of the number sequence up to 25:

H̄	8	K̄	20
Θ̄	9	K̄A	21
Ī	10	K̄B	22
ĪA	11	K̄Γ	23
ĪB	12	K̄Δ	24
ĪΓ	13	K̄ε	25

FIG. 17.28.

Guéraud & Jouguet (1938) note that the list is an elementary one, and does not even include all the symbols the pupil would need to understand the table of squares given at the end of the primer (Fig. 17.31). However, the table of squares itself, besides giving the young reader some basic ideas of arithmetic, also served to show the sequence of numbers beyond those given at the start of the scroll and to familiarise the learner with the principles of Greek numbering from 1 to 640,000, and that may have been its real purpose.

How could the scribe represent numbers from 1 to 640,000 when the highest numeral in the alphabet was only 900? For numbers up to 9,000, he just added a distinctive sign to the letters representing the units, thus*:

ʹA	ʹB	ʹΓ	ʹΔ	ʹE	ʹϛ	ʹZ	ʹH	ʹΘ
1,000	2,000	3,000	4,000	5,000	6,000	7,000	8,000	9,000

FIG. 17.29.

* Printed Greek usually puts the distinctive sign (a kind of *iota*) as a subscript, to the lower left corner of the character.

When he got to 10,000, otherwise called the myriad (Μύριοι)*, the second "base" of Greek numerals, he put an M (the first letter of the Greek word for "ten thousand") with a small *alpha* over the top. All following multiples of the myriad could therefore be written in the following way:

α	β	γ	δ	ϵ		ια	ιβ		χξθ	
M	M	M	M	M	...	M	M	...	M	...
10,000	20,000	30,000	40,000	50,000		110,000	120,000		6,690,000	

FIG. 17.30.

As he gave these numbers in the form 1 myriad, 2 myriads, 3 myriads, etc. the scribe could reach 640,000 without any difficulty. He could obviously have continued the sequence up to the 9,999th myriad, which he would have written thus:

$$\overset{'\theta\vartheta\varrho\theta}{M} \quad \overset{9999}{(M} = 99{,}990{,}000)$$

TRANSCRIPTION	TRANSLATION

Left-hand column					Right-hand column				
Δ			$\overset{\circ}{4}$						
ἐ	Κέ		$\overset{\circ}{5}$	25	P̣	P	M̊		
								$\overset{\circ}{90}$	8,100
ϛ̓	ϛ̄	Λ⊂	6	6	36	Σ	Σ	M̊	100	100	10,000
Z	Z	MΘ	7	7	79	T	T	M̊	200	200	40,000
H	H	Ξ Δ	8	8	64	Y	Y	M̅	300	300	90,000
Θ	Θ	ΠΑ	9	9	81	Φ	Φ	M	400	400	160,000
I	I	P	10	10	100	X	X	M̅	500	500	250,000
K	K	Y	20	20	400	Ψ	Ψ	M̅	600	600	360,000
Λ	Λ	ϒ	30	30	900	Ω	Ω	M̅	700	700	490,000
M	M	ΛΧ	40	40	1,600				800	800	640,000

FIG. 17.31. *Fragment of a Greek papyrus, third quarter of the the third century BCE (Cairo Museum, inv. 65 445). See Guéraud & Jouguet (1938), plate X. The papyrus gives a table of squares, from 1 to 10 and then in tens to 40 (left-hand column), and from 50 to 800 (right-hand column). The squares of 1, 2 and 3 are missing from the start of the table.*

*When the accent is on the first syllable, the word means "ten thousand"; when the accent is on the second syllable, it has the meaning "a very large number".

These kinds of notation for very large numbers were frequently used by Greek mathematicians. For example, Aristarch of Samos (?310–?230 BCE) wrote the number 71,755,875 in the following way, according to P. Dedron & J. Itard (1959), p. 278:

$$'\zeta \; \rho \; o \; \epsilon \; M \; '\epsilon \; \omega \; o \; \epsilon$$

$$\xrightarrow{\hspace{5cm}}$$

$$7{,}175 \; \times \; 10{,}000 \; + \; 5{,}875$$

FIG. 17.32.

We find a different system in Diophantes of Alexandria (c. 250 CE): he separates the myriads from the thousands by a single point. So for him the following expression meant 4,372 myriads and 8,097 units, or 43,728,097 [from C. Daremberg & E. Saglio (1873), p. 426]:

$$\delta \tau o \beta \; '\eta \; \mathsf{?} \; \zeta$$

$$\xrightarrow{\hspace{4cm}}$$

$$4{,}372 \; \times \; 10{,}000 \; + \; 8{,}097$$

FIG. 17.33.

The mathematician and astronomer Apollonius of Perga (c. 262–c. 180 BCE) used a different method of representing very large numbers, and it has reached us through the works of Pappus of Alexandria (third century CE). This system was based on the powers of the myriad and used the principle of dividing numbers into "classes". The first class, called the elementary class, contained all the numbers up to 9,999, that is to say all numbers less than the myriad. The second class, called the class of *primary myriads*, contained the multiples of the myriad by all numbers up to 9,999 (that is to say the numbers 10,000, 20,000, 30,000, and so on up to 9,999 × 10,000 = 99,990,000). To represent a number in this class, the number of myriads in the number is written after the sign $\overset{\alpha}{\mathrm{M}}$. A reconstructed example:

$$\overset{\alpha}{\mathrm{M}} \chi \xi \delta$$

$$\xrightarrow{\hspace{1.5cm}} \text{means } 664 \times 10{,}000 = 6{,}640{,}000$$

$$664$$

FIG. 17.34.

Next comes the class of *secondary myriads*, which contains the multiples of a myriad myriads by all the numbers between 1 and 9,999 (that is to say, the numbers 100,000,000, 200,000,000, 300,000,000, and so on up to

9,999 × 100,000,000 = 999,900,000,000. A number in this range is expressed by writing *beta* over M before the number (written in the classical letter-number system) of one hundred millions that it contains. A reconstructed example:

$$\overset{\beta}{\mathrm{M}}{}'\epsilon\omega\xi\gamma$$

-----------→

5,863

FIG. 17.35.

This notation thus means: 5,863 × 100,000,000 = 586,300,000,000, and is "read" as 5,863 *secondary myriads*.

Next come the tertiary myriads, signalled by *gamma* over M, which begin at 100,000,000 × 10,000 = 1,000,000,000,000; then the quaternary myriads (signalled by *delta* over M), and so on.

The difference between the system used in the papyrus of Fig. 17.31 and the system of Apollonius is that whereas for the papyrus the superscribed letter over M is a *multiple* of 10,000, for Apollonius the superscript represents a *power* of 10,000.

In the Apollonian system, intermediate numbers can be expressed by breaking them down into a sum of numbers of the consecutive classes. Pappus of Alexandria [as quoted in P. Dédron & J. Itard (1959) p. 279] gave the example of the number 5,462,360,064,000,000, expressed as 5,462 tertiary myriads, 3,600 secondary myriads, and 6,400 primary myriads (in which the Greek word καɩ can be taken to mean "plus"):

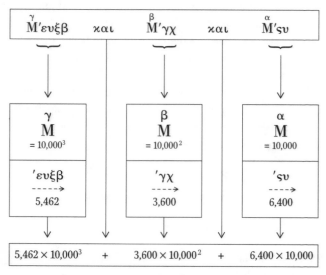

FIG. 17.36.

Archimedes (?287– 212 BCE) proposed an even more elaborate system for expressing even higher magnitudes, and laid it out in an essay on the number of grains of sand that would fill a sphere whose diameter was equal to the distance from the earth to the fixed stars. Since he had to work with numbers larger than a myriad myriads, he imagined a "doubled class" of numbers containing eight digits instead of the four allowed for by the classical letter-number system, that is to say octets. The first octet would contain numbers between 1 and 99,999,999; the second octet, numbers starting at 100,000,000; and so on. The numbers belonged to the first, second, etc. class depending on whether they figure in the first, second, etc. octet.

As C. E. Ruelle points out in DAGR (pp. 425–31), this example suffices to show just how far Greek mathematicians developed the study and applications of arithmetic. Archimedes's conclusion was that the number of grains of sand it would take to fill the sphere of the world was smaller than the eighth term of the eighth octet, that is to say the sixty-fourth power of 10 (1 followed by 64 zeros). However, Archimedes's system, whose purpose was in any case theoretical, never caught on amongst Greek mathematicians, who it seems preferred Apollonius's notation of large numbers.

From classical times to the late Middle Ages, Greek alphabetic numerals played almost as great a role in the Middle East and the eastern part of the Mediterranean basin as did Roman numerals in Western Europe.

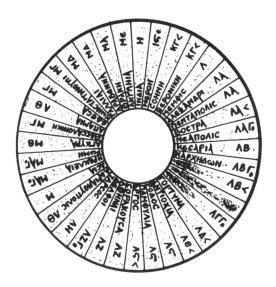

FIG. 17.37A. *Part of a portable sundial from the Byzantine era (Hermitage Museum, St Petersburg). This disc gives the names of the regions where it can be used, with latitudes indicated in Greek alphabetical numerals in ascending clockwise order.*

TRANSCRIPTION		TRANSLATION	
INΔIA	H	India	8
MEPOH	Iϛ<	Meroe	$16\,^1/_2$
COHNH	KΓ<	Syena	$23\,^1/_2$
BEPONIKH	KΓ<	Beronika	$23\,^1/_2$
MEMΦIC	Λ	Memphis	30
AΛEΞANΔPI	ΛA	Alexandria	31
ΠENTAΠOΛIC	ΛA	Pentapolis	31
BOCTPA	ΛA<	Bostra	$31\,^1/_2$
NEAΠOΛIC	ΛA Γo	Neapolis	$31\,^2/_3$
KECARIA	ΛB	Caesaria	32
KAPXHΔΩN	ΛB Γo	Carthage	$32\,^2/_3$
....	ΛB<	....	$32\,^1/_2$
....	...	....	
....	ΛΓ Γo	....	$33\,^2/_3$
ΓOPTYNA	ΛΔ<	Gortuna	$34\,^1/_2$
ANTIOXIA	ΛE<	Antioch	$35\,^1/_2$
POΔOC	Λϛ	Rhodes	36
ΠAMΦYΛIA	Λϛ	Pamphilia	36
APΓOC	Λϛ<	Argos	$36\,^1/_2$
COPAKOYCA	ΛZ	Syracuse	37
AΘHNAI	ΛZ	Athens	37
ΔEΛΦOI	ΛZ Γo	Delphi	$37\,^2/_3$
TAPCOC	ΛH	Tarsus	38
AΔPIANOYΠOΛIC	ΛΘ	Adrianopolis	39
ACIA	M	Asia	40
HPAKΛEIA	MA Γo	Heraklion	$41\,^2/_3$
PΩMH	MA Γo	Rome	$41\,^2/_3$
AΓKYPA	MB	Ankara	42
ΘECCAΛONIKH	MΓ	Thessalonika	43
AΠAMIA	ΛΘ	Apamea	39
EΔECA	MΓ	Edessa	43
KΩNCTA'TINOYΠI	MΓ	Constantinople	43
ΓAΛΛIAI	MΔ	Gaul	44
APABENNA	MΔ	Aravenna	44
ΘPAKH	MA	Thrace	41 (?44)
AKYΛHIA	ME	Aquileia	45

$< = ^1/_2$ $Γo = ^2/_3$

Fig. 17.37B.

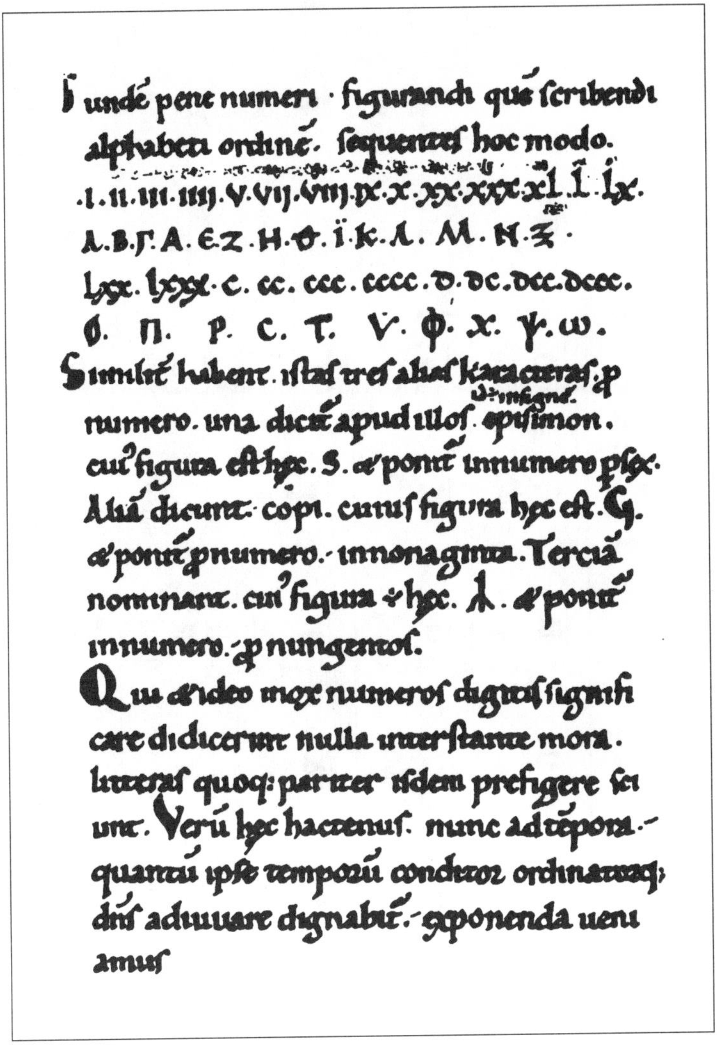
s undē pene numeri · figurandi quē scribendi
alphabeti ordinē. sequentes hoc modo.
·I·II·III·IIII·V·VI·VII·VIII·IX·X·XX·XXX·XL·L·LX·
A·B·Γ·Δ·E·Z·H·Θ·Ï·K·Λ·M·N·Ξ·
LXX·LXXX·C·CC·CCC·CCCC·D·DC·DCC·DCCC·
O· Π· P· C· T· V· Φ· X· Ψ· Ω·
Similr̄ habent. istas tres alias karacteras. p
id est insigne.
numero. una dicit apud illos. episimon.
cui figura est hec. S. et ponit innumero p sex.
Alia dicunt: copi. cuius figura hec est. Ϙ.
et ponit p numero· innonaginta. Tercia
nominant. cui figura ÷ hec. ϡ. et ponit
innumero· p nungentos.
Qui et ideo mox numeros digitis signifi
care didicerint nulla interstante mora.
litteras quoq: pariter isdem prefigere sci
unt. Verū hec hactenus. nunc ad tēpora·
quantū ipse temporū conditor ordinatorq;
dns adiuuare dignabit· exponenda ueni
amus

FIG. 17.38. *Fragment of a Spanish manuscript concerning the Venerable Bede's finger-counting system, copied in c. 1130 CE, probably at Santa Maria de Ripoll (Catalonia). Madrid, National Library, Cod. A 19 folio 2 (top left). To explain the finger diagrams given on the following pages, the scribe uses two different numerical notations – Roman numerals and the Greek alphabetic system, with their correspondence.*

1	10	100	1,000	10,000	100,000
2	20	200	2,000	20,000	200,000
3	30	300	3,000	30,000	300,000
4	40	400	4,000	40,000	400,000
5	50	500	5,000	50,000	500,000
6	60	600	6,000	60,000	600,000
7	70	700	7,000	70,000	700,000
8	80	800	8,000	80,000	800,000
9	90	900	9,000	90,000	900,000

FIG. 17.39. *Coptic numerals. [From Mallon, (1956); Till, (1955)]. The script of Egyptian Christians has 31 letters, of which 24 derive directly from Greek, and the others from demotic Egyptian writing. However, Coptic numerals use the same signs as the Greek system (that is to say, the 24 signs of the classical alphabet plus the three obsolete letters,* digamma, koppa *and* san, *with the same values as in Greek). In Coptic, letters used as numbers have a single superscripted line up to 999, and a double superscript for 1,000 and above.*

ARMENIAN LETTERS		NAMES OF THE LETTERS	SOUNDS		NUMERICAL VALUES
UPPER-CASE	LOWER-CASE		WESTERN ARMENIAN	EASTERN ARMENIAN	
Ա	ա	ayp/ayb	a	a	1
Բ	բ	pén/bén	p	b	2
Գ	գ	kim/gim	k	g	3
Դ	դ	ta/da	t	d	4
Ե	ե	yétch	é	ye/e	5
Զ	զ	za	z	z	6
Է	է	é	é	é	7
Ը	ը	et	e	e	8
Թ	թ	to	t	t/th	9
Ժ	ժ	jé	j	j	10
Ի	ի	ini	i	i	20
Լ	լ	lyoun	l	l	30
Խ	խ	khé	kh	kh	40
Ծ	ծ	dza/tsa	dz	ts	50
Կ	կ	gen/ken	g	k	60
Հ	հ	ho	h	h	70
Ձ	ձ	tsa/dza	tz	dz	80
Ղ	ղ	ghad	gh	gh	90

FIG. 17.40. *Armenian numerals. Armenian uses an alphabet of 32 consonants and 6 vowels, designed specifically for this language in the fifth century CE by the priest Mesrop Machtots (c.362–440CE). The alphabet was based on Greek and Hebrew.*

ARMENIAN LETTERS		NAMES OF THE LETTERS	SOUNDS		NUMERICAL VALUES
UPPER-CASE	LOWER-CASE		WESTERN ARMENIAN	EASTERN ARMENIAN	
Ճ	ճ	djé/tché	dj	tch	100
Մ	մ	mén	m	m	200
Յ	յ	hi	y	y/h	300
Ն	ն	nou	h	n	400
Շ	շ	cha	ch	ch	500
Ո	ո	vo	o	o	600
Չ	չ	tcha	tch	tch	700
Պ	պ	bé/pé	b	p	800
Ջ	ջ	tché/djé	tch	dj	900
Ռ	ռ	ra	r	rr	1,000
Ս	ս	sé	s	s	2,000
Վ	վ	vév	v	v	3,000
Տ	տ	dyoun/tyoun	d	t	4,000
Ր	ր	ré	r	r	5,000
Ց	ց	tso	ts	ts	6,000
Ւ	ւ	hyoun	u	iu	7,000
Փ	փ	pyour	p	p	8,000
Ք	ք	ké	k	k	9,000
Օ	o	o	ô	o	
Ֆ	ֆ	fé	f	f	

Fig. 17.40(*continued*). *Like Greek, Armenian uses the first 9 letters to represent the units, the second 9 for the tens, the third 9 for the hundreds. However, as it has more letters than Greek, it can use the fourth set of 9 letters to represent the thousands. Note than only 36 of the 38 letters are used for numerical purposes.*

GEORGIAN LETTERS		VALUES		GEORGIAN LETTERS		VALUES	
UPPER-CASE	LOWER-CASE	PHONETIC	NUMERICAL	UPPER-CASE	LOWER-CASE	PHONETIC	NUMERICAL
Ⴀ	ⴀ	a	1	Ⴐ	ⴐ	r	100
Ⴁ	ⴁ	b	2	Ⴑ	ⴑ	s	200
Ⴂ	ⴂ	g	3	Ⴒ	ⴒ	t	300
Ⴃ	ⴃ	d	4	Ⴓ	ⴓ	u	400
Ⴄ	ⴄ	e	5	Ⴔ	ⴔ	vi	500
Ⴅ	ⴅ	v	6	Ⴕ	ⴕ	p'	600
Ⴆ	ⴆ	z	7	Ⴖ	ⴖ	k'	700
Ⴡ	ⴡ	h	8	Ⴗ	ⴗ	γ	800
Ⴇ	ⴇ	t'	9	Ⴘ	ⴘ	q	900
Ⴈ	ⴈ	i	10	Ⴙ	ⴙ	š	1,000
Ⴉ	ⴉ	k	20	Ⴚ	ⴚ	tš	2,000
Ⴊ	ⴊ	l	30	Ⴛ	ⴛ	ts	3,000
Ⴋ	ⴋ	m	40	Ⴜ	ⴜ	dz	4,000
Ⴌ	ⴌ	n	50	Ⴝ	ⴝ	ts'	5,000
Ⴢ	ⴢ	ï	60	Ⴞ	ⴞ	tš'	6,000
Ⴍ	ⴍ	o	70	Ⴟ	ⴟ	h̭	7,000
Ⴎ	ⴎ	p	80	Ⴠ	ⴠ	h	8,000
Ⴏ	ⴏ	ž	90	Ⴣ	ⴣ	dž	9,000
				Ⴥ	ⴥ	h	10,000

FIG. 17.41. *Georgian alphabetic numerals. An example of a script and numeral system influenced by Greek in the Christian era. There are two distinct styles of writing the Georgian alphabet: the "priestly" script, or* khoutsouri, *reproduced above, and the "military", or* mkhedrouli. *Both have 38 letters.*

GOTHIC LETTERS	VALUES		GOTHIC LETTERS	VALUES		GOTHIC LETTERS	VALUES	
	PHONETIC	NUMERICAL		PHONETIC	NUMERICAL		PHONETIC	NUMERICAL
Ᏸ	a	1	I	i	10	Ᏸ	r	100
Ᏸ	b	2	K	k	20	S	s	200
Γ	g	3	λ	l	30	T	t	300
ᕷ	d	4	M	m	40	Ᏹ	w	400
Ɛ	e	5	N	n	50	Ᏺ	f	500
ᴜ	q	6	Ϛ	y	60	X	ch	600
Z	z	7	ᴨ	u	70	⊙	hw	700
ᴛ	h	8	ᴨ	p	80	Ᏸ	o	800
Φ	th	9	Ϥ		90	✝		900

FIG. 17.42. *Gothic: Another alphabetical numeral system influenced by Greek in the Christian era. The Goths – a Germanic people living on the northeastern confines of the Roman Empire, were Christianised by Eastern (Greek-speaking) priests in the second and third centuries CE. Wulfila (311–384 CE), a Christianised Goth who became a bishop, translated the Bible into his own tongue, and invented the Gothic alphabet, based on Greek together with some additional characters, in order to do this. The Goths eventually merged into other peoples, from Crimea to North Africa, and disappeared, leaving only the term "Gothic" with its various acquired meanings.*

A	1	K	10	T	100
B	2	L	20	V	200
C	3	M	30	X	300
D	4	N	40	Y	400
E	5	O	50	Z	500
F	6	P	60		
G	7	Q	70		
H	8	R	80		
I	9	S	90		

FIG. 17.43. *Numeral alphabet used by some mediaeval and Renaissance mystics. This adaptation of the Greek system to the Latin alphabet is described by A. Kircher in* Oedipi Aegyptiaci, *vol. II/1, p. 488 (1653).*

CHAPTER 18

THE INVENTION OF
ALPHABETIC NUMERALS

Greek alphabetic numerals were, as we have seen, pretty much identical to the system of Hebrew numerals, save for a few details. The similarity is such as to prompt the question: which came first?

What follows is an attempt to answer the question on the basis of what is currently known.

First of all, though, we have to clear away a myth that has been handed down uncritically as the truth for more than a hundred years.

THE MYTH OF PHOENICIAN LETTER-NUMBERS

It has long been asserted that, long before the Jews and the Greeks, the Phoenicians first assigned numerical values to their alphabetic signs and thus created the first alphabetic numerals in history.

However, this assumption rests on no evidence at all. No trace has yet been discovered of the use of such a system by the Phoenicians, nor by their cultural heirs, the Aramaeans.

The idea is in fact but a conjecture, devoid of proof or even indirect evidence, based solely on the fact that the Phoenicians managed to simplify the business of writing down spoken language by inventing an alphabet.

As we shall see, Phoenician and Aramaic inscriptions that have come to light so far, including the most recent, show only one type of numerical notation – which is quite unrelated to alphabetic numerals.

In the present state of our knowledge, therefore, we can consider only the Greeks and the Jews as contenders for the original invention of letter-numerals.

THE NUMERALS OF THE NORTHWESTERN SEMITES

The numerical notations used during the first millennium BCE by the various northwestern Semitic peoples (Phoenicians, Aramaeans, Palmyreneans, Nabataeans, etc.) are very similar to each other, and manifestly derive from a common source.

Leaving aside the cases of Hebrew and Ugaritic, the earliest instance of "numerals" found amongst the northwestern Semites dates from no earlier than the second half of the eighth century BCE. It is in an inscription on a monumental statue of a king called Panamu, presumed to have come from Mount Gercin, seven km northeast of Zencirli, Syria (not far from the border with Turkey). Semites generally liked to "write out" numbers, that is to say to spell out number-names, and this tradition, which continued for many centuries, no doubt explains why specific number-signs made such a late appearance. But that does not mean to say that their system of numerals is at all obscure.

The Aramaeans were traders who, from the end of the second millennium BCE, spread all across the Middle East; their language and culture were adopted in cities and ports from Palestine to the borders of India, from Anatolia to the Nile basin, and of course in Mesopotamia and Persia, over a stretch of time that goes from the Assyrian Empire to the rise of Islam. Thanks to the economic and legal papyri that constitute the archives of an Aramaic-speaking Jewish military colony established in the fifth century BCE at Elephantine in Egypt, we can easily reconstruct the Aramaeans' numeral system.

Aramaic numerals were initially very simple, using a single vertical bar to represent the unit, and going up to 9 by repetition of the strokes. To make each numeral recognisable at a glance, the strokes were generally written in groups of three (Fig. 18.1 A). A special sign was used for 10, and also (oddly enough) for 20 (Fig. 18.1 B and 18.1 C), whereas all other numbers from 1 to 99 were represented by the repetition of the basic signs. Aramaic numerals to 99 were thus based on the principle that any number of signs juxtaposed represented the sum of the values of those signs. As we shall see (Fig. 18.2), Aramaic numerals up to this point were thus identical to those of all other western Semitic dialects, namely:

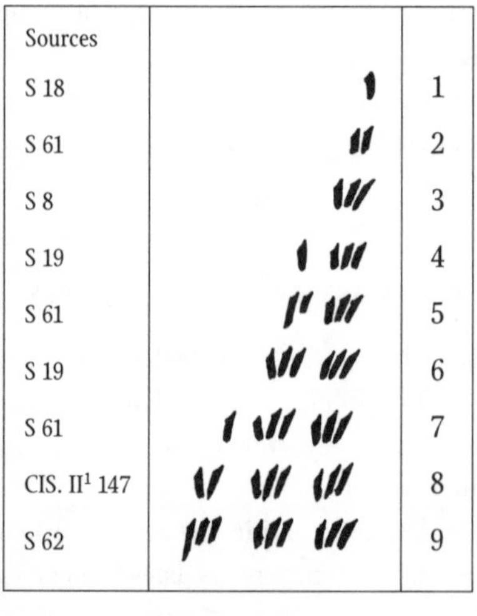

Sources		
S 18	**1**	1
S 61	**11**	2
S 8	**111**	3
S 19	**1 111**	4
S 61	**1′ 111**	5
S 19	**111 111**	6
S 61	**1 111 111**	7
CIS. II¹ 147	**11 111 111**	8
S 62	**111 111 111**	9

FIG. 18.1A. *Aramaic figures for the numbers 1 to 9. Copied from Sachau (1911), abbreviated as S, from fifth century BCE papyri from Elephantine (same source for Fig. 18.1 B – E)*

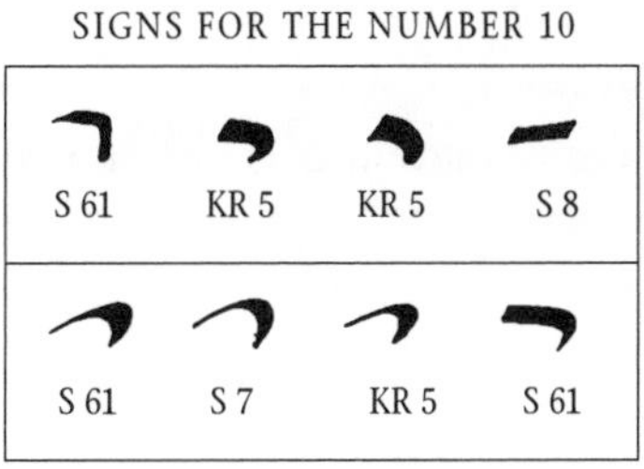
SIGNS FOR THE NUMBER 10
S 61
KR 5
KR 5
S 8
S 61
S 7
KR 5
S 61

FIG. 18.1B.

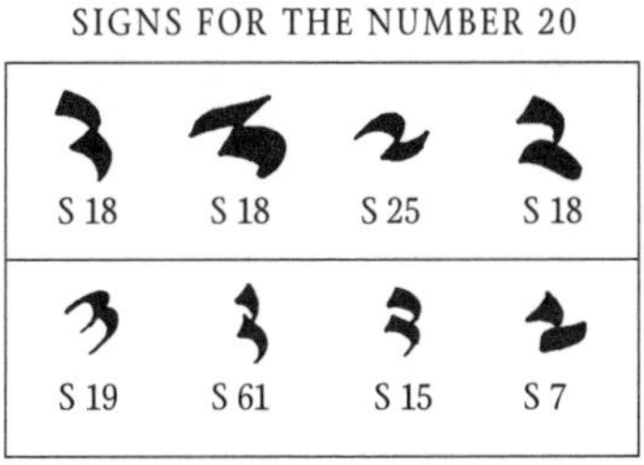
SIGNS FOR THE NUMBER 20
S 18
S 18
S 25
S 18
S 19
S 61
S 15
S 7

FIG. 18.1C.

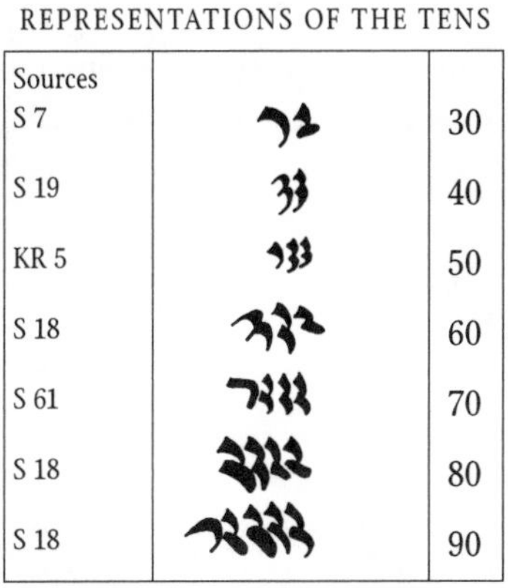
REPRESENTATIONS OF THE TENS
Sources
S 7
30
S 19
40
KR 5
50
S 18
60
S 61
70
S 18
80
S 18
90

FIG. 18.1D.

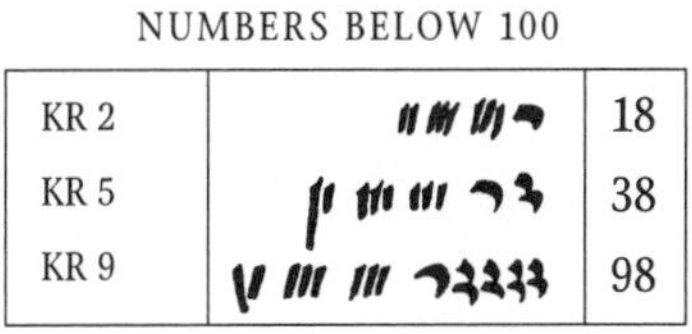
NUMBERS BELOW 100
KR 2
18
KR 5
38
KR 9
98

FIG. 18.1E.

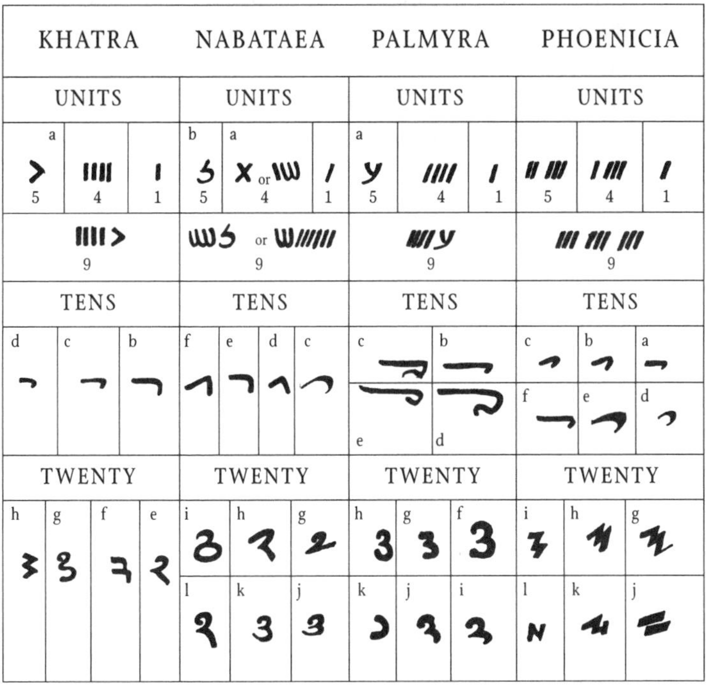
KHATRA
NABATAEA
PALMYRA
PHOENICIA
UNITS
UNITS
UNITS
UNITS
a
5
4
1
b
a
5
or
4
1
a
5
4
1
5
4
1
9
or
9
9
9
TENS
TENS
TENS
TENS
d
c
b
f
e
d
c
c
b
e
d
c
b
a
f
e
d
TWENTY
TWENTY
TWENTY
TWENTY
h
g
f
e
i
h
g
l
k
j
h
g
f
k
j
i
i
h
g
l
k
j

FIG. 18.2.

KHATRA	NABATAEA	PALMYRA	PHOENICIA
References: B. Aggoula (1972); Milik (1972); Naveh (1972)	References: G. Cantineau (1930)	References: M. Lidzbarski (1962)	References: M. Lidzbarski (1962)
a Khatra no. 65	a CIS II¹, 161	a CIS II³, 3 913	a CIS I¹, 165
b Khatra no. 65	b CIS II¹, 212	b CIS II³, 3 952	b CIS I¹, 165
c Khatra no. 62	c CIS II¹, 158	c CIS II³, 4 036	c CIS I¹, 93
d Abrat As-Saghira	d CIS II¹, 147 B	d CIS II³, 3 937	d CIS I¹, 88
e Abrat As-Saghira	e CIS II¹, 349	e CIS II³, 3 915	e CIS I¹, 165
f Khatra no. 62	f CIS II¹, 163 D	f CIS II³, 3 937	f CIS I¹, 3 A
g Abrat As-Saghira	g CIS II¹, 354	g CIS II³, 4 032	g CIS I¹, 87
h Khatra nos. 34, 65, 80	h CIS II¹, 211	h CIS II³, 3 915	h CIS I¹, 93
i Doura-Europos	i CIS II¹, 161	i CIS II³, 3 969	i CIS I¹, 7
j Ashoka	j CIS II¹, 213	j CIS II³, 3 969	j CIS I¹, 86 B
k Ostraca nos. 74 & 113 from Nisa	k CIS II¹, 204	k CIS II³, 3 935	k CIS I¹, 13
l Khatra nos. 62 & 65	l CIS II¹, 204	l CIS II³, 3 915	l CIS I¹, 165
	m N, II, 12	m CIS II³, 3 917	m CIS I¹, 143
	n CIS II¹, 163D		n CIS I¹, 65
	o CIS II¹, 161		o IS I¹, 7
			p CIS I¹, 217

• Phoenician, the language of a people of traders and sailors who settled, from the third millennium BCE, in Canaan (on the Mediterranean shore of Syria and Palestine); but Phoenician numerals are not found earlier than the sixth century BCE;

• Nabataean, spoken by people who, from the fourth century BCE, were settled at Petra, a city (now in Jordan) at the crossroads of trails leading from Egypt and Arabia to Syria and Palestine, and whose numeral system is attested from the second century BCE;

• Palmyrenean, spoken at Palmyra (east of Homs, in the Syrian desert), from around the beginning of the Common Era;

• Syriac, in use from the beginning of the Common Era;

• the dialect of Khatra, spoken in the early centuries of the Common Era by the inhabitants of the city of Khatra, in upper Mesopotamia, southwest of Mosul;

• Indo-Aramaic, a numeral system found in Kharoshthi inscriptions in the former province of Gandhara (on the borders of present-day Afghanistan and the Punjab), from the fourth century BCE to the third century CE;

• Pre-Islamic Arabic, in the fifth and sixth centuries CE.

However, despite affirmations to the contrary, the existence in these systems of a special sign for 20 is not a trace of an underlying vigesimal system borrowed by the Semites from a prior civilisation. The Semitic sign for 10 was originally a horizontal stroke or bar, and the tens were represented by repetitions of these bars, two by two:

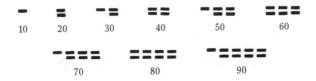

FIG. 18.3. *Figures for the tens on the Aramaic inscription at Zencirli (eighth century BCE). Donner & Röllig, Inscr. 215*

By a natural process of graphical development, which is found in all cursive scripts written with a reed brush on papyrus or parchment, the stroke became a line rounded off to the right. The double stroke for the number 20 developed into a ligature in rapid notation, and that "joined-up" form then gave rise to a whole variety of shapes, all deriving simply from writing two strokes without raising the reed brush.

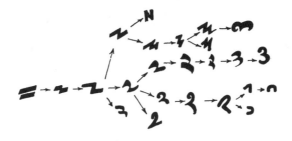

FIG. 18.4. *Origin and development of the figure for 20*

Aramaic numerals are thus strictly decimal, and do not have any trace of a vigesimal base. It was identical in principle to the Cretan Linear system for numbers below 100 – but that does not mean that it was a "primitive" form of number-writing nor that it lacked ways of coping with numbers above the square of its base. In fact, the system had a very interesting device for representing higher numbers which makes it significantly more sophisticated than many numeral systems of the Ancient World.

The Elephantine papyrus shows that Semitic numbering possessed distinctive signs for 100, 1,000 and 10,000 (though this last is not found on Phoenician or Palmyrenean inscriptions). What is more, the system did not require these higher signs to be repeated on the additive principle, but put unit expressions to the right of the higher numeral, that is to say used the multiplicative principle for the expression of large numbers (see Fig. 18.7 and 18.8).

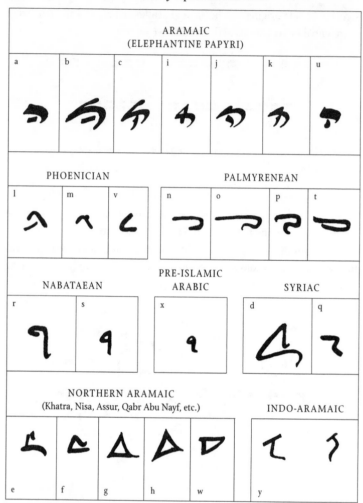

FIG. 18.5. *Variant forms of the Semitic numeral 100*

SOURCES

a CIS II 147	h Khatra	o CIS II 4 021	u CIS II 147
b S 19	i S 15	p CIS II 3 935	v CIS I 7
c S 61	j KR 4	q Sumatar Harabesi	w Assur
d Sari inscription	k S 61	r CIS II 161	x En-Namara
e Nisa ostracon 113	l CIS I 165	s CIS II 163 D	(Cantineau, p. 49)
f Qabr Abu Nayf	m CIS I 143	t CIS II 3 915	y Bühler, p. 77
g Khatra	n CIS II 3 999		

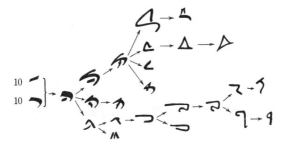

FIG. 18.6. *Origin and development of the figure 100. All these signs derive from placing two variants of the sign for 10 one above the other. This multiplicative combination has a kind of additional superscript to avoid confusing it with the sign for 20, and produced widely different graphical representations of the number 100.*

ARAMAIC (ELEPHANTINE PAPYRI)

S 61	100×5	500	S 19	100×1	100
CIS II¹	100×8	800	S fragm. 3	100×2	200
S 61	100×9	900	S 19	100×4	400

KHATRA			NABATAEA	PALMYRA	PHOENICIA		
k	j	i			o	n	m
100×1	100×1	100×1	100×1	100×1	100×1	100×1	100×1
k				m			
100×2			100×2	100×2	100×2		100×2
j			n	l			
100×3			100×3	100×3	100×3		100×3
l			o	m	p		
100×4			100×4	100×4	100×4		100×4

FIG. 18.7. *Semitic representations of the number 100. Attested examples are given in solid lines; reconstructed examples in outline. For sources, see list of references in Fig. 18.2 and 18.5.*

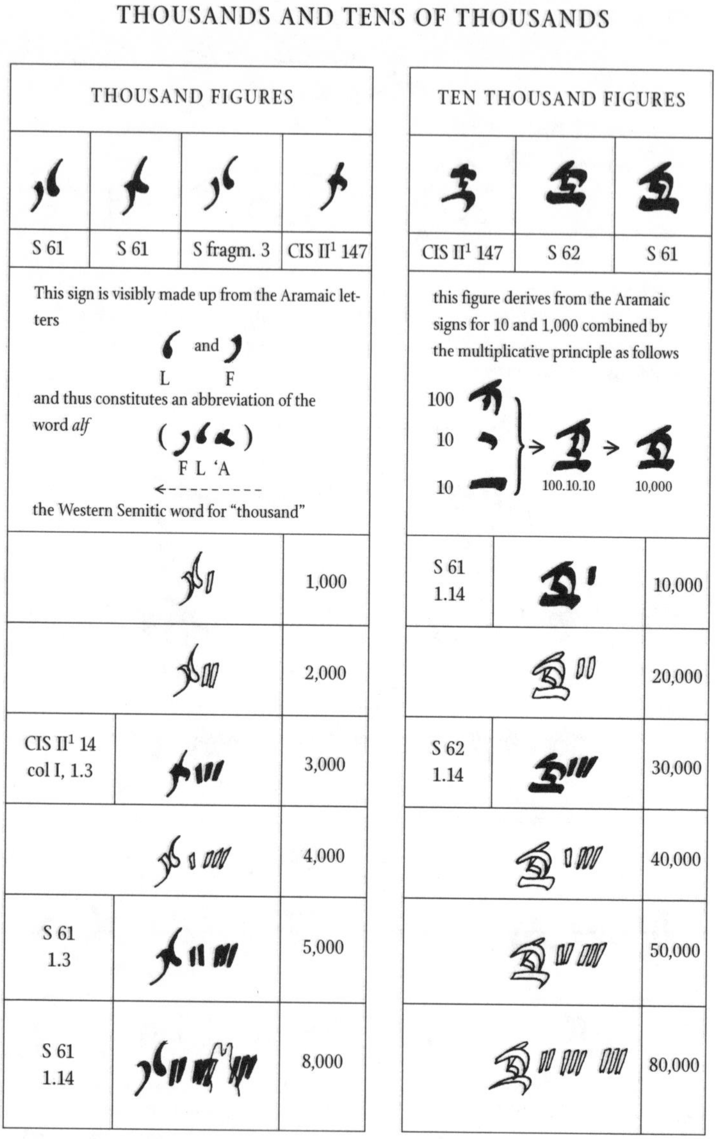

THOUSANDS AND TENS OF THOUSANDS

THOUSAND FIGURES				TEN THOUSAND FIGURES		
S 61	S 61	S fragm. 3	CIS II¹ 147	CIS II¹ 147	S 62	S 61

This sign is visibly made up from the Aramaic letters

     L   F

and thus constitutes an abbreviation of the word *alf*

     F L ʽA

the Western Semitic word for "thousand"

this figure derives from the Aramaic signs for 10 and 1,000 combined by the multiplicative principle as follows

100
10
10
   100.10.10   10,000

	1,000	S 61 1.14	10,000
	2,000		20,000
CIS II¹ 14 col I, 1.3	3,000	S 62 1.14	30,000
	4,000		40,000
S 61 1.3	5,000		50,000
S 61 1.14	8,000		80,000

FIG. 18.8. *Aramaic representations of the numbers 1,000 and over. Figures for these numbers have not been found in other northwestern Semitic numeral systems.*

In other words, the Semites used the additive principle for numbers from 1 to 99, but for multiples of 100, 1,000 and 10,000, they adopted the multiplicative principle by writing the numbers in the form 1 x 100, 2 x 100, 3 × 100, etc.; 1 × 1,000, 2 × 1,000, 3 × 1,000, etc. So for intermediate numbers above 100 they used a combination of the additive and multiplicative principles.

This corresponds with the general traditions of numbering amongst Semitic peoples. It is found amongst all the northwestern Semites (Phoenicians, Palmyreneans, Nabataeans, etc.) who used, as we have seen, numerical notations of the same kind as the Aramaic system of Elephantine. But it is also found amongst the so-called eastern Semites. The Assyrians and the Babylonians certainly inherited the additive sexagesimal system of the Sumerians, but they modified it completely even whilst adopting the cuneiform script for writing it down. Precisely because of their tradition of counting in hundreds and thousands, and finding no numeral for 100 or 1,000 in the Sumerian system, their scribes wrote those two numbers in phonetic script and represented their multiples not by addition of a sequence of signs, but by multiplication (Fig. 18.9).

So we can say that with the obvious exception of late Hebrew, none of the Semitic numeral systems had anything to do with the use of letters as numbers.

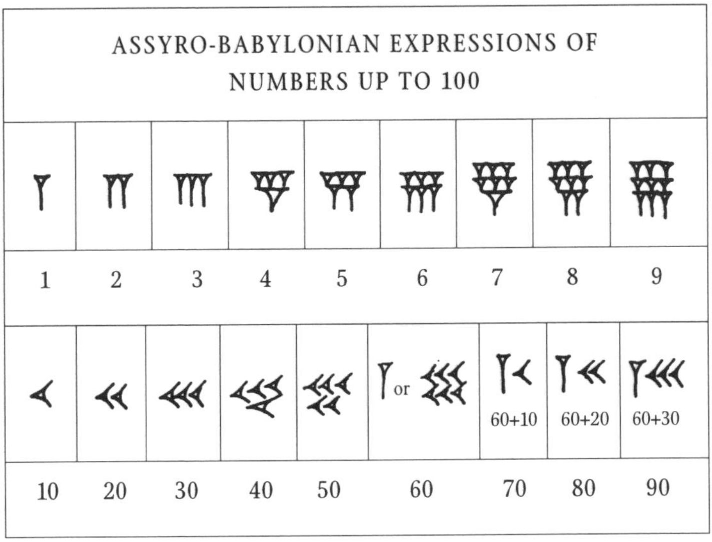

FIG. 18.9. *Assyro-Babylonian "ordinary" numerals – an adaptation of Sumerian numerals to Semitic numbering traditions*

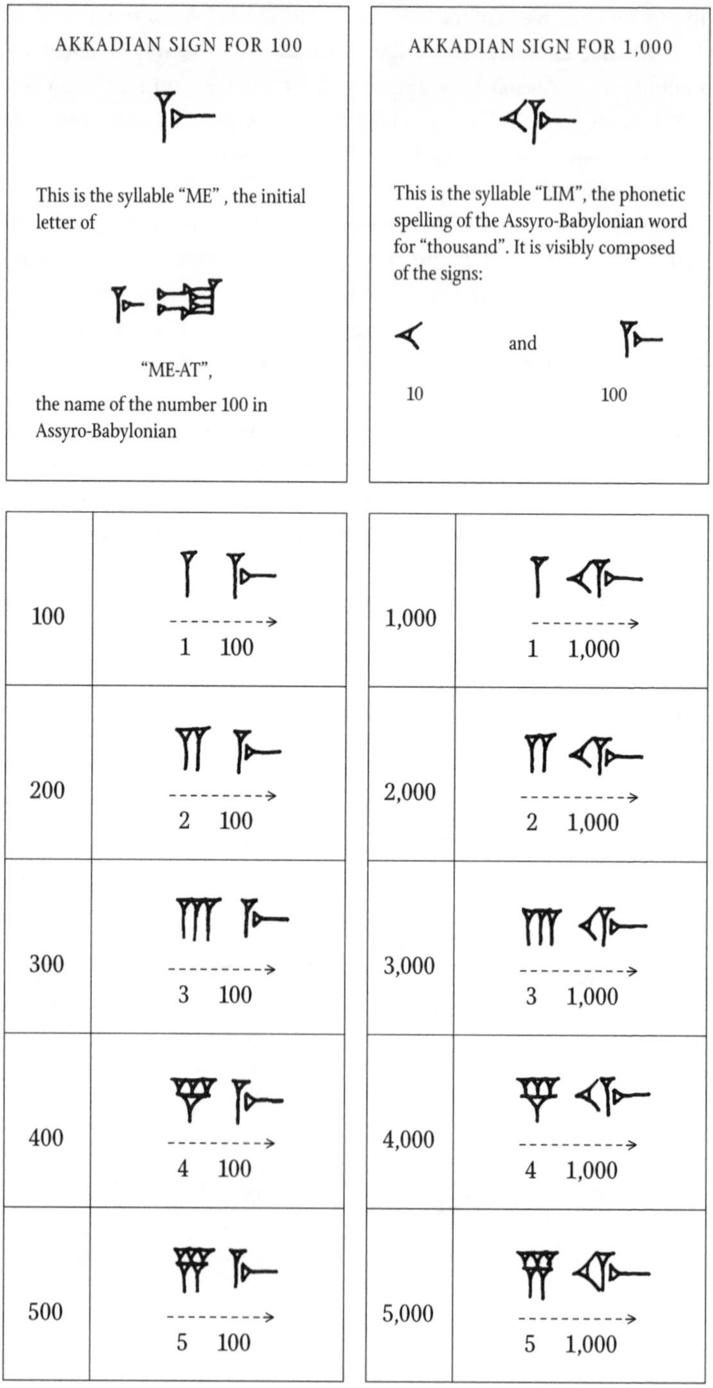

AKKADIAN SIGN FOR 100	AKKADIAN SIGN FOR 1,000
This is the syllable "ME", the initial letter of	This is the syllable "LIM", the phonetic spelling of the Assyro-Babylonian word for "thousand". It is visibly composed of the signs:
	◅ and ͳ⟞
"ME-AT",	10 100
the name of the number 100 in Assyro-Babylonian	

100	ͳ ͳ⟞ - - - - - - - -→ 1 100	1,000	ͳ ◅ͳ⟞ - - - - - - - -→ 1 1,000
200	ͳͳ ͳ⟞ - - - - - - - -→ 2 100	2,000	ͳͳ ◅ͳ⟞ - - - - - - - -→ 2 1,000
300	ͳͳͳ ͳ⟞ - - - - - - - -→ 3 100	3,000	ͳͳͳ ◅ͳ⟞ - - - - - - - -→ 3 1,000
400	ͳͳͳͳ ͳ⟞ - - - - - - - -→ 4 100	4,000	ͳͳͳͳ ◅ͳ⟞ - - - - - - - -→ 4 1,000
500	ͳͳͳͳͳ ͳ⟞ - - - - - - - -→ 5 100	5,000	ͳͳͳͳͳ ◅ͳ⟞ - - - - - - - -→ 5 1,000

Fig. 18.9 *(Continued)*.

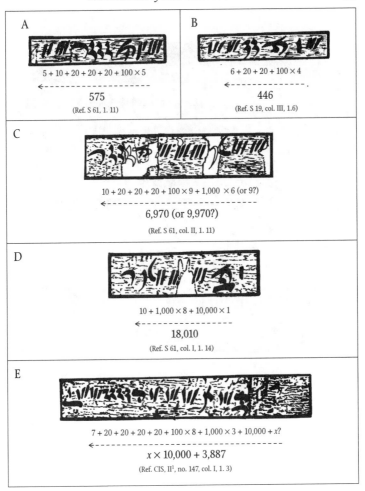

A

$5 + 10 + 20 + 20 + 20 + 100 \times 5$

575

(Ref. S 61, l. 11)

B

$6 + 20 + 20 + 100 \times 4$

446

(Ref. S 19, col. III, 1.6)

C

$10 + 20 + 20 + 20 + 100 \times 9 + 1{,}000 \times 6 \text{ (or 9?)}$

6,970 (or 9,970?)

(Ref. S 61, col. II, l. 11)

D

$10 + 1{,}000 \times 8 + 10{,}000 \times 1$

18,010

(Ref. S 61, col. I, l. 14)

E

$7 + 20 + 20 + 20 + 20 + 100 \times 8 + 1{,}000 \times 3 + 10{,}000 + x?$

$x \times 10{,}000 + 3{,}887$

(Ref. CIS, II^1, no. 147, col. I, l. 3)

FIG. 18.10. *Facsimile and interpretation of numerical expressions in the Elephantine papyrus*

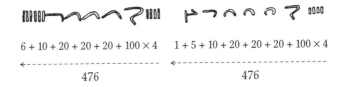

$6 + 10 + 20 + 20 + 20 + 100 \times 4$

476

$1 + 5 + 10 + 20 + 20 + 20 + 100 \times 4$

476

FIG. 18.11. *Tracing and interpretation of two examples from Syriac inscriptions at Sumatar Harabesi, dated 476 of the Seleucid era (165–166 CE). Source: Naveh*

FIG. 18.12. *Phoenician inscription, fifth century BCE. Source: CIS I', 7*

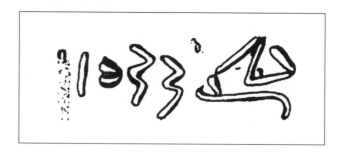

FIG. 18.13. *The number 547 on a Syriac inscription at Sari. Source: Naveh*

THE OLDEST ARCHAEOLOGICAL EVIDENCE OF GREEK ALPHABETIC NUMERALS

Amongst the oldest known uses of Greek alphabetical numerals are those to be found on coins minted in the reign of Ptolemy II (286–246 BCE), the second of the Macedonian kings who ruled over Egypt after the death of Alexander the Great (Fig. 18.14).

Coin inventory numbers	Date symbols	Transcription and translation		Coin inventory numbers	Date symbols	Transcription and translation	
CGC 44	К	K	20	CGC 61	Λ	Λ	30
CGC 45	Ϗ	KA	21	CGC 63	ΛΑ	ΛA	31
CGC 46	Ѱ	KA	21	CGC 68	ΛΒ	ΛB	32
CGC 48	К В	KB	22	CGC 70	ΛΓ	ΛΓ	33
CGC 49	Ϝ	KΓ	23	CGC 73	ΛΔ	ΛΔ	34
CGC 50	К	KΔ	24	CGC 99	ΛΕ	ΛE	35
CGC 53	Ϟ	KE	25	CGC 100	ΛC	Λ	36
CGC 57	Ж	KZ	27	CGC 101	ΛΣ	ΛZ	37
CGC 50	Ͷ	KH	28	CGC 77	ΛΗ	ΛH	38

FIG. 18.14. *Coins from the British Museum, catalogued by R. S. Poole*

Even earlier, in a Greek papyrus from Elephantine, we find a marriage contract that states that it was drawn up in the seventh year of the reign of Alexander IV (323–311 BCE), that is to say in 317–316 BCE, in which the dowry is expressed as *alpha drachma*, thus:

 (transcription: Ͱ A
translation: *drachma* A)

FIG. 18.15.

The alphabetic numeral *alpha* probably means 1,000 in this case, unless the father of the bride was a real miser, since *alpha* could either mean 1,000 – or 1!

It therefore seems that the use of Greek alphabetic numerals was common by the end of the fourth century BCE.

Moreover, relatively recent excavations of the agora and north slope of the Acropolis in Athens prove that the system arose even earlier, in the fifth century BCE, since it is found on an inscription on the Acropolis that is assumed to date from the time of Pericles (see N. M. Tod, in ABSA, 45/1950).

THE OLDEST ARCHAEOLOGICAL EVIDENCE OF HEBREW ALPHABETIC NUMERALS

Amongst the earliest instances of Hebrew alphabetic numerals are those found on coins struck in the second century CE by Simon Bar Kokhba, who seized Jerusalem in the Second Jewish Revolt (132–134 CE). The *shekel* coin shown in Fig. 18.16 bears an inscription in what were already the obsolete forms of the palaeo-Hebraic alphabet* that gives the date as *bet*, that is to say "Year 2", in alphabetic numerals, which corresponds (as Year 2 of the Liberation of Israel) to 133 CE.

"YEAR 2 OF THE LIBERATION OF ISRAEL"

FIG. 18.16. *Coin from the Second Jewish Revolt (132–134 CE). Kadman Numismatic Museum, Israel*

Other earlier instances are found on coins from the First Jewish Revolt in 66–73 CE (Fig. 18.17), and Hasmonaean coins dating from the end of the first century CE. These inscriptions, such as the one reproduced as Fig. 18.18 (from a coin minted in 78 CE), are in the Aramaic language but written in palaeo-Hebraic script.

A	B	C
"SHEKEL [OF] ISRAEL YEAR 2"	"SHEKEL [OF] ISRAEL YEAR 3"	"SHEKEL [OF] ISRAEL YEAR 5"

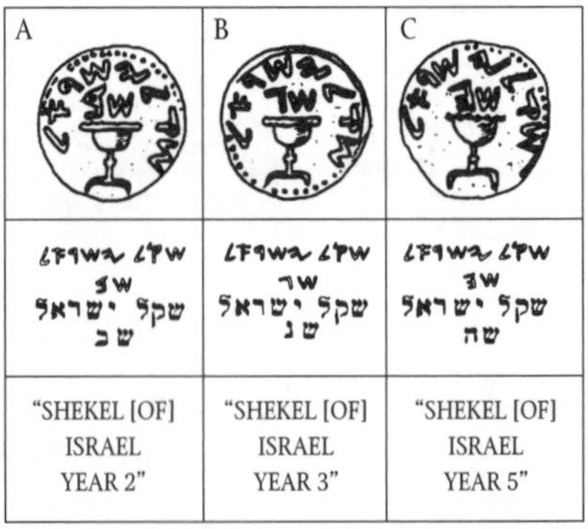

FIG. 18.17. *Coins struck during the First Jewish Revolt (66–73 CE):* shekels *dated Year 2 (A: 67 CE), Year 3 (B: 68 CE), and Year 5 (C: 70 CE) with alphabetic numerals in palaeo-Hebraic script. Kadman Numismatic Museum, Israel. See Kadman (1960), plates I–III.*

* Palaeo-Hebraic letters are close to Phoenician script. They were replaced by Aramaic script (which gave rise to modern square-letter Hebrew around the beginning of the CE) in the fifth century BCE (see Fig. 17.2 above). However, the archaic forms of the letters continued to be used sporadically up to the second century CE, most particularly by the leaders of the two Jewish revolts, to signify a return to the "true traditions of Israel". The alphabet of the present-day Samaritans is derived directly from palaeo-Hebraic script.

A B C

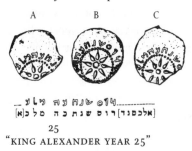

... ע ל מ הש ה ש ⊙ ש ...
[אלכסנד]רוס שנת כה מל כ[א]
25
"KING ALEXANDER YEAR 25"

FIG. 18.18. *Coins struck in 78 BCE under Alexander Janneus. Kadman Numismatic Museum, Israel. See Naveh (1968), plate 2 (nos. 10 & 12) and plate 3 (no. 14).*

We must also mention a clay seal in the Jerusalem Archaeological Museum which must have originally served to fix a string around a papyrus scroll (Fig. 18.19). The seal bears an inscription in palaeo-Hebraic characters which can be translated as: "Jonathan, High Priest, Jerusalem, M". The letter *mem* at the end is still a puzzle, but it could be a numeral, with a value (= 40) referring to the reign of Simon Maccabeus, recognised by Demetrius II in 142 BCE as the "High Priest, leader and ruler of the Jews". If this were so, then the seal would date from 103 BCE (the "fortieth year" of Simon Maccabeus) and thus constitute the oldest known document showing the use of Hebrew alphabetic numerals.

Length: 13 mm
Width: 12 mm
Thickness: 2–3 mm

יהונתן כהן גדל
ירשלם
מ

"JONATHAN HIGH PRIEST
JERUSALEM M"

FIG. 18.19. *Bulla of "Jonathan the High Priest", probably from the Hasmonaean period (second century BCE). Israel Museum, Jerusalem, item 75.35. See Avigad (1975), Fig. 1 and Plate I-A.*

Finally, there is this fragment of a parchment scroll from Qumran (one of the "Dead Sea Scrolls"):

FIG. 18.20. *Fragment of a parchment scroll, recently found at Khirbet Qumran. Scroll 4QSd, no. 4Q 259. See Milik (1977).*

The scroll contains a copy of the Rule of the Essene community, written in square-letter Hebrew of a style that dates from the first century BCE at the earliest. The fragment comes from the first column of the third sheet of the scroll as it was found in the caves at Qumran. In the top right-hand corner there is a letter, *gimmel*: since this is the third letter of the Hebrew alphabet, people have assumed that the letter gives the sheet number, 3. However, the *gimmel* was not written by the same hand as the rest of the scroll; J. T. Milik has explained that the page-numbering was probably the work of an apprentice, using what was then a novel procedure for numbering manuscripts by the letters of the alphabet, whereas the main scribe used an older form of writing.

JEWISH NUMERALS FROM THE PERSIAN TO THE HELLENISTIC PERIOD

The preceding section shows that in Palestine Hebrew letters were only just beginning to be used as numerals at the start of the Common Era.

This is confirmed by the discovery, in the same caves at Qumran, of several economic documents belonging to the Essene sect and dating from the first century BCE. One of them, a brass cylinder-scroll (Fig. 18.21), uses number-signs that are quite different from Hebrew alphabetic numerals.

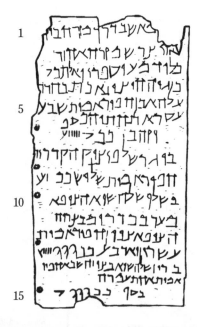

FIG. 18.21A. *Fragment of a brass cylinder-scroll, first century BCE, from the third of the Qumran caves. See DJD III, 3Q, plate LXII, column VIII.*

Lines	NUMERALS FOUND ON THE DOCUMENT SHOWN IN FIG. 21A	VALUES	HAD THE SCRIBE USED LETTER-NUMERALS, HE WOULD HAVE WRITTEN:
7	 2 + 5 + 10 ←---------	17	 (יז) 7 + 10 ←-----
13	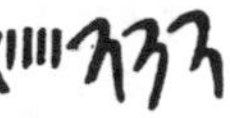 2 + 4 + 20 + 20 + 20 ←--------------	66	 (סו) 6 + 60 ←----

FIG. 18.21B.

Further confirmation is provided by the many papyri from the fifth century BCE left by the Jewish military colony at Elephantine (near Aswan and the first cataract of the Nile). These consist of deeds of sale, marriage contracts, wills, and loan agreements, and they use numerals that are identical to those of the Essene scroll. For example, one such papyrus [E. Sachau (1911), papyrus no. 18] uses the following representations of 80 and 90, which are obviously unrelated to the Semitic letter-numbers *pe* (for 80) and *tsade* (for 90).

20 + 20 + 20 + 20
←------------
80

10 + 20 + 20 + 20 + 20
←---------------
90

FIG. 18.22.

An even more definitive piece of evidence comes from the archaeological site of Khirbet el Kom, not far from Hebron, on the West Bank (Israel). It is a flat piece of stone that was used, at some point in the third century BCE, for writing a receipt for the sum of 32 *drachmas* loaned by a Semite called Qos Yada to a Greek by the name of Nikeratos – and is thus written in both Aramaic and Greek.

TRANSCRIPTION

Greek Text	Aramaic Text

Greek Text:

6 12
↑ ↑
L Ϛ ΙΒΜΗΝΟΣ ΠΑ
ΝΗΜΟΥ ΕΧΕΙ ΝΙ
ΚΗΡΑΤΟΣ ΣΟΒΒΑ
ΘΟ ΠΑΡΑ ΚΟΣΙΔΗ ΚΑ
ΠΗΛΟΥ ⊢ΛΒ
- - - - →
30.2

Aramaic Text:

6   12

‏**III ﬡ** ‏לתמוז שנת‏ ‏ב ר II
‏קוסידע בר חנא קפילס‏
‏הו נתן [ﬡ] ניקרתם זוזן‏
‏ﬤ רII‏

← - - - - - - - -
2 10 20

TRANSLATION

6th year, the 12th of the month of Panemos, Nikeratos, son of Sobbathos, received from Koside the moneylender [the sum of] 32 drachma

The 12th [of the month] of Tammuz [of] the 6th year Qos Yada son of Khanna the trader gave Nikeratos in "Zuz": 32.

FIG. 18.23. *Bilingual ostracon from Khirbet el Kom (Israel), probably dating from 277 BCE (Year 6 of Ptolemy II). See Geraty (1975), Skaist (1978).*

Close scrutiny of the inscription shows first of all that the two languages are written by different hands: probably the moneylender wrote the Aramaic and the borrower wrote the Greek. Moreover, we can see that Nikeratos the Greek wrote the sum he had borrowed and the date of the loan ("6th year, on the 12th of the month of Panemos") using Greek alphabetic numerals: **⊂** *digamma* (= 6), ιβ *iota-beta* (= 12), and λβ *lambda-beta* (=32). On the other hand, Qos Yada the Semite wrote the sum of the loan (32 *zuz*) using the numeral system we have seen on the Essene scroll above, broken down as:

$$20 + 10 + 1 + 1$$

It seems indisputable that if Hebrew alphabetic numerals had been in use in Palestine at this time, then Qos Yada would have used them, and written the number 32 much more simply as

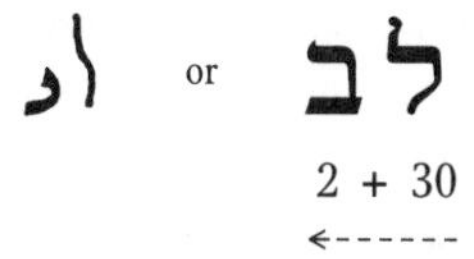

בל   or   נו

2 + 30

←------

FIG. 18.24.

We can therefore conclude that in all probability the inhabitants of Judaea did not use alphabetic numerals in ordinary transactions until the dawn of the Common Era.

The numeral system we have found in use amongst Jews from the Persian to the Hellenistic period (fifth to second centuries BCE) is in fact nothing other than the old western Semitic system, borrowed by the Hebrews from the Aramaeans together with their language (Aramaic) and script. Because the Aramaeans were very active in trade and commerce – their role across the land-mass of the Middle East was similar to that of the Phoenicians around the shores of the Mediterranean Sea – Aramaic script spread more or less everywhere. It finally killed off the cuneiform writing of the Assyro-Babylonians, and became the normal means of international correspondence.

ACCOUNTING IN THE TIME OF THE KINGS OF ISRAEL

How did the Jews do their accounting in the age of the Kings, roughly from the tenth to the fifth centuries BCE? In the absence of archaeological evidence, it was long thought that numbers were simply written out as words, for the numeral system explained below remained undiscovered until less than a hundred years ago.

That was when excavations in Samaria uncovered a hoard of *ostraca* in palaeo-Hebraic script in the storerooms of the palace of King Omri. An *ostracon* is a flat piece of rock, stone or earthenware used as a writing surface. (The use of *ostraca* as "scribble-pads" for current accounts, lists of workers, messages and notes of every kind was very common amongst the Ancient Egyptians, the Phoenicians, the Aramaeans, and the Hebrews.) The Samarian *ostraca* consist of bills and receipts for payments in kind to the stewards of the King of Israel, and reveal that the Jews wrote out their numbers as words and also used a real system of numerals.

Subsequent discoveries confirmed the existence of these ancient Hebrew numerals. They have been found on a hoard of about a hundred *ostraca* unearthed at a site at Arad (in the Negev Desert, on the trail from Judaea to Edom); on another score of *ostraca* found at Lakhish in 1935, which contain messages from a Jewish military commander to his subordinates, written in the months prior to the fall of Lakhish to Nebuchadnezzar II in 587 BCE; numerous Jewish weights and measures; and on various similar discoveries made at the Ophel in Jerusalem, at Murabba'at and at Tell Qudeirat.

Although it took a long time to decipher these inscriptions, there is no longer any doubt (Fig. 18.26) but that these number-signs are Egyptian hieratic numerals in their fully developed form from the New Empire (shown in Fig. 14.39 and 14.46 above). This incidentally provides additional confirmation of the significant cultural relations between Egypt and Palestine which historians have revealed in other ways. In other words, in the period of the Kings of Israel, the Jews were influenced by the civilisation of the Pharaohs to the extent of adopting from it Egyptian cursive hieratic numerals (Fig. 18.25 and 18.27).

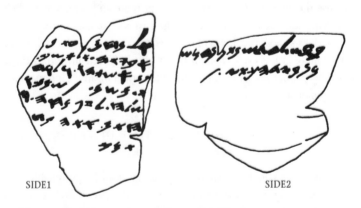

SIDE1   SIDE2

FIG. 18.25. *Hebrew* ostracon *from Arad, sixth century BCE (ostracon no 17). Written in palaeo-Hebraic script, side 2 has the number 24 written as:* ●●●● ∑ *See Aharoni (1966).*

4   20

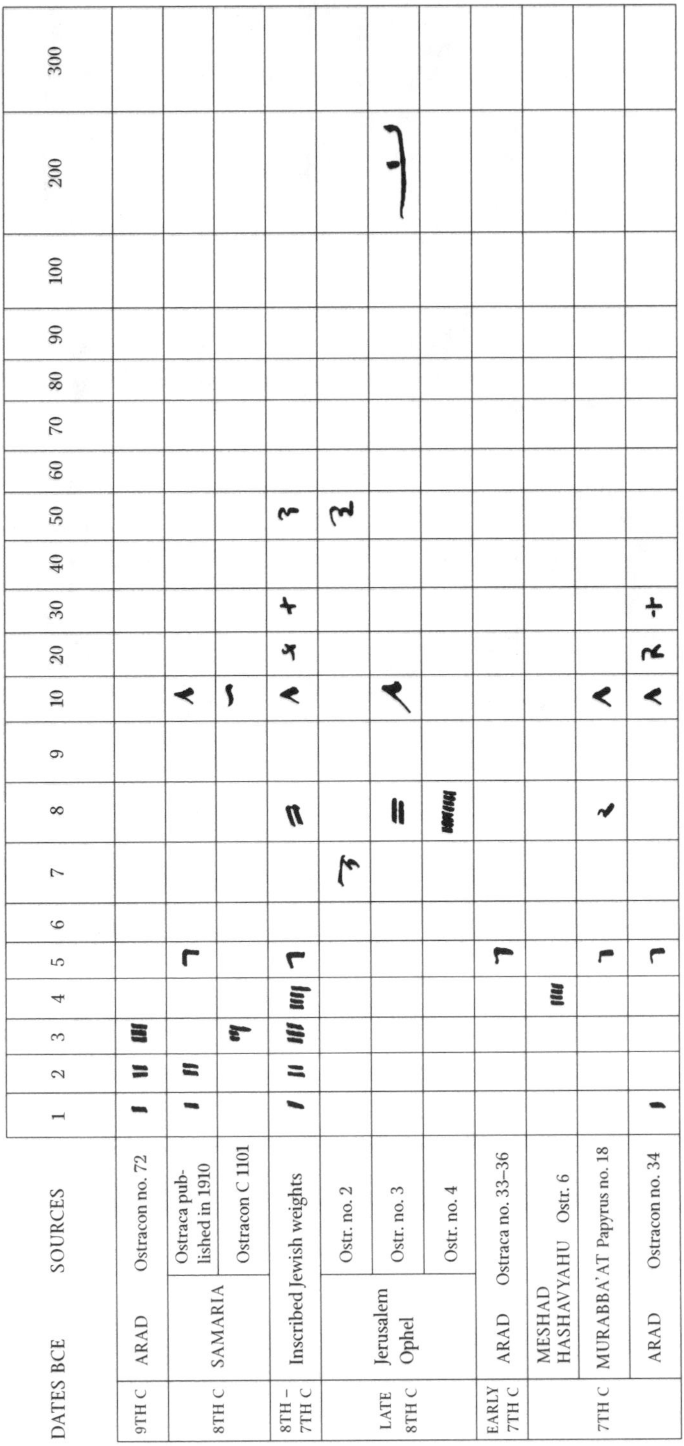

DATES BCE	SOURCES		1	2	3	4	5	6	7	8	9	10	20	30	40	50	60	70	80	90	100	200	300
9TH C	ARAD	Ostracon no. 72	𐤖	𐤖𐤖	𐤖𐤖𐤖																		
8TH C	SAMARIA	Ostraca published in 1910	𐤖	𐤖𐤖								𐤶											
		Ostracon C 1101			⟨glyph⟩	⟨glyph⟩																	
8TH – 7TH C		Inscribed Jewish weights	𐤖	𐤖𐤖	𐤖𐤖𐤖	⟨glyph⟩	⟨glyph⟩			⟨glyph⟩		⟨glyph⟩	⟨glyph⟩	⟨glyph⟩		⟨glyph⟩							
LATE 8TH C	Jerusalem Ophel	Ostr. no. 2							⟨glyph⟩							⟨glyph⟩							
		Ostr. no. 3								⟨glyph⟩		⟨glyph⟩										⟨glyph⟩	
		Ostr. no. 4					⟨glyph⟩			⟨glyph⟩													
EARLY 7TH C	ARAD	Ostraca no. 33–36					⟨glyph⟩																
	MESHAD HASHAVYAHU	Ostr. 6				⟨glyph⟩																	
	MURABBA'AT	Papyrus no. 18					⟨glyph⟩			⟨glyph⟩		⟨glyph⟩		⟨glyph⟩									
7TH C	ARAD	Ostracon no. 34	𐤖				⟨glyph⟩					⟨glyph⟩	⟨glyph⟩	⟨glyph⟩									

FIG. 18.26. *Table showing the identity of numerals used in Palestine under the Jewish Kings with Egyptian hieratic numerals*

DATES BCE	SOURCES	1	2	3	4	5	6	7	8	9	10	20	30	40	50	60	70	80	90	100	200	300
LAKHISH	Ostr. no. 9	I	II								⌄											
LAKHISH	Ostr. no. 19	I									⌄	⌄⌄⌄										
6TH C · ARAD	Ostr. no. 1–4	I	II	III																		⟋
6TH C · ARAD	Ostr. no. 16–18	I			IIII				∩			⌄										
6TH C · ARAD	Ostr. no. 24–29					⌐					⌄											

EGYPTIAN HIERATIC NUMERALS (NEW KINGDOM, CURSIVE). FROM MÖLLER (1911).

	1	2	3	4	5	6	7	8	9	10	20	30	40	50	60	70	80	90	100	200	300

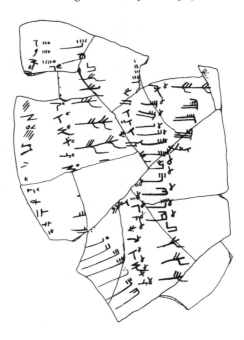

FIG. 18.27. Ostracon *no. 6 from Tell Qudeirat, late seventh century BCE, the largest known palaeo-Hebraic ostracon, found by R. Cohen in 1979. This text confirms the results of Fig. 18.26, since it gives almost the whole series of the hundreds and thousands in Egyptian hieratic script.*

TRANSJORDAN PALESTINE	Rosette from Hösn, east of the Jordan. See RB, 1900, p. 119

Great Theatre of Gerasa. See RB, 1895/3, pp. 373–400

CIB	CIe	CK	CKA
- - →	- - →	- - →	- - →
212	215	220	221
ƎΙΦ	ΔΙΦ	ΠΙΦ	ВΙΦ
← - -	← - -	← - -	← - -
515	514	513	512

EGYPT	*Coptic* inscription concerning Luke and two of his works. See ASAE, X, 1909, p.51
	K̄H **K̄Δ** **K̄Z** ----→ ----→ ----→ 28 24 27
	Jewish funerary *stelae* from Tell el Yahudieh (10 km north of Cairo), dating from the first century CE. See CII 1454, 1458 and 1460
	ĪB ĪΓ K̄Γ Λ̄E N P̄B --→ --→ --→ --→ --→ --→ 12 13 23 35 50 102
PHRYGIA	*Jewish* inscription dated 253–254 CE. See CII 773
	TΛH -----→ 338
ETHIOPIA	Aksum inscription, third century CE. See DAE 3 and 4
	KΔ **ΓPIB** **ʾƷCKΔ** ----→ ----→ -------→ 24 3112 6224
LATIUM	*Jewish* catacombs on the Via Nomentana, Via Labicana and Via Appia Pignatelli. See CII 44, 78, 79
	ΛΓ **KA** **ΞΘ** ----→ ----→ ----→ 33 21 69
NORTHERN SYRIA	Synagogue mosaic. *Jewish* inscription dated 392 CE. See CII 805
	ΨΓ̄ ----→ 703
SOUTH OF THE DEAD SEA	*Jewish* grave marking dated 389–390 CE. See CII 1209
	ΣΠ **CΠΓ** -----→ ------→ 86 283

FIG. 18.28B.

JEWISH LAPIDARY NUMERALS AT THE DAWN
OF THE COMMON ERA

There is a final curiosity to add to this story. From the first century BCE to the seventh century CE, the use of Hebrew alphabetic numerals grew ever more common amongst Jews all over the Mediterranean basin, from Italy to Palestine and northern Syria, from Phrygia to Egypt and even Ethiopia. However, during this period, Jewish stone-carvers, who could write just as well in Hebrew as in Greek or Latin, most often put dates and numbers not in Hebrew, but in Greek alphabetic numerals, as the examples reproduced in Fig. 18.28 show.

THE JEWS: NATIONAL IDENTITY AND
CULTURAL COMPLEXITY

The people of Israel certainly played a major role in the history of the world's religions; but at the same time, Jewish culture has, throughout its history, accepted and adopted influences of the most diverse kinds.

The most notable of these "foreign influences" include:

- the adoption of the Phoenician alphabet in the period of the Kings;
- the adoption of the Assyro-Babylonian sexagesimal system for weights and measures (see Ezekiel XLV:12, where the *talent* is set at 60 *maneh*, and the *maneh* at 60 *shekels*);
- the presumed adoption of the Canaanites' calendar, in which each month starts with the appearance of the new moon;
- the borrowing of the names of the months from the ancient calendar of Nippur, used throughout Mesopotamia from the time of Hammurabi (*Nisân, Ayar, Siwan, Tammuz, Ab, Elul, Teshrêt, Arashamna, Kisilimmu, Tebet, Shebat,* and *Adar).* In Modern Hebrew, the names are still almost identical;
- the adoption of Aramaic and its script (the only ones in general use in Judaea at the time of Jesus).

What is remarkable about Jewish culture is that despite these numerous borrowings it retained a separate identity. Since the expulsion of the Jews from Palestine in the first century CE, and for the following 1,800 years, it has not ceased to adapt itself to the most diverse situations and to incorporate new elements, whilst also exercising a determining influence over developments in Western and Islamic culture. As Jacques Soustelle sees it, this long history of a cultural identity within a complex of cultural influences is what accounts for the successful re-founding of a Jewish nation-state in the twentieth century: Israel today is made of more than a score of distinct ethnic groups with many different mother-tongues, but sharing a common cultural identity.

SUMMARY

From the tenth to the sixth century BCE (the era of the Kingdom of Israel), the Hebrews used Egyptian hieratic numerals; from the fifth to the second century BCE, they used Aramaic numerals; and from around the start of the common era, many Jews used Greek alphabetical numerals.

In the present state of knowledge, it seems that Greek alphabetic numerals go back at least as far as the fifth century BCE; whereas Hebrew alphabetic numerals are not found before the second century BCE.

Does that mean to say that the Greeks invented the idea of representing numbers by the letters of their alphabet, and that the Jews copied it during the Hellenistic period? It seems very likely, and all the more plausible in the light of the Jews' adoption of numerous other "outside" influences.

However, this is not the only possible conclusion. Many passages in the Torah (the Old Testament) suggest very strongly that the scribes or authors of these ancient texts were familiar with the art of coding words according to the numerical value of the letters used (see further explanations in Chapter 20 below). It is currently reckoned that the oldest biblical texts were composed in the reign of Jeroboam II (eighth century BCE) and that the definitive redaction of the main books of the Torah took place in the sixth century BCE, around the time of the Babylonian exile.

Do Hebrew alphabetic numerals go so far back in time? Or are the passages showing letter-number coding later additions?

If the system is as old as it seems, and which would imply that Hebrew letter-numbers were invented independently of the Greek model, we would still have to explain why they had no use in everyday life until the Common Era. One plausible answer to that question would be that since the letters of the Hebrew alphabet acquired a sacred character very early on, the Jews avoided using sacred devices for profane purposes.

In conclusion, let us say that the "Greek hypothesis" seems to have most of the actual evidence on its side; but that the possibility of an independent origin for Hebrew alphabetic numerals and of their restriction over several centuries to religious texts alone is not to be rejected out of hand.

CHAPTER 19

OTHER ALPHABETIC
NUMBER-SYSTEMS

SYRIAC LETTER-NUMERALS

The Arabic-speaking Christians of the Maronite sect have maintained, mainly for liturgical use, a relatively ancient writing system which is known as *serto* or *Jacobite* script.

Christians of the Nestorian sect, who are found mainly in the region of Lake Urmia (near the common frontier between Iraq, Turkey, Iran and the former Soviet Union), still speak a dialect of Aramaic which they write in a graphical system called *Nestorian* writing.

Each of these two writing systems has an alphabet of twenty-two letters, and is derived from a much older script called *estranghelo*, formerly used to write Syriac, a ancient Semitic language related to Aramaic.

Graphically, the Nestorian form, which is more rounded than the *estranghelo*, is intermediate between this and *serto* which in turn has a more developed and cursive form (Fig. 19.1). The letters themselves are written from right to left, are joined up, and, as in the writing of Arabic, undergo various modifications according to their position within a word, i.e. according to whether they stand alone or are in the initial, medial, or final position (Fig. 19.1 only shows the independent forms of Syriac letters).

The oldest known Syriac inscriptions seem to date from the first century BCE. *Estranghelo* writing seems to have been used only up to the sixth or seventh century. As used by the Nestorian Christians, fairly numerous in Persia in the period of the Sassanid Dynasty (226–651 CE), it gradually evolved until, around the ninth century, it attained its canonical Nestorian form. With the Jacobites, who mainly lived in the Byzantine Empire, it seems to have evolved more rapidly towards the *serto* form, since it was gradually replaced by this after the seventh or eighth century.

Finally, *estranghelo* (which is simply a variant of Aramaic script and therefore ultimately derives from the Phoenician alphabet) has preserved in its entirety the order of the original twenty-two Phoenician letters (the same order which is to be found with all the western Semites).

In *serto*, however, as in Nestorian, letters have been used (and still are used) as number-signs. This is confirmed by the fact that in all Syriac manuscripts (at least those later than the ninth century), codices are made up of serially numbered quires, ensuring the correct order of composition of the bound book. (The manuscript folios, however, were only numbered later, often using Arabic numerals.)

The numerical values of the Syriac letters are assigned as follows. The first nine letters are assigned to the units, the next nine letters to the tens, and the remaining four are assigned to the first four hundreds. Also, as in Hebrew, the numbers from 500 to 900 are written as additive combinations of the sign for 400 with the signs for the other hundreds, according to the schema:

$$500 = 400 + 100$$
$$600 = 400 + 200$$
$$700 = 400 + 300$$
$$800 = 400 + 400$$
$$900 = 400 + 400 + 100$$

The thousands are represented by a kind of accent mark placed beneath the letters representing the units, and the tens of thousands by a short horizontal mark beneath these same letters:

! 10,000	! 1,000	?1
20,000	2,000	2
30,000	3,000	
! 40,000	! 4,000	?4

Similar conventions allowed the Maronites to represent numbers greater than the tens of thousands. With a few exceptions, this number-system is quite analogous to that of the Hebrew letter-numerals. It is however a relatively late development in Syriac writing, since the oldest documents show that it does not go back earlier than the sixth or seventh centuries. Older Syriac inscriptions only reveal a single kind of numerical notation related to the "classical" Aramaic system.

HEBRAIC LETTERS	ARCHAIC PHOENICIAN	PALMYRENEAN	ESTRANGHELO	NESTORIAN	SERTO	NAMES TRANSCRIPTIONS AND NUMERICAL VALUES OF SYRIAC LETTERS		
Aleph	א	א	ܐ	?	? ?	Ōlap	'	1
Bet	בבב	9	ܒ	ܒ	?	ܒ	Bét	b bh	2
Gimmel	גגג	ܓ	ܓ	ܓ	ܓ	ܓ	Gómal	g gh	3
Dalet	דדדד	ܕ	ܕ	ܕ	?	?	Dólat	d dh	4
He	הה	ܗ	ܗ	ܗ	ܗ	?	Hé	h	5
Vov	ו	ܘ	?	ܘ	ܘ	ܘ	Waw	w	6
Zayin	ז	ܙ	ܙ	ܙ	?	?	Zayn	z	7
Het	ח	ܚ	ܚ ܚ	?	?	?	Ḥét	ḥ	8
Tet	טט	ܛ	6	?	?	?	Ṭét	ṭ	9
Yod	יי	ܝ	?	?	?	?	Yud	y	10
Kof	ךכ	ܟ	?	?	?	?	Kóp	k kh	20
Lamed	ךלס	ܠ	?	?	?	?	Lómad	l	30
Mem	ם	ܡ	?	?	?	?	Mim	m	40
Nun	ן	ܢ	?	?	?	?	Nun	n	50
Samekh	ס	ܣ	?	?	?	?	Semkat	ṣ	60
Ayin	עע	ܥ	?	?	?	?	'E	'	70
Pé	ף	ܦ	3	?	?	?	Pé	p ph	80
Tsade	ץץצ	ܨ	?	?	?	?	Ṣódé	ṣ	90
Quf	ףקק	?	?	?	?	?	Quf	q	100
Resh	ר	?	?	?	?	?	Rish	r	200
Shin	שש	w	?	?	?	?	Shin	sh	300
Tav	ת	ܬ ܬ	?	?	?	?	Taw	t	400

FIG. 19.1. *Syriac alphabets compared with Phoenician, Aramaean (from Palmyra) and Hebraic alphabets. The use of Syriac letters as number-signs is attested notably in a manuscript in the British Museum (Add. 14 620) which features the above order. (See M. Cohn, Costaz, Duval, Février, Hatch, Pihand, W. Wright)*

When did letter-numerals in Syriac writing first arise? In the absence of documents, it is hard to say. But there are several good reasons to suppose that the introduction of this system owed much to Jewish influence on the Christian and Gnostic communities of Syria and Palestine.

One final question: a Syriac manuscript, now in the British Museum (reference Add. 14 603), which probably dates from the seventh or eighth century [W. Wright (1870), p. 587a], reveals some interesting information. Its quires are numbered in the usual way, with Syriac letters according to their numerical values; but these have alongside them the corresponding older number-signs. Should we conclude that, at the date of this manuscript, the system of letter-numerals had not been universally adopted? Or, taking the question in the other sense, should we conclude that at that time the use of the old system was already a traditional but archaic usage, and the letter-numerals were by then not only widespread but considered by the majority of Syrians to be the only normal and official system of notation? The documentation which we have to hand does not give us an answer.

ARABIC LETTER-NUMERALS

Arabic has a number-system modelled not only on the Hebrew system, but also on the Greek system of letter-numerals. But first we need to look at a curious problem.

The order of the twenty-eight letters of the Arabic alphabet, in its Eastern usage, is quite different from the order of the letters in the Phoenician, Aramaic or Hebrew alphabets.

A glance at the names of the first eight Arabic letters compared with the first eight Hebrew letters shows this straight away:

ARABIC	HEBREW
'alif	'aleph
ba	bet
ta	gimmel
tha	dalet
jim	he
ḥa	vov
kha	zayin
dal	het

We would expect to find the twenty-two western Semitic letters in the Arabic alphabet, and in the same order, since Arabic script derives from archaic Aramaic script. So how did the traditional order of the Semitic letters get changed in Arabic? The answer lies in the history of their system for writing numbers.

The Arabs have frequently used a system of numerical notation in which each letter of their own alphabet has a specific numerical value (Fig. 19.3); according to F. Woepke, they "seem to have considered [this system] as uniquely and by preference their own".

They call this *ḥurūf al jumal*, which means something like "totals by means of letters".

But, if we look closely at the numerical value which this system assigns to each letter, we are bound to note that the method used by the Arabs of the East is not quite the same as the one adopted, later, by western (North African) Arabs, since the values for six of the letters differ in the two systems.

LETTER		ITS VALUE	
		IN THE MAGHREB	IN THE EAST
س	sin	300	60
ص	ṣad	60	90
ش	shin	1,000	300
ض	ḍad	90	800
ظ	ḍha	800	900
غ	ghayin	900	1,000

Fig. 19.2.

Now, let us first note that the numerical values of the Arabic letters can be arranged into a regular series, as follows:

1; 2; 3; 4; ... 10; 20; 30; 40; ... 100; 200; 300; 400; ... ; 1,000,

and if we set out, according to this sequence, the letter-numerals of the eastern Arabic system (the more ancient of the two) we obtain the order of the western Semitic letters of which we have just written (Fig. 17.2 and 17.4 above). Furthermore, if we tabulate the letter-numerals of the Arabic system (as in Fig. 19.4) and compare this with the Hebrew letter-numerals (Fig. 17.10) and also with the Syriac system of alphabetic numbering (Fig. 19.1), then it is easy to see that for the numbers below 400 all three systems agree perfectly. This shows that "in the initial system of numeration, the order of the northern Semitic alphabet was preserved, and additional letters from the Arabic alphabet were added later in order to go up to 1,000" [M. Cohen (1958)].

LETTERS						NUMERICAL VALUES	
LETTERS ON THEIR OWN	LETTER-NAMES	PHONETIC VALUES OF LETTERS	LETTERS IN INITIAL POSITION	LETTERS IN MEDIAN POSITION	LETTERS IN END POSITION	IN THE EAST	IN THE MAGHREB
ا	'Alif	'	ا	ـا	ـا	1	1
ب	Ba	b	بـ	ـبـ	ـب	2	2
ت	Ta	t	تـ	ـتـ	ـت	400	400
ث	Tha	th	ثـ	ـثـ	ـث	500	500
ج	Jim	j	جـ	ـجـ	ـج	3	3
ح	Ḥa	ḥ	حـ	ـحـ	ـح	8	8
خ	Kha	ḫ	خـ	ـخـ	ـخ	600	600
د	Dal	d	د	ـد	ـد	4	4
ذ	Dhal	dh	ذ	ـذ	ـذ	700	700
ر	Ra	r	ر	ـر	ـر	200	200
ز	Zay	z	ز	ـز	ـز	7	7
س	Sin	s	سـ	ـسـ	ـس	60	300
ش	Shin	sh	شـ	ـشـ	ـش	300	1,000
ص	Ṣad	ṣ	صـ	ـصـ	ـص	90	60
ض	Ḍad	ḍ	ضـ	ـضـ	ـض	800	90
ط	Ṭa	ṭ	طـ	ـطـ	ـط	9	9
ظ	Ḍha	ḍh	ظـ	ـظـ	ـظ	900	800
ع	'Ayin	'	عـ	ـعـ	ـع	70	70
غ	Ghayin	gh	غـ	ـغـ	ـغ	1,000	900
ف	Fa	f	فـ	ـفـ	ـف	80	80
ق	Qaf	q	قـ	ـقـ	ـق	100	100
ك	Kaf	k	كـ	ـكـ	ـك	20	20
ل	Lam	l	لـ	ـلـ	ـل	30	30
م	Mim	m	مـ	ـمـ	ـم	40	40
ن	Nun	n	نـ	ـنـ	ـن	50	50
ه	Ha	h	هـ	ـهـ	ـه	5	5
و	Wa	w	و	ـو	ـو	6	6
ي	Ya	y	يـ	ـيـ	ـي	10	10

FIG. 19.3. *The Arabic alphabet, in its modern representation*

We may therefore conclude that the use of alphabetic numerals by the Arabs was introduced in imitation of the Jews and the Christians of Syria for the first twenty-two letters (numbers below 400), and according to the example of the Greeks for the remaining six (values from 400 to 1,000).

ا	'Alif	'	1	س	Sin	s	60
ب	Ba	b	2	ع	'Ayin	'	70
ج	Jim	j	3	ف	Fa	f	80
د	Dal	d	4	ص	Ṣad	ṣ	90
ه	Ha	h	5	ق	Qaf	q	100
و	Wa	w	6	ر	Ra	r	200
ز	Zay	z	7	ش	Shin	sh	300
ح	Ḥa	ḥ	8	ت	Ta	t	400
ط	Ṭa	ṭ	9	ث	Tha*	th	500
ي	Ya	y	10	خ	Kha*	kh	600
ك	Kaf	k	20	ذ	Dhal*	dh	700
ل	Lam	l	30	ض	Ḍad*	ḍ	800
م	Mim	m	40	ظ	Ḍha*	ḍh	900
ن	Nun	n	50	غ	Ghayin*	gh	1,000
					* subsequently added		

FIG. 19.4. *The order of Arabic letters as ordained according to the regular development of the values of the alphabetic number-system of eastern Arabs*

In fact, "following the conquest of Egypt, Syria and Mesopotamia, numbers were habitually written, in Arabic texts, either spelled out in full or by means of characters borrowed from the Greek alphabet" [A. P. Youschkevitch (1976)].

Thus we find in an Arabic translation of the Gospels, the manuscript verses have been numbered with Greek letters:

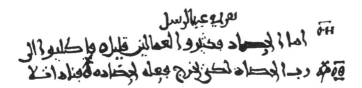

FIG. 19.5. *Excerpt from a Christian ninth-century manuscript. In this manuscript, which gives a translation from the Gospels, the corresponding verses have been numbered by reference to Greek letter-numbers, (first line, right: ŌH = v. 78; second line, right: OΘ = v. 79. Vatican Library, Codex Borghesiano arabo 95, folio 173. (See E. Tisserant, pl. 55)*

Similarly, in a financial papyrus written in Arabic and dating from the year 248 of the Hegira (862–863 CE), the sums were written exclusively according to the Greek system. [This document is, along with others of similar kind, in the Egyptian Library, inventory number 283; cf. A. Grohmann (1962)].

This usage persisted in Arabic documents for several centuries, but disappeared completely in the twelfth century. For all that, we should not conclude that Arabic letter-numerals were introduced only at that time. The system certainly first arose before the ninth century. We have, in fact, a mathematical manuscript copied at Shiraz between the years 358 and 361 of the Hegira (969–971 CE) in which all of the Arabic alphabetic numerals are used according to the Eastern system.*

Likewise, there is an astrolabe[†] dating from year 315 of the Hegira (927–928 CE) where this date is expressed in a palaeographic style known as Kufic script (Fig. 19.10). Other older documents indicate that the introduction of this system to the Arabs occurred as early as the eighth century, or, at the earliest, the end of the seventh.

From then on, all becomes clear. After adding six letters to the western Semitic alphabet which they inherited, and having established their system of alphabetic numerals preserving the traditional order of the letters, the Arab grammarians of the seventh or eighth century, apparently for pedagogical reasons, completely changed the original order of the letters by bringing together letters which had much the same graphical forms. At that time these grammarians "worked mainly in Mesopotamia where Jewish and Christian studies flourished, with Greek influences" (M. Cohen).

Thus it was that, from that time on, letters such as *ba*, *ta*, and *tha*, or *jim*, *ḥa*, and *kha* were placed in sequence in the Arabic alphabet (Fig. 19.3).

خ	ح	ج	ث	ت	ب
kha	ḥa	jim	tha	ta	ba
600	8	3	500	400	2

FIG. 19.6.

* "Treatise by Ibrahim ibn Sinan on the Methods of Analysis and Synthesis in Problems of Geometry", a tenth-century copy of fifty-one works on mathematics (BN Ms.arab. 2457; see for example ff. 53v and 88).
† A scientific instrument for observing the position of stars and their height above the horizon. It was used in particular by Arabian astrologers, but some examples have been found from the Graeco-Byzantine era.

The better to establish the order of the alphabetic numerals, the eastern Arabs invented eight mnemonics which every user had to learn by heart in order to be able to recall the number-letters according to their regular arithmetic sequence (Fig. 19.7).

This clearly shows that the "ABC" order, pronounced *Abajad* (or *Abjad, Abujad, Aboujed*, etc. depending on accent), which sometimes governs the order of letters in the Arabic alphabet, does not correspond to their phonetic value nor to their graphical form, but to their respective numerical values according to the eastern Arabian system (Fig. 19.4).

In the usage of the Maghreb, it should be noted, the numerical values given to six of the twenty-eight Arabic letters are different from those in the Eastern system; also, the grouping of the number-letters is different, being done according to nine mnemonics which yield the following groups of values: (1; 10; 100; 1,000); (2; 20; 200); (3; 30; 300); etc. (Fig. 19.11).

MNEMONIC WORDS		DECOMPOSED AS	
Abajad	ابجد	اب ج د ←‐‐‐‐‐‐‐ d j b 'a	4. 3. 2. 1
Hawazin	هوز	ه و ز ←‐‐‐‐‐ z w h	7. 6. 5.
Ḥuṭiya	حطي	ح ط ي ←‐‐‐‐‐ y ṭ ḥ	10. 9. 8
Kalamuna	كلمن	ك ل م ن ←‐‐‐‐‐‐‐ n m l k	50. 40. 30. 20
Saʻfaṣ	سعفص	س ع ف ص ←‐‐‐‐‐‐‐ ṣ f ʻ s	90. 80. 70. 60
Qurshat	قرشت	ق ر ش ت ←‐‐‐‐‐‐‐ t sh r q	400. 300. 200. 100.
Thakhudh	ثخذ	ث خ ذ ←‐‐‐‐‐‐ dh kh th	700. 600. 500.
Ḍaḍhugh	ضظغ	ض ظ غ ←‐‐‐‐‐‐ gh dh ḍ	1,000. 900. 800

FIG. 19.7. *Mnemonic words enabling eastern practitioners to find the order of numerical values associated with Arabic letters*

604 خ د **خد** 4 600 ←----		12 ي ب **يب** 2 10 ←----	
472 ب ع ت **تعب** 2 70 400 ←-------		58 ح ن **نح** 8 50 ←----	
1,283 ج ف ر غ **غرفج** 3 80 200 1,000 ←--------		96 و ص **صو** 6 90 ←----	
1,631 ا ل خ غ **غخلا** 1 30 600 1,000 ←--------		169 ط س ق **قسط** 9 60 100 ←------	
1,629 ط ك خ غ **غخكط** 9 20 600 1,000 ←--------		315 ي ي ش **شيه** 5 10 300 ←--------	

Fig. 19.8. *The writing of numbers by reference to the number-letters of the eastern Arabic system (transcribed into current characters) is always from right to left in descending order of values, starting with the highest order. Moreover, these number-signs (as with ordinary Arabic letters) have an inter-relationship generally by undergoing slight graphic modifications according to the position they occupy within the body of the number- or word-combinations. Examples reconstituted from an Arabic manuscript copied at Shiraz c. 970. Paris: Bibliothèque nationale, Ms. ar. 2457*

On the other hand, this same order occurs not only with the Jews, but also with all the northwestern Semites, as well as the Greeks, the Etruscans and the Armenians, to cite but a few. It is a very ancient ordering since, more than twenty centuries earlier than the Arabs, the inhabitants of Ugarit were familiar with it.

Nonetheless the Arabs, lacking knowledge of the other Semitic languages ... sought other explanations for the mnemonics *abjad*, etc. which had come down to them by tradition but which they found incomprehensible. The best that they could propose on this subject, interesting though it is, is pure fable. According to some, six kings of Madyan arranged the Arabic letters according to their own names. According to a different tradition, the first six mnemonics were the names of six demons. According to a third, it was the names of the days of the week. ... We may none the less discern an interesting detail amongst these fables. One of the six kings of Madyan had supremacy over the others (*ra'īsuhum*): this was Kalaman, whose name bears perhaps some relation with the Latin *elementa**. In North Africa, the adjective *bujadi* is still used to mean beginner, novice literally, someone who is still on his ABC. [G. S. Colin]

---

* According to M. Cohen (p. 137), the Latin word *elementum* goes back to an earlier alphabet that began in the middle, with the letters L, M, N. So giving the LMN (*elemen-tum*) of a matter was the same as "saying the ABC of it all".

FIG. 19.9. *Seventeenth-century Persian astrolabe inscribed by Mohannad Muqim (Delhi, Red Fort, Isa 8). Note that the rim is marked in fives to 360 degrees by means of Arabic letter-numbers. (See B. von Dorn)*

"WORK OF BASTULUS
YEAR 315"

FIG. 19.10. *Detail from an early oriental astrolabe, ostensibly once the property of King Farouk of Egypt, inscribed by Bastulus and dating from 315 of the Hegira (927–928 CE). The date is expressed by means of letter-numbers from the eastern number-system ("Kufic" characters with diacritics). (Personal communication from Alain Brieux)*

				MNEMONIC WORDS RETAINED BY THIS USAGE
1 ا 'alif	10 ي ya	100 ق qaf	1,000 ش shin	Ayqash ايقش ←-------
2 ب ba	20 ك kaf	200 ر ra		Bakar بكر ←-------
3 ج jim	30 ل lam	300 س sin		Jalas جلس ←-------
4 د dal	40 م mim	400 ت ta		Damat دمت ←-------
5 ه Ha	50 ن Nun	500 ث Tha		Hanath هنث ←-------
6 و Wa	60 ص Ṣad	600 خ Kha		Waṣakh وصخ ←-------
7 ز Zay	70 ع 'Ayin	700 ذ Dhal		Za'adh زعذ ←-------
8 ح Ḥa	80 ف Fa	800 ظ Dha		Ḥafaḍh حفظ ←-------
9 ط Ṭa	90 ض Ḍad	900 غ Ghayin		Taḍugh طضغ ←-------

FIG. 19.11. *Numeral alphabet used by African Arabs. (For mnemonic words see Fig. 19. 7 and footnote [same page])*

The eastern Arabs represented thousands, tens of thousands and hundreds of thousands by the multiplicative method. For this purpose, they adopted the convention of putting the letter associated with the corresponding numbers of units, of tens or of hundreds to the right of the Arab letter *ghayin*, whose value was 1,000 (Fig. 19.12).

Arabic letter-number attributed to 1,000*					
غ isolated form					ـغ final form
1,000 × 8 ←--------	حغ gh Ḥ	8,000	1,000 × 2 ←--------	بغ gh B	2,000
1,000 × 9 ←--------	طغ gh Ṭ	9,000	1,000 × 3 ←--------	جغ gh J	3,000
1,000 × 10 ←---------	يغ gh Y	10,000	1,000 × 4 ←--------	دغ gh D	4,000
1,000 × 20 ←---------	كغ gh K	20,000	1,000 × 5 ←--------	هغ gh H	5,000
1,000 × 30 ←---------	لغ gh L	30,000	1,000 × 6 ←--------	وغ gh W	6,000
1,000 × 40 ←---------	مغ gh M	40,000	1,000 × 7 ←--------	زغ gh Z	7,000
* i.e. the letter *ghayin*, twenty-eighth in the Abjad system (Fig. 19.4)					

FIG. 19.12. *Eastern Arabic notation for numbers above 1,000*

THE ETHIOPIAN NUMBER-SYSTEM

The Ethiopians borrowed the Greek alphabetical numbering system during the fourth century CE, no doubt under the influence of Christian missionaries who came from Egypt, Syria and Palestine.*

Starting, however, with 100, they radically altered the Greek system. Having adopted the first nineteen Greek alphabetic numerals to represent the first hundred whole numbers, they decided to indicate the hundreds and thousands by putting the letters for the units and tens to the left of the sign P (Greek *rho*) whose value was 100. That is to say that instead of representing the numbers 200, 300, ... 9,000 after the Greek fashion:

Σ	T	Y	...	Ꝺ	'A	'B	...	'Θ
200	300	400		900	1,000	2,000		9,000

they expressed them as follows (see Fig. 19.13 A):

BP	...	HP	...	KP	...	ΠP
2 × 100		8 × 100		20 × 100		80 × 100
- - - - - -→		- - - - - - -→		- - - - - - -→		- - - - - - -→
200		800		2,000		8,000

They denoted 10,000 by marking a ligature between two identical P signs (making a composite sign equivalent to multiplying 100 by itself, which we shall transcribe as P-P. Then multiples of 10,000 were expressed by placing the symbol for the multiplier to the left of this symbol P-P for 10,000.

BP-P	...	HP-P	...	KP-P	...	ΠP-P
2 × 10,000		8 × 10,000		20 × 10,000		80 × 10,000
- - - - - - - -→		- - - - - - - -→		- - - - - - - -→		- - - - - - - -→
20,000		80,000		200,000		800,000

* The numerals that Ethiopians still sometimes use today are actually much more rounded stylisations of the numerical signs found on the Aksum inscriptions (Aksum, near the modern port of Adowa, was the capital of the Kingdom of Abyssinia from the fourth century CE). The modern signs follow the same principles as the ancient ones, which are themselves derived from the first nineteen letter-numerals of the Greek alphabet. Since the fifteenth century Ethiopian numerals have always been written inside two parallel bars with a curlicue at either end, signifying that they are to be taken as numbers, not as letters.

VALUES	GREEK LETTER-NUMBERS	ETHIOPIAN INSCRIPTIONS AT AKSUM (4th century CE) DAE no. 7, 10, 11	MODERN ETHIOPIAN NUMBERS
1	A	*0*	6
2	B	*B* *13*	E
3	Γ	*Γ* *⌐*	Γ
4	Δ	▽	O
5	E	*Ł* *Ł*	E
6	Ϛ	*?* *Ⴑ* *Ⴑ*	Z
7	Z	*z* *Ʒ*	Z
8	H	Ⲉ	A
9	Θ	*Ꮵ* *H* *ϑ*	Ꮎ
10	I	*ı*	Ꭲ
20	K	*Ⴆ* *ⲕ*	Ⴕ
30	Λ	*Λ* *ᴎ*	ᴎ
40	M	*Ꮃ*	Ꮃ
50	N	*ϟ* *Ꮞ* *Ꮞ*	Ꮞ
60	Ξ	*Ⴑ*	Ⴑ
70	O		Ⴑ
80	Π	*Π*	Π
90	Ϟ		Ꭹ
100	P	*P* *Ⴗ* *Ⴗ*	Ⴗ

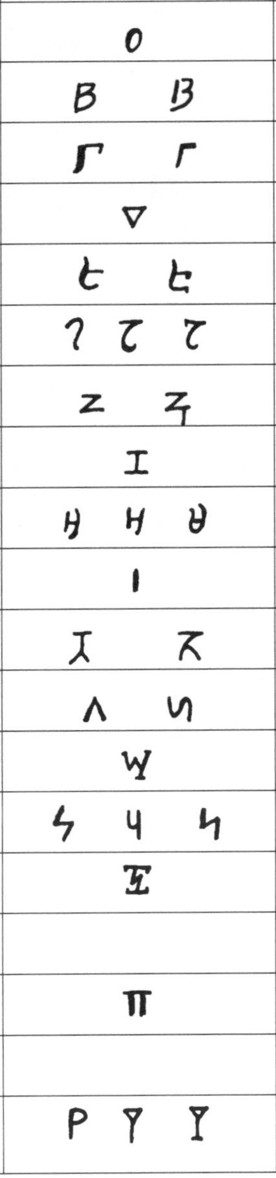

FIG. 19.13A *Ethiopian numbering*

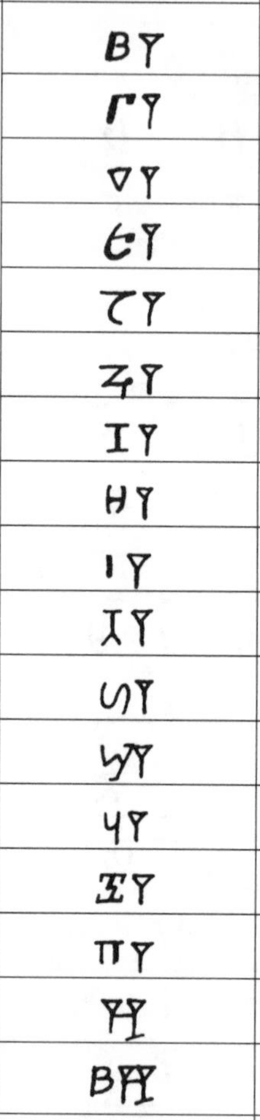

VALUES AND GREEK LETTER-NUMBERS		ETHIOPIAN INSCRIPTIONS AT ASKUM (4th century CE) DAE no. 7, 10, 11	MODERN ETHIOPIAN NUMBERS AND ARITHMETICAL TRANSLATIONS	
200	Σ	B Υ	𐌺𐌺	2×100
300	T	Γ Υ	𐌺𐌺	3×100
400	Υ	▽ Υ	𐌺𐌺	4×100
500	Φ	Ϭ Υ	𐌺𐌺	5×100
600	X	ᛐ Υ	𐌺𐌺	6×100
700	Ψ	ᛉ Υ	𐌺𐌺	7×100
800	Ω	Ⅱ Υ	𐌺𐌺	8×100
900	⟋	H Υ	𐌺𐌺	9×100
1,000	'A	ꟾ Υ	𐌺𐌺	10×100
2,000	'B	人 Υ	𐌺𐌺	20×100
3,000	'Γ	ᛁ Υ	𐌺𐌺	30×100
4,000	'Δ	Ϟ Υ	𐌺𐌺	40×100
5,000	'E	Ϥ Υ	𐌺𐌺	50×100
6,000	'Ϛ	Ξ Υ	𐌺𐌺	60×100
8,000	'H	π Υ	𐌺𐌺	80×100
10,000	M$^{\alpha}$	Ϥ٣	𐌺	100×100
20,000	M$^{\beta}$	BϤ٣	𐌺𐌺	$2 \times 10,000$
31,900		Γ٣ ꟾᛐ ϑ Υ	𐌺𐌺 𐌺𐌺 𐌺𐌺	
		$\xrightarrow{\hspace{2cm}}$		
		$3 \times 100 \times 100 + 10 \times 100 + 9 \times 100$		
25,140		B Ϥ٣ Ϥρ ρᛁ	𐌺 𐌺𐌺 𐌺𐌺 𐌺 𐌺	
		$\xrightarrow{\hspace{2cm}}$		
		$2 \times 100 \times 100 + 50 \times 100 + 100 + 40$		

FIG. 19.13B *Ethiopian numbering for numbers above 100*

CHAPTER 20

MAGIC, MYSTERY, DIVINATION, AND OTHER SECRETS

SECRET WRITING AND SECRET NUMBERS IN THE OTTOMAN EMPIRE

We shall close our account of alphabetic numerals with an examination of the secret writing and secret numerals used until recently in the Middle East and, especially, in the official services of the Ottoman Empire.*

> The Turks used cryptography with abandon. Documents on Mathematics, Medicine and the occult, written or translated by the Turks, teem with secret alphabets and numerals, and they made use of every alphabet they knew. Usually they adopted such alphabets in the form in which they came across them, but sometimes they changed them; either deliberately, or as a result of the mutations which attend repeated copying. [M. J. A. Decourdemanche (1899)]

Fig. 20.1 shows secret numerals which were used for a long time in Egypt, Syria, North Africa, and Turkey. At first sight these would seem to have been made up throughout. However, if they are put alongside the Arabic letters which have the same numerical values, and then we put alongside these the corresponding Hebrew and Palmyrenean characters, we can at once see that the figures of these secret numerals are simply survivals of the ancient Aramaic characters in their traditional *Abjad* order (Fig. 20.2; see also Fig. 17.2, 17.4, 17.10 and 19.4).

---

* Such esoteric writing was used in a great variety of contexts: occultism, divination, science, diplomacy, military reports, business letters, administrative circulars, etc. Until the beginning of this century, the Turkish and Persian offices of the Ministry of Finance used a system of numerals known as *Siyaq*, whose figures were used in balance sheets and business correspondence. These figures were abbreviations of the Arabic names of the numbers, and their purpose was both to keep the sums of money secret from the public and also to prevent fraudulent alteration (see Chapter 25).

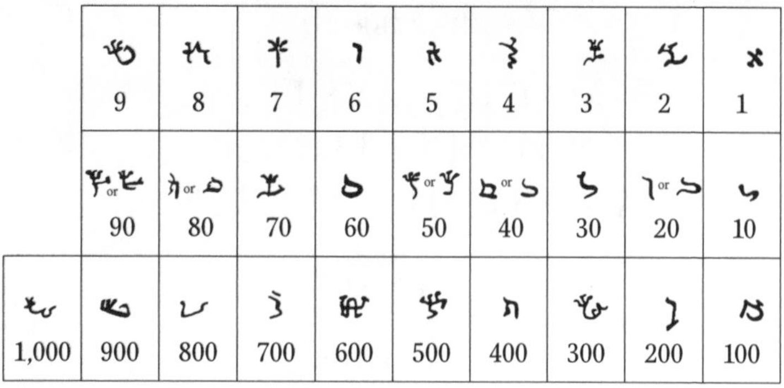

9	8	7	6	5	4	3	2	1
90	80	70	60	50	40	30	20	10
1,000	900	800	700	600	500	400	300	200	100

FIG. 20.1.

Among these secret numerals there were alternative forms for the values 20, 40, 50, 80 and 90. These are in fact the final forms of the Hebrew and Palmyrenean letters *kof, mem, nun, pe* and *tsade*. The correspondences noted here are confirmed by treatises on arithmetic. The Egyptian treatises refer to this system as *al Shāmisī* ("sunlit"), which was used in those parts to designate things related to Syria. The Syrian documents themselves called it *al Tadmurī* ("from Tadmor"), which was the former Semitic name of Palmyra, an ancient city on the road linking Mesopotamia to the Mediterranean via Damascus to the south and via Homs to the north.

The people who had devised these secret writings had therefore taken the twenty-two Aramaic letters as they found them and (as has been explicitly mentioned by Turkish writers) they added six further conventional signs in order to complete a correspondence with the Arabic alphabet and to achieve a system of numerals which was complete from 1 to 1,000. This system was used until recent times, not only for writing numbers, but also as secret writing:

> In 1869, in order to draw up for French military officers a comparison between the abortive expedition of Charles III of Spain against Algiers, and the French expedition of 1830, the Ministry of War brought to Paris the original military report on the expedition of Charles III, which had been written in Turkish by the Algerian Regency at the Porte. This document was given to a military interpreter to be summarised. The manuscript, which I have seen, carried the stamp of a library in Algiers. After a whole wad of financial accounts came the report from the Regency. Following this came a series of annexes amongst which is an espionage report written as a long letter in the Hebraic script called *Khat al barāwāt*.

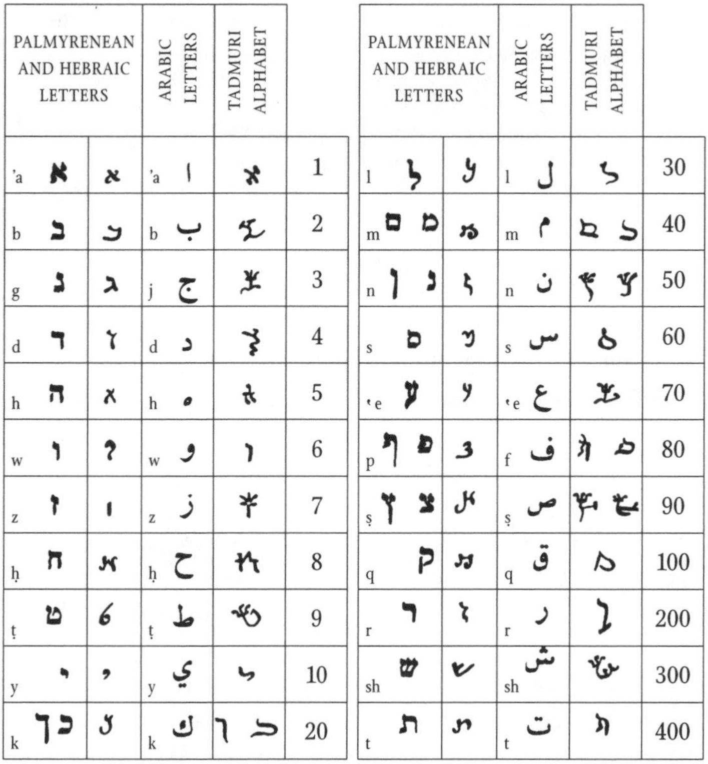

PALMYRENEAN AND HEBRAIC LETTERS		ARABIC LETTERS	TADMURI ALPHABET		PALMYRENEAN AND HEBRAIC LETTERS		ARABIC LETTERS	TADMURI ALPHABET	
'a		'a		1	l		l		30
b		b		2	m		m		40
g		j		3	n		n		50
d		d		4	s		s		60
h		h		5	'e		'e		70
w		w		6	p		f		80
z		z		7	ṣ		ṣ		90
ḥ		ḥ		8	q		q		100
ṭ		ṭ		9	r		r		200
y		y		10	sh		sh		300
k		k		20	t		t		400

FIG. 20.2. *Secret alphabet (still used in Turkey, Egypt, and Syria in the nineteenth century) compared with the Arabic, Palmyrenean, and Hebraic alphabets*

The signature was written in Tadmuri characters, not Latin: *Felipe, rabbina Yusuf ben Ezer, nacido en Granada.*

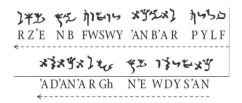

RZ'E   NB FWSWY 'AN B'A R   PYLF
←- -

'AD'AN'A R Gh   N'E WDY S'AN
←- -

FIG. 20.3.

Then, on exactly the same kind of paper as the letter, is a detailed analysis of the Spanish land and sea forces, again written in Tadmuri characters. Since this analysis is also reproduced line for line in normal Turkish characters in the Regency report, it was easy for me to discern the value of each of the Tadmuri signs.

As an example, here in reproduction is the first line of the analysis, possibly for the army, possibly for the navy:

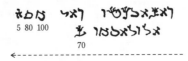

5 80 100

70

FIG. 20.4.

in which the following Spanish expressions are written in Tadmuri script:

Regimento (del) Rey,	185	(hombres)
"King's Army"	185	men
El Velasco,	70	(cañones)
"Navy"	70	guns

[M. J. A. Decourdemanche (1899)]

We have no intention of presenting a general survey of the very many clandestine systems of the East; nonetheless we shall discuss two other systems of secret numerals which were used until recent times in the Ottoman army.

We begin with the simplest case. This is a system of numerals used in Turkish military inventories of provisions, supplies, equipment, and so on.

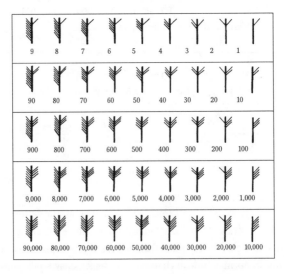

FIG. 20.5.

Here the numbers 1, 10, 100, 1,000 and 10,000 are represented by a vertical stroke with, on the right, one, two, three, four, or five upward oblique strokes. Adding one upward oblique stroke on the left of each of these gives the figures for 2, 20, 200, 2,000, and 20,000; two strokes on the left gives the figures for 3, 30, 300, 3,000, and 30,000; and so on, until with eight oblique strokes on the left we have the figures for 9, 90, 900, 9,000, and 90,000.

The above system is very straightforward, which is not the case for the next one. This was used in the Turkish army for recording the strengths of their units.

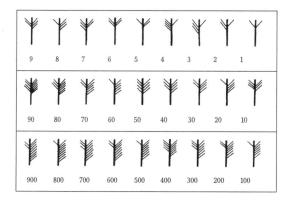

FIG. 20.6.

To the uninitiated, this system follows no obvious pattern. However, it was used both for writing numbers and also as a means of secret writing, which leads us to suppose that each of these signs corresponded to the Arabic letter corresponding to the numeral in question.

Proceeding as we did before, placing each of these numerals beside the Arabic letter corresponding to the same numerical value (see Fig. 19.4 above), we now consider the eight mnemonics for the letters of the Arabic numerals (Fig. 19.7), and it becomes clear how the figures of this system were formed.

For the numbers 1, 2, 3, 4 (corresponding to the first mnemonic, *ABJaD*), we take a vertical stroke with one oblique upward stroke on its right, and adjoin successively one, two, three, or four oblique upward strokes on its left.

Then, for the second mnemonic, *HaWaZin*, we take a vertical stroke with two upward oblique strokes on the right, and add successively one, two, or three upward oblique strokes on the left, and so on (Fig. 20.7).

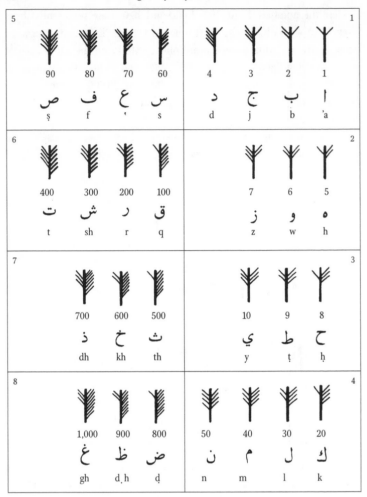

FIG. 20.7. *Secret numerical notation based on the succession of eight mnemonic words in the eastern Arab alphabetical numbering*

THE ART OF CHRONOGRAMS

Jewish and Muslim writings since the Middle Ages abound in what are called "chronograms": these correspond to a method of writing dates, but – like calligraphy or poetry – are an art form in themselves.

This is the *Ramz* of the Arab poets, historians and stone-carvers in North Africa and in Spain, the *Tarikh* of the Turkish and Persian writers, which "consists of grouping, into one meaningful and characteristic word or short phrase, letters whose numerical values, when totalled, give the year of a past or future event." [G. S. Colin]

The following example occurs on a Jewish tombstone in Toledo [IHE, inscr. 43]:

שנת אגלי טל על חמשת אלפם

THOUSAND   FIVE   ON   DEW   DROP   YEAR

←- -

YEAR: "ONE DROP OF DEW ON FIVE THOUSAND"

FIG. 20.8.

If we take it literally, the phrase is meaningless. But if we add up the numerical values of the letters in the phrase translated as "drop of dew", we discover that this phrase represents, according to the Hebrew calendar, the date of death of the person buried here:

"ONE DROP OF DEW"   ל ט י ל ג א

30   9   10   30   3   1

←- - - - - - - - - - - - - - - -

83

FIG. 20.9.

This person died, in fact, in the year "eighty-three [= *drop of dew*] on five thousand," or, in plain language, in the year 5083 of the Hebrew era, i.e. 1322–1323 CE.

In the following two further examples from the Jewish cemetery in Toledo we find the years 5144 (Fig. 20.10) and 5109 (Fig. 20.11, in two different forms) shown in the chronograms: but note that the "5000" is not indicated, since it would have been implicitly understood, much as we understand "1974" when someone says "I was born in seventy-four". Also, note that in these examples the words whose letters represent numerals have been marked with three dots.

אב   אין   היינה   שנת

2 1   50 1   5 10 5   YEAR
         10   50 10

←- -

YEAR: "WE HAVE BEEN MADE FATHERLESS"

144

FIG. 20.10.

החיים   ל ו   מן וחה

40 10 10 8 5   6   30   5 8 6 50 40

←- - - - - - - - - - - - - -   ←- - - - - - - - -

"FOR LIFE"   "REST"

109   109

FIG. 20.11.

The same procedure is found in Islamic countries, especially Turkey, Iraq, Persia, and Bīhar (in northwest India); but, like the oriental art of calligraphy, it seems to go no further back than the eleventh century.* The dreadful death of King Sher of Bīhar in an explosion occurred in the year 952 of the Hegira (1545 CE), which is recorded in the following chronogram [CAPIB, vol. X, p. 368]:

د	ر	م	ش	ت أ	ز
4	200	40	300	400 1	7

زأتش مرد

"DIED OF BURNS"

←--------------------------

952 (of the Hegira)

FIG. 20.12.

Another interesting chronogram was made by the historian, mathematician and astronomer Al Biruni (born 973 CE at Khiva, died 1048 at Ghazni) in his celebrated *Tarikh ul Hind*. This learned man accused the Jews of deliberately changing their calendar so as to diminish the number of years elapsed since the Creation, in order that the date of birth of Christ should no longer agree with the prophecies of the coming of the Messiah; he boldly asserted that the Jews awaited the Messiah for the year 1335 of the Seleucid era (1024 CE), and he wrote this date in the following form:

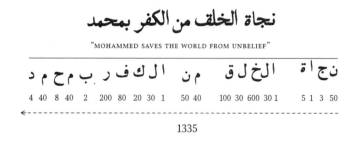

نجاة الخلق من الكفر بمحمد

"MOHAMMED SAVES THE WORLD FROM UNBELIEF"

د م ح م ب . ر ف ك ل ا ن م	ق ل خ ل ا	ة ا ج ن
4 40 8 40 2 200 80 20 30 1 50 40	100 30 600 30 1	5 1 3 50

←--

1335

FIG. 20.13.

---

* In Persian and in Turkish certain letters have exactly the same numerical values as the equivalent Arabic letters according to the Eastern usage. For instance:

the letter پ , or P, has the same value as ب , or B;

the letter چ , or Ch, has the same value as ج , or J;

the letter گ , or G, has the same value as ک , or K.

Chronograms were also common in Morocco, but only from the seventeenth century CE (possibly the sixteenth, or earlier, according to recent documentation). They were often used in verse inscriptions commemorating events or foundations, and by writers, poets, historians and biographers, including the secretary and court poet Muhammad Ben Ahmad al Maklati (died 1630), and also the poets Muhammed al Mudara (died 1734) and 'Abd al Wahab Adaraq (died 1746) who both composed instructional historical synopses on the basis of chronograms, which in one case referred to the notabilities of Fez, and in the other to the saints of Maknez.*

The following example comes from an Arabic inscription discovered by Colin in the Kasbah of Tangier over fifty years ago, in the south chamber of the building known as *Qubbat al Bukhari*, in the old Sultan's Palace. We make a brief detour in time so as to stand in the period when this building was constructed.

The inscription was written to the glory of Ahmad ibn 'Ali ibn 'Abdallah. This notable person was:

> the son of the famous 'Ali ibn 'Abdallah, governor (*qā'id*) of Tetuan and chief of the Rif contingents destined for holy war (*mujāhidīn*) who, after a long siege, entered Tangier in 1095 of the Hegira (1684 CE) after its English occupiers had abandoned it ...

> When *Qa'id* 'Ali ibn 'Abdallah, commandant (*amir*) of all the people of the Rif, died in year 1103 of the Hegira (1691–1692), Sultan Isma'il gave to them as chief the dead man's son, *basa* Ahmad ibn 'Ali; henceforth, almost all the history of northwest Morocco can be found in this man's biography ... After 1139 of the Hegira (1726–1727), following the death of Sultan Isma'il, he took the opportunity provided by the weakness of his successor, Ahmad ad Dahabi, to try to seize Tetuan which was administered by another, almost independent, governor (*amīr*), Muhammed al Waqqas, but he was repulsed with loss.

> In 1140 of the Hegira (1727–1728), when Sultan Ahmad ad Dahabi (who had been overturned by his brother 'Abd al Malik) was restored to the throne, Ahmad ibn 'Ali refused to recognise him and declined to send him a deputation (a snub which was imitated by the town of Fez). The enmity between the Rif chieftain and the 'Alawite kings waxed from then on, and an impolitic gesture by Sultan 'Abdallah, successor of Ahmad ad Dahabi, transformed this into overt hostility ...

---

* In epigraphic texts, chronograms were often written in a contrasting colour, and sometimes also in manuscripts where, however, we also find them written with thicker strokes. Arab chronograms, like those in Hebraic inscriptions, were always preceded by the preposition *fi* or by *Sanat 'ama* in the year, etc.

In 1145 of the Hegira, when a delegation of 350 holy warriors from the Rif came from Tangier to Sultan 'Abdallah to try to resolve the differences between him and *basa* Ahmad ibn 'Ali, he had them killed. The Rif chieftain distanced himself from the King and came closer to his brother and rival Al Mustadi. Thenceforth, until his unfortunate death in 1156 of the Hegira (1743), he did not cease from fighting with 'Abdallah, son of Sultan Isma'il, and to support his rivals against him. [G. S. Colin]

Returning now to our inscription, the date 1145 is given in the following verse (in which the numerical values have been calculated from the Arabic alphabetic numerals according to the Maghreb usage; see Fig. 19.11).

YEAR:"THE FULL MOON OF MY BEAUTY HAS ENTERED THE CHAMBER OF HAPPINESS"

جم الدي	ر دب د ع	ال س ع دب د ر	حل ب ي ت
10 30 1 40 3		200 4 2 4 70 300 30 1	400 10 2 30 8

1145

Fig. 20.14.

In other words, the *Qubbat al Bukhari* in the Kasbah of Tangier was constructed in the year 1145 of the Hegira, the very time when *basa* Ahmad ibn 'Ali broke away from Sultan 'Abdallah.

We find in this chronogram, therefore, testimony to an art in which one's whole imagination is deployed to create a phrase which is both eloquent and, at the same time, has a numerical value that reveals the date of an event which one wishes to commemorate.

GNOSTICS, CABBALISTS, MAGICIANS, AND SOOTHSAYERS

Once the letters of an alphabet have numerical values, the way is open to some strange procedures. Take the values of the letters of a word or phrase and make a number from these. Then this number may furnish an interpretation of the word, or another word with the same or a related numerical value may do so. The Jewish *gematria*,* the Greek *isopsephy* and the Muslim *khisab al jumal* ("calculating the total") are examples of this kind of activity.

---

* Possibly a corruption of the Greek *geometrikos arithmos*, geometrical number

Especially among the Jews, these calculations enriched their sermons with every kind of interpretation, and also gave rise to speculations and divinations. They are of common occurrence in Rabbinic literature, especially the Talmud* and the Midrash.† But it is chiefly found in esoteric writings, where these cabbalistic procedures yielded hidden meanings for the purposes of religious dialectic.

Though not adept in the matter, we would here like to describe some examples of religious, soothsaying or literary practices which derive from such procedures.

The two Hebrew words *Yayin*, meaning "wine", and *Sod*, meaning "secret", both have the number 70 in the normal Hebrew alphabetic numerals (Fig. 20.15), and for this reason some rabbis bring these words together: *Nichnas Yayin Yatsa Sod*: "the secret comes out of the wine" (Latin: *in vino veritas*, the drunken man tells all).

יַיִן   סוד

50 10 10   4   6   60

←-----   ←------

YAYIN   SOD

70   70

FIG. 20.15.

In *Pardes Rimonim*, Moses Cordovero gives an example which relates *gevurah* ("force") to *arieh* ("lion"), which both have value 216. The lion, traditionally, is the symbol of divine majesty, of the power of *Yahweh*, while *gevurah* is one of the Attributes of God.

גבורה   אריה

5   200   6   2   3   5   10   200   1

←----------   ←----------

GEVURAH   ARIEH

216   216

FIG. 20.16.

The Messiah is often called *Tzemach*, "seed", or *Menakhem*, "consoler", since these two words have the same value:

צמח   מנחם

8   40   90   40   8   50   40

←-------   ←--------

TZEMACH   MENAKHEM

138   138

FIG. 20.17.

---

* The Rabbinic compilation of Jewish laws, customs, traditions and opinions which forms the code of Jewish civil and canon law

† Hebrew commentaries on the Old Testament

The letters of *Mashiyakh*, "Messiah", and of *Nakhash*, "serpent", give the same value:

FIG. 20.18.   NAKHASH   MASHIYAKH

and this gives rise to the conclusion that "When the Messiah comes upon earth, he shall measure himself against Satan and shall overcome him."

We may also conclude that the world was created at the beginning of the Jewish civil year, from the fact that the two first words of the Torah (*Bereshit Bara*, "in the beginning [God] created") have the same value as *Berosh Hashanah Nibra*, "it was created at the beginning of the year":

BERESHIT BARA   BEROSH HASHANAH NIBRAH
1116   1116

FIG. 20.19.

In Genesis XXXII:4, Jacob says "I have sojourned with Laban" (in Hebrew, *'Im Laban Garti*). According to the commentary by Rashi* on this phrase (*Bereshit Rabbati*, 145), this means that "during his sojourn with Laban the impious, Jacob did not follow his bad example but followed the 613 commandments of the Jewish religion"; for, as he explains, *Garti* ("I have sojourned") has the value 613:

FIG. 20.20.   GARTI

Genesis recounts elsewhere (XIV:12–14) how, in the battle of the kings of the East in the Valley of Siddim, Lot of Sodom, the kinsman of Abraham, was captured by his enemies: "When Abraham heard that his brother was taken captive, he armed his trained servants, born in his own house, three hundred and eighteen, and pursued them unto Dan", where he smote his adversaries with the help of "God Most High" (XIV:20). Then he addresses God in these words: "Lord GOD [*Yahweh*], what wilt thou give me, seeing I go childless and the steward of my house is this Eliezer of Damascus?" (Genesis XV:2).

* Rabbenu Shelomoh Yishakhi (1040–1105)

The *barayta* of the thirty-two Haggadic rules (for the interpretation of the Torah) gives the following interpretation (rule 29): the 318 servants are none other than the person of Eliezer himself. In other words, Abraham smote his enemies with the help of Eliezer alone, his trusted servant who was to be his heir; and whose name in Hebrew means "My God is help". The argument put forward for this brings together the two verses

> his trained servants, born in his own house, *three hundred and eighteen*

and

> the steward of my house is this *Eliezer of Damascus*

and the fact that the numerical value of the name *Eliezer* is 318:

$$\text{אליעזר}$$

200 7 70 10 30 1
←--------------

FIG. 20.21.   ELI'EZER
                            318

Another concordance which the exegetes have achieved brings *Ahavah* ("Love") together with *Ekhad* ("One"):

$$\text{אחד} \qquad \text{אהבה}$$

4 8 1   5 2 5 1
←------   ←---------

FIG. 20.22.   EKHAD   AHAVAH
              13   13

As well as their numerical equivalence, it is explained that these two terms correspond to the central concept of the biblical ethic, that "God is Love", since on the one hand "One" represents the One God of Israel and, on the other hand, "Love" is supposed to be at the very basis of the conception of the Universe (Deuteronomy: V 6–7; Leviticus XIX:18). At the same time, the sum of their values is 26, which is the number of the name *Yahweh* itself:

$$\text{יהוה}$$

5 6 5 10
←---------

FIG. 20.23.   YHWH
              26

The common Semitic word for "God" is *El*, but in the Old Testament this only occurs in compounds (*Israel*, *Ismael*, *Eliezer*, etc.). To refer to God, the Torah uses *Elohim* (which in fact is plural), and is the word which is supposed to express all the force and supernatural power of God. The Torah refers also to the attributes of God, such as *khay* ("living"), *Shadai* ("all-powerful"), *El Ilyion* ("God Most High") and so on. But *YHWH*, "Yahweh", is the only true Name of God: it is the *Divine Tetragram*. It is supposed to incorporate the eternal nature of God since it embraces the three Hebrew tenses of the verb "to be", namely:

היה   הוה   יהיה

HaYaH "He was"   HoWeH "He is"   YiHYeH "He shall be"

FIG. 20.24.

To invoke God by this name is therefore to appeal to His intervention and His concern for all things. But this name may be neither written nor spoken casually, and in order not to violate what is holy and incommunicable, in common use it must be read as *Adonai* ("My Lord").

Every kind of speculation has been founded on the numerical value of 26 which the Tetragram assumes according to the classical system of alphabetic numerals. Some adept writers have thereby been led to point out that in Genesis I:26, God says: "Let us make man in our image"; that 26 generations separate Adam and Moses; that 26 descendants are listed in the genealogy of Shem, and the number of persons named in this is a multiple of 26; and so on. According to them, the fact that God fashioned Eve from a rib taken from Adam is to be found in the numerical difference (= 26) between the name of Adam (= 45) and the name of Eve (= 19):

חוה   אדם

5 6 8   40 4 1
←- - - - - -   ←- - - - - - - -
KHAWAH   ADAM
19   45

FIG. 20.25.

The usual alphabetic numerals were not the only basis adopted by the rabbis and Cabbalists for this kind of interpretation. A manuscript in the Bodleian Library at Oxford (Ms. Hebr. 1822) lists more than seventy different systems of *gematria*.

One of these involves assigning to each letter the number which gives its position in the Hebrew alphabet but with reduction of numbers above 9, that is to say with the same units figure as in the usual method, but ignoring tens and hundreds. The letter מ (*mem*), for example, which traditionally has the value 40, is given the value 4 in this system.* Similarly, the

---

* This can be found by the alternative method of noting that *Mem* is in the thirteenth place, so its value is equal to $1 + 3 = 4$.

letter ש (*shin*), whose usual value is 300, has value 3 in this system.*From this, some have concluded that the name *Yahweh* can be equated to the divine attribute *Tov* ("Good"):

טוב   יהוה

2 6 9   5 6 5 1

←------   ←-------

TOV   YHWH
"Good"

17   17

FIG. 20.26.

Another method gives to the letters values equal to the squares of their usual values, so that *gimmel*, for example, which usually has value 3, is here assigned the value 9 (Fig. 20.29, column B). According to a further system, the value 1 is assigned to the first letter, the sum (3) of the first two to the second letter, the sum (6) of the first three to the third, and so on. The letter *yod* , which is in the tenth position, therefore has a value equal to the sum of the first ten natural numbers: $1 + 2 + 3 + ... + 9 + 10 = 55$ (Fig. 20.29, column C).

Yet another system assigns to each letter the numerical value of the word which is the name of the letter. Thus *aleph* has the value $1 + 30 + 80 = 111$:

טוב

80 30 1

←------

111

FIG. 20.27.

With these starting points, one can make a concordance between two words by evaluating them numerically according to either the same numerical system, or two different numerical systems. For instance, the word *Maqom* ("place"), which is another of the names of God, can be equated to *Yahweh* because in the traditional system the word *Maqom* has value 186, and *Yahweh* also has value 186 if we use the system which gives each letter the square of its usual value:

מקום   יהוה

40 6 100 40 5^2 6^2 5^2 10^2

←---------   ←---------

MAQOM   YHWH

186   186

FIG. 20.28.

* *Shin* is in the twenty-first place, so its value is $2 + 1 = 3$.

Order number and normal values of the letters			A	B	C	D	
1	א	1	1	1^2	1	111	value of אלף ALEPH
2	ב	2	2	2^2	$1+2$	412	" בית BET
3	ג	3	3	3^2	$1+2+3$	73	" גמל GIMMEL
4	ד	4	4	4^2	$1+2+3+4$	434	" דלת DALET
5	ה	5	5	5^2	$1+2+3+4+5$	6	" הא HE
6	ו	6	6	6^2	$1+2+3+4+5+6$	12	" וו VOV
7	ז	7	7	7^2	$1+2+3+4+5\ldots+7$	67	" זין ZAYIN
8	ח	8	8	8^2	$1+2+3+4+5\ldots+8$	418	" חית HET
9	ט	9	9	9^2	$1+2+3+4+5\ldots+9$	419	" טית TET
10	י	10	1	10^2	$1+2+3+4+5\ldots+10$	20	" יוד YOD
11	כ	20	2	20^2	$1+2+3+4+5\ldots+11$	100	" כף KOF
12	ל	30	3	30^2	$1+2+3+4+5\ldots+12$	74	" למד LAMED
13	מ	40	4	40^2	$1+2+3+4+5\ldots+13$	90	" מים MEM
14	נ	50	5	50^2	$1+2+3+4+5\ldots+14$	110	" נון NUN
15	ס	60	6	60^2	$1+2+3+4+5\ldots+15$	120	" סמך SAMEKH
16	ע	70	7	70^2	$1+2+3+4+5\ldots+16$	130	" עין AYIN
17	פ	80	8	80^2	$1+2+3+4+5\ldots+17$	85	" פה PE
18	צ	90	9	90^2	$1+2+3+4+5\ldots+18$	104	" צדי TSADE
19	ק	100	1	100^2	$1+2+3+4+5\ldots+19$	104	" קוף QUF
20	ר	200	2	200^2	$1+2+3+4+5\ldots+20$	510	" ריש RESH
21	ש	300	3	300^2	$1+2+3+4+5\ldots+21$	360	" שין SHIN
22	ת	400	4	400^2	$1+2+3+4+5\ldots+22$	406	" תו TAV

Fig. 20.29. *Some of the many systems for the numerical evaluation of Hebraic letters. They are used by rabbis and Cabbalists for the interpretation of their homilies.*

This, it is emphasised, is confirmed by Micah I:3.

> For, behold, the LORD [*Yahweh*] cometh forth out of his place [*Maqom*].

This selection of examples – which could easily be much extended – gives a good idea of the complexities of Cabbalistic calculations and investigations which the exegetes went into, not only for the purpose of interpreting certain passages of the Torah but for all kinds of speculations.*

The Greeks also used similar procedures. Certain Greek poets, such as Leonidas of Alexandria (who lived at the time of the Emperor Nero), used them to create distichs and epigrams with the special characteristic of being *isopsephs*. A distich (consisting of two lines or two verses) is an isopseph if the numerical value of the first (calculated from the sum of the values of its letters) is equal to that of the second. An epigram (a short poem which might, for example, express an amorous idea) is an isopseph if all of its distichs are isopsephs, with the same value for each.

More generally, isopsephy consists of determining the numerical value of a word or a group of letters, and relating it to another word by means of this value.

At Pergamon, isopseph inscriptions have been found which, it is believed, were composed by the father of the great physician and mathematician Galen, who, according to his son, "had mastered all there was to know about geometry and the science of numbers."

At Pompeii an inscription was found which can be read as "I love her whose number is 545", and where a certain Amerimnus praises the mistress of his thoughts whose "honourable name is 45."

In the *Pseudo-Callisthenes* † (I, 33) it is written that the Egyptian god Sarapis (whose worship was initiated by Ptolemy I) revealed his name to Alexander the Great in the following words:

> Take two hundred and one, then a hundred and one, four times twenty, and ten. Then place the first of these numbers in the last place, and you will know which god I am.

Taking the words of the god literally, we obtain

$$200 \ 1 \ 100 \ 1 \ 80 \ 10 \ 200$$

---

* We claim no competence to make the slightest commentary on these matters, neither on the delicate questions of the historical origins of *Gematria* in the Hebrew texts, nor on its evolution, nor on the extent to which it was regarded (or discredited) in Rabbinic and Cabbalistic writings throughout the centuries and in various countries. The reader who is interested in these questions may consult F. Dornseiff (1925) or G. Scholem.

† A spurious work associated with the name of Callisthenes, companion of Alexander in his Asiatic expedition.

which corresponds to the Greek name

$$\Sigma A P A \Pi I \Sigma$$

200 1 100 1 80 10 200

-------------------→

SARAPIS

FIG. 20.30.

In recalling the murder of Agrippina, Suetonius (*Nero*, 39) relates the name of Nero, written in Greek, to the words *Idian Metera apekteine* ("he killed his own mother"), since the two have exactly the same value according to the Greek number-system:

$$N E P \Omega N \quad I \Delta I A N \quad M H T E P A \quad A \Pi E K T E I N E$$

50 5 100 800 50   10 4 10 1 50   40 8 300 5 100 1   1 80 5 20 300 5 10 50 5

----------→   --→

"NERO"   "HE KILLED HIS OWN MOTHER"

1005   1005

FIG. 20.31.

The Greeks apparently came rather late to the practice of speculating with the numerical values of letters. This seems to have occurred when Greek culture came into contact with Jewish culture. The famous passage in the Apocalypse of Saint John clearly shows how familiar the Jews were with these mystic calculations, long before the time of their Cabbalists and the *Gematria*. Both Jews and Greeks were remarkably gifted for arithmetical calculation and also for transcendental speculation; every form of subtlety was apt to their taste, and number-mysticism appealed to both predilections at the same time. The Pythagorean school, the most superstitious of the Greek philosophical sects, and the most infiltrated by Eastern influence, was already addicted to number-mysticism. In the last age of the ancient world, this form of mysticism experienced an astonishing expansion.

It gave rise to arithmomancy; it inspired the Sybillines, the seers and soothsayers, the pagan *Theologoi*; it troubled the Fathers of the Church, who were not always immune to its fascination. Isopsephy is one of its methods. [P. Perdrizet (1904)]

Father Theophanus Kerameus, in his *Homily* (XLIV) asserts the numerical equivalence between *Theos* ("God"), *Hagios* ("holy") and *Agathos* ("good") as follows:

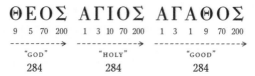

FIG. 20.32.

He likewise saw in the name *Rebecca* (wife of Isaac and mother of the twins Jacob and Esau) a figure of the Universal Church. According to him, the number (153) of great fish caught in the "miraculous draught of fishes" is the same as the numerical value of the name Rebecca in Greek (*Homily* XXXVI; John XXI).

FIG. 20.33.

In the New Testament, the phrase *Alpha and Omega* (Apocalypse XXII: 13) is a symbolic designation of God: formed from the first and last letters of the Greek alphabet, in the Gnostic and Christian theologies it corresponded to the "Key of the Universe and of Knowledge" and to "Existence and the Totality of Space and Time". When Jesus declares that he is *the Alpha and the Omega*, he therefore declares that he is the beginning and the end of all things. He identifies himself with the "Holy Ghost" and therefore, according to Christian doctrine, with God Himself. According to Matthew III:16, the Holy Ghost appeared to Jesus at the moment of his birth in the form of a dove; the Greek word *Peristera* for "dove" has the value 801; and this is also the value of the letters of the phrase "Alpha and Omega" which, therefore, is no other than a mystical affirmation of the Christian doctrine of the Trinity.

FIG. 20.34.

In another conception much exercised in the Middle Ages, numbers were given a supernatural quality according to the graphical shape of their symbols.

In a manuscript which is in the Bibliothèque nationale in Paris (Ms. lat. 2583, folio 30), Thibaut of Langres wrote as follows, about the number 300 represented by the Greek letter T (*tau*), which is also the sign of the Cross:

> The number is a secret guarded by writing, which represents it in two ways: by the letter and by its pronunciation. By the letter, it is represented in three ways: shape, order, and secret. By shape, it is like the 300 who, from the Creation of the World, were to find faith in the image of the Crucifix since, to the Greeks, these are represented by the letter T which has the form of a cross.

Which is why, according to Thibaut, Gideon conquered Oreb, Zeeb, Zebah, and Zalmunna with only the three hundred men who had drunk water "as a dog lappeth" (Judges VII:5).

A similar Christian interpretation is to be seen in the *Epistle of Barnabas*. In the patriarch Abraham's victory over his enemies with the help of 318 circumcised men, Barnabas finds a reference to the cross and to the two first letters of the name of Jesus (Ιησους)

$$T + IH = 318$$
$$\quad 300 \qquad 10+8$$

Fig. 20.35.

He considers that the number 318 means that these men would be saved by the crucifixion of Jesus.

In the same fashion, according to Cyprian (*De pascha computus*, 20), the number 365 is sacred because it is the sum of 300 (T, the symbol of the cross), 18 (IH, the two first letters of the name of Jesus), 31 (the number of years Christ is supposed to have lived, in Cyprian's opinion) and 16 (the number of years in the reign of Tiberius, within which Jesus was crucified). This may well also explain why certain heretics believed that the End of the World would occur in the year 365 of the Christian era.*

---

* "But because this sentence is in the Gospel, it is no wonder that the worshippers of the many and false gods ... invented I know not what Greek verses, ... but add that Peter by enchantments brought it about that the name of Christ should be worshipped for three hundred and sixty-five years, and, after the completion of that number of years, should at once take end. Oh the hearts of learned men!" [Augustine, *The City of God*, Book 18, Chapter 53]

TRANSCRIPTION

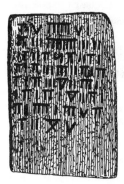

I V IIIIII V I

II III IIIIIII III II

I I II II VI II II I I

II I II III III II II I II

I I II III V II III I

I III I II I V III II I

II I IIII I III V II

I II X V

FIG. 20.36. *Wooden tablet found in North Africa, dating from the late fifth century CE. Note that on each line the Roman numerals total 18 (the overline denotes a part-total). It is not known whether this is a mathematical (indeed a teaching) document or a " magic" tablet relating to speculations on the numerical value of Greek or Hebraic letters. (See TA, act XXXIV, tabl. 3a)*

Clearly, all possible resources have been exploited for these purposes. The Christian mystics, who wished to support the affirmation that Jesus was the Son of God, often equated the Hebrew phrase *'Ab Qal* which Isaiah used to mean "the swift cloud" on which "the Lord rideth" (XIX: 1) and the word *Bar* ("son"):

FIG. 20.37.

For their part, the Gnostics* were able to draw almost miraculous consequences from the practice of isopsephy. P. Perdrizet (1904) explains:

A text, which is probably by Hippolytus, says that in certain Gnostic sects isopsephy was a normal form of symbolism and catechesis. It did not serve only to wrap a revelation in a mystery: if in certain cases it served to conceal, in others it served to reveal, throwing light on things which otherwise would never have been understood ...

Gnosticism seems loaded with a huge burden of Egyptian superstitions. It purported to rise to knowledge of the Universal Principle;

---

* Gnosticism (from the Greek *gnosis*, "knowledge") is a religious doctrine which appeared in the early centuries of our era in Judaeo-Christian circles, but was violently opposed by rabbis and by the New Testament apostles. It is based essentially on the hope that salvation may be attained through an esoteric knowledge of the divine, as transmitted through initiation.

in fact it was preoccupied with the quest to know the name of God and thence, with the aid of magic (the ancient magic of Isis), the means to induce God to allow Man to raise himself to God's own level. The name, like the shadow or the breath, is a part of the person: more, it is identical with the person, it is the person himself.

To know the name of God, therefore, was the problem which Gnosticism addressed. At first it seems insoluble: how can we know the Ineffable? The Gnostics did not pretend to know the name of God, but they believed it possible to learn its formula; and for them this was sufficient, since for them the formula of the divine name contained its complete magical virtue: and this formula was the number of the name of God.

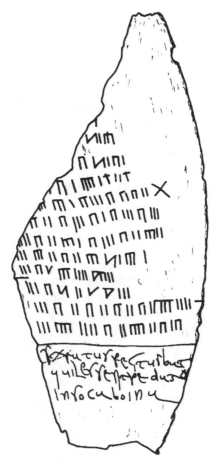

FIG. 20.38. *One of many slates found in the region of Salamanca. This one was discovered at Santibanez de la Sierra and dates from about the sixth century. It is a document similar to the previous one; each line that has remained intact shows a total count of 26. (See G. Gomez-Moreno, pp.24, 117)*

The supreme God of the Gnostics united in himself, according to Basilides the Gnostic, the 365 minor gods who preside over the days of the year ... and so the Gnostics referred to God as "He whose number is 365" (ου εστιν η ψηφος ΤΞΕ). From God, on the other hand, proceeded the magical power of the seven vowels, the seven notes of the musical scale, the seven planets, the seven metals (gold, silver, tin, copper, iron, lead, and mercury); and of the four weeks of the lunar month. Whatever was the name of the Ineffable, the Gnostic was sure it involved the magic numbers 7 and 365. We may not know the unknowable name of God, so instead we seek a designation which would serve as its formula, and we only have to combine the mystic numbers 7 and 365. Thus Basilides created the name *Abrasax*, which has seven letters whose values add up to 365:

$$\text{Α Β Ρ Α Σ Α Ξ}$$

Α	Β	Ρ	Α	Σ	Α	Ξ
1	2	100	1	200	1	60

---------------------->

365

FIG. 20.39.

God, or the name of God (for they are the same) has first the character of holiness. Αγιος ο Θεος (*Hagios o Theos*) says the seraphic hymn; "hallowed be thy name" says the Lord's Prayer, that is "let the holiness of God be proclaimed."

Though the name of God remained unknown, it was known that it had the character to be the ideal holy name. Nothing therefore better became the designation of the Ineffable than the locution *Hagion Onoma* ("Holy Name") which the Gnostics indeed frequently employed. But this was not only for the above metaphysical or theological reason, nor because they had borrowed this same appellation from the Jews, but for a more potent mystical reason peculiar to them. By a coincidence of which Gnosticism had seen a revelation, the biblical phrase *Hagion Onoma* had the same number (365) as *Abrasax*.

$$\text{Α Γ Ι Ο Ν} \quad \text{Ο Ν Ο Μ Α}$$

Α	Γ	Ι	Ο	Ν		Ο	Ν	Ο	Μ	Α
1	3	10	70	50		70	50	70	40	1

-->

365

FIG. 20.40.

Once embarked on this path, Gnosticism made other discoveries no less gripping.

Mingled as it was with magic, Gnosticism had a fatal tendency to syncretism. In isopsephy it had the means to identify with its own supreme God the national god of Egypt. The Nile, which for the Egyptians was the same as Osiris, was a god of the year, for the regularity of its floods followed the regular course of the years; and now, the number of the name of the Nile, *Neilos*, is 365:

FIG. 20.41.

By isopsephy, Gnosticism achieved another no less interesting syncretism. The Mazdean cult of Mithras underwent a prodigious spread in the second and third centuries of our era. The Gnostics noticed that *Mithras*, written ΜΕΙΘΡΑΣ, has the value

FIG. 20.42.

Therefore the Sun God of Persia was the same as the "Lord of the 365 Days".

As Perdrizet says, the Christians often put new wine in old bottles, and they found that this kind of practice offered ample scope for fantasy. When the scribes and stone-carvers wished to preserve the secret of a name, they wrote only its number instead.

In Greek and Coptic Christian inscriptions, following an imprecation or an exhortation to praise, we sometimes come across the sign ϟ Θ made up of the letters *Koppa* and *Theta*. This cryptogram remained obscure until the end of the nineteenth century, when J. E. Wessely (1887) showed that it was simply a mystical representation of *Amen* ('Αμην), since both have numerical value 99:

$$
\begin{array}{ccccc}
\mathrm{A} & \mathrm{M} & \mathrm{H} & \mathrm{N} & \qquad \text{ϟΘ} \\
1 & 40 & 8 & 50 & 90 \quad 9 \\
\multicolumn{4}{c}{\text{- - - - - - - - - - - →}} & \text{- - - →} \\
\multicolumn{4}{c}{99} & 99
\end{array}
$$

FIG. 20.43.

Similarly, the dedication of a mosaic in the convent of Khoziba near Jericho begins:

$$\Phi\ \Lambda\ E\ MNH\Sigma\Phi HTI\ TOY\ \Delta OY\Lambda OY\Sigma OY$$

ΦΛΕ REMEMBER YOUR SERVANT

FIG. 20.44.

What does the group *Phi-Lambda-Epsilon* stand for? The problem was solved by W. D. Smirnoff (1902). These letters correspond to the Greek word for "Lord", Κυριε, whose numerical value is 535:

$$\Phi\ \Lambda\ E \qquad K\ Y\ P\ I\ E$$

| 500 | 30 | 5 | | 20 | 400 | 100 | 10 | 5 |

--------->   -------------------->

535   535

FIG. 20.45.

Much more significant are the speculations of the Christian mystics surrounding the number 666, which the apostle John ascribed to the *Beast of the Apocalypse*, a monster identified as the Antichrist, who shortly before the end of time would come on Earth to commit innumerable crimes, to spread terror amongst men, and raise people up against each other. He would be brought down by Christ himself on his return to Earth.

16 And he shall make all, both little and great, rich and poor, freemen and bondsmen, to have a character in their right hand, or on their foreheads.

17 And that no man might buy or sell, but he that hath the character, or the name of the beast, or the number of his name.

18 Here is wisdom. He that hath understanding, let him count the number of the beast. For it is the number of a man: and the number of him is six hundred and sixty-six. [Apocalypse, XIII:16–18]

We clearly see an allusion to isopsephy here, but the system to be used is not stated. This is why the name of the Beast has excited, and continues to excite, the wits of interpreters, and many are the solutions which have been put forward.

Taking 666 to be "the number of a man", some have searched amongst the names of historical figures whose names give the number 666. Thus Nero, the first Roman emperor to persecute the Christians, has been identified as the Beast of the Apocalypse since the number of his name, accompanied by the title "Caesar", makes 666 in the Hebraic system:

קסר נרון

50 6 200 50   200 60 100

QSAR NERO

666

FIG. 20.46.

On the same lines, others have found that the name of the Emperor Diocletian (whose religious policies included the violent persecution of Christians), when only the letters that are Roman numerals are used, also gives the number of the Beast:

(Diocletian Augustus)

D I o C L E s   A V G V s T V s

500 1   100 50   5   5   5

666

FIG. 20.47.

Yet others, reading the text as "the number of a *type of man*", saw in 666 the designation of the Latins in general since the Greek word *Lateinos* gives this value:

Λ ΑΤΕΙΝΟ Σ

30   1 300 5 10 50 70 200

666

FIG. 20.48.

Much later, at the time of the Wars of Religion, a Catholic mystic called Petrus Bungus, in a work published in 1584–1585 at Bergamo, claimed to have demonstrated that the German reformer Luther was none other than the Antichrist since his name, in Roman numerals, gives the number 666:

L	V	T	H	E	R	N	V	C
30	200	100	8	5	80	40	200	3

- ->

666

FIG. 20.49.

But the disciples of Luther, who considered the Church of Rome as the direct heir of the Empire of the Caesars, lost no time in responding. They took the Roman numerals contained in the phrase *VICARIUS FILII DEI* ("Vicar of the Son of God") which is on the papal tiara, and drew the conclusion that one might expect:

V	I	C	A	R	I	V	s	F	I	L	I	I	D	E	I
5	1	100			1	5			1	50	1	1	500		1

- ->

666

FIG. 20.50.

The numerical evaluation of names was also used in times of war by Muslim soothsayers, under the name of *khisab al nim*, to predict which side would win. This process was described as follows by Ibn Khaldun in his "Prolegomena" (*Muqāddimah*, I):

> Here is how it is done. The values of the letters in the name of each king are added up, according to the values of the letters of the alphabet; these go from one to 1,000 by units, tens, hundreds and thousands. When this is done, the number nine is subtracted from each as many times as required until what is left is less than nine. The two remainders are compared: if one is greater than the other, and if both are even numbers or both odd, the king whose name has the smaller number will win. If one is even and the other odd, the king with the larger number will win. If both are equal and both are even numbers, it is the king who has been attacked who will win; if they are equal and odd, the attacking king will win.

Since each Arabic letter is the first letter of one of the attributes of Allah (*Alif*, the first letter of *Allah*; *Ba*, first letter of *Baqi*, "He who remains", and so on), the use of the Arabic alphabet led to a "Most Secret" system. In this, each letter is assigned, not its usual value, but instead the number of the divine attribute of which it is the first letter. For instance, the letter *Alif*, whose usual value is 1, is given the value 66 which is the number of the name of Allah calculated according to the *Abjad* system. This is the system used in the symbolic theology called *da'wa*, "invocation", which allowed mystics and soothsayers to make forecasts and to speculate on the past, the present and the future.

LETTERS		VALUES	ASSOCIATED DIVINE ATTRIBUTES		VALUES
			NAMES	MEANING	
ا	'alif	1	الله ALLAH	Allah	66
ب	ba	2	باقي BĀQĪ	He who remains	113
ج	jim	3	جامع JĀMI'	He who collects	114
د	dal	4	ديّان DAYĀN	Judge	65
ه	ha	5	هادي HĀDĪ	Guide	20
و	wa	6	ولي WALĪ	Master	46
ز	zay	7	زكي ZAKĪ	Purifier	37
ح	ḥa	8	حق ḤAQ	Truth	108
ط	ṭa	9	طاهر ṬĀHIR	Saint	215
ي	ya	10	يسين YASSĪN	Chief	130
ك	kaf	20	كافي KĀFĪ	Sufficient	111
ل	lam	30	لطيف LAṬĪF	Benevolent	129
م	mim	40	ملك MALIK	King	90
ن	nūn	50	نور NŪR	Light	256
س	sin	60	سميع SAMĪ'	Listener	180
ع	'ayin	70	علي 'ALĪ	Raised up	110
ف	fa	80	فتاح FATĀḤ	Who opens	489
ص	ṣad	90	صمد ṢAMAD	Eternal	134
ق	qaf	100	قادر QĀDIR	Powerful	305
ر	ra	200	رب RAB	Lord	202
ش	shin	300	شفيع SHAFĪ'	Who accepts	460
ت	ta	400	توب TAWAB	Who restores to the good	408
ث	tha	500	ثابت THĀBIT	Stable	903
خ	kha	600	خالق KHĀLIQ	Creator	731
ذ	dhal	700	ذاكر DHĀKIR	Who remembers	921
ض	ḍad	800	ضار ḌĀR	Chastiser	1,001
ظ	dha	900	ظاهر DḤĀHIR	Apparent	1,106
غ	gha	1,000	غفور GHAFŪR	Indulgent	1,285

FIG. 20.51. *The Da'wa system, after the tabulation made by Sheikh Abu'l Muwwayid of Gujarat in* Jawahiru'l Khamsah

The same type of procedure allowed magicians to contrive their talismans, and to indulge in the most varied practices. In order to give their co-religionists the means to get rich quickly, to preserve themselves from evil and to draw down on themselves every grace of God, some *tolba* of North Africa offered their clients a *kherz* ("talisman") containing:

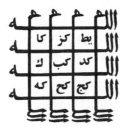

FIG. 20.52A.

This is a "magic square" whose value is 66, which can be obtained as the sum of every row, of every column, and of each diagonal:

21	26	19
20	22	24
25	18	23

FIG. 20.52B.

and is itself the number of the name of Allah according to the *Abjad*:

5 30 30 1
←---------
ALLAH

FIG. 20.53.

We can see, therefore, to what lengths the soothsayers, seers and other numerologists were prepared to go in applying these principles of number to the enrichment of their dialectic.

CHAPTER 21

NUMBERS IN CHINESE CIVILISATION

THE THIRTEEN FIGURES OF THE TRADITIONAL CHINESE NUMBER-SYSTEM*

The Chinese have traditionally used a decimal number-system, with thirteen basic signs denoting the numbers 1 to 9 and the first four powers of 10 (10, 100, 1,000, and 10,000). Fig. 21.1 shows the simplest representations of these, which is the one most commonly used nowadays.

1	一	10	十
2	二	100	百
3	三	1,000	千
4	四	10,000	萬
5	五		
6	六		
7	七		
8	八		
9	九		

FIG. 21.1.

To an even greater extent than in the ancient Semitic world, this written number-system corresponds to the true type of "hybrid" number-system, since the tens, the thousands, and the tens of thousands are expressed according to the multiplicative principle (Fig. 21.2).

* I wish to express here my deep gratitude to my friends Alain Briot, Louis Frédéric and Léon Vandermeersch for their valuable contributions, and for their willing labour in reading this entire chapter.

TENS		HUNDREDS		THOUSANDS		TENS OF THOUSANDS	
10	一十 1 × 10	100	一百 1 × 100	1,000	一千 1 × 1,000	10,000	一萬 1 × 10,000
20	二十 2 × 10	200	二百 2 × 100	2,000	二千 2 × 1,000	20,000	二萬 2 × 10,000
30	三十 3 × 10	300	三百 3 × 100	3,000	三千 3 × 1,000	30,000	三萬 3 × 10,000
40	四十 4 × 10	400	四百 4 × 100	4,000	四千 4 × 1,000	40,000	四萬 4 × 10,000
50	五十 5 × 10	500	五百 5 × 100	5,000	五千 5 × 1,000	50,000	五萬 5 × 10,000
60	六十 6 × 10	600	六百 6 × 100	6,000	六千 6 × 1,000	60,000	六萬 6 × 10,000
70	七十 7 × 10	700	七百 7 × 100	7,000	七千 7 × 1,000	70,000	七萬 7 × 10,000
80	八十 8 × 10	800	八百 8 × 100	8,000	八千 8 × 1,000	80,000	八萬 8 × 10,000

FIG. 21.2. *The modern Chinese notation for consecutive multiples of the first four powers of 10.*

For intermediate numbers, the Chinese used a combination of addition and multiplication, so that the number 79,564, for example, is decomposed as:

七　萬　九　千　五　百　六　十　四

7 × 10,000 + 9 × 1,000 + 5 × 100 + 6 × 10 + 4

79,564

FIG. 21.3.

NUMERALS OCCURRING IN THE DOCUMENT IN FIGURE 21.5

Col. VIII	Col. VII	Col. IV	Col. I
1 × 100 + 6 × 10 + 1 → 一百六十一 161	3 × 100 + 4 × 10 + 5 → 三百四十五 345	2 × 100 + 4 × 10 → 二百四十 240	1 × 10,000 + 6 × 1,000 + 3 × 100 + 4 × 10 + 3 → 一萬六千三百四十三 16,343
3 × 10 + 2 → 三十二 32	1 × 10 + 2 → 一十二 12	1 × 1,000 + 3 × 100 + 2 × 10 + 8 → 一千三百二十八 1,328	

Traditionally, these numbers, as in Chinese generally, would be placed vertically from top to bottom of columns which would be placed from right to left. However, in the People's Republic of China it is now preferred to write them horizontally from left to right.

FIG. 21.4. *Examples of numbers written with Chinese numerals*

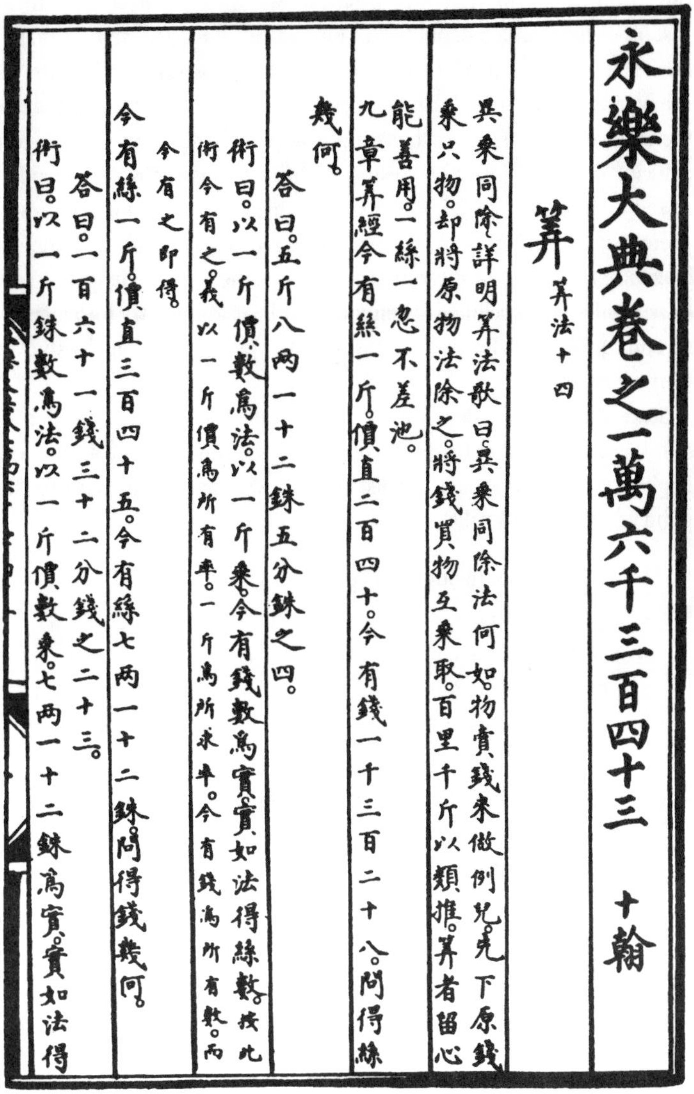

永樂大典卷之一萬六千三百四十三　十翰

算　算法十四

其乘同除詳明算法歌曰其乘同除法何如物貴錢賤來做例兒先下原錢乘只物。卻將原物法除之。將錢買物互乘取。百里千斤以類推算者留心

能善用一絲一忽不差池。

九章算經今有絲一斤。價直二百四十。今有錢一十三百二十八。問得絲

幾何

答曰五斤八兩一十二銖五分銖之四。

術曰。以一斤價數為法。以一斤乘今有錢數為實。實如法得絲數挨九絲二兩一斤為所求率。今有錢數為所有數。而

今有之義以一斤價為所有率。一斤為所求率。今有錢數為所有數。而

今有之即得

今有絲一斤。價直三百四十五。今有絲七兩一十二銖問得錢幾何。

答曰一百六十一錢三十二分錢之二十三。

術曰。以一斤錢數為法。以一斤價數乘七兩一十二銖為實。實如法得

FIG. 21.5. *A page from a Chinese mathematical document dating from the beginning of the fifteenth century. Cambridge University Library [Ms. Yong-le da dian, chapter 16 343, introductory page. From Needham (1959), III, Fig. 54].*

TRANSCRIPTION OF CHINESE CHARACTERS

To transcribe Chinese characters into the Latin alphabet, we shall adopt the so-called *Pinyin* system in what follows. This has been the official system of the People's Republic of China since 1958. "This transcription", according to D. Lombard (1967), "was developed by Chinese linguists for use by the Chinese people and especially to assist schoolchildren to learn the language and its characters, and it is based mainly on phonological principles. The majority of Western Chinese scholars nowadays tend to abandon the older transcription systems (which sought in vain to represent pronunciation in terms of the spelling conventions of various European languages) in favour of this one. The reader is therefore no longer obliged to remember any spelling conventions, but instead must try to remember certain equivalences between sound and letter (as in beginning the study of German or Italian)."

Since the *Pinyin* system was not conceived with European readers in mind, it is natural that the values of its letters do not always coincide with English pronunciation. Here is a list of the most important aspects from the point of view of the English reader.

b   corresponds to our letter "p"

c   corresponds to our "ts"

d   corresponds to our "t"

g   corresponds to our "k"

u   corresponds to the standard English pronunciation of "u" as in "bull" (except after j, q or x)

ı   corresponds to the pronunciation of "u" as, for instance, in Scotland or in French

z   corresponds to our "dz"

zh   corresponds to "j" as in "join"

ch   corresponds to "ch" as in "church"

h   in initial position, corresponds to the hard German "ch" (as in "Bach")

x   in initial position, corresponds to the soft "ch" (as in German "Ich")

i   corresponds to our "i" (as in "pin"); but, following z, c, s, sh, sh or r it is pronounced like "e" (in "pen") or like "u" in "fur"; following a or u, it is pronounced like the "ei" in "reign".

q   stands for a complex sound consisting of "ts" with drawing-in of breath

r   in initial position is like the "s" in pleasure; in other cases it is like the "el" in "channel".

THE CHINESE ORAL NUMERAL SYSTEM

The number-signs shown above are in fact ordinary characters of Chinese writing. They are therefore subject to the same rules as govern the other Chinese characters. These are, in fact, "word-signs" which express in graphical form the ideographic and phonetic values of the corresponding numbers. In other words, they constitute one of the graphical representations of the thirteen monosyllabic words which the Chinese language possesses to denote the numbers from 1 to 9 and the first four powers of 10.

Having a decimal base, the oral Chinese number-system gives a separate name to each of the first ten integers:

yī	èr	sān	sì	wǔ	liù	qī	bā	jiǔ	shí
1	2	3	4	5	6	7	8	9	10

The numbers from 11 to 19 are represented according to the additive principle:

11	shí yī	ten-one	= 10 + 1
12	shí èr	ten-two	= 10 + 2
13	shí sān	ten-three	= 10 + 3
14	shí sì	ten-four	= 10 + 4
...	....	....	....

The tens are represented according to the multiplicative principle:

20	èr shí	two-ten	$= 2 \times 10$
30	sān shí	three-ten	$= 3 \times 10$
40	sì shí	four-ten	$= 4 \times 10$
50	wǔ shí	five-ten	$= 5 \times 10$
60	liù shí	six-ten	$= 6 \times 10$
...	....	....	....

For 100 (= 10^2), 1,000 (= 10^3) and 10,000 (= 10^4), the words bǎi, qiān and wàn are used; for the various multiples of these the multiplicative principle is used:

100	yī bǎi	one-hundred	
200	èr bǎi	two-hundred	$= 2 \times 100$
300	sān bǎi	three-hundred	$= 3 \times 100$
400	sì bǎi	four-hundred	$= 4 \times 100$
....	....	....	....
1,000	yī qiān	one-thousand	
2,000	èr qiān	two-thousand	$= 2 \times 1,000$
3,000	sān qiān	three-thousand	$= 3 \times 1,000$
4,000	sì qiān	four-thousand	$= 4 \times 1,000$
....	....	....	....
10,000	yī wàn	one-myriad	
20,000	èr wàn	two-myriad	$= 2 \times 10,000$
30,000	sān wàn	three-myriad	$= 3 \times 10,000$
40,000	sì wàn	four-myriad	$= 4 \times 10,000$
....	....	....	....

Starting with these, intermediate numbers can be represented very straightforwardly:

53,781	wǔ wàn	sān qiān	qī bǎi	bā shí	yī
	(five-myriad	three-thousand	seven-hundred	eight-ten	one)

$$(= 5 \times 10,000 + 3 \times 1,000 + 7 \times 100 + 8 \times 10 + 1)$$

Thus the Chinese number-signs are a very simple way of writing out the corresponding numbers "word for word".

Finally, note that such a system has no need of a zero. For the numbers 504, 1,058, or 2,003, for example, one simply writes (or says):

五 百 四
wǔ bǎi sì $(= 5 \times 100 + 4)$

一 千 五 十 八
yī qiān wǔ shí bā $(= 1 \times 1,000 + 5 \times 10 + 8)$

二 千 三
èr qiān sān $(= 2 \times 1,000 + 3)$

FIG. 21.6.

Note, however, that in current usage the word 零, *ling* (which means "zero"), is mentioned whenever any power of 10 is not represented in the expression of the number. This is done in order to avoid any ambiguity. But this usage was only established late in the development of the Chinese number-system.

504 五 百 零 四
    5   100   0   4
wǔ bǎi ling sì
("five hundred zero four")

1,058 一 千 零 五 十 八
    1 1,000 0 5 10 8
yī qiān ling wǔ shí bā
("one thousand zero five ten eight")

2,003 二 千 零 三
    2 1,000 0 3
èr qiān ling sān
("two thousand zero three")

FIG. 21.7.

CHINESE NUMERALS ARE DRAWN
IN MANY WAYS

Even today, the thirteen basic number-signs are drawn in several different ways. Obviously they are spoken in the same way, but are a result of the many different ways of writing Chinese itself.

The forms we have considered so far, which may be called "classical", is the one in common use nowadays, especially in printed matter. It is also the simplest. Some of these signs are among the "keys" of Chinese writing: they are used in the elementary teaching of Chinese, at the stage of learning the Chinese characters.

They are part of the now standard *kăishū* notation, a plain style in which the line segments making up each character are basically straight, but of varying lengths and orientations; they are to be drawn in a strict order, according to definite rules (Fig. 21.8).

FIG. 21.8. *The basic strokes of Chinese writing in the standard style called* kăishū, *and the order in which they are to be written in composing certain characters*

It is also the oldest of the common contemporary forms, having been used as early as the fourth century CE, and it is derived from the ancient writing called *lìshū** ("the writing of clerks") which was used in the Han Dynasty (Fig. 21.9).

* The *lìshū* style of notation is the earliest of the modern forms: it is the first "line writing" in Chinese history. However, "in seeking the maximum enhancement of the precision of the *lìshū* an even more geometrical style resulted, the inflexibly regular *kăishū*." [V. Alleton (1970)]. This regular style became fixed as the standard for Chinese writing in the earliest centuries of the current era: administrative documents, official and scientific writings, were usually written in this style from that time on, when most such works were printed and the fonts for the characters had been made. When, below, we refer to "Chinese writing" without further qualification, it is this style which is meant.

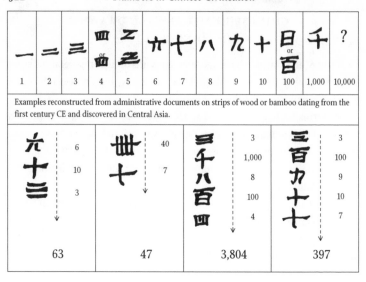

FIG. 21.9. *The earliest of the modern Chinese numerical notations. This is of the* lìshū *type and was in use during the Han Dynasty (206 BCE to 220 CE). The documents used for this diagram were written by scribes of the first century CE. [See de Chavannes (1913); Maspéro; Guitel]*

The second form of the Chinese numerals is called *guān zí* ("official writing"). It is used mainly in public documents, in bills of sale, and to write the sums of money on cheques, receipts or bills. Although still written like the classic *kǎishū*, it is somewhat more complicated, having been made more elaborate in order to avoid fraudulent amendments in financial transactions (Fig. 21.10).

Classical notation	一 萬 三 千 六 百 八 十 四
Guàn zí notation	壹 萬 參 仟 陸 佰 捌 拾 肆

yǐ　　wàn　　sán　qián　　liù　　bǎi　　bá　　shí　　sì
1 × 10,000　+　3 × 1,000 + 6 × 100 + 8 × 10　+ 4

FIG. 21.10.

The third style of writing the numerals is a cursive form of the classical numerals, which is routinely used in handwritten letters, personal notes, drafts, and so on. It belongs to the *xíngshū* style of writing, a cursive style which was developed to meet the need for abbreviation without detracting from the structure of the characters; the changes lay in the manner of

drawing the characters more rapidly and flexibly using upward and down-
ward brushstrokes. (Fig. 21.11).

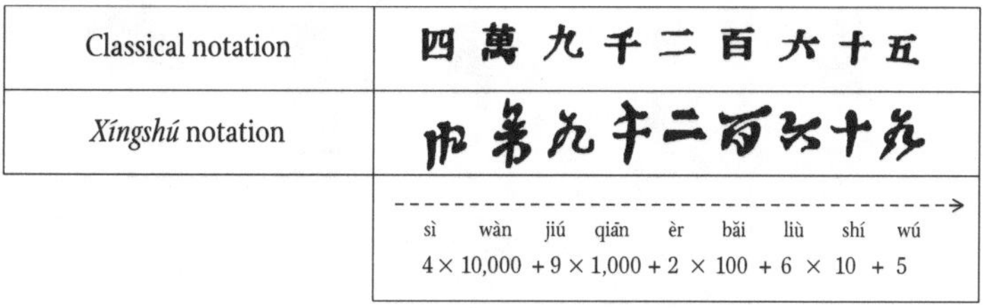

| Classical notation | 四 萬 九 千 二 百 六 十 五 |
| Xíngshú notation | 《cursive characters》 |

sì   wàn   jiú   qián   èr   bǎi   liù   shí   wú

$4 \times 10{,}000 + 9 \times 1{,}000 + 2 \times 100 + 6 \times 10 + 5$

FIG. 21.11.

A combination of exaggerated abbreviation with virtuosity and imagination on the part of calligraphers rapidly brought these cursive forms, which
still resembled the classical style, into an exaggeratedly simplified style
which the Chinese call *cǎoshū* (literally, "plant-shaped"). It can only be deciphered by initiates, with the result that nowadays it is used only in painting
and in calligraphy* (Fig. 21.12 and 21.13).

七	7 ×
萬	10,000 +
五	5 ×
千	1,000 +
六	6 ×
百	100 +
九	9 ×
十	10 +
六	6

FIG. 21.12. *Example of 75,696*

* "Chinese writing underwent two transformations in the *cǎoshū* :

a. Lines and elements of characters were suppressed; save for characters with a small number of strokes,
almost all elements are represented by symbols, leading to a kind of "writing of writing".

b. The strokes lose their individuality and join up: eventually a character is written in one movement; then
the characters themselves join up, and even a whole column may be written without lifting brush from
paper." [V. G. Alleton (1970)]

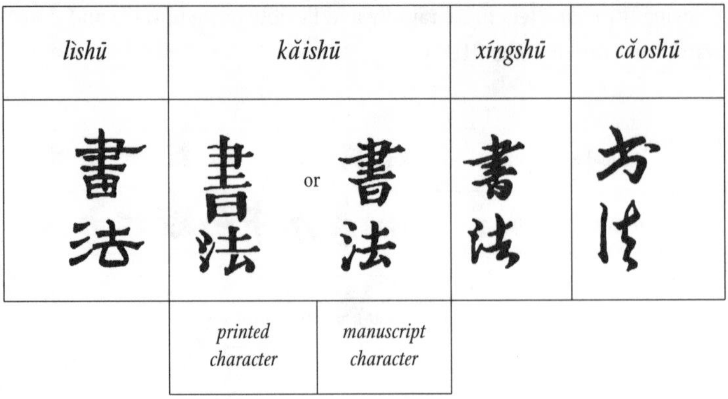

lìshū	*kǎishū*	*xíngshū*	*cǎoshū*
書法	書法 or 書法	書法	方法
	printed character manuscript character		

FIG. 21.13. *The difference between the principal styles of modern Chinese writing, as shown in writing the word* shūfǎ *("calligraphy") in the styles of* lìshū *("official writing", used in the Han period),* kǎishū *("standard style", which replaced the* lìshū *and has been used since the fourth century CE),* xíngshū *(the current cursive style) and* cǎoshū *(a cursive style which has been reduced to maximum abbreviation and is now used only in calligraphy).* [Alleton (1970)]

Yet another form corresponds to a curiously geometrical way of drawing the numerals and characters, called *sháng fāng dà zhuàn*, which is still employed on seals and signatures (Fig. 21. 14).

1	2	3	4	5	6	7	8	9	10	100	1,000	10,000

FIG. 21.14. *Example of the singular* sháng fàng dà zhuàn *calligraphy as used for the thirteen basic characters of the Chinese number-system on seals and in signatures. [See Perny (1873); Pihan (1860)]*

As well as the forms already mentioned, there is the form used by traders to display the prices of goods. This is called *gán mà zí* ("secret marks"). Anyone who travels to the interior of China should be sure of knowing these numerals by heart, if he wishes to understand his restaurant bill (Fig. 21.15).

There are so many different styles for writing numerals in China that we should stop at this point, having described the important ones; to describe them all would be self-indulgent, and little to our purpose.

VALUES	guān zí			gán mà zí	TRANSCRIPTIONS
	1st form	2nd form	3rd and 4th forms	5th form	
	Classical forms	Elaborate augmented forms used in finance	Cursive forms of the classical signs	Cursive forms currently used in business and calculation	
1	一	壹 or 弌	一 一	一	**ι**	yī
2	二	貳 or 弍	二 or 乙 二	二	刂	èr
3	三	叁 or 弎	三 or 乞 三	三	川	sān
4	四	肆	四	囚	メ	sì
5	五	伍	孖	匇	彡 or 彡	wǔ
6	六	陸	孖	彡	山	liù
7	七	柒	仏	仏	亠	qī
8	八	捌	八	八	亠	bā
9	九	玖 or 久	九	九	攵	jiǔ
10	十	拾 or 什	十 or 十	十	十	shí
100	百	佰	百	百	百 or 3	bǎi
1,000	千	仟	千	千	千	qiān
10,000	萬	萬	萬	萬	万	wàn
	Standard kǎishū style	Xíngshū style	Cǎoshū style		

FIG. 21.15. *The principal graphic styles for the thirteen basic signs of the modern Chinese number-system. [Giles (1912); Mathews (1931); Needham (1959); Perny (1873); Pihan (1860)]*

THE ORIGINS OF THE CHINESE NUMBER-SYSTEM

Several thousand bones and tortoise shells: these are the most ancient evidence we have of Chinese writing and numerals. They have for the most part been found since the end of the nineteenth century at the archaeological site of Xiao dun;* called *jiaguwen* ("oracular bones"), they date from around the Yin period (fourteenth–eleventh centuries BCE). On one side they bear inscriptions graven with a pointed instrument, on the other the surface is a maze of cracks due to heat. They would once have belonged to soothsayer-priests attached to the court of the Shang kings (seventeenth–eleventh centuries BCE) and would have been used in divination by fire.[†]

The writing on them is probably pictographic in origin, and seems to have reached a well-developed stage since it is no longer purely pictographic nor purely ideographic. The basis of the ancient Chinese writing in fact consists of a few hundred basic symbols which represent ideas or simple objects, and also of a certain number of more complicated symbols composed of two elements, of which one relates to the spoken form of a name and the other is visual or symbolic.[‡] It represents a rather advanced stage of graphical representation (Fig. 21.16). "The stylisation and the economy of means are so far advanced in the oldest known Chinese writings that the symbols are more letters than drawings" [J. Gernet (1970), p. 31].

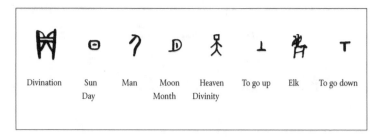

| Divination | Sun | Man | Moon | Heaven | To go up | Elk | To go down |
| | Day | | Month | Divinity | | | |

FIG. 21.16. *Some archaic Chinese characters*

---

* Village in the northwest of the An'yang district in the province of Henan

† According to H. Maspéro, this ritual took place as follows. Ancestor worship was of great importance in Chinese religion, and the priests consulted the royal ancestors on a great diversity of subjects. They first inscribed their questions on the ventral side of a tortoise shell which had been previously blessed (or on one side of the split shoulder-blade of a stag, of an ox or of a sheep). They then brought the other side towards the fire and the result of the divination was supposed to be decipherable from the patterns of cracks produced by the fire.

‡ "The peculiarities of the Chinese language may possibly explain the creation and persistence of this very complicated writing system. In ancient times, the language seems to have consisted of monosyllables of great phonemic variety, which did not allow the sounds of the language to be analysed into constituents, so Chinese writing could not evolve towards a syllabic notation, still less towards an alphabetic one. Each written sign could correspond to a single monosyllable and a single linguistic unit." (J. Gernet).

Gernet continues: "Moreover, this writing abounds in its very constitution with abstract elements (symbols reflected or rotated, strokes that mark this or that part of a symbol, representations of gestures, etc.) and with compounds of simpler signs with which new symbols are created."

The numerals, in particular, seem to have already embarked on the road towards abstract notation and appear to reflect a relatively advanced intellectual perspective.

In this system, unity is represented by a horizontal line, and 10 by a vertical line. Their origin is clear enough, since they reflect the operation of the human mind in given conditions: we know, for instance, that the people of the ancient Greek city of Karystos, and the Cretans, the Hittites and the Phoenicians, all used the same kind of signs for these two numerals. A hundred is denoted by what Joseph Needham called a "pine cone", and a thousand by a special character which closely resembles the character for "man" in the corresponding writing.

The figures 2, 3 and 4 are represented each by a corresponding number of horizontal strokes: an old ideographic system which is not used for the figures from 5 onwards. Like all the peoples who have used a similar numerical notation, the Chinese also stopped at 4; in fact few people can at a glance (and therefore without consciously counting) recognise a series of more than four things in a row. The Egyptians continued the series from 4 by using parallel rows, and the Babylonians and Phoenicians had a ternary system, but the Chinese introduced five distinct symbols for each of the five successive numbers: symbols, apparently, devoid of any intuitive suggestion. The number 5 was represented by a kind of X closed above and below by strokes; the number 6, by a kind of inverted V or by a design resembling a pagoda; 7, by a cross; 8, by two small circular arcs back to back; and the number 9 by a sign like a fish-hook (Fig. 21.17).

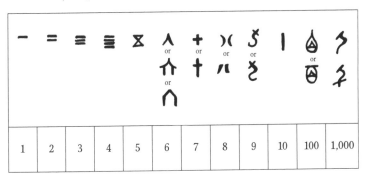

FIG. 21.17. *The basic signs of archaic Chinese numerals. They have been found on divinatory bones and shells from the Yin period (fourteenth to eleventh centuries BCE), and also on bronzes from the Zhou period (tenth to sixth centuries BCE). [Chalfant (1906); Needham (1959); Rong Gen (1959); Wieger (1963)]*

Now, did these number-signs evolve graphically from forms which originally consisted of groupings of corresponding numbers of identical elements? Or are they original creations? The history of Chinese writing leads us to form two hypotheses about these questions, both of them plausible, and not incompatible with each other.

We may in fact suppose that, for some of these numbers, their signs were, more or less, "phonetic symbols" which were used for the sake of the sounds they stood for, independently of their original meaning just as, indeed, was the case for Chinese writing. Such, for example, may well be why the number 1,000 has the same representation as "man", since the two words were probably pronounced in the same way at the time in question.

Another possible explanation may be of religious or magical origin, and may have determined the choice of the other symbols. Gernet (*EPP*) writes: "From the period of the inscriptions on bones and tortoiseshells at the end of the Shang Dynasty until the seventh century BCE, writing remained the preserve of colleges of scribes, adepts in the arts of divination and, by the same token, adepts also in certain techniques which depended on number, who served the princes in their religious ceremonies. Writing was therefore primarily a means of communication with the world of gods and spirits, and endowed its practitioners with the formidable power, and the respect mingled with dread, which they enjoyed. In a society so enthralled to ritual in behaviour and in thought, its mystical power must have preserved writing from profane use for a very long period."

Therefore it is by no means impossible that certain of the Chinese number-signs may have had essentially magical or religious roots, and were directly related to an ancient Chinese number-mysticism. Each number-sign, according to its graphical form, would have represented the "reality" of the corresponding number-form.

Whatever the case may be, the system of numerals which may be seen in the divinatory inscriptions on the bones and tortoise shells from the middle of the second millennium BCE is, intellectually speaking, already well on the way to the modern Chinese number-notation.

FIG. 21.18A. *Copy of a divinatory inscription on the ventral surface of a tortoise shell discovered at Xiao dun, which dates from the Yin period (fourteenth to eleventh centuries BCE). [Diringer (1968) plate 6-4: Yi 2908, translated and interpreted by L. Vandermeersch]*

TRANSLATION

MARGINAL NOTES*

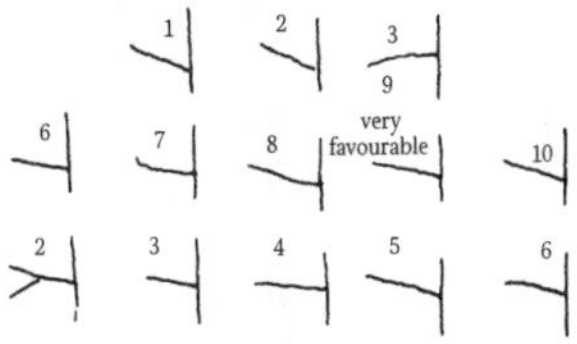

THE QUESTION

Augury of the day Wuwu by the god Ke on the question:
– Should we go hunting at Gui [a place-name]?
– What success shall we have?

THE ANSWER

– Today (after consultation with the ancestors) we have been hunting and we have taken the following prey: 1 tiger, 40 stags, 164 foxes (?), 159 fawns and 18 pheasants with a double pair of red streaks (?)

* The figures number the different parts of the tortoise shell, no doubt to show the order in which to read the cracks made by the heat. The character shown on the ninth corresponds to good prospects.

TALLY OF CAPTURES (LITERALLY)

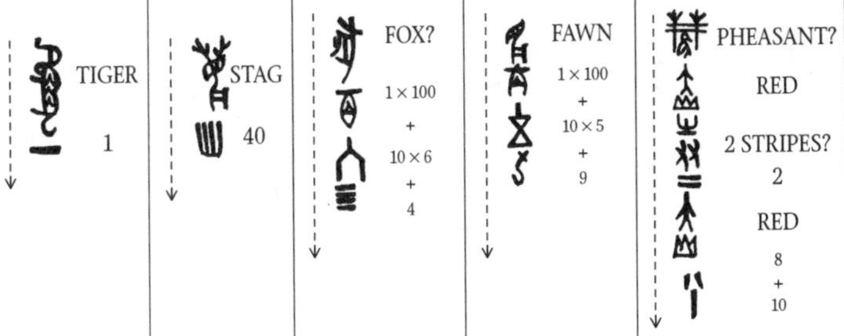

TIGER	STAG	FOX?	FAWN	PHEASANT?
1	40	1×100 + 10×6 + 4	1×100 + 10×5 + 9	RED 2 STRIPES? 2 RED 8 + 10
1 Tiger	**40** Stag	**164** Fox	**159** Fawn	**18** Pheasant with a double pair of red stripes (?)

FIG. 21.18B.

Leaving aside the numbers 20, 30 and 40 (to which we shall shortly return), the tens, hundreds and thousands are in fact represented according to the multiplicative principle by combining the signs corresponding to the units associated with them: in other words, the numbers from 50 to 90, for instance, are represented by superpositions according to the principle:

10	10	10	10	10
$\times$	$\times$	$\times$	$\times$	$\times$
5	6	7	8	9

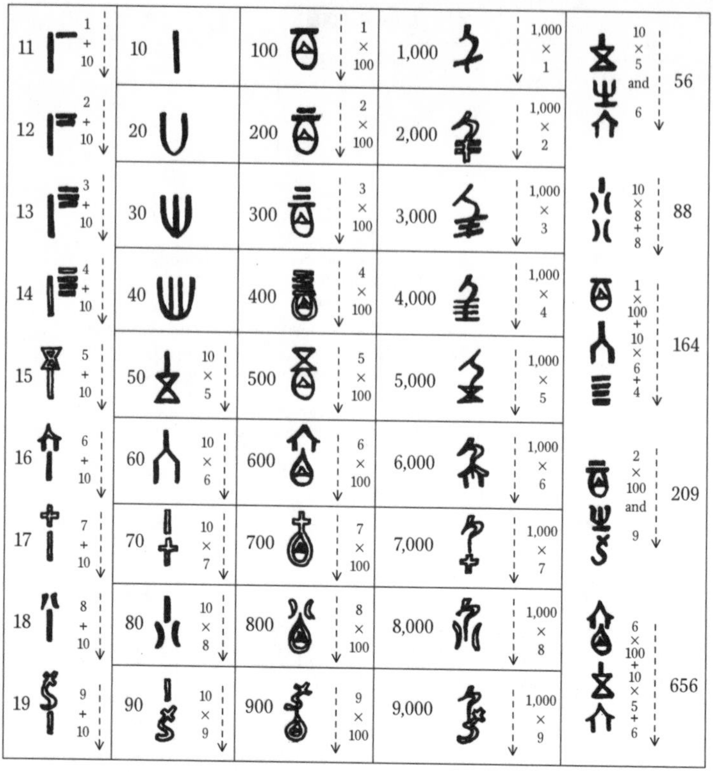

FIG. 21.19. *Principle of archaic Chinese numbering*

This representation should not be confused with the one used for the numbers 15 to 19, which was:

5	6	7	8	9
+	+	+	+	+
10	10	10	10	10

The numbers from 100 to 900 were written by placing the symbols for the successive units above the symbol for 100, and the thousands were written in a similar way to the tens (Fig. 21.19). Intermediate numbers were usually written by combining the additive and multiplicative methods.

We therefore see that, since the time of the very earliest known examples, the Chinese system was founded on a "hybrid" principle. That the numbers 20, 30 and 40 were often written as requisite repetitions of the symbol for 10 is quite simply due to the fact that the use of the multiplicative method would not have made the result any simpler. This kind of ideographic notation, natural though it was, was nevertheless limited, for psychological reasons, to a maximum of four identical elements.

The structure of the Chinese numerals stayed basically the same throughout its long history, even though the arrangement of the signs changed somewhat and their graphical forms underwent some variations (see Fig. 21.17, 21.21, then 21.9 and finally 21.15).

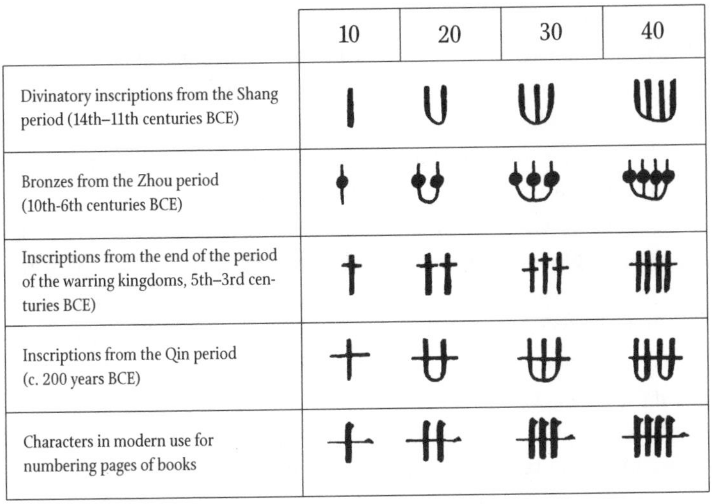

	10	20	30	40
Divinatory inscriptions from the Shang period (14th–11th centuries BCE)				
Bronzes from the Zhou period (10th-6th centuries BCE)				
Inscriptions from the end of the period of the warring kingdoms, 5th–3rd centuries BCE)				
Inscriptions from the Qin period (c. 200 years BCE)				
Characters in modern use for numbering pages of books				

FIG. 21.20. *The durability of the ideographic forms of the first four numbers, as seen throughout the history of Chinese numerals*

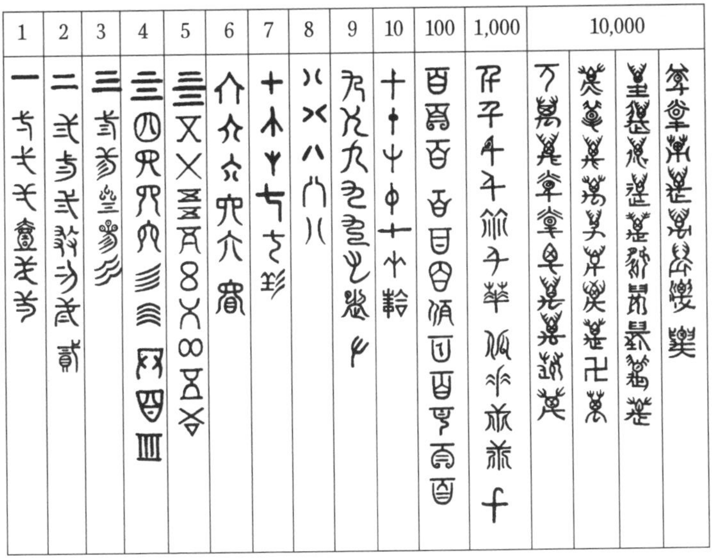

FIG. 21.21. *Variations in the graphical forms of the Chinese numerals, as found on inscriptions from the end of the period of the warring kingdoms (fifth to third centuries BCE). [Perny (1873); Pihan (1860)]*

THE SPREAD OF WRITING THROUGHOUT
THE FAR EAST

Over all the centuries, the structure of the Chinese characters has not fundamentally changed at all. The Chinese language is split into many regional dialects, and the characters are pronounced differently by the people of Manchuria, of Hunan, of Peking, of Canton, or of Singapore. Everywhere, however, the characters have kept the same meanings and everyone can understand them.

For example, the word for "eat" is pronounced *chi* in Mandarin and is written with a character which we shall denote by "A". In Cantonese, this character is pronounced like *hek* but the Cantonese word for "eat" is pronounced *sik* and itself is represented by a character which we shall denote by "B". Nevertheless, all educated Chinese – even if in their dialect the word for "eat" is pronounced neither *chi* nor *sik* – readily understand the characters "A" and "B", which both mean "eat". [V. Alleton (1970)]

Chinese writing is therefore, in the words of B. Karlgren, a visual Esperanto: "The fact that people who are unable to communicate by the spoken word can understand each other when each writes his own language in Chinese characters has always been seen as one of the most remarkable features of this graphical system." [V. Alleton (1970)] We can easily understand why it is that some of China's neighbours have adopted this writing system for their own languages.

NUMERALS OF THE FORMER KINGDOM OF ANNAM

This last was especially the case for the literate people of Annam (now Vietnam). They considered that the Chinese language was superior to their own, richer and more complete, and they adopted the Chinese characters as they stood but pronounced them in their own way (called "Sino-Annamite"). This gave rise to the Vietnamese writing called *chữ' nôm* (meaning "letter writing").

The Chinese numerals were also borrowed at the same time, and were read as follows in the Sino-Annamite pronunciation (*sô dêm táu*) which derived from an ancient Chinese dialect (Fig. 21.22).

一	二	三	四	五	六	七	八	九	十	百	千	萬
nhât	*nhi*	*tam*	*tír*	*ngũ*	*luc*	*thàt*	*bát*	*ciru*	*thập*	*bách*	*thiên*	*vạn*
1	2	3	4	5	6	7	8	9	10	100	1,000	10,000

Fig. 21.22. *The Chinese numerals and the Sino-Annamite names of the numbers [Dumoutier (1888)]*

In the present day, and since a date usually taken as the end of the thirteenth century CE, for most purposes (including letters, contracts, deeds and popular literature) the numerals are made in the *chũ' nôm* writing which is perfectly adapted to the Annamite number-names (the *sô dêm annam* system) (Fig. 21.23).

没	仁	巴	罘	南	叟	毕	釟	尬	进	扉	新	闎
một	*hai*	*ba*	*bôn*	*năm*	*sáu*	*bảy*	*tám*	*chín*	*muòi*	*trăm*	*nghìn*	*muôn*
1	2	3	4	5	6	7	8	9	10	100	1,000	10,000

FIG. 21.23. Chũ' nôm *numerals and the Annamite names of the numbers. [Dumoutier (1888); Fossey (1948)]*

Although they look different from their Chinese prototypes, these numerals are in fact made up by combining a Chinese character (generally one of the Chinese numerals) as an ideogram, with some element of a character (or the whole character) chosen to represent the pronunciation of the pure Annamite number which is to be written (Fig. 21.24).

figures	2	3	4	5	6	7	8	9	10	100	1,000	10,000
Chinese	二	三	四	五	六	七	八	九	十	百	千	万
chũ' nôm	仁	巴	罘	南	叟	毕	釟	尬	进	扉	新	闎

FIG. 21.24.

This changed nothing in the number-system itself, which continued to follow the Chinese rule of alternating digit and decimal order of magnitude, as in Fig. 21.25.

叟	*sáu*	6
		×
新	*nghìn*	1,000
		+
罘	*bôn*	4
		×
扉	*trăm*	100
		+
尬	*chín*	9
		×
进	*muòi*	10
		+
釟	*tám*	8

FIG. 21.25.

Chinese characters were however abandoned in Vietnam at the start of the twentieth century in favour of an alphabetic system of Latin origin. The Annamite number-names (which are the only ones in current use) are either spelled out using Latin letters or are represented by Arabic numerals.

JAPANESE NUMERALS

The Japanese also borrowed Chinese writing. However, according to M. Malherbe (1995), this was ill-adapted to the multiple grammatical suffixes of Japanese which are intrinsically incapable of ideographic representation. Therefore the Japanese early adopted (around the ninth century) a mixed system based on the following principle:

> Whatever corresponds to an idea is rendered by one of the Chinese *kanji* ideograms [the *kanji* system has been simplified to the point that there now remain only 1,945 official *kanji* characters, plus 166 for personal names, of which 996 are considered essential and are taught as part of primary education]. The more complicated ideograms have fallen into disuse and have been replaced by the *hiragana* characters.
>
> *Hiragana* is a syllabary: there are fifty-one signs, each of which represents a syllable, and not a letter as in the case of our alphabet. This can represent all the grammatical inflections and endings and, indeed, anything which cannot be written using ideograms.
>
> *Katakana* is a syllabary which exactly matches the *hiragana* but is used for recently imported foreign words, geographical names, foreign proper names, and so on.
>
> Finally, the *rômaji*, that is to say our own Western alphabet, is used in certain cases where using the other systems would be too complicated. For example, in a dictionary it is much more convenient to arrange the Japanese words according to the alphabetical order of their transcriptions into Latin characters.
>
> This writing system, which is the most complicated in the world, is regarded as inviolable by the Japanese who would consider themselves cut off from their culture if they gave themselves over to the use of *rômaji*, even though this would cause no practical difficulties nor inconvenience. [M. Malherbe (1995)]

The traditional Japanese numerals continue to be used despite the growing importance of Arabic numerals; they are the same as the Chinese numerals, in all their diverse forms (classical, cursive, commercial, etc.).

However, they are not pronounced as in Chinese. There are two different pronunciations: one is the "Sino-Japanese" which is derived from their

Chinese pronunciation at the time when these characters were borrowed into Japanese; the other is "Pure Japanese".

The Japanese language therefore has two completely different series of number-names which still exist side by side.

The "Pure Japanese" system is a vestige of the ancient indigenous number-system. It consists of an incomplete list of names, which have short forms and complete forms (Fig. 21.26).

	Short forms			Full forms	
1	*hi-*	or	*hito-*	*hitotsu*[a]	*hitori*[b]
2	*fu-*	or	*futa-*	*futatsu*[a]	*futari*[b]
3	*mi-*			*mitsu*[a]	*mitari*[b]
4	*yo-*			*yotsu*[a]	*yotari*[b]
5	*itsu-*			*itsutsu*	
6	*mu-*			*mutsu*	
7	*nana-*			*nanatsu*	
8	*ya-*			*yatsu*	
9	*kokono-*			*kokonotsu*	
10	*tō*				

a. The number-names ending in -tsu are only used to refer to objects
b. The number-names ending in -tari are only used to refer to persons

FIG. 21.26. *The Pure Japanese names of numbers. [Frédéric (1994 and 1977–87); Haguenauer (1951); Miller (1967); Plaut (1936)]*

Only the first four number-names have the ending -*tari* when applied to persons. From five persons upwards the base forms are used, which have neither inflection nor gender. This provides another instance of the psychological phenomenon described in Chapter 1, that only four items can be directly perceived.

The name of the number 8 also means "big number" and occurs in numerous locutions which express great multiplicity. So, where we for instance would say "break into a thousand pieces", the Japanese say

literally: "break into 8 pieces"

yatsuzaki

FIG. 21.27.

A market greengrocer – who sells every kind of fruit and vegetable – is likewise called

literally: [the man who sells] 800 kinds of produce

yaoya

FIG. 21.28.

The city of Tokyo, which is of enormous extent, used to be called

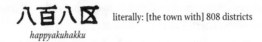

八百八区   literally: [the town with] 808 districts

happyakuhakku

Fig. 21.29.

And to indicate the innumerable gods of their Shintô religion, the Japanese say

八百万の神   literally: 8 million gods

happyakuman no kami

Fig. 21.30.

As C. Haguenauer (1951) points out for the Pure Japanese number-names, there is a clear relation between the odd forms and the even forms, in the series "one–two" [*hito–futa*] and "three–six" [*mi–mu*], and an equally clear one between the even numbers four and eight [*yo–ya*]. The even numbers 2 and 6 have been obtained from the corresponding odd numbers by simple sound changes. In the latter case, a mere change of vowel makes the difference between "four" [*yo*] and "eight" [*ya*]. At first sight, only *i.tsu*, "five", and *tô*, "ten" are exceptions – as well, of course, as the odd numbers greater than 5. (Fig. 21.31)

1	*hito* ≈ *hi*	←----------------→ 2×1	2	*futa* ≈ *fu*
3	*mi*	←----------------→ 2×3	6	*mu*
4	*yo*	←----------------→ 2×4	8	*ya*

Fig. 21.31.

This could indicate that long ago, among the indigenous peoples of Japan, the series of numbers came to a second break at 8 (the sequence 1, 2, 3, 4 being extended up to 8 by the additive principle: $5 = 3 + 2$, $6 = 3 + 3$, $7 = 4 + 3$, $8 = 4 + 4$).

In the aboriginal Japanese number-system there were also special names for some orders of magnitude above 10: a word for 20 (whose root is *hat'*) and individual names for 100 (*momo*), 1,000 (*chi*) and 10,000 (*yorozu*).

Nowadays, however, this system has been reduced to the barest minimum and is only now used for numbers between 1 and 10. The words for higher numbers have mostly fallen out of use except for the word for 20 (still used for lengths of time) and the word for 10,000 (sometimes used for the number itself, but most often simply to mean a boundless number).

The second of the Japanese number-systems has considerably greater capability than the one we have just looked at. It has a complete set of names for numbers, as follows:

1	*ichi*	10		*jû*
2	*ni*	100	(= 10^2)	*hyaku*
3	*san*	1,000	(= 10^3)	*sen*
4	*shi*	10,000	(= 10^4)	*man*
5	*go*			
6	*roku*			
7	*shichi*			
8	*hachi*			
9	*ku*			

FIG. 21.32. *The Sino-Japanese number-names. [Haguenauer (1951); Miller (1967); Plaut (1936)]*

The numbers from 11 to 19 are represented according to the additive principle:

11	*jû.ichi*	ten-one	= 10 + 1
12	*jû.ni*	ten-two	= 10 + 2
13	*jû.san*	ten-three	= 10 + 3

For the tens, hundreds and thousands, and so on, it used the multiplicative principle:

20	*ni.jû*	two-ten	= 2 × 10
30	*san.jû*	three-ten	= 3 × 10

....

100	*hyaku*	hundred= 10^2	
200	*ni.hyaku*	two-hundred	= 2 × 100
300	*san.hyaku*	three-hundred	= 3 × 100

....

1,000	*sen*	thousand	= 10^3
2,000	*ni.sen*	two-thousand	= 2 × 1,000
3,000	*san.sen*	three-thousand	= 3 × 1,000

....

10,000	*ichi.man*	myriad	= 10^4
20,000	*ni.man*	two-myriad	= 2 × 10,000

五 萬		三 千		六 百		八 十		一
go.man		*san.sen*		*roku.hyaku*		*hachi.jû*		*ichi*
("five-myriad		three-thousand		six-hundred		eight-ten		one")
(= 5 × 10,000	+	3 × 1,000	+	6 × 100	+	8 × 10	+	1)

53,681

FIG. 21.33.

The word for 10,000 in Sino-Japanese is *man*. Previously, *ban* was also used but nowadays it is only used in the sense of "unlimited number" or, rather, "maximum". While *sen.man* means "a thousand myriad", namely 10,000,000, its obsolete homologue *sen.ban* nowadays means "in the highest degree" or "extremely". The famous Japanese war-cry *banzai*, "long life (to) ..." (to the Emperor, is understood), is made up of *ban*, "10,000", and *zai*, a modification of *sai*, "life". On its own, the word also means "bravo", in the sense that "for what you are doing you deserve to live ten thousand years!"

This oral number-system is of Chinese origin and so it is called the Sino-Japanese system. It long ago displaced the old Pure Japanese system whose structure was rather complicated. The changeover took place under the influence of Chinese culture and manifested itself not only in the disappearance of the number-names for the indigenous numbers above 10, but also by the adoption of the Chinese characters which express the names of these numbers; these characters are, of course, pronounced in the Japanese way. This is the reason why there are two systems in use together.

Two parallel systems are also used in Korea. In the aboriginal, true Korean system it is only possible to count up to 99, and it is only written in *hangŭl* (a Korean alphabet which has nothing to do with Chinese or Japanese writing and was created in 1443 by King Sezhong of the Yi Dynasty). The second, Sino-Korean system was derived from Chinese and allows arbitrarily large numbers; it is written with characters of Chinese origin or by means of Arabic numerals [see J. M. Li (1987)].

CUSTOM AND SUPERSTITION:
LINGUISTIC TABOOS

For numbers from 1 to 10, the Sino-Japanese system is only used in special circumstances, but is used without exception for larger numbers. In conversation, however, the Japanese often use both systems at the same time.

The main reason for this is the speaker's desire to make sure that the listener does not misunderstand. Since different words often sound alike in Japanese, ambiguity can only be avoided by careful choice of words.

This can be seen in the following examples (Fig. 21.34 and 21.35).

The word for "evening" is *ban*. For "one evening" one would say *hito.ban* and not *ichi.ban* since the latter spoken words may also mean "ordinal number" or "first number".

Similarly, *jû.nana* (combining the Sino-Japanese for 10 with the Pure Japanese for 7) can be heard more clearly than *jû.shichi* (in which both elements are Sino-Japanese) and so is more commonly used for 17; and for the same reason 70 is pronounced *nana.jû* and not *shichi.jû*. For 4,000, the indigenous word *yon* for 4 is combined with the Sino-Japanese *sen* for 1,000 in saying *yon.sen* rather than *shi.sen*. C. Haguenauer (1951) also gives the following examples:

To say:	A Japanese would never use the form:	He would rather use the word or expression:
4	shi	yo
7	shichi	nana
9	ku	kokono
14	jû.shi	jû.yon
17	jû.shichi	jû.nana
40	shi.jû	yon.jû
42	shi.jû.ni	yon.jû.ni
47	shi.jû.shichi	yon.jû.nana
70	shichi.jû	nana.jû
400	shi.hyaku	yon.hyaku
4,000	shi.sen	yon.sen
7,000	shichi.sen	nana.sen

FIG. 21.34.

However, concern for clarity is not the whole story. Another reason is that the Japanese have always had scrupulous respect for certain linguistic taboos imposed by mystical fears.

In Japan, a "name" (in the widest sense of the term) has a very special significance. The sound of the name, it is held, is produced by the action of motive forces which, indeed, are the very essence of the name, so to pronounce a name is not merely to utter some expression but also – and above all – is to set in motion forces which may have malign powers. This is an ancient and universal belief: to name a being or a thing is to assume power over it; to pronounce a name, or even to utter a sound resembling the name of some malevolent spirit, is to risk awakening its powers and suffering their evil effects. We can therefore understand why the Japanese have attached such importance to precision of utterance and why they take such trouble to avoid using a name which might resemble the sound of a name of evil import.

In addition to this, there are mystical reasons. Numbers, in Japan as elsewhere, have hidden meanings. The Japanese even today still have a degree of numerical superstition, manifest as a respect or even an instinctive fear for certain numbers such as 4 or 9. Try to park your car in bays 4, 9, 14, 19, or 24 of a Tokyo car park: you may locate these places, perhaps, if the secret of perpetual motion is ever discovered. Seat number 4 in a plane of Japanese Airlines, rooms 304 or 309 of a hotel – these can hardly ever be found (still less in a hospital!). Simply because the number in "Renault 4" has always been one of the most menacing, the Japanese launch of this car failed miserably.

This superstition originates in an unfortunate coincidence of sound (resulting from the adoption of the Chinese number-system and its development according to the rules for reading and writing Sino-Japanese). In the Sino-Japanese system, the word for 4 is *shi* which has the same sound as the word for death. Therefore the Japanese recoil from using the Sino-Japanese word for 4, usually using the Pure Japanese word *yo-*. For 9, the Sino-Japanese word is *ku*, with the same sound as the word for pain. Throughout the Far East, including Japan, the ills of the human race are popularly attributed to Spirits of Evil which breathe their poisoned breath all round. Always meticulous about their health, the Japanese therefore sought to avoid attracting the malign attention of these spirits by avoiding the use of this word for 9, using instead the indigenous word *kokono-*.

For exactly the same reason, 4,000 is spoken as *yon.sen* rather than *shi.sen* which has the same sound as the expression for "deadly line"; for "four men" they say *yo.nin* and not *shi.nin* which also means "death" or "corpse". The indigenous word *nana* for "seven" is preferred to the numeral *shichi* (7) because the latter might be mistaken for *shitou* which means death, or loss. Finally, 42 is never spoken as *shi.ni* (a simplified expression: "four-two") nor *shi.jû.ni* (= 4 × 10 + 2), because of the dread presence of " death" in the name of the number 4 as *shi* in each case. There is a further reason: in the first form, the listener may hear *shin.i* – "occurrence of death"; in the second form we also have the name of "42 years of age" which is held to be an especially dangerous age for a man. This number is therefore usually expressed as *yon.jû.ni*.

It is a strange paradox that a civilisation which is at the forefront of science and technology has preserved the fears and superstitions of thousands of years ago, and that there is no thought that these should be overturned.

	NUMERALS OF CHINESE ORIGIN				READ AS:		
	Standard forms	Cursive forms	Calligraphic forms	Commercial forms	Sino-Japanese	Pure Japanese short	Pure Japanese complete
1	一	一	一	〡	*ichi*	*hi-, hito-*	*hitotsu*
2	二	二 or [illegible]	二	〢	*ni*	*fu-, futa-*	*futatsu*
3	三	三 or [illegible]	三	〣	*san*	*mi-*	*mitsu*
4	四	四	四	〤	*shi*	*yo-*	*yotsu*
5	五	五	五	〥 or [illegible]	*go*	*itsu-*	*itsutsu*
6	六	六	六	〦	*roku*	*mu-*	*mutsu*
7	七	七	七	〧	*shichi*	*nana-*	*nanatsu*
8	八	八	八	〨	*hachi*	*ya-*	*yatsu*
9	九	九	九	〩	*ku*	*kokono-*	*kokonotsu*
10	十	十 or 十	十	十	*jû*	*tō*	
100	百	百	百	[illegible] or [illegible]	*hyaku*		
1,000	千	千	千	千	*sen*		
10,000	萬 or 万	萬	萬	万	*man*		

FIG. 21.35. *Number-names and numerals in current use in Japan*

WRITING LARGE NUMBERS

In everyday use, neither the Chinese nor the Japanese have need of special signs for very large numbers. Using only the thirteen basic characters of their present-day number-system they can write down any number, up to at least a hundred billion (10^{11}).

Although usually only used for numbers up to 10^8, the method they use is a simple extension of their ordinary number-system, namely introducing ten thousand (10^4) as an additional counting unit. The following shows how the Chinese represent consecutive powers of 10 (Fig. 21.36):

10,000 :	yī wàn (=	1 × 10,000)
100,000 :	shí wàn (=	10 × 10,000)
1,000,000 :	yī bǎi wàn (=	1 × 100 × 10,000)
10,000,000 :	yī qiān wàn (=	1 × 1,000 × 10,000)
100,000,000 :	yī wàn wàn (=	1 × 10,000 × 10,000)
1,000,000,000 :	shí wàn wàn (=	10 × 10,000 × 10,000)
10,000,000,000 :	yī bǎi wàn wàn (=	1 × 100 × 10,000 × 10,000)
100,000,000,000 :	yī qiān wàn wàn (=	1 × 1,000 × 10,000 × 10,000)

FIG. 21.36A. *The usual Chinese notation for the successive powers of 10. [Guitel; Menninger (1957); Ore (1948); Tchen Yon-Sun (1958)]*

10^4	一萬 *yī wàn*	1×10^4	10^8	一萬萬 *yī wàn wàn*	$1 \times 10^4 \times 10^4$
10^5	十萬 *shí wàn*	10×10^4	10^9	十萬萬 *shí wàn wàn*	$10 \times 10^4 \times 10^4$
10^6	一百萬 *yī bǎi wàn*	$1 \times 10^2 \times 10^4$	10^{10}	一百萬萬 *yī bǎi wàn wàn*	$1 \times 10^2 \times 10^4 \times 10^4$
10^7	一千萬 *yī qiān wàn*	$1 \times 10^3 \times 10^4$	10^{11}	一千萬萬 *yī qiān wàn wàn*	$1 \times 10^3 \times 10^4 \times 10^4$

FIG. 21.36B.

For a very large number such as 487,390,629, therefore, they would write:

四　萬　八　千　七　百　三　十　九　萬　六　百　二　十　九

sí　wàn　bā　qiān　qī　bǎi　sān　shí　jiǔ　wàn　liù　bǎi　èr　shí　jiǔ

---→

$(4 \times 10^4 + 8 \times 10^3 + 7 \times 10^2 + 3 \times 10 + 9) \times 10^4 + (6 \times 10^2 + 2 \times 10 + 9)$

FIG. 21.37.

decomposing it as

$(4 \times 10,000 + 8 \times 1,000 + 7 \times 100 + 3 \times 10 + 9) \times 10,000 + 6 \times 100 + 2 \times 10 + 9$

or $48,739 \times 10,000 + 629$.

The system just described is in practice the only one used for ordinary purposes. However, though only in scientific and especially astronomical texts, one may encounter special characters for higher orders than 10^4 which can therefore be used to express much larger numbers than are possible with the usual system. However, the signs used have meanings which vary according to which of three value conventions is being used. Each sign may have one of three different values depending on whether it is used on the *xià deng* system ("lower degree"), the *zhōng deng* system ("middle degree") or the *shàng deng* system ("higher degree").

The character 兆, *zhào*, therefore, may represent a million (10^6) in the lower degree, a thousand billion (10^{12}) in the middle degree, and 10^{16} in the higher degree.

In the lower degree (*xià deng*) the system is a direct continuation of the ordinary number-system since the ten successive additional characters are simply the ten consecutive powers of 10 following 10^4, namely

$10^5, 10^6, 10^7, 10^8 \dots, 10^{13}, 10^{14}$

which are represented by the characters

yì, zháo, jing, gai, ... , zheng, zài.

So, written in the lower degree, one million and three million would be written as follows:

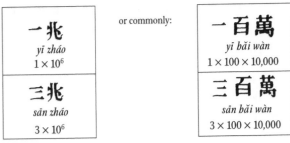

一兆 *yī zháo* 1×10^6	or commonly:	一百萬 *yī bǎi wàn* $1 \times 100 \times 10{,}000$
三兆 *sān zháo* 3×10^6		三百萬 *sān bǎi wàn* $3 \times 100 \times 10{,}000$

FIG. 21.38.

The *xià deng* system therefore allows any number less than 10^{15} to be written down straightforwardly. For example, the number 530,010,702,000,000 would be written as

五	載	三	正	一	壤	七	垓	二	兆
wǔ	*zái*	*sān*	*zheng*	*yī*	*ràng*	*qī*	*gai*	*èr*	*zháo*

$$5 \times 10^{14} + 3 \times 10^{13} + 1 \times 10^{10} + 7 \times 10^{8} + 2 \times 10^{6}$$

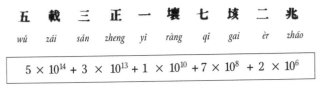

FIG. 21.39.

In the middle system the same ten consecutive characters represent increasing powers of 10 greater than 10^4, but they now increase, not by a factor of 10 each time, but by a factor of 10,000, namely

$10^8, 10^{12}, 10^{16}, \ldots, 10^{40}, 10^{44}$ (Fig. 21.42).

With the convention that two of these characters should never occur consecutively, this system can be used to represent all the numbers less than 10^{48}. For example:

三	百	五	十	壤	七	千	三	百	兆	二	十	六	億
sān	*bǎi*	*wǔ*	*shí*	*ràng*	*qī*	*qiān*	*sān*	*bǎi*	*zháo*	*èr*	*shí*	*liù*	*yì*

$$(3 \times 10^2 + 5 \times 10).\ 10^{28} + (7 \times 10^3 + 3 \times 10^2).\ 10^{12} + (2 \times 10^6 + 6).\ 10^8$$

3,500,000,000,000,007,300,002,600,000,000

FIG. 21.40.

In the higher degree system, only the first three of these ten characters are used, namely *yì*, *zhǎo* and *jing*. These are given the values 10^8, 10^{16} and 10^{32} respectively. With these, it is possible to represent all numbers less than 10^{64}. For example:

三 京 五 千 三 百 一 億 二 百 七 萬 六 千 一 百 八 十 五 兆 三 億 一 萬

sān jing wǔ qiān sān bǎi yì yì èr bǎi qī wàn liù qiān yì bǎi bā shí wǔ zhǎo sān yì yì wàn

$(3 \times 10^{32} + [[5 \times 10^3 + 3 \times 10^2 + 1] \cdot 10^8 + [2 \times 10^2 + 7] \cdot 10^4 + 6 \times 10^3 + 1 \times 10^2 + 8 \times 10 + 5] \cdot 10^{16} + 3 \times 10^8 + 1 \times 10^4$
300,005,301,020,761,850,000,000,300,010,000

FIG. 21.41.

		Xià deng LOWER DEGREE SYSTEM	Zhōng deng MIDDLE DEGREE SYSTEM	Shàng deng HIGHER DEGREE SYSTEM
萬	*wàn*	10^4	10^4	10^4
億	*yì*[a]	10^5	10^8	10^8
兆	*zhǎo*	10^6	10^{12}	10^{16}
京	*jing*	10^7	10^{16}	10^{32}
垓	*gai*	10^8	10^{20}	10^{64}
補	*bù*[b]	10^9	10^{24}	10^{128}
壤	*ràng*	10^{10}	10^{28}	10^{256}
蕫	*gou*[c]	10^{11}	10^{32}	10^{512}
澗	*jiǎn*	10^{12}	10^{36}	10^{1024}
正	*zheng*	10^{13}	10^{40}	10^{2048}
載	*zái*	10^{14}	10^{44}	10^{4096}

THEORETICAL VALUES

[a] Graphical variant 亿　　[b] Equivalent word 溝　　[c] Graphical variant 枺

FIG. 21.42. *Chinese scientific notation for large numbers [Giles (1912); Mathews (1931); Needham (1959)]*

Such very large numbers are, however, very infrequently used: "in mathematics, business or economics numbers greater than 10^{14} are very rare;

only in connection with astronomy or the calendar do we sometimes find larger numbers" [R. Schrimpf (1963–64)].

Finally, let us draw attention to a very interesting notation which Chinese and Japanese scientists have used to express negative powers of 10:

$$10^{-1} = 1/10, \ 10^{-2} = 1/100, \ 10^{-3} = 1/1,000, \ 10^{-4} = 10,000, \text{ etc.}$$

They especially find mention in the arithmetical treatise *Jinkoki* published in 1627 by the Japanese mathematician Yoshida Mitsuyoshi (Fig. 21.43).

分	*fēn*	10^{-1}
厘	*lí*	10^{-2}
毛	*máo*	10^{-3}
糸	*mi*	10^{-4}
忽	*hū*	10^{-5}
微	*wēi*	10^{-6}
纖	*xiān*	10^{-7}
沙	*shā*	10^{-8}
塵	*chén*	10^{-9}
埃	*āi*	10^{-10}

FIG. 21.43. *Sino-Japanese scientific notation for negative powers of 10 [Yamamoto (1985)]*

THE CHINESE SCIENTIFIC POSITIONAL SYSTEM

Further evidence of advanced intellectual development in the Far East comes from the written positional notation formerly used by Chinese, Japanese, and Korean mathematicians.

Though we only know examples of this system dating back to the second century BCE, it seems probable that it goes back much further.

Known by the Chinese name *suan zí* (literally, "calculation with rods"), and by the Japanese name *sangi*, this system is similar to our modern number-system not only by virtue of its decimal base, but also because the values of the numerals are determined by the position they occupy. It is therefore a strictly positional decimal number-system.

However, whereas our system uses nine numerals whose forms carry no intrinsic suggestion of value, this system of numerals makes use of systematic combinations of horizontal and vertical bars to represent the first nine units. The symbols for 1 to 5 use a corresponding number of vertical strokes, side by side, and the symbols for 6, 7, 8, and 9 show a horizontal bar capping 1, 2, 3, or 4 vertical strokes:

FIG. 21.44.

Examples of numbers written in this system are given by Cai Jiu Feng, a Chinese philosopher of the Song era who died in 1230 [in *Huang ji*, in the chapter *Hong fan* of his "Book of Annals", cited by A.Vissière (1892)]. Example:

FIG. 21.45.

Ingenious as it was, this system lent itself to ambiguity.

For one thing, people writing in this system tended to place the vertical bars for the different orders of magnitude side by side. So the notation for the number 12 could be confused with that for 3 or for 21; 25 could be confused with 7, 34, 43, 52, 214, or 223, and so on (Fig. 21.45).

However, the Chinese found a way round the problem, by introducing a second system for the units, analogous to the first but made up of horizontal bars rather than vertical. The first five digits were represented by as many horizontal bars, and the numbers 6, 7, 8, 9 by erecting a vertical bar (with symbolic value 5) on top of one, two, three, or four horizontal bars:

FIG. 21.46.

Then, to distinguish between one order of magnitude and the next, they alternated figures from one series with figures from the other, therefore alternately vertical and horizontal. The units, hundreds, tens of thousands, millions, and so on (of odd rank) were drawn with "vertical" symbols (Fig. 21.44), whereas the tens, thousands, hundreds of thousands, tens of millions, etc. (of even rank) were drawn with "horizontal" symbols (Fig. 21.46), by which means the ambiguities were elegantly resolved (Fig. 21.48).

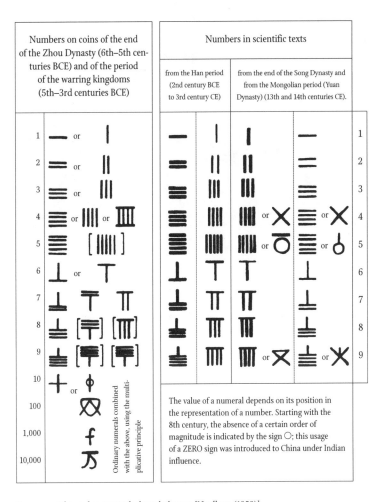

FIG. 21.47. *Chinese bar numerals through the ages [Needham (1959)]*

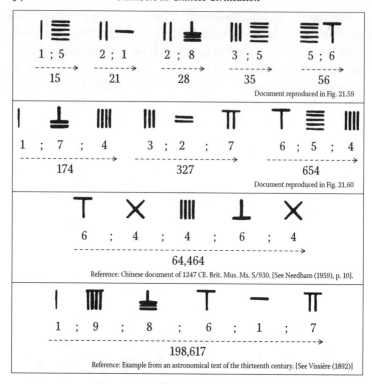

FIG. 21.48. *Examples of numbers written in the Chinese bar notation* (suan zí)

This step was taken at the time of the Han Dynasty (second century BCE to third century CE). This did not solve all the problems there and then, however, since the Chinese mathematicians were to remain unaware of zero for several centuries yet. The following riddle bears witness to this, in the words of the mathematician Mei Wen Ding (1631–1721):

> The character *hai* 亥 has 2 for its head and 6 for its body. Lower the head to the level of the body, and you will find the age of the Old Man of Jiangxian.

In the above, the character playing the main role in the riddle has been written in the *kǎishū* style:

FIG. 21.49.   hai

and the riddle remains obscure since the modern character is not the same shape as it was before. According to Chinese sources, however, the riddle dates from long before the Common Era, originating in the middle of the

Zhou era (seventh to sixth centuries BCE; see Needham (1959), p. 8). And since at that time Chinese characters were drawn in the *dà zhuàn* ("great seal") style, we must therefore see the character in question drawn in this style if we are to solve the riddle.

In this style, the word was written:

FIG. 21.50.   hai

Its "head", therefore, is indeed the figure 2 ▅ , and its lower part is a "body" consisting of three identical signs 佾 each of which resembles the "vertical" symbol for the figure 6 (Fig. 21.47). Arrange the two horizontal lines of the head vertically and on the left-hand side of the body, and you find

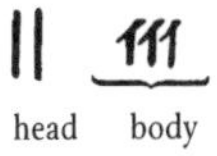

‖ 佾   ‖ T T T
head body   or, nearly enough,   2   6   6   6

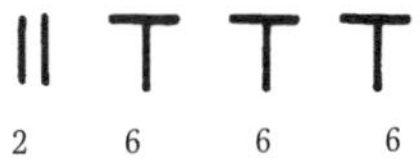

FIG. 21.51.   FIG. 21.52.

The Chinese system being decimal and strictly positional, this represents the number

$$2 \times 1{,}000 + 6 \times 100 + 6 \times 10 + 6 = 2{,}666$$

so the solution of the riddle is the number 2,666. But this cannot be an age in years, unless the Old Man of Jiangxian was a Chinese Methuselah. To consider them as 2,666 days would give an absurd answer, since the "Old Man" would then only be seven and a half years old. In fact, this number system had no zero until much later, so the answer can only be one of the numbers 26,660, 266,600, 2,666,000, etc. But since 266,600 or any higher number is out of the question, we are left with 26,660 days. In the riddle, the number sought does not represent days but tens of days: the Old Man of Jiangxian had lived 2,666 tens of days, or about 73 years.

The lack of a sign to represent missing digits also gave rise to confusion. In the first place, a blank space was left where there was no digit, but this was inadequate since numbers like 764, 7,064, 70,640 and 76,400 could easily be confused:

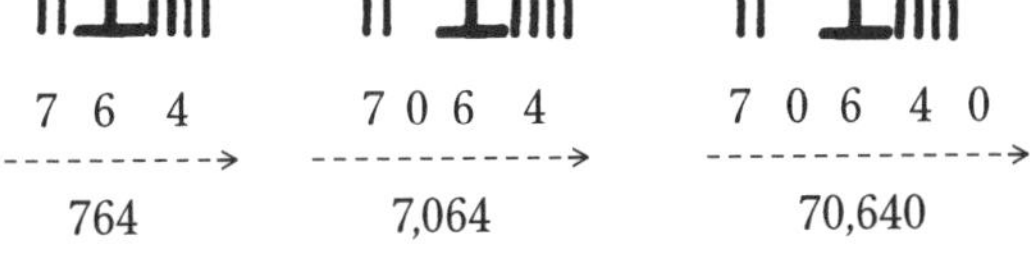

TⅢ   T Ⅲ   T Ⅲ

7 6 4   7 0 6 4   7 0 6 4 0
- - - - - - →   - - - - - - - →   - - - - - - - - →
FIG. 21.53.   764   7,064   70,640

To avoid such ambiguities, some used signs indicating different powers of 10 from the traditional number-system, so that numbers such as 70,640 and 76,400 would be written as:

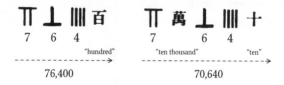

FIG. 21.54.

Others used the traditional expression, therefore writing out in full:

76,400

七	7
	×
萬	10,000
	+
六	6
	×
千	1,000
	+
四	4
	×
百	100

FIG. 21.55.

Yet others placed their numerals in the squares of a grid, leaving an empty square for each missing digit:

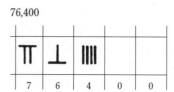 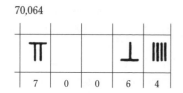

FIG. 21.56.

Only since the eighth century CE did the Chinese begin to introduce a special positional sign (drawn as a small circle) to mark a missing digit (Fig. 21.57); this idea no doubt reached them through the influence of Indian civilisation.

Once this had been achieved, all of the rules of arithmetic and algebra were brought to a degree of perfection similar to ours of the present day.

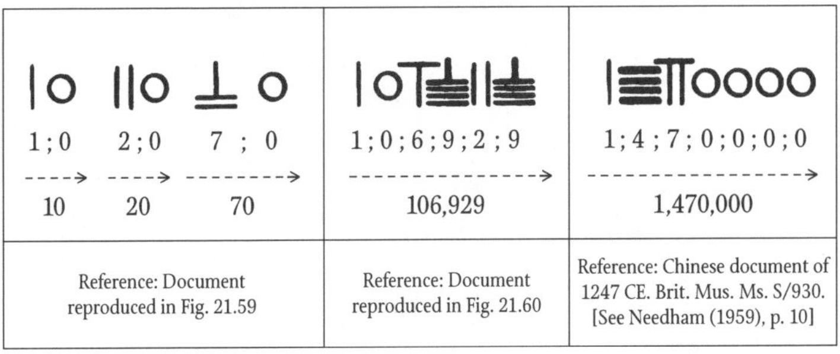

｜○ ‖○ ⊥ ○ 1 ; 0 2 ; 0 7 ; 0 ---→ ---→ --------→ 10 20 70 Reference: Document reproduced in Fig. 21.59	**｜○⊤≣‖≣** 1 ; 0 ; 6 ; 9 ; 2 ; 9 ----------------→ 106,929 Reference: Document reproduced in Fig. 21.60	**｜≣⫪○○○○** 1 ; 4 ; 7 ; 0 ; 0 ; 0 ; 0 -----------------→ 1,470,000 Reference: Chinese document of 1247 CE. Brit. Mus. Ms. S/930. [See Needham (1959), p. 10]

FIG. 21.57. *The use of zero in the Chinese bar numerals*

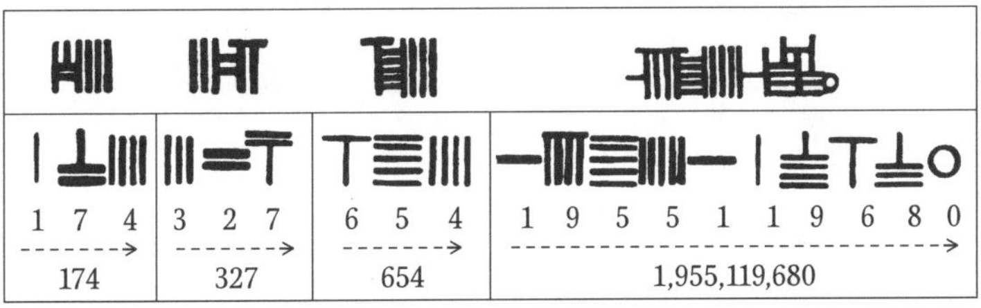

⋕‖‖ ‖⊓⊤ ⊤‖‖‖ ⫪⊤≣‖‖⊢≣			
｜⊥‖‖‖ ‖‖=⊤ ⊤≣‖‖‖ —⫪≣‖‖—｜≟⊤⊥○ 1 7 4	3 2 7	6 5 4	1 9 5 5 1 1 9 6 8 0
------→ 174	------→ 327	------→ 654	--------------------------→ 1,955,119,680

FIG. 21.58. *As a rule, in Chinese manuscripts or printed documents, numbers written in the bar notation are written as monograms, i.e. in a condensed form in which the horizontal strokes are joined to the vertical ones. (Examples taken from the document reproduced in Fig. 21.60)*

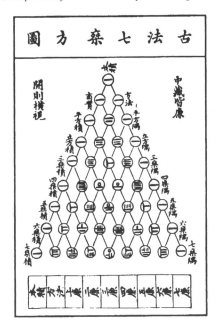

FIG. 21.59A. *Page from a text entitled* Su Yuan Yu Zhian, *published in 1303 by the Chinese mathematician Zhu Shi Jie (see the commentary in the text). [Reproduced from Needham (1959), III, p. 135, Fig. 80]*

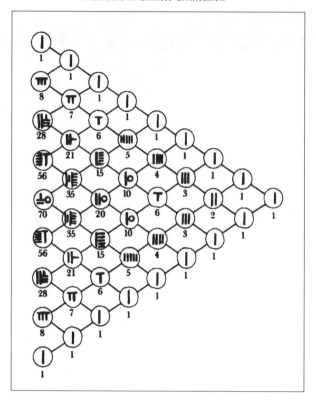

Fig. 21.59B.

Blaise Pascal was long believed in the West to have been the first to discover the famous "Pascal triangle" which gives the numerical coefficients in the expansion of $(a + b)^m$, where m is zero or a positive integer:

BINOMIAL EXPANSIONS	PASCAL'S TRIANGLE
$(a+b)^0 = 1$	1
$(a+b)^1 = a + b$	1　1
$(a+b)^2 = a^2 + 2ab + b^2$	1　2　1
$(a+b)^3 = a^3 + 3a^2b + 3ab^2 + b^3$	1　3　3　1
$(a+b)^4 = a^4 + 4a^3b + 6a^2b^2 + 4ab^3 + b^4$	1　4　6　4　1
$(a+b)^5 = a^5 + 5a^4b + 10a^3b^2 + 10a^2b^3 + 5ab^4 + b^5$	1　5　10　10　5　1
$(a+b)^6 = a^6 + 6a^5b + 15a^4b^2 + 20a^3b^3 + 15a^2b^4 + 6ab^5 + b^6$	1　6　15　20　15　6　1
- →	- - - - - - - - - - - - - - →

In fact, as we can see from Fig. 21.59A, which is schematically redrawn on its side in Fig. 21.59B (to be read from right to left), the Chinese had known of this triangle long before the famous French mathematician.

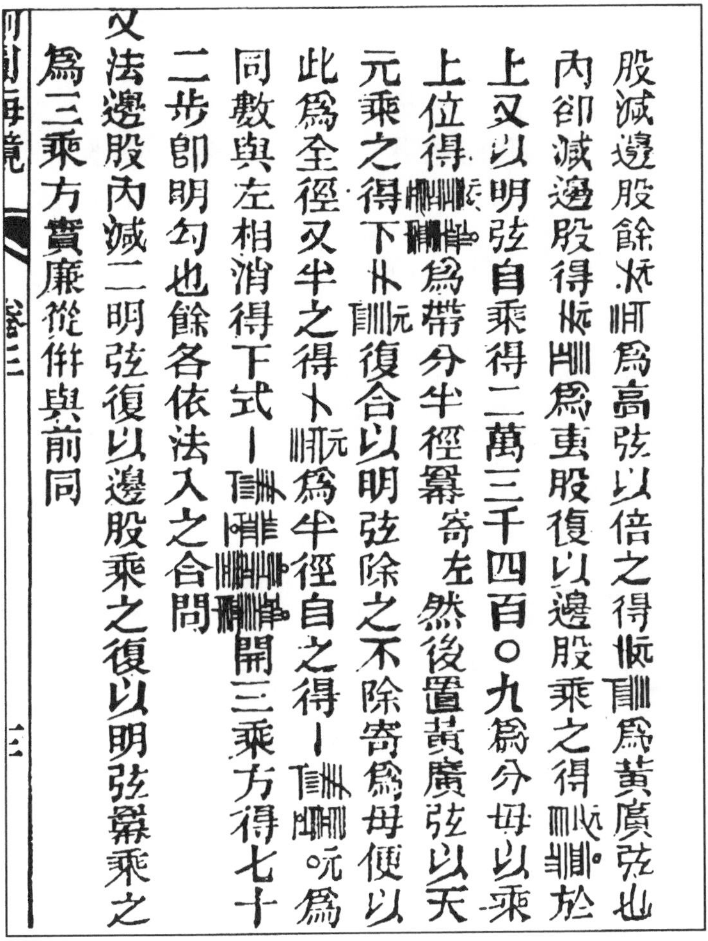

股減邊股餘▯▯為高弦以倍之得▯▯為黃廣弦也
內卻減邊股得▯▯為亩股復以邊股乘之得▯▯於
上又以明弦自乘得二萬三千四百○九為分母以乘
元乘之得▯▯下▯▯復合以明弦除之不除寄為母便以
此為全徑又半之得▯為半徑自之得▯▯元為
同數與左相消得下式▯▯▯開三乘方得七十
二步即明勾也餘各依法入之合問
又法邊股內減二明弦復以邊股乘之復以明弦冪乘之
為三乘方實廉從併與前同

FIG. 21.60. *Extract from* Ce Yuan Hai Jing, *published in 1248 by the mathematician Li Ye.*
[Reproduced from Needham (1959), III, page 132, Fig. 79]

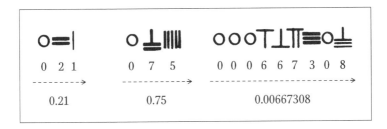

FIG. 21.61. *How Chinese mathematicians extended their positional notation to decimal fractions.*
Reconstructed examples based on a text from the Mongol period: Biot (1839)

EXAMPLES FROM A 13TH-CENTURY CHINESE TREATISE (cf. Fig. 21.60)				EXAMPLES FROM AN 18TH-CENTURY JAPANESE TEXT
𝕂	T≣ℍℍ	|≡T⍺	⌇≣||⌐	⊣|||||T∘∘⊞⊞⊤⊤
	654	1360	1536	152710100928
	------>	----->	------>	------------>
− 2	− 654	− 1,360	− 1,536	− 152,710,100,928

FIG. 21.62A. *Extension of scientific numerical notation to negative numbers. To indicate a negative number, the Chinese and Japanese mathematicians often drew an oblique stroke through the rightmost symbol of the written number. [Menninger (1957); Needham (1959)]*

Polynomial $P(x) = 2x + 654$ cf. Fig. 21.60, col. I		
𝕂元 冒|||	− 2 元 Character representing the variable 654	X 1

Polynomial $P(x) = 2x^2 + 654x$ cf. Fig. 21.60, col. V		
𝕂 冒||元	− 2 654 元 "variable"	X^2 X

Polynomial $P(x) = x^4 - 654x^3 + 106{,}924x^2$ cf. Fig. 21.60, col. VI		
		1	X^4
T≣ℍℍ	− 654	X^3
|∘⊥⊞⊞|||	106,924	X^2
∘ 元 "variable"	0	X
	0	1

Equation $2x^3 + 15x^2 + 166x - 4460 = 0$ cf. J. Needham III, p. 45		
			X^4
||	2	X^3
⊣||||	15	X^2
|⌐T 元 "unknown"	166	X
≣|||⊥⍺太 Character which means "the centre of the earth"	− 4,460	1

FIG. 21.62B. *Notation for polynomials and for equations in one unknown, used by Li Ye (1178–1265)*

THE CHINESE VERSION OF THE RODS ON
THE CHECKERBOARD

Although the numerals discussed above served for writing, they were not used for calculation. For arithmetical calculation, the Chinese used little rods made of ivory or bamboo which were called *chóu* ("calculating rods") which were placed on the squares of a tiled surface or a table ruled like a checkerboard.

FIG. 21.63. *Model of a Chinese checkerboard used for calculation*

The following story from the ninth century CE is evidence in point. It tells how the Emperor Yang Sun selected his officials for their skill and rapidity in calculation.

Once two clerks, of the same rank, in the same service, and with the same commendations and criticisms in their records, were candidates for the same position. Unable to decide which one to promote, the superior officer called upon Yang Sun, who had the candidates brought before him and announced: Junior clerks must know how to calculate at speed. Let the two candidates listen to my question. The one who solves it first will have the promotion. Here is the problem:

A man walking in the woods heard thieves arguing over the division of rolls of cloth which they had stolen. They said that, if each took six rolls there would be five left over; but if each took seven rolls, they would be eight short. How many thieves were there, and how many rolls of cloth?

Yang Sun asked the candidates to perform the calculation with rods upon the tiled floor of the vestibule. After a brief moment, one of the clerks gave the right answer and was given the promotion, and all then departed without complaining about the decision. (See J. Needham in HGS 1, pp. 188–92).

Fig. 21.64. *A Chinese Master teaches the arts of calculation to two young pupils, using an abacus with rods. Reproduced from the* Suan Fa Tong Zong, *published in 1593 in China: [Needham (1959) III, p. 70]*

FIG. 21.65. *An accountant using the arithmetic checkerboard with rods. Reproduced from the Japanese* Shojutsu Sangaka Zue *of Miyake Kenriyû, 1795: (D. E. Smith)*

On an abacus of this kind, each column corresponds to one of the decimal orders of magnitude: from right to left, the first is for the units, the second for the tens, the third for the hundreds, and so on. A given number, therefore, is represented by placing in each column, along a chosen line, a number of rods equal to the multiplicity of the corresponding decimal order of magnitude. For the number 2,645, for example, there would be 5 rods in the first column, 4 in the second, 6 in the third and 2 in the fourth.

For the sake of simplicity, Chinese calculators adopted the following convention (in the words of the old Chinese textbooks of arithmetic): "Let the units lie lengthways and the tens crosswise; let the hundreds be upright and the thousands laid down; let the thousands and the hundreds be face to face, and let the tens of thousands and the hundreds correspond."

The mathematician Mei Wen Ding explains that there was a fear that the different groups might get muddled because there were so many of them. Numbers such as 22 or 33 were therefore represented by two groups of rods, one horizontal and the other vertical, which allowed them to be differentiated. To prevent errors of interpretation, the rods were laid down vertically in the odd-numbered columns (counting from the right), and horizontally in the even-numbered columns (Fig. 21.67).

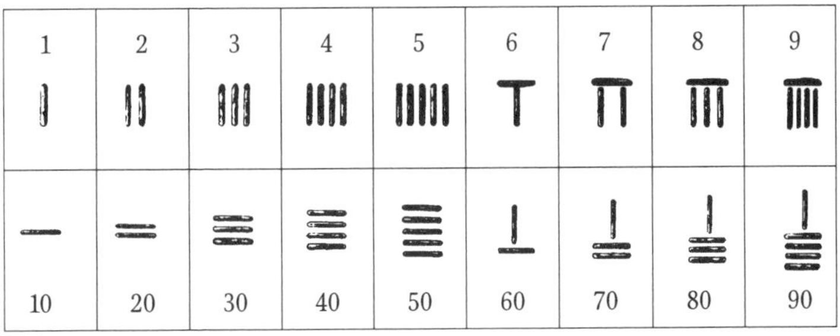

1	2	3	4	5	6	7	8	9
10	20	30	40	50	60	70	80	90

FIG. 21.66. *How the units and tens are represented by rods on the arithmetical checkerboard*

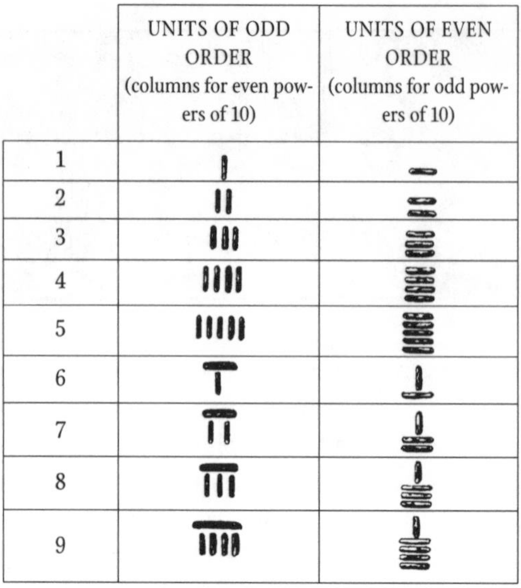

UNITS OF ODD ORDER (columns for even powers of 10)
UNITS OF EVEN ORDER (columns for odd powers of 10)
1
2
3
4
5
6
7
8
9

FIG. 21.67. *The rods are laid vertically for the units, the hundreds, the tens of thousands, and so on; they are laid horizontally for the tens, the thousands, the hundreds of thousands, and so on.*

myriads
thousands
hundreds
tens
units
8 1 2 2 1
81,221
1 1 1 1
1,111
3 (0) 1 (0)
3,010
6 (0) (0) (0)
6,000

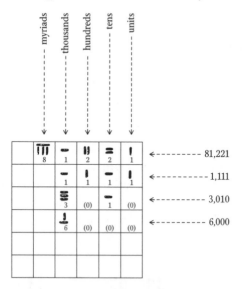

FIG. 21.68. *How certain numbers are represented by laying rods on the checkerboard*

From antiquity until recent times, the Chinese were able to perform every kind of arithmetical operation by means of this device: addition, subtraction, multiplication, division, raising to a power, extraction of square and cube roots, and so on.

The methods used for addition and subtraction were straightforward.

The numbers to be added or subtracted were represented in the squares, and rods were added or removed column by column. Multiplication was almost as simple: the multiplier was placed at the top of the board, with the number to be multiplied placed a few rows lower down. The partial products were then set out on an intermediate line and added in as they were obtained.

For example, to work out the product 736 × 247 (as set out by Yang Hui in the thirteenth century), first of all the two numbers are set out on the board as follows, keeping two empty squares at the right of the multiplier:

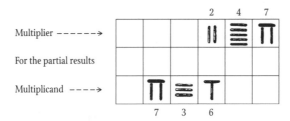

FIG. 21.69A.

Since the multiplier contains three figures, the method proceeds in three stages.

First stage: multiplying 736 by 200

Mentally multiply the 2 of the multiplier by the 7 of the multiplicand, and place the result 14 (in fact 140,000) in the middle line, taking care to place the units of the result above the hundreds of the multiplicand:

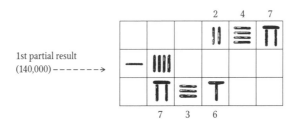

FIG. 21.69B.

Then multiply the 2 of the multiplier by the 3 of the multiplicand, and add the result 6 (in fact 6,000) to the partial result already obtained, placing it on the square to the right of the 4 in 14:

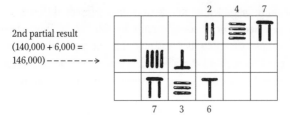

2nd partial result
(140,000 + 6,000 =
146,000) – – – – – – –→

Fig. 21.69c.

Then multiply the 2 of 247 by the 6 of 736, and add this result 12 (in fact 1,200) to the partial result already obtained: in this case, the 2 is placed on the square to the right of the 6 from the preceding stage, and the 1 is placed on the next square to the left thereby being added to the number already there:

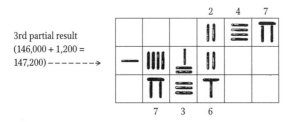

3rd partial result
(146,000 + 1,200 =
147,200) – – – – – – –→

Fig. 21.69d.

Second stage: multiplying 736 by 40

The 2 of the multiplier has now done its work, so it is removed, and the multiplicand is moved bodily one square to the right:

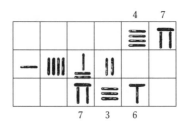

Fig. 21.69e.

Now multiply the 4 of the multiplier by the 7 of the multiplicand, place the result 28 (in fact 28,000) to the partial result in the middle row, and complete the addition:

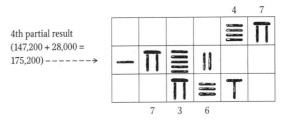

4th partial result
(147,200 + 28,000 =
175,200) ------->

Fig. 21.69F.

Now multiply the 4 by the 3 of 736, and add the result 12 (in fact 1,200) to the middle line:

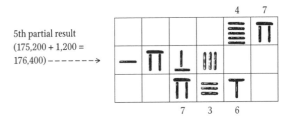

5th partial result
(175,200 + 1,200 =
176,400) ------->

Fig. 21.69G.

Now multiply the 4 by the 6 of 736 and add the result 24 (in fact 240) to the middle line:

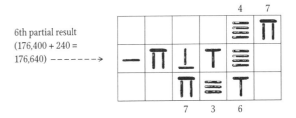

6th partial result
(176,400 + 240 =
176,640) ------->

Fig. 21.69H.

Third stage: multiplying 736 by 7

The 4 of the multiplier has done its work and it too is now removed, and the multiplicand again moved bodily one square to the right:

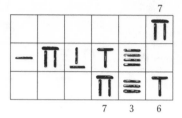

Fɪɢ. 21.69I.

The remaining 7 of the multiplier is now multiplied by the 7 of the multiplicand, and the result 49 (in fact 4,900) is added to the middle line:

7th partial result
(176,640 + 4,900 =
181,540) – – – – – – →

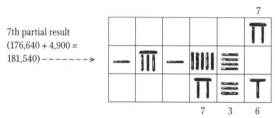

Fɪɢ. 21.69J.

Now multiply the 7 by the 3 of 736, and add the result 21 (in fact 210) to the middle line:

8th partial result
(181,540 + 210 =
181,750) – – – – – – →

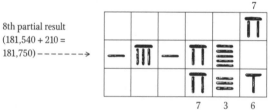

Fɪɢ. 21.69K.

Finally multiply the 7 by the 6 of 736, and add the result 42 to the middle line. This gives the following tableau, where the middle line shows the result of the multiplication (736 × 247 = 181,792):

Final result
(181,750 + 42 =
181,792) – – – – – – →

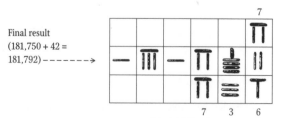

Fɪɢ. 21.69L.

Division was carried out by placing the divisor at the bottom and the dividend on the middle line. The quotient, which was placed at the top, was built up by successively removing partial products from the dividend.

On this numerical checkerboard it was also possible to solve equations, and systems of algebraic equations in several unknowns. The *Jiu Zhang Suan Shu* ("Art of calculation in nine chapters"), an anonymous work compiled during the Han Dynasty (206 BCE to 220 CE), gives much detail about the latter. Each vertical column is associated with one of the equations, and each horizontal row is associated with one of the unknowns, with the co-efficient of an unknown in an equation being placed in the square where the row intersects the column. Also, for this purpose, as well as the ordinary rods (reserved for "true" (*zheng*) numbers, i.e. positive numbers), black rods were used for negative numbers (*fu*: "false" numbers).

A system of equations such as the following, for example:

$$2x - 3y + 8z = 32$$
$$6x - 2y - z = 62$$
$$3x + 21y - 3z = 0$$

was therefore represented as:

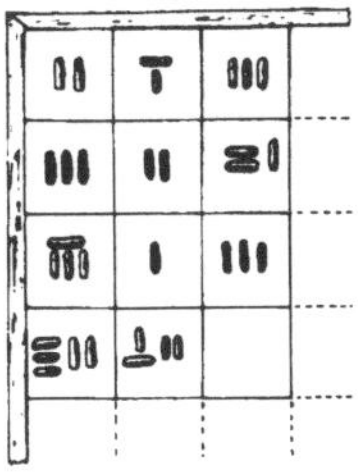

Fig. 21.70A.

The representation of a system of three equations in three unknowns on the arithmetical checkerboard. (From a treatise on mathematics of the Han period: 206 BCE to 220 CE): The first column on the left represents
$2x - 3y + 8z = 32;$
the second column represents
$6x - 2y - z = 62;$
the third column represents
$3x + 21y - 3z = 0.$

x	2	6	3
y	-3	-2	21
z	8	-1	-3
	32	62	0

Fig. 21.70B.

It could be solved quite easily by skilful manipulation of the rods.

This system of numerals is of particular interest for the history of numerical notation, since it is what led to the discovery of the principle of position by the Chinese.

Their system of writing numbers with vertical and horizontal strokes was simply the written copy of the way numbers were represented by rods on the abacus, where the different decimal orders of magnitude progressed in decreasing order from left to right. Once a calculation had been completed on the abacus by manipulation of the rods, their disposition on the abacus could be copied in writing, ignoring the lines dividing the abacus into squares. However, the rods were arranged on the abacus according to the principle of position, for the purposes of calculation, and so this principle was carried over into the written copy.

REPRESENTATION OF THE NUMBER 3,764

with rods on the abacus	using bar numerals combined according to the positional principle

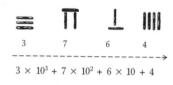

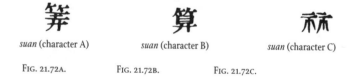

3   7   6   4

$$3 \times 10^3 + 7 \times 10^2 + 6 \times 10 + 4$$

10^3 10^2 10   1

FIG. 21.71. *Origin of the Chinese bar numerals: how a manual calculating aid led to a written positional number-system*

The system of rods on the abacus was the practical means of performing arithmetic calculations, and the *suan zí* notation was used to transcribe the results into their mathematical texts.

The earliest known examples of the use of this abacus date from the second century BCE, but it is very likely that it goes much further back in time.

In any case, the characters used today for the Chinese word *suan*, which means "calculation", have a suggestive etymology. This word may be written using three apparently quite different characters, namely:

suan (character A)   *suan* (character B)   *suan* (character C)

FIG. 21.72A.   FIG. 21.72B.   FIG. 21.72C.

Derived from the following archaic form A', the first character is an ideogram expressing two hands, a ruled table and a bamboo rod:

 *suan* (archaic character A')

FIG. 21.73A.

The second character is derived from the following archaic form B' which expresses two hands and a ruled table:

suan (archaic character B')

FIG. 21.73B.

and the third comes from the following ancient form C' which clearly evokes the representation of numbers on the checkerboard by means of rods vertically and horizontally oriented:

suan (archaic character C')

FIG. 21.73C.

THE CHINESE ABACUS:
THE CALCULATOR OF MODERN CHINA

The celebrated "Chinese abacus" is, therefore. neither the first nor the only calculating device which has been used in China in the course of her long history. It is in fact of relatively recent creation, the earliest known examples being not older than the fourteenth century CE.

Amongst all the calculating devices which the Chinese have used, however, the *suan pan* (meaning "calculating board") is the only one with which all the arithmetical procedures can be performed simply and quickly. In fact almost everyone in China uses it: illiterate trader or accountant, banker, hotelier, mathematician, or astronomer. The most Westernised Chinese or Vietnamese, whether in Bangkok, Singapore, Taiwan, Polynesia, Europe, or the United States, carry out every kind of calculation using the abacus despite having ready access to electronic calculators, so deeply ingrained in their culture is its use. Even the Japanese, major world manufacturers of pocket calculators, still consider the *soroban* (the Japanese word for the abacus) as the principal calculating device and the one item that every schoolchild, businessman, peddler or office-worker should carry with them.

Likewise in the former Soviet Union the *schoty* (счёты), as the abacus is called, may be seen alongside the cash register and will be used to calculate the bill, in boutiques and hotels, department stores and banks.

A friend of mine, on a visit to the former Soviet Union, changed some French francs into roubles. The cashier first worked out the amount on an electronic calculator, and then checked the result on his abacus.

FIG. 21.74. *A Chinese shopkeeper doing his accounts with an abacus. (Reproduced from an illustration in the Palais de la Découverte in Paris)*

FIG. 21.75. *A Japanese accountant working with a* soroban. *From an eighteenth-century Japanese book,* Kanjô Otogi Zôshi *by Nakane Genjun, 1741: [Smith and Mikami (1914)]*

Westerners are invariably astonished at the speed and dexterity with which the most complicated calculations can be done on an abacus. Once, in Japan, there was even a contest between the Japanese Kiyoshi Matzusaki (*soroban* champion of the Post Office Savings Bank – a significant title, given what it means to be champion of anything in Japan) and the American Thomas Nathan Woods, Private Second Class in the 240th Financial Section of US Army HQ in Japan, the acknowledged "most expert electric calculator operator of the American forces in Japan". It took place in November 1945, just after the end of the Second World War, and the men of General MacArthur's army were eager to show the Japanese the superiority of modern Western methods.

The match took place over five rounds involving increasingly complicated calculations. And who won, four rounds out of five with numerous mistakes on the part of the loser? Why, the Japanese with the abacus! (Fig. 21.76)

RESULTS OF THE MATCH

KIYOSHI MATSUZAKI   versus   THOMAS NATHAN WOODS

Soroban champion of the Japanese Post Office Savings Bank

Private 2nd class in the 240th financial section of the US Forces HQ in Japan. The "top expert with the calculator in Japan"

Contested on 12 November 1945 under the auspices of the US Army daily *Stars and Stripes*

1st round	2nd round	3rd round	4th round	Composite round
Additions of numbers with 3 to 6 figures	Subtractions of numbers with 6 to 8 figures	Multiplications of numbers with 5 to 12 figures	Divisions of numbers with 5 to 12 figures	30 additions 3 subtractions 3 multiplications 3 divisions (Numbers with from 6 to 12 figures)
Matsuzaki beat Woods	Matsuzaki beat Woods	Woods beat Matsuzaki	Matsuzaki beat Woods	Matsuzaki beat Woods
1'14"8 / 2'00"2 1'16"0 / 1'53"0	1'04"0 / 1'20"0 1'00"8 / 1'36"0 1'00"0 / 1'22"0 (with mistakes)	(with mistakes by the loser)	1'36"6 / 1'48"0 1'23"4 / 1'19"0 1'21"0 / 1'26"6	1'21"0 / 1'26"6 (with mistakes by the loser)

Overall: Woods on the calculator is beaten 4 to 1 by Matsuzaki on the *soroban*

FIG. 21.76. Reader's Digest *no. 50, March 1947, p. 47*

The match, contested on 12 November 1945 under the auspices of the American Army daily *Stars and Stripes*, was a sensation. Their reporter wrote that: "Machinery suffered a setback yester-day in the Ernie Pyle theatre in Tokyo, when an abacus of centuries-old design crushed the most modern electrical equipment of the United States Government." The *Nippon Times* was exultant at this modest intellectual revenge for military defeat: "In the dawn of the atomic age, civilisation reeled under the blows of the 2,000-year-old *soroban*." An exaggeration, of course – above all concerning the age of the *soroban* – but one which must be viewed in the context of a Japan which, less than three months earlier, had seen two of its greatest cities destroyed by unprecedented military force. But anyone who has watched a Japanese of any competence operate the abacus would have no doubt that the same result could be obtained even today, with electronic instead of electrical calculators, at any rate for additions and subtractions. The keyboard speeds of most of us would be no match for the dexterity of the *soroban* operator. (*Science et Vie*, no. 734, November 1978, pp. 46–53).

The Chinese form of the instrument has a hardwood frame which holds a number of metal rods upon each of which slide wooden (or plastic) beads which may be of somewhat flattened shape. The beads are on either side of a wooden partition, two beads above and five below, and the beads may be slid towards the partition. Each of the metal rods corresponds to one of the decimal orders of magnitude, the value of a bead increasing by a factor of 10 as one moves from one rod to the rod on its left. (In theory, a base different from 10 may be used – 12 or 20 for example – provided each rod carries a sufficient number of beads.)

The normal abacus will have between eight and twelve rods, but the number may be fifteen, twenty, thirty, or even more, according to need. The more rods there are, the larger the numbers that the abacus can handle. With fifteen rods, for example, it can handle up to $10^{15}-1$ (a thousand million million, minus one!)

As a rule, the first two rods on the right are reserved for decimal fractions of first and second order, i.e. for the first two decimal places, and it is the third rod which is used for the units, the fourth for the tens, the fifth for the hundreds, and so on.

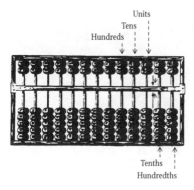

FIG. 21.77. *The representation of numbers on the Chinese* suan pan

The Russian abacus is somewhat different in design from the Chinese *suan pan* (Fig. 21.78). It has ten beads on each rod, of which two (the fifth and the sixth) are usually of a different colour, which makes it easier for the eye to recognise the numbers from 1 to 10. To represent a number the corresponding number of beads are slid towards the top of the frame.

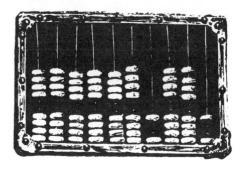

FIG. 21.78. *Russian abacus* (schoty). *It generally has four white beads, then two black and then four white. This type of instrument is still in use in Iran, Afghanistan, Armenia and Turkey*

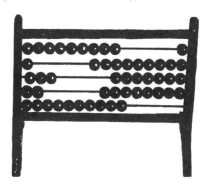

FIG. 21.79. *French abacus used for teaching arithmetic in municipal schools in the nineteenth century*

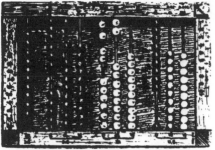

Fig. 21.80. *Abacus marketed by Fernand Nathan at the beginning of the twentieth century as a teaching aid*

On the Chinese abacus, each of the five beads on the lower part is worth one unit, and each of the two on the upper part is worth five. Arithmetical operations involve sliding beads from either side towards the central partition.

To place the number 3 on the abacus, slide three of the five beads on the lower part of the first rod upwards towards the partition. To place the number 9, slide four of the five lower beads upwards towards the partition, and one of the two upper beads downwards towards the partition:

Fig. 21.81.   9

For a larger number such as 4,561,280, the same principle is adopted for each digit: since the first digit is zero, the beads on the first rod are not displaced (denoting absence of number in this position), giving the result shown:

10^{11} 10^{10} 10^9 10^8 10^7 10^6 10^5 10^4 10^3 10^2 10 1

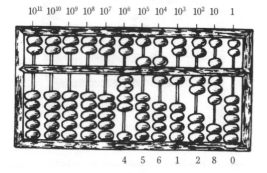

Fig. 21.82.   4   5   6   1   2   8   0

To place the number 57.39, which has a decimal fraction part, the same principle is used for the hundredths, then the tenths, and then the units, tens and hundreds (Fig. 21.83):

HUNDREDTHS 1st rod on the right	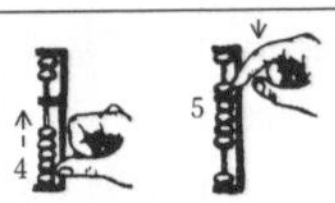	9/100
TENTHS 2nd rod		3/10
UNITS 3rd rod	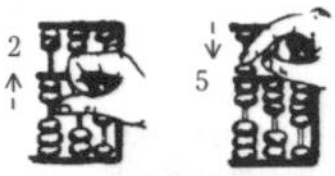	7
TENS 4th rod	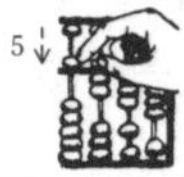	50
Result		57.39

FIG. 21.83.

It is therefore a very simple matter to enter a number onto the Chinese abacus. Actual arithmetic is hardly any more complicated, provided one has learned the addition and multiplication tables by heart for the numbers from 1 to 9.

For convenience of exposition, we shall only consider whole numbers, and therefore we can allocate the first rod to the units, the second rod to the tens, and so on. Now consider addition of the three numbers 234, 432 and 567.

First of all we "clear" the abacus by sliding all the beads to the top and bottom extremities of the rods, leaving the central partition clear. To enter the number 234, first on the third rod from the right (for the hundreds), we slide two beads upwards; then, on the second rod (for the tens) three beads upwards; and finally, on the first rod (for the units), four beads upwards:

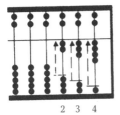

FIG. 21.84A.

Next, to add to this the number 432, we move the corresponding number of beads towards the centre in a similar way. However, on the hundreds rod there are already two beads touching the partition so we do not have four beads available to slide; but we can bring down one bead (representing 5) from the top against the partition and slide one of the lower beads back down away from the centre, since 5 − 1 = 4. On the tens rod, where three beads have already been moved upwards leaving two, in order to add in the 3 of 432 we again slide down one of the upper beads (for 5) and retract two of the lower beads (since 5 − 2 = 3). Finally, on the units rod, we slide down one of the upper beads (for 5) and retract three of the lower beads (since 5 − 3 = 2):

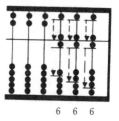

6   6   6

FIG. 21.84B.

As the third and final stage, to add the number 567 to this result, we start by sliding one of the upper beads (for 5) downwards on the hundreds rod. Then, on the tens rod, we slide down one of the upper beads (for 5) and we slide up one of the lower beads (for 1), since 5 + 1 = 6. Finally, on the units rod, we slide downwards one of the upper beads (for 5) and we slide upwards two of the lower beads (for 2), since 5 + 2 = 7. Our abacus now looks like the following:

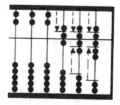

FIG. 21.84C.

But it is not yet all over: what is represented on each rod is no longer a decimal digit, and some further reduction is required before the result can be announced. Therefore, on the third (hundreds) rod, we slide the two upper beads away upwards: each counts for five hundreds, and so we then slide one lower bead (for 1) of the thousands rod towards the centre. Next, in a similar way, the two upper beads of the tens rods are slid upwards away from the centre and one lower bead of the hundreds rod is slid towards the centre; and, finally, the two upper beads of the units rod are replaced by a single bead on the tens rod. When this has been done, the abacus looks like the following, and the result can be read off from it: 234 + 432 + 567 = 1,233.

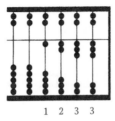

1 2 3 3

FIG. 21.84D.

Subtraction is carried out by the reverse process, multiplication by repeated addition of the multiplicand for as many times as each digit in the multiplier, and division by repeated subtraction of the divisor from the dividend as many times as possible, this number then being the quotient.

Suppose we want to evaluate the product 24 × 7.

We first note that the method is independent of the overall order of magnitude of the result: technically, the procedure is identical whether we want 24 × 7, 24,000 × 7, 24 × 700, 0.24 × 7 or 24 × 0.007, and the digits in the result will be the same; to get the correct result it is enough to keep the order of magnitude in mind.

To work out the above calculation, we start by placing the multiplier (7) on a rod at the left, and the multiplicand (24) towards the right, making sure to leave a few empty rods between them.

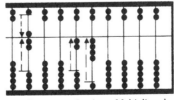

Multiplier - - → 7   2   4 ← - - Multiplicand

FIG. 21.85A.

Now we mentally multiply 7 by 4, getting 28, and we place this result immediately to the right of the multiplicand:

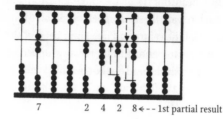

7 2 4 2 8 ⟵ -- 1st partial result

FIG. 21.85B.

Now the 4 of the multiplicand is eliminated by sliding its four units beads back downwards:

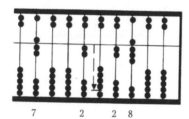

FIG. 21.85C. 7 2 2 8

Next we mentally multiply 7 by 2, getting 14, and we now enter this result as before but at one place further left. Adding it to what is already there, we therefore slide one lower bead upwards on the hundreds rod, and on the tens rod we slide one upper bead downwards and one lower bead upwards:

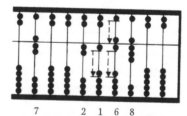

FIG. 21.85D. 7 2 1 6 8

The 2 of the multiplicand is now eliminated, and the multiplier also, and all that remains is to read off the result (168):

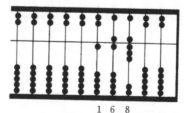

FIG. 21.85E. 1 6 8

So it is not very complicated to do arithmetic on the Chinese abacus. Even square roots or cube roots, or more complicated problems still, can be worked out by operators who know how to use it well. (Our intention here is only to give a general idea of how to use the abacus; we therefore abstain from describing the detailed technique for manipulating it, and we do not discuss its general arithmetical or algebraic applications.)

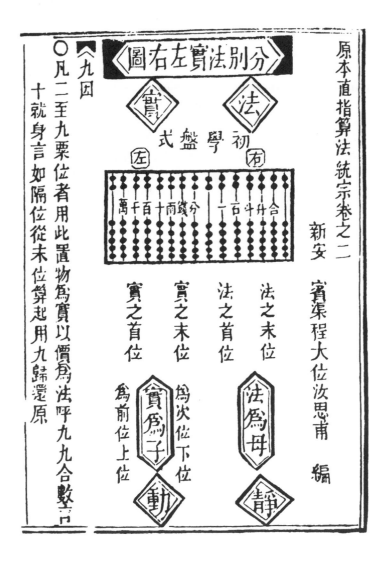

FIG. 21.86. *Instructions for using the* suan pan *in the Chinese* Suan Fa Tong Zong *printed in 1593. [Reproduced from Needham (1959), III, p. 76]*

For all its convenience, this aid to calculation has a number of disadvantages. It takes a long time, and thorough training, to learn how to use it. The finger-work must be extremely accurate, and the abacus must rest on a very solid support. Moreover, if one single error is made the whole procedure must be restarted from scratch, since the intermediate results (partial products, etc.) disappear from the scene once they have been used. None of this, however, detracts from the ingenious simplicity of the device.

After a little thought, however, we are led to ask a question touching on the basic concept of the Chinese abacus. We have seen that on each rod nine units are represented by one upper bead (worth five) and four lower beads (worth one each). Therefore five beads (one upper and four lower) always suffice to represent any number from 1 to 9. Why, therefore, do we find seven beads, whose total value is 15? The answer lies in the fact that (as we have seen in some of the above examples), it is often useful to represent on one rod, temporarily, an intermediate result whose value exceeds 9.

In this connection we may note that the Japanese *soroban* began to do away with the second upper bead, from around the middle of the nineteenth century (Fig. 21.87), and that since the end of the Second World War it has definitively lost the fifth lower bead. This change has obliged the Japanese abacists to undergo an even longer and more arduous training, and it has obliged them to acquire a finger technique even more elaborate and precise than that of the operators of the Chinese *suan pan* (Fig. 21.88).

The post-war Japanese abacus is therefore the fully perfected state of the instrument and marks the close of an evolution in the techniques of calculation which derive from arithmetical manipulations of pebbles, an evolution which has largely been independent of the development of written number-systems.

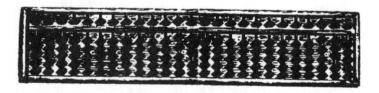

FIG. 21.87. *Pre-war Japanese* soroban *with a single upper bead and five lower beads*

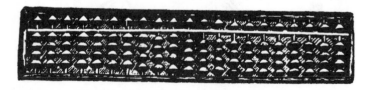

FIG. 21.88. *Post-war Japanese* soroban *with a single upper bead and four lower beads*

NUMBER-GAMES AND WORD-PLAYS

We should not bid farewell to the Far Eastern civilisations without enjoying some examples of their wit.

Both the Chinese and the Japanese have always had a great weakness for plays on words and characters. Since their numerals correspond both to words and to characters, they have taken every opportunity to indulge it. Here are some examples.

The first example (noted by Mannen Veda) bears on the character for the figure 8. For the age of a 16-year-old girl, the Chinese use the expression *pogua*, which literally means "to cut the watermelon in two":

破 瓜

po ("*cut into two*") *gua* ("*watermelon*")

This is a number-play on the form of the character *gua* ("watermelon") which seems to be composed of two characters identical to the figure 8 side by side, representing an addition:

$$瓜 = 八 + 八 \quad = 8 + 8 = 16$$

Furthermore, the pun involves the fact that "watermelon" can also mean virginity (much as we use the word "flower"), which means that *pogua* is also an erotic image of the "defloration" of the young girl.

Other examples (noted by Masahiro Yamamoto) concern the names given to the various major anniversaries of old age in Japan.

1. The 77th birthday is the "happy anniversary". In Japanese it is called *kiju* and written

喜 寿

kiju

Graphically this yields the 77, since the word *ki* ("happy") is written, in the cursive style, as

ki

namely as a character which can be decomposed as follows:

$$七 十 七 \quad = 7 \times 10 + 7 = 77$$

2. The 88th birthday is the "rice anniversary". In Japanese it is called *beiju* and written

$$米 寿$$
beiju

Graphically this yields the 88, since the word *bei* ("rice") is written using a character which can be decomposed as follows:

$$米 \; = \; 八十八 \quad = 8 \times 10 + 8 = 88$$

3. The 90th birthday is the "accomplished anniversary". In Japanese it is called *sotsuju* and is written

$$卒 寿$$
sotsuju

Graphically this yields the 90, since the word *sotsu* ("accomplished") is written using a character which in turn may be replaced by an abbreviation which itself may be decomposed as follows:

$$卒 \; = \; 卆 \; = \; 九十 \quad = 9 \times 10 = 90$$
sotsu   *sotsu*

4. The 99th birthday is the "white-haired anniversary". In Japanese it is called *hakuju* and is written

$$白寿$$
hakuju

Graphically this yields the 99, since the word *haku* ("white") is written using a character which is none other than the character for 100 from which one unit (the horizontal line) has been removed:

$$白 \; = \; 百 \; - \; 一 \quad = 100 - 1 = 99$$
haku   *hyaku*   *ichi*

5. Finally, the 108th birthday is the "tea anniversary". In Japanese it is called *chaju* and is written

$$茶 寿$$
chaju

Graphically this yields the 108, since the word *cha* ("tea") is written using a character which can be decomposed as

$$茶 \; = \; 十 十 八 十 八 \quad = 10 + 10 + (8 \times 10 + 8) = 108$$
10   10   8   10   8

We may also note the strange number-names used by Zen monks to express sums of money in the Edo period (eighteenth century). For these monks, anything to do with money was considered vulgar and not to be mentioned directly. Therefore, to express numerical sums of money, they euphemistically made use of plays on characters (Fig. 21.89).

	ZEN NUMERALS	LITERAL MEANING	EXPLANATION OF NUMERICAL INTERPRETATION
1	大無人 *dai ni jin nashi*	"size without man"	= 大 without 人 ---→ 一 = 1
2	天無人 *ten ni jin nashi*	"heaven without man"	= 天 without 人 ---→ 二 = 2
3	王無中 *ô ni chû nashi*	"king without centre"	= 王 without ｜ ---→ 三 = 3
4	罪無非 *zai ni hi nashi*	"fault without evil"	= 罪 without 非 ---→ 四 = 4
5	吾無口 *go ni kuchi nashi*	"myself without mouth"	= 吾 without 口 ---→ 五 = 5
6	交無人 *kô ni jin nashi*	"exchange without man"	= 交 without 人 ---→ 六 = 6
7	切無刀 *setsu ni to nashi*	"cutting without a knife"	= 切 without 刀 ---→ 七 = 7
8	分無刀 *bun ni to nashi*	"dividing without a knife"	= 分 without 刀 ---→ 八 = 8
9	丸無点 *gan ni chu nashi*	"circle without accent"	= 丸 without ヽ ---→ 九 = 9
10	針無金 *shin ni kin nashi*	"needle without metal"	= 針 without 金 ---→ 十 = 10

FIG. 21.89. *Esoteric numerals of the Zen monks (eighteenth century) (M. Yamamoto. Personal communication from Alain Birot)*

We close with the following Japanese verses, attributed to Kôbô Daishi (775–835)*:

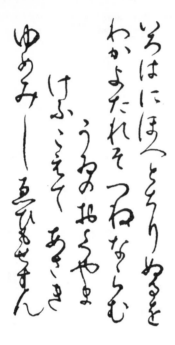

TRANSCRIPTION

I-ro-ha-ni-ho-he-to-
Chi-ri-nu-ru-wo
Wa-ka-yo-ta-re-so
Tsu-ne-na-ra-mu
U-i-no-o-ku-ya-ma
Ke-fu-ko-e-te
A-sa-ki-yu-me-mi-shi
E-hi-mo-se-su-n'

TRANSLATION

Though pretty be its colour,
The flower alas will fade;
What is there in this world
That can forever stay?
As I go forward from today,
To the end of the visible world,
I shall see no more dreams drift by
And I shall not be fooled by them.

This poem contains every sound of the Japanese language with no repetitions. It is therefore often used in teaching Japanese.

However, number has never been far from poetry in the oriental cultures: these same syllables which have been so to speak frozen into a given order by this poem, have finally acquired numerical values. Which is why the Japanese often count using the syllables of the poem:

I-ro-ha-ni-ho-he-to-chi-ri- ...
1 2 3 4 5 6 7 8 9 ...

CHAPTER 22

THE AMAZING ACHIEVEMENTS
OF THE MAYA

The civilisation of the Maya was without question the most glorious of all the pre-Columbian cultures of Central America. Its influence over the others, particularly over Aztec culture, can be likened to the influence of Greece over Rome in European antiquity.

SIX CENTURIES OF INTELLECTUAL AND ARTISTIC CREATION

In the course of the first millennium CE the Maya people produced art, sculpture, and architecture of the highest quality and made great strides in education, trade, mathematics, astronomy, etc.

Maya builders discovered cement, learned how to make arches, built roads, and, of course, they put up vast and complex cities whose buildings were heavily decorated with sculpture and painting. Surprisingly, all this was done with tools that had not developed since the Stone Age: the Maya did not discover the wheel, nor use draught animals, nor any metals. The Mayas' true glory rests on their abstract, intellectual achievements.

They were, in the first place, astronomers of far greater precision than their European contemporaries. As C. Gallenkamp (1979) tells us, the Maya used measured sight-lines, or alignments of buildings that served the same purpose, to make meticulous records of the movements of the sun, the moon, and the planet Venus. (They may also have observed the movements of Mars, Jupiter and Mercury.) They studied solar eclipses in sufficient detail to be able to predict their recurrence. They were acutely aware that apparently small errors could lead in time to major discrepancies; the care they took with their observations allowed them to reduce margins of error to almost nothing. For example, the Maya calculation of the synodic revolution of Venus was 584 days, compared to the modern calculation of 583.92.

FIG. 22.1. *The Great Jaguar Temple at Tikal, constructed in c. 702 CE. Copy by the author from Gendrop (1978), p. 72*

The Maya also made their own very accurate measurement of the solar year, putting it at 365.242 days.* The latest computations give us the figure of 365.242198: so the Maya were actually far nearer the true figure than the current Western calendar of 365 days (which, with leap years, gives a true average of 365.2425).

They were no less precise in their measurement of the lunar cycle. Modern measuring devices of the most sophisticated kind allow us to fix the average length of a lunar cycle at 29.53059 days. Using only their eyes and their brains, the Mayan astronomers of Copán found that 149 new moons occurred in 4,400 days, which gives an average for each lunar month of 29.5302. At Palenque, the same calculation was made over 81 new moons and produced the even more accurate figure of 2,392 days, or 29.53086 per cycle.

---

* The Maya did not express the figure in this way of course, since they could only operate arithmetically in integers.

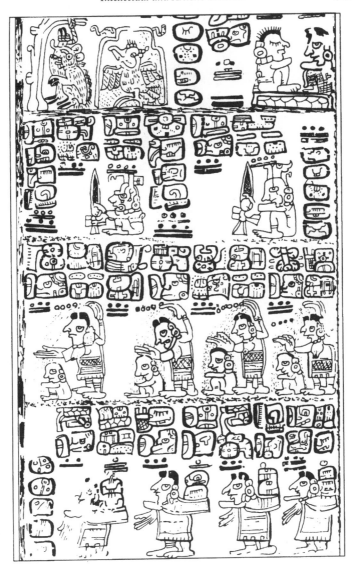

FIG. 22.2. *Extract from a Maya manuscript (lower part of p. 93 of the Codex Tro-Cortesianus, from the American Museum, Madrid). It shows a kind of memorandum for prophet-priests, part of a treatise on ritual magic which includes some astronomical observations.*

Even more fascinating is the Mayas' use of very high numbers for the measurement of time. On a *stela* at Quiriguà, for instance, there is an inscription that mentions the last 5 *alautun*, a period of no less than 300,000,000 years, and gives the precise start and end of the period according to the ritual calendar. Why did they count in terms so far beyond any human experience of life? Perhaps that will always remain a mystery; but it suggests that the Maya had a concept if not of infinity, then of a boundless, unending stretch of time.

FIG. 22.3. *Alone in the darkness of the night, a Maya astronomer observes the stars. Detail from the Codex Tro-Cortesianus. Copied from Gendrop (1978), p. 41, Fig. 2*

It is even more puzzling that the Maya measurements were done without any tools to speak of. They had not discovered glass, so there were no optical instruments. They had no clockwork, no hour-glasses, no idea of water-clocks (*clepsydras*), no means at all of measuring time in units less than a day (such as hours, minutes, seconds, etc.); nor did they have any concept of fractions. It is hard to imagine how to measure time without at least basic measuring devices.

The tool that the Maya used for measuring the true solar day was the very simple but utterly reliable device called a *gnomon*. It consists of a rigid stick or post fixed at the centre of a perfectly flat area. The stick's shadow alters as the day progresses. When the shadow is at its shortest, then the sun is at its meridian: that is to say, the sun has reached its highest point above the horizon, and it is "true noon".

As for astronomical observations, according to P. Ivanoff (1975), these were done by means of a jadeite tube placed over a wooden cross-bar, as shown in codices, thus:

FIG. 22.4. *Astronomical observations, as shown in the Mexican manuscripts, Codex Nuttall and Codex Selden. Copied from Morley (1915). In the left-hand drawing, an astronomer seen in profile watches the sky through a wooden X; the right-hand drawing shows an eye looking through the angle of the X.*

The Maya also developed an elaborate writing system, consisting of intricate signs known as glyphs. These include numerals (as we shall see below) and many names or "emblem glyphs" associated with the main cities in the central Mayan area. The decipherment of Maya glyphs is currently the subject of intense and recently successful research.*

MAYA GODS

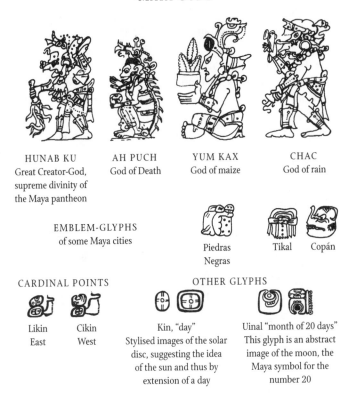

HUNAB KU	AH PUCH	YUM KAX	CHAC
Great Creator-God, supreme divinity of the Maya pantheon	God of Death	God of maize	God of rain

EMBLEM-GLYPHS
of some Maya cities

Piedras Negras   Tikal   Copán

CARDINAL POINTS

OTHER GLYPHS

Likin   Cikin
East   West

Kin, "day"
Stylised images of the solar disc, suggesting the idea of the sun and thus by extension of a day

Uinal "month of 20 days"
This glyph is an abstract image of the moon, the Maya symbol for the number 20

FIG. 22.5. *Some of the Maya hieroglyphs deciphered to date*

MAYA CIVILISATION

Several dozen abandoned cities buried in the tropical jungles and savannah of Central America bear witness to one of the most mysterious episodes of human history.

With their stately temples perched atop pyramids up to 170 feet high, with their intricately carved pillars and altars and brightly painted earthenware vessels, these forgotten cities are all that is left of a sophisticated civilisation that is thought to have begun in the jungles of Peten. At the height of its glory, Maya civilisation covered the area shown in Fig. 22.6, and included:

* See Michael D. Coe, *Breaking the Maya Code* (London: Thames and Hudson, 1992), for a fascinating account of recent breakthroughs.

- the present-day Mexican provinces of Tabasco, Campeche, and Yucatan, the region of Quintana Roo and a part of Chiapas province;
- the Peten region and almost all the uplands of present-day Guatemala;
- the whole of Belize (formerly British Honduras);
- parts of Honduras;
- the western half of Salvador;

making an area of about 325,000 km².

There are reckoned to be about two million direct descendants of the Maya alive today, most of whom are in Guatemala, and the remainder spread around Honduras and the Mexican provinces of Yucutan, Tabasco, and Chiapas.

Maya civilisation was fully developed at least as early as the third century CE and reached its greatest heights of artistic and intellectual creation long before the discovery of the New World by Christopher Columbus.

FIG. 22.6. *Map by the author, after P. Ivanoff*

It is widely assumed that there was an early period of Maya civilisation dating from about the fifth century BCE, during which the Maya differentiated themselves from other Amerindian cultures; but of this era of formation, there remain few traces apart from shards of pottery, and little can be known of it.

The period from the third to the tenth century CE is the "classical" period of Maya civilisation, and it is in these centuries that the Maya developed their arts and sciences to their highest point. But at some point in the ninth or tenth centuries there occurred an unexpected and mysterious event which Americanists have not yet fully explained: the Maya began to abandon their ritual centres and cities in the central area of the "Old Empire". Their departure was so sudden in some places that buildings were left half-finished.

It was long thought that what had happened was an exodus of the entire population, but recent excavations have shown this not to be true. Various theories have been put forward to explain this resettlement of the Maya to the north and the south – epidemics, earthquakes, climate change, invasion, perhaps even their priests' interpretation of the wishes of the gods. The most plausible of these hypotheses are those that see the main cause of the exodus in the exhaustion of the soil. Mayan agriculture was based on the use of burnt clearings, which created ever more extensive infertile areas. In addition, there may well have been a peasant revolt, provoked by the vast inequality between the classes of Maya society.

Whatever the real cause, large sections of the Maya people left the central area, leaving a much reduced population which gave up the traditional rituals in the cities and allowed the religious monuments to fall into decay.

There was also an invasion of a different people, from the west. To judge by the ruins of Chichén Itzá (Yucatan), these invaders were probably Toltecs, who came from an area north of present-day Mexico City. After the "interregnum" (925–975 CE), the period following the fall of classical Maya culture is called the "Mexican period", and it lasted until 1200 CE.

The Maya accepted Toltec domination and adopted some of the Mexican gods, including Quetzalcoatl, the plumed serpent. The Maya also became more warlike, in line with the traditions of the Mexicans, whose gods required countless human sacrifices. However, even if the Maya of the Mexican period tore the hearts out of their human sacrificial victims, they were never as bloodthirsty as their neighbours, the Aztecs, whose religious rituals were frenetically violent.

Toltec and Maya civilisations gradually merged into one. The language, religion and even the physical characteristics of the Maya changed so much that it is hard to compare Maya civilisation before and after the Mexican invasion.

Between 1200 and 1540, the course of Maya history changed completely once again. Mexican civilisation was rejected, and the invaders adopted Maya customs. This period is called the age of "Mexican absorption". Maya civilisation continued to decline, as can be seen in the art and architecture of the period. Wars of annihilation broke out, and Maya civilisation soon came to an end. Only a small group from Chichén Itzá managed to escape and resettle on the island of Tayasal, in Lake Peten, where they maintained their independence until 1697.

THE DOCUMENTARY SOURCES
OF MAYA HISTORY

The first light to be shed on the civilisation of the Maya was the work of the famous American diplomat and traveller, John Lloyd Stephens, who explored the jungles of Guatemala and southern Mexico with the English artist Frederick Catherwood in 1839. A more detailed survey of Maya sites and buildings was carried out from 1881 by Alfred Maudslay, which marked the true beginning of scholarly research on the world of the Maya. But most of the knowledge we now have of this lost civilisation has been gained in the last few decades.

When the Spaniards conquered Central America in the sixteenth century, Maya civilisation had been all but extinct for several generations, and most of its magnificent cities were but inaccessible ruins in the midst of the jungle. This explains why the early Spanish chroniclers were bedazzled by the Aztecs and hardly mentioned the Maya at all.

Pre-Columbian cultures, moreover, were systematically suppressed by the conquistadors. Deeply shocked by the bloodthirstiness of Aztec and Maya rituals, and believing that their mission was to convert the natives to Christianity, the Spaniards sought to eradicate all traces of the devilish practices that they came across. In order to ensure that such abominable religions would never re-emerge, they burnt everything they could find in *autos-da-fé*.

Nonetheless it is to a Spaniard that we owe a significant part of our present knowledge of the history, customs and institutions of the Maya. In 1869, the colourful and indefatigable French monk Brasseur de Bourbourg unearthed in the Royal Library of Madrid a manuscript entitled *Relación de las Cosas de Yucatán* by the first bishop of Merida (Yucatan), Diego de Landa. Written shortly after the Spanish conquest, the *Relación* is full of priceless ethnographic information, including descriptions and drawings of the glyphs used by the indigenous population of Yucatan in the sixteenth century. Ironically, Landa was proud of having burned all the texts using this writing, the better to bring the natives into the embrace of the Catholic

Church. He wrote his chronicle in order to explain why he had destroyed all those precious painted codices – but thereby unwittingly preserved the basic elements of one of the most important pre-Columbian civilisations of the Americas.

The discovery of this sixteenth-century manuscript aroused great interest, because the glyphs copied down by Landa were similar to the carved shapes on the ruins found in the virgin jungle of Central America by Stephens and later explorers. It provided solid evidence of the cultural connection between the sixteenth-century population of the Yucatan peninsula and the builders of the lost cities of the jungle, both in Yucatan and further south.

Landa's manuscript is a major source for the history of the Maya, but it is not the only one. Much was also written down by the natives themselves, who were taught by Spanish missionaries to read and write in the Latin alphabet, which they then also used for writing in their own tongue. Although the teaching was intended to support the spread of Christianity, it was also used – inevitably – to set down the fast-disappearing oral traditions of the local populations.

A good number of anonymous accounts of this kind have survived, and give a reflection of the history, traditions and customs of the indigenous peoples of Spanish Central America. From the Guatemalan uplands comes the manuscript known as *Popol Vuh*, which contains fragments of the mythology, cosmology and religious beliefs of the Quiché Maya; and it was in the same area that the *Annals of the Cakchiquels* were found, which provide in addition the story of the tribe of that name during the Spanish conquest. The *Books of Chilám Balám* are a collection of native chronicles from Yucatan, and are named after a class of "Jaguar Priests", famed for their prophetic powers and their mastery of the supernatural. Fourteen of these manuscripts go a long way back in history; though they deal mostly with traditions, calendars, astrology, and medicine, three of them mention historical events that can be precisely situated in the year 1000 CE. Some parts of the *Chilám Balám* may even have been copied directly from ancient codices.

The ancient Maya codices used parchment, tree bark, or mashed vegetable fibres strengthened with glue to provide a writing surface. The glyphs were written with a brush pen dipped in wood ash, and then coloured with dyes from various animal and vegetable sources. The pages were glued together, then folded like a concertina and bound between wood or leather covers, much like a book. Three of them miraculously escaped the attention of the conquistadors, and found their way back to Europe, where they are now known by the names of the cities where they are kept: the Dresden Codex (in the Sächsische Landesbibliothek, Dresden,

Germany) is an eleventh-century copy of an original text drafted in the classical period, and deals with astronomy and divination; the Codex Tro-Cortesianus (American Museum, Madrid) is less elaborate and was probably composed no earlier than the fifteenth century; and the Paris Codex (Bibliothèque nationale, Paris), likewise from the late period, gives illustrations of ceremonies and prophecies.

Despite these various documentary sources, much of Maya civilisation remains mysterious and unexplained to this day.

AZTEC CIVILISATION

The legendary homeland of the Aztecs, according to the few manuscripts that have survived and the tales of Spanish conquerors, was called Aztlan and was located somewhere in northwestern Mexico, maybe in Michoacan. In a cave in Aztlan they are supposed to have found the "colibri wizard", Huitzilopochtli, who gave such good advice that he became the Aztecs' tribal god. Then began their long migration, by way of Tula and Zumpango (on the high plateau), and the Chapultepec, where they lived peaceably for more than a generation. Thereafter, they were defeated in battle and exiled to the infertile lands of Tizapan, infested with poisonous snakes and insects. A group of rebels took refuge on the islands in Lake Texcoco, where, in 1325 CE (or 1370, according to more recent calculations), they founded the city of Tenochtitlán, which has become present-day Mexico City.

Within a century Tenochtitlán became the centre of a vast empire. The Aztec King Itzcoatl subdued and enslaved most of the tribes in the valley; then under Motecuhzoma I (1440–1472) they battled on into the Puebla region in the south. Axayacatl, son of Motecuhzoma, led the Aztec armies even further south, as far as Oaxaca; he also attacked, but failed to conquer the Matlazinca and Tarasques in the west.

By the time the Spaniards arrived in 1519, the Aztecs possessed most of Mexico, and their language and religion held sway over a vast territory stretching from the Atlantic to the Pacific Oceans and from the northern plains to Guatemala. The name of the king, Motecuhzoma (Europeanised as "Montezuma") struck fear from one end to another of the empire; Aztec traders, with great caravans of porters, scoured the entire kingdom; and taxes were levied everywhere by the king's administrators. It was a relatively recent civilisation, at the height of its wealth and glory.

FIG. 22.7. *Page 1 of the Codex Mendoza (post-conquest). Through a number of Aztec hieroglyphs, this illustration sums up Aztec history and relates the founding of the city of Tenochtitlán.*

It was also a very violent civilisation. The continual military campaigns were for the most part undertaken in the service of the Aztec gods – for every aspect of Aztec history, culture, and society can only be understood in terms of a tyrannical religion which left no space for anything resembling hope or even virtue in the Christian sense. The main purpose of war-making was to seize prisoners who could be used in the ritual sacrifices. About 20,000 people were thus slaughtered every year in the service of magic. The Aztecs believed that the Sun and the Earth (both considered gods) required constant replenishment with human blood, or else the world's mechanism would cease to function. The slaughter also had a straightforward nutritional use, for only the victims' hearts were reserved for the gods' consumption. Human legs, arms and rumps were treated much as we treat butcher's meat, and sold retail at Aztec markets, for ordinary consumption.

Beside the priestly and the warrior castes, there were also castes of artisans and traders, organised into a set of guilds. The main market of the empire was at Tlatelolco, Tenochtitlán's twin town, founded in 1358, where merchandise of every sort, brought from the four corners of the Aztec empire, was traded. The records of the taxes levied by the imperial administrators of the Tlatelolco market have survived, and give a good picture of the wealth and variety of trade in the Aztec empire: gold, silver, jade, shells, feathers for ceremonial wear, ceremonial garb, shields, raw cotton for spinning, cocoa beans, coats, blankets, embroidered cloth, etc.

The empire and the whole of Aztec civilisation collapsed in the early sixteenth century. "Stout" Cortez, accompanied by a mere handful of men armed with guns, landed at Vera Cruz and marched towards the highlands. He gained the support of tribes that were the Aztecs' enemies or their subjects, and from them acquired supplies and reinforcements. After a violent struggle, Cortez seized Tenochtitlán on 13 August 1521, and destroyed Aztec civilisation for ever.

AZTEC WRITING

At the time of the Spanish Conquest, Mexican script was a mixture of ideographic and phonetic representation, with some more or less "pictorial" signs designating directly the beings, objects or ideas that they resembled, and others (including the same ones) standing for the sound of the thing that they represented. Names of people and places were written in the manner of a rebus or puzzle (rather approximate ones, in fact, since the writing took no account of case-endings). For example, the name of the city of Coatlan (literal meaning: "snake-place") was represented by the drawing of a snake (= *coatl*) together with the sign for "teeth", pronounced *tlan*. The name of the city of Coatepec (literal meaning: "snake-mountain-place") was represented similarly by a snake (= *coatl*) together with the sign for "mountain" (*tepetl*).

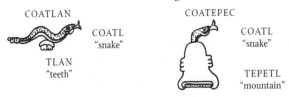

COATLAN

COATL
"snake"

TLAN
"teeth"

COATEPEC

COATL
"snake"

TEPETL
"mountain"

FIG. 22.8. *Examples of Aztec names written in the form of a rebus*

Aztec script is used in a number of Mexican documents written just before and just after the Spanish conquest. Some of these deal with matters of religion, ritual, prophecy, and magic; others are narratives of real or mythical history (tribal migrations, foundations of cities, the origins and history of different dynasties, etc.); and others are registers of the vast taxes paid in kind (goods, food supplies, and men) by the subject cities to the lords of Tenochtitlán.

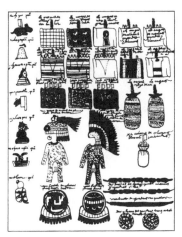

FIG. 22.9. *Codex Mendoza (folio 52 r), showing the tributes to be paid by seven Mexican cities to Tenochtitlán*

The most important by far of these Aztec documents in the Codex Mendoza, drawn up by order of Don Antonio de Mendoza, the first Viceroy of New Spain, and sent to the court of Spain. It contains three parts, dealing respectively with the conquests of the Aztecs, the taxes that they levied on each of the conquered towns, and with the life-cycle of an Aztec, from birth through education, punishment, recreation, military insignia, battles, the genealogy of the royal family, and even the ground-plan of Motecuhzoma's palace ... It was written in a period of ten days (since the fleet was about to put to sea) in the native language and script, but with a simultaneous commentary on the meaning of every detail in Spanish. And it is largely thanks to the Spanish commentary that we can now seek to understand Aztec numerals ...

HOW THE MAYA DID THEIR SUMS

Most of what could be "read" in Maya texts and inscriptions until very recently consists of numerical, astronomical and calendrical information. However, before we can approach Mayan arithmetic, we need to know what their oral numbering system was.

Like all the other peoples of pre-Columbian Central America, the Maya counted not to base 10, but to base 20. As we now know, this was due to their ancestors' habits of using their toes as well as their fingers as a model set.

The language of the Maya and various dialects of it are still in use nowadays in the Mexican states of Yucatan, Campeche, and Tabasco, in a part of the Chiapas and the region of Quintana Roo, in most of Guatemala and in parts of Salvador and Honduras. The names of the numbers are as follows:

1	*hun*	11	*buluc*	
2	*ca*	12	*lahca*	(*lahun + ca* = 10 + 2)
3	*ox*	13	*ox-lahun*	(3 + 10)
4	*can*	14	*can-lahun*	(4 + 10)
5	*ho*	15	*ho-lahun*	(5 + 10)
6	*uac*	16	*uac-lahun*	(6 + 10)
7	*uuc*	17	*uuc-lahun*	(7 + 10)
8	*uaxac*	18	*uaxac-lahun*	(8 + 10)
9	*bolon*	19	*bolon-lahun*	(9 + 10)
10	*lahun*			

FIG. 22.10A.

The units up to and including 10 thus have their own separate names, and above that number are made of additive compounds that rely on 10 as an auxiliary base. The one exception is the name of the number 11, *buluc*, which was probably invented to avoid confusion of a regular form *hun-lahun*, "one + ten" with *hun-lahun*, in the meaning "a ten".

Numbers from 20 to 39 are expressed as follows:

20	*kun kal*	score (*hun uinic*, "one man", in some dialects)
21	*hun tu-kal*	one (after) twentieth
22	*ca tu-kal*	two (after) twentieth
23	*ox tu-kal*	three (after) twentieth
24	*can tu-kal*	four (after) twentieth
25	*ho tu-kal*	five (after) twentieth
26	*uac tu-kal*	six (after) twentieth
27	*uuc tu-kal*	seven (after) twentieth
28	*uaxac tu-kal*	eight (after) twentieth
29	*bolon tu-kal*	nine (after) twentieth
30	*lahun ca kal*	ten-two-twenty
31	*buluc tu-kal*	eleven (after) twentieth
32	*lahca tu-kal*	twelve (after) twentieth
33	*ox-lahun tu-kal*	thirteen (after) twentieth

34	*can-lahun tu-kal*	fourteen (after) twentieth
35	*holhu ca kal*	fifteen-two-twenty
36	*uac-lahun tu-kal*	sixteen (after) twentieth
37	*uuc-lahun tu-kal*	seventeen (after) twentieth
38	*uaxac-lahun tu-kal*	eighteen (after) twentieth
39	*bolon-lahun tu-kal*	nineteen (after) twentieth

Fig. 22.10B.

So, as a general rule, these numbers are formed by inserting the ordinal prefix *tu* between the name of the unit and the name of the base, 20. But there are two exceptions:

30 "ten-two-twenty", instead of "ten (after) twentieth"
35 "fifteen-two-twenty", instead of "fifteen (after) twentieth"

These two anomalies cannot be explained by addition or by subtraction, since 35 is neither $15 + (2 \times 20)$ nor $(2 \times 20) - 15$. Moreover, the irregularity is repeated in the next sequence of numbers, which begin "one-three-twenty", "two-three-twenty" and so on.

40	*ca kal*	two score
41	*hun tu-y-ox kal*	one – third score
42	*ca tu-y-ox kal*	two – third score
43	*ox tu-y-ox kal*	three – third score
44	*can tu-y-ox kal*	four – third score
	....	
58	*uaxac-lahun tu-y-ox kal*	eighteen – third score
59	*bolon-lahun tu-y-ox kal*	nineteen – third score
60	*ox kal*	three score
61	*hun tu-y-can kal*	one – fourth score
62	*ca tu-y-can kal*	two – fourth score
	....	
78	*uaxac-lahun tu-y-can kal*	eighteen – fourth score
79	*bolon-lahun tu-y-can kal*	nineteen – fourth score
80	*can kal*	four score
81	*hun tu-y-ho-kal*	one – fifth score
82	*ca tu-y-ho-kal*	two – fifth score
	....	
98	*uaxac-lahun tu-y-ho-kal*	eighteen – fifth score
99	*bolon-lahun tu-y-ho-kal*	nineteen – fifth score
100	*ho kal*	five score
	....	
400	*hun bak*	one four-hundreder
8,000	*hun pic*	one eight-thousander
160,000	*hun calab*	one hundred-and-sixty-thousander

Fig. 22.10C.

To work out how such a numbering system might have come into being, we have to imagine something like the following scene taking place several thousand years ago somewhere in Central America.

FIG. 22.11.

As they prepare to set off to fight a skirmish, warriors line up a few men to serve as "counting machines" or model sets, and one of the men proceeds to check off the number of warriors in the group. As the first one files past, the checker touches the first finger of the first "counting machine", then for the second he touches the second finger, and so on up to the tenth. The "accountant" then moves on to the toes of the first model set, up to the tenth, which matches the twentieth warrior that has filed past. For the next man, the accountant proceeds in exactly the same way using the second of the "counting machines", and when he gets to the last toe of the second man, he will have checked off forty warriors. He moves on to the third man, which would take him up to sixty, and so on until the count is finished.

Let us suppose that there are 53 men in the group. The accountant will reach the third toe of the first foot of the third man, and will announce the result of the count in something like the following manner: "There are as many warriors as make three toes on the first foot of the third man". But the result could also be expressed as: "Two hands and three toes of the third man" or even "ten-and-three of the third twenty". If applied to English, such a system would produce a set of number-names of the following sort:

1	one		11	ten-one
2	two		12	ten-two
3	three		13	ten-three
4	four		14	ten-four
5	five		15	ten-five
6	six		16	ten-six
7	seven		17	ten-seven
8	eight		18	ten-eight
9	nine		19	ten-nine
10	ten		20	one man

Style A		Style B
one after the first man	21	one of the second man
two after the first man	22	two of the second man
three after the first man	23	three of the second man
four after the first man	24	four of the second man
five after the first man	25	five of the second man
six after the first man	26	six of the second man
seven after the first man	27	seven of the second man
eight after the first man	28	eight of the second man
nine after the first man	29	nine of the second man
ten after the first man	30	ten of the second man
ten-one after the first man	31	ten-one of the second man
ten-two after the first man	32	ten-two of the second man
ten-three after the first man	33	ten-three of the second man
ten-four after the first man	34	ten-four of the second man
ten-five after the first man	35	ten-five of the second man
....		
ten-nine after the first man	39	ten-nine of the second man
two men	40	two men
one after the second man	41	one of the third man
two after the second man	42	two of the third man
three after the second man	43	three of the third man
....		
ten-one after the second man	51	ten-one of the third man
ten-two after the second man	52	ten-two of the third man
ten-three after the second man	53	ten-three of the third man
....		
ten-nine after the second man	59	ten-nine of the third man
three men	60	three men
one after the third man	61	one of the fourth man
two after the third man	62	two of the fourth man
....		
nineteen after the third man	79	nineteen of the fourth man
four men	80	four men

FIG. 22.12.

It is now easy to see how the irregularities of the Maya number-names arose. The numbers 21 to 39 (except 30 and 35) are expressed in terms of Style A: 21 = *hun tu-kal* = "one (after) the twentieth" or "one (after the) first twenty", 39 = *bolon-lahun tu-kal* = "nine-ten (after the) twentieth" or "nine-ten (after the) first twenty"; whereas the numbers from 41 to 59, 61 to 79, etc. as well as the numbers 30 and 35, are expressed in terms of Style B: 30 = *lahun ca kal* = "ten-two-twenty" or "ten of the second twenty", and so forth. The Maya were not alone in counting in this way. The number 53, for instance, is expressed as follows:

- by the Inuit of Greenland, as *inûp pingajugsane arkanek pingasut*, literally, "of the third man, three on the first foot";
- by the Ainu of Japan and Sakhalin, as *wan-re wan-e-rehotne*, literally "three and ten of the third twenty" [see K. C. Kyosuke and C. Mashio (1936)];
- by the Yoruba (Senegal and Guinea) as *eeta laa din ogota*, literally "ten and three before three times twenty" [see C. Zaslavsky (1973)];
- and other instances of similar systems can be found amongst the Yedo (Benin) and the Tamanas of the Orinoco (Venezuela).

THE "ORDINARY" NUMBERS OF THE MAYA

Now that we can see the reasons for the irregularities of the Maya number-name system, we can try to grasp their written numerals. Or rather, we would have been able to, had the Spanish Inquisition not stupidly destroyed almost every trace of it. So we are forced to take a step backwards.

Amongst the cultures of pre-Columbian Central America there are four main types of writing system: Maya, Zapotec (in the Oaxaca Valley), Mixtec (southwest Mexico), and Aztec (around Mexico City). Zapotec is the oldest, probably dating from the sixth century BCE, and Aztec is the most recent (see above). Now, although these scripts served to represent languages belonging to quite different linguistic families, they possess a number of graphical features in common, including (as far as Aztec, Mixtec and Zapotec are concerned) the basic features of numerical notation.

In Aztec vigesimal numerals, for instance, the unity was represented by a dot or circle, the base by a hatchet, the square of the base ($20 \times 20 = 400$) by a sign resembling a feather, the cube of the base ($20 \times 20 \times 20 = 8,000$) by a design symbolising a purse.

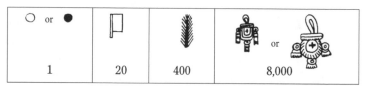

FIG. 22.13. *Aztec numerals*

The numeral system relied on addition: that is to say, numbers were expressed by repeating the component figures as many times as necessary. To express 20 shields, 100 sacks of cocoa beans, or 200 pots of honey, for example, one, five or ten "hatchets" were attached to the pictogram for the relevant object:

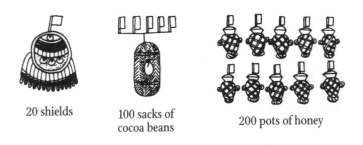

20 shields   100 sacks of cocoa beans   200 pots of honey

FIG. 22.14.

To record 400 embroidered cloaks, 800 deerskins or 1,600 cocoa bean-pods, one, two or four "feather" signs were similarly attached to the respective object-sign:

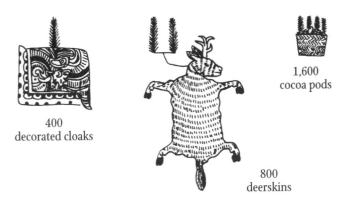

400 decorated cloaks

1,600 cocoa pods

800 deerskins

FIG. 22.15.

This was the way that the scribe of the Codex Mendoza recorded the taxes that were paid once, twice or four times a year by the subject-cities to the Aztec lords of Tenochtitlán. The page shown in Fig. 22.9 above gives the taxes due from seven cities in one province, and expresses them as follows:

1. *Left column*: the names of the seven cities, expressed by combinations of signs in the manner of a rebus:

| Tochpan | Tlaltiçapan | Civateopan | Papantla | Ocelotepec | Miaua apan | Mictlan |

FIG. 22.16A.

2. *Line 1, horizontally:*

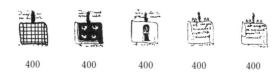

400   400   400   400   400

FIG. 22.16B.

- 400 cloaks of black-and-white chequered cloth
- 400 cloaks of red-and-white embroidered cloth (worn by the lords of Tenochtitlán)
- 400 loincloths
- 2 sets of 400 white cloaks, size 4 *braza* (a unit of length indicated by the finger-sign)

3. *Line 2*

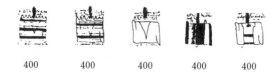

400   400   400   400   400

FIG. 22.16C.

- 2 sets of 400 orange-and-white-striped cloaks, size 8 *braza*
- 400 white cloaks, size 8 *braza*
- 400 polychrome cloaks, size 2 *braza*
- 400 women's skirts and tunics

4. *Line 3*

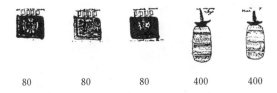

     80   80   80   400   400

FIG. 22.16D.

- 3 sets of 80 coloured and embroidered cloaks (as worn by the leading figures of the capital)
- 2 sets of 400 bundles of dried peppers (used amongst other things to punish young people for breaking rules)

5. *Line 4*

20

FIG. 22.16E.

- 2 ceremonial costumes, 20 sacks of down, and 2 strings of jade pearls

6. *Last line*

FIG. 22.16F.

- 2 shields, a string of turquoise, and 2 plates with turquoise incrustation

The Codex Telleriano Remensis, another post-conquest document in Aztec script, also provides examples of numerals:

FIG. 22.17. *Detail from a page of the Aztec Codex Telleriano Remensis*

What this page says in effect is that 20,000 men from the subject provinces were sacrificed in 1487 CE to consecrate a new building. The number was written by the native scribe thus:

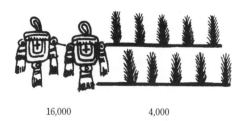

16,000   4,000

FIG. 22.18.

The Spanish annotator, however, made a mistake in transcribing this number: as he did not know the meaning of the two purses worth 8,000 each, he "translated" only the ten feathers, giving a total of 4,000.

Aztec numerals were identical to those of the Zapotecs and Mixtecs, as the following painting shows. It was done in Zapotec by order of the Spanish colonial authorities in Mexico in 1540 CE and shows the numbering conventions common to Zapotec, Mixtec and Aztec cultures:

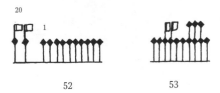

So it seems certain that "ordinary" Maya numerals must also have been strictly vigesimal and based on the additive principle. It can be safely assumed that a circle or dot was used to represent the unity (the sign is common to all Central American cultures, and derives from the use of the cocoa bean as the unit of currency), that there was a special sign, maybe similar to the "hatchet" used by other Central American cultures, for the base (20), and other specific signs for the square of the base (400) and the cube (8,000), etc.

As we shall see below, it is also quite probable that, like the Zapotecs, the Maya introduced an additional sign for 5, in the form of a horizontal line or bar.

Even though no trace of it remains, we can reasonably assume that the Maya had a numeral system of this kind, and that intermediate numbers were figured by repeating the signs as many times as was needed. But that kind of numeral system, even if it works perfectly well as a recording device, is of no use at all for arithmetical operations. So we must assume that the Maya and other Central American civilisations had an instrument similar to the abacus for carrying out their calculations.

The Inca of South America certainly did have a real abacus, as shown in Fig. 22.20. The Spaniards were amazed at the speed with which Inca accountants could resolve complex calculations by shifting ears of maize, beans or pebbles around twenty "cups" (in five rows of four) in a tray or table, which could be made of stone, earthenware or wood, or even just laid out in the ground. Inca civilisation was obviously quite different from the Maya world, but it did have one thing in common: a method of recording numbers and tallies (the *quipus*, or knotted string) that was entirely unsuitable for performing arithmetical operations. For that reason the Inca were obliged to devise a different kind of operating tool.

FIG. 22.20. *Document proving the use of the abacus amongst the Peruvian and Ecuadorian Incas. It shows a* quipucamayoc *manipulating a* quipu *and on his right a counting table. From the Peruvian Codex of Guaman Poma de Ayala (16th century), Royal Library, Copenhagen*

THE PLACE-VALUE SYSTEM OF "LEARNED" MAYA NUMERALS

The only numerical expressions of the Maya that have survived are in fact not of the ordinary or practical kind, but astronomical and calendrical calculations. They are to be found in the very few Maya manuscripts that exist, and most notably in the Dresden Codex, an astronomical treatise copied in the eleventh century CE from an original that must have been three or four centuries older.

What is quite remarkable is that Maya priests and astronomers used a numeral system with base 20 which possessed a true zero and gave a specific value to numerical signs according to their position in the written expression. The nineteen first-order units of this vigesimal system were represented by very simple signs made of dots and lines: one, two, three and four dots for the numbers 1 to 4; a line for 5, one, two, three and four dots next to the line for 6 to 9; two lines for 10, and so on up to 19:

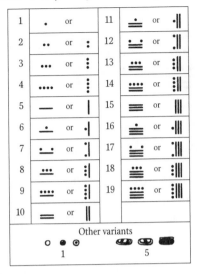

FIG. 22.21. *The first nineteen units in the numeral system of the Maya priests*

Numbers above 20 were laid out vertically, with as many "floors" as there were orders of magnitude in the number represented. So for a number involving two orders, the first order-units were expressed on the first or "bottom floor" of the column, and the second-order units on the "second floor". The numbers 21 (= 1 × 20 + 1) and 79 (3 × 20 + 19) were written thus:

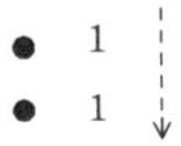

FIG. 22.22.

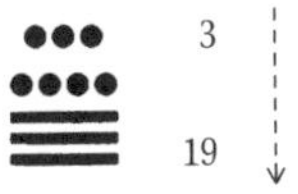

FIG. 22.23.

The "third floor" should have been used for values twenty times as great as the "second floor" in a regular vigesimal system. Just as in our decimal system the third rank (from the right) is reserved for the hundreds (10 × 10 = 100), so in Maya numbering the third level should have counted the "four hundreds" (20 × 20 = 400). However, in a curious irregularity that we will explain below, the third floor of Mayan astronomical numerals actually represented multiples of 360, not 400. The following expression:

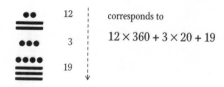

corresponds to

$$12 \times 360 + 3 \times 20 + 19$$

FIG. 22.24.

actually meant $12 \times 360 + 3 \times 20 + 19 = 4,399$, and not $12 \times 400 + 3 \times 20 + 19 = 4,879$!

Despite this, higher floors in the column of numbers were strictly vigesimal, that is to say represented numbers twenty times as great as the immediately preceding floor. Because of the irregularity of the third position, the fourth position gave multiples of 7,200 (360×20) and the fifth gave multiples of 144,000 ($20 \times 7,200$) – and not of 8,000 and 160,000.

A four-place expression can thus be resolved by means of three multiplications and one addition, thus:

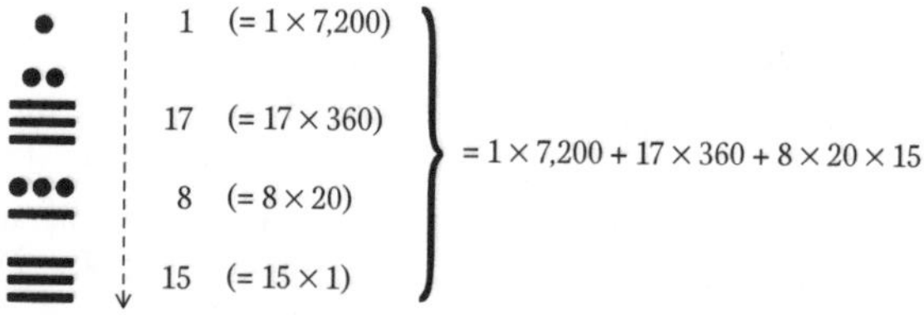

$$1 \quad (= 1 \times 7,200)$$
$$17 \quad (= 17 \times 360)$$
$$8 \quad (= 8 \times 20)$$
$$15 \quad (= 15 \times 1)$$

$$= 1 \times 7,200 + 17 \times 360 + 8 \times 20 \times 15$$

FIG. 22.25.

So that each numeral would be in its right place even when there were no units to insert in one or another of the "floors", Mayan astronomers invented a zero, a concept which they represented (for reasons we cannot pierce) by a sign resembling a snail-shell or sea-shell.

For instance, a number which we write as 1,087,200 in our decimal place-value system and which corresponds in Mayan orders of magnitude to $7 \times 144,000 + 11 \times 7,200$ and no units of any of the lower orders of 360, 20 or 1, would be written in Maya notation thus:

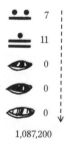

1,087,200

FIG. 22.26.

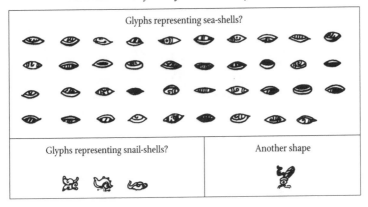

Glyphs representing sea-shells?

Glyphs representing snail-shells?	Another shape

FIG. 22.27.

We can see the system in operation in these very interesting numerical expressions in the Dresden Codex:

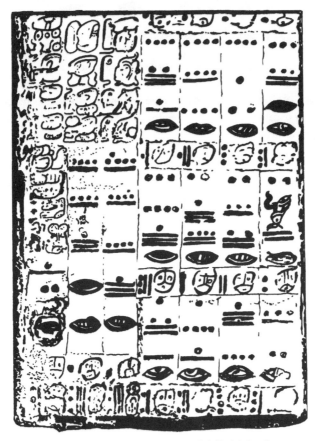

FIG. 22.28. *The Dresden Codex, p. 24 (part). Sächsische Landesbibliothek, Dresden*

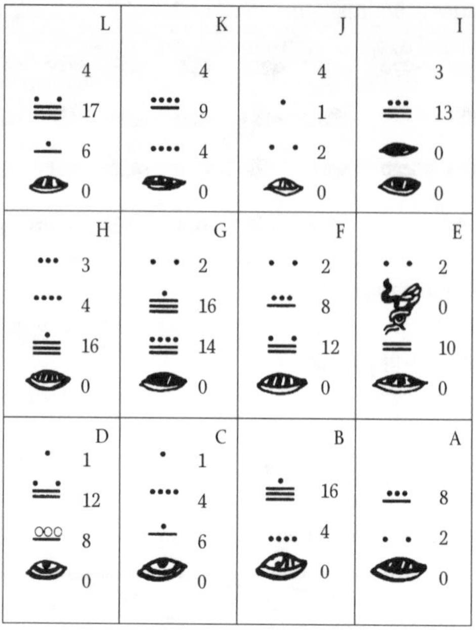

L		K		J		I	
4		4		4		3	
17		9		1		13	
6		4		2		0	
0		0		0		0	
H		**G**		**F**		**E**	
3		2		2		2	
4		16		8		0	
16		14		12		10	
0		0		0		0	
D		**C**		**B**		**A**	
1		1					
12		4		16		8	
8		6		4		2	
0		0		0		0	

FIG. 22.29. *Transcriptions of the numerals on the right-hand side of Fig. 22.28*

Each of these expressions in Mayan astronomical notation refers to a number of days (we know this from the context) and gives the following set of equivalences:

$$
\begin{aligned}
A &= [8;\ 2;\ 0] = 2{,}920 = 1 \times 2{,}920 = 5 \times 584 \\
B &= [16;\ 4;\ 0] = 5{,}840 = 2 \times 2{,}920 = 10 \times 584 \\
C &= [1;\ 4;\ 6;\ 0] = 8{,}760 = 3 \times 2{,}920 = 15 \times 584 \\
D &= [1;12;\ 8;\ 0] = 11{,}680 = 4 \times 2{,}920 = 20 \times 584 \\
E &= [2;\ 0;10;\ 0] = 14{,}600 = 5 \times 2{,}920 = 25 \times 584 \\
F &= [2;\ 8;12;\ 0] = 17{,}520 = 6 \times 2{,}920 = 30 \times 584 \\
G &= [2;16;14;\ 0] = 20{,}440 = 7 \times 2{,}920 = 35 \times 584 \\
H &= [3;\ 4;16;\ 0] = 23{,}360 = 8 \times 2{,}920 = 40 \times 584 \\
I &= [3;13;\ 0;\ 0] = 26{,}280 = 9 \times 2{,}920 = 45 \times 584 \\
J &= [4;\ 1;\ 2;\ 0] = 29{,}200 = 10 \times 2{,}920 = 50 \times 584 \\
K &= [4;\ 9;\ 4;\ 0] = 32{,}120 = 11 \times 2{,}920 = 55 \times 584 \\
L &= [4;17;\ 6;\ 0] = 35{,}040 = 12 \times 2{,}920 = 60 \times 584
\end{aligned}
$$

So this series is nothing other than a table of the synodic revolutions of Venus, calculated by Mayan astronomers as 584 days.

This gives us two indisputable proofs of the mathematical genius of Maya civilisation:

- it shows that they really did invent a place-value system;
- it shows that they really did invent zero.

These are two fundamental disoveries that most civilisations failed to make, including especially Western European civilisation, which had to wait until the Middle Ages for these ideas to reach it from the Arabic world, which had itself acquired them from India.

One problem remains: why was this system not strictly vigesimal, like the Mayas' oral numbering? For instead of using the successive powers of 20 (1, 20, 400, 8,000, etc.), it used orders of magnitude of 1, 20, $18 \times 20 = 360$, $18 \times 20 \times 20 = 7,200$, etc. In short, why was the third "floor" of the system occupied by the irregular number 360?

If Maya numerals had been strictly vigesimal, then its zero would have acquired operational power: that is to say, adding a zero at the end of a numerical string would have multiplied its value by the base. That is how it works in our system, where the zero is a true operational sign. For instance, the number 460 represents the product of 46 multiplied by the base. For the Maya, however, [1; 0; 0] is not the product of [1; 0] multiplied by the base, as the first floor gives units, the second floor gives twenties, but the third floor gives 360s. [1; 0] means precisely 20; but [1; 0; 0] is not 400 $(20 \times 20 + 0 +)$, but 360. The number 400 had to be written as [1; 2; 0] or $(1 \times 360 + 2 \times 20 + 0)$:

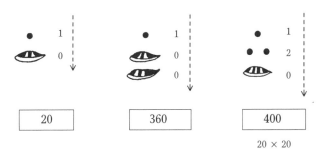

20   360   400

20 × 20

Fig. 22.30.

This anomaly deprived the Maya zero of any operational value, and prevented Mayan astronomers from exploiting their discovery to the full. We must therefore not confuse the Maya zero with our own, for it does not fulfil the same role at all.

A SCIENCE OF THE HIGH TEMPLES

To understand the odd anomaly of the third position in the Maya place-value system we have to delve deep into the very sources of Maya mathematics, and make a long but fascinating detour into Maya mysticism and its reckoning of time.

Maya learned numerals were not invented to deal with the practicalities of everyday reckoning – the business of traders and mere mortals – but to meet the needs of astronomical observation and the reckoning of time. These numerals were the exclusive property of priests, for Maya civilisation made the passing of time the central matter of the gods.

Maya science was practised in the high temples: astronomy was what the priests did. Mayan achievements in astronomy, including the invention of one of the best calendars the world has ever seen, were part and parcel of their mystical and religious beliefs.

The Maya did not think of time as a purely abstract means of ordering events into a methodical sequence. Rather, they viewed it as a super-natural phenomenon laden with all-powerful forces of creation and destruction, directly influenced by gods with alternately kindly and wicked intentions. These gods were associated with specific numbers, and took on shapes which allowed them to be represented as hieroglyphs. Each division of the Maya calendar (days, months, years, or longer periods) was thought of as a "burden" borne on the back of one or another of the divine guardians of time. At the end of each cycle, the "burden" of the next period of time was taken over by the god associated with the next number. If the coming cycle fell to a wicked god, then things would get worse until such time as a kindly god was due to take over. These curious beliefs supported the popular conviction that survival was impossible without learned mediators who could interpret the intentions of the irascible gods of time. The astronomer-priests alone could recognise the attributes of the gods, plot their paths across time and space, and thus determine times that would be controlled by kindly gods, or (as was more frequent) times when the number of kindly gods would exceed that of evil gods. It was an obsession for calculating periods of luck and good fortune over long time-scales, in the hope that such foreknowledge would enable people to turn circumstances to their advantage. [See C. Gallenkamp (1979)]

FIG. 22.31. *The cyclical conception of events in the Mayas' mystical thinking. The inexorable cycle of Chac, god of rain, planting a tree, followed by Ah Puch, god of death, who destroys it, and by Yum Kax, god of maize and of agriculture, who restores it. From the Codex Tro-Cortesianus, copy from Girard (1972), p. 241, Fig. 61*

The priests were thus the possessors of the arcana of time and of the foretelling of the gods bearing the burden of particular times. Mysticism, religion and astronomy formed a single, unitary sphere which gave the priestly caste enormous power over the people, who needed priestly mediation in order to learn of the mood of the gods at any given moment. So despite its amazing scientific insights, Mayan astronomy was very different from what we now imagine science to be: as Girard puts it [R. Girard (1972)], its main purpose was to give mythical interpretations of the magical powers that rule the Universe.

THE MAYA CALENDAR

The Maya had two calendars, which they used simultaneously: the *Tzolkin* – the "sacred almanac" or "magical calendar" or "ritual calendar", used for religious purposes; and the *Haab*, which was a solar calendar.

The religious year of the Maya consisted of twenty cycles of thirteen days, making 260 days in all. It had a basic sequence of twenty named days in fixed order:

Imix	*Cimi*	*Chuen*	*Cib*
Ik	*Manik*	*Eb*	*Caban*
Akbal	*Lamat*	*Ben*	*Etznab*
Kan	*Muluc*	*Ix*	*Cauac*
Chicchan	*Oc*	*Men*	*Ahau*

Each day had its distinct hieroglyph, which also represented directly the corresponding deity or sacred animal or object. As J. E. Thompson explains, prayers were addressed to the days, each of which was the incarnation of a divinity, such as the sun, the moon, the god of maize, the god of death, the Jaguar, etc.

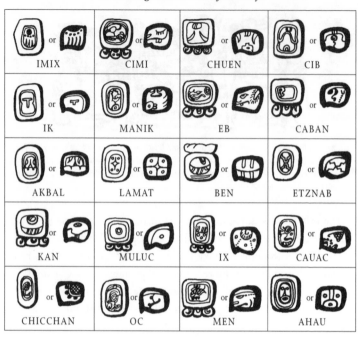

FIG. 22.32. *Hieroglyphs for the twenty days of the Maya calendar, with their names in the Yucatec language. [See Gallenkamp (1979), Fig. 9; Peterson (1961), Fig. 55]*

Each of the days was also associated with a number-sign, in the range 1 to 13 (itself associated with thirteen Maya gods of the "upper world" or *Oxlahuntiku*).

In the first cycle, the first day was associated with the number 1, the second day with the number 2, and so on to the thirteenth day. The numbering then started over, so that the fourteenth day was associated with the number 1, the fifteenth with the number 2, and the last day of the first cycle had number 7.

The second cycle thus began with 8 and reached 13 with the sixth day, so that the numbering began again at 1 with the seventh day of the second cycle.

Thus it took thirteen cycles for the numbering to come back to where it started, with day one counting once again as 1. As there are 13×20 possible pairings of the sets 1–13 and 1–20, the whole series of cycles lasted 260 days.

Each day of the religious year therefore had a unique name consisting of its hieroglyph together with its number resulting from the cyclical recurrence explained above. So a day-hieroglyph plus number gives an unambiguous identification of any day in the religious year. The following expressions, for instance:

FIG. 22.33.

13 CHUEN   4 IMIX

specify the 91st and 121st days of a religious year that begins on *1 Imix*.
(Fig. 22.34 below shows the whole cycle.)

	I	II	III	IV	V	VI	VII	VIII	IX	X	XI	XII	XIII
IMIX	1	8	2	9	3	10	4	11	5	12	6	13	7
IK	2	9	3	10	4	11	5	12	6	13	7	1	8
AKBAL	3	10	4	11	5	12	6	13	7	1	8	2	9
KAN	4	11	5	12	6	13	7	1	8	2	9	3	10
CHICCHAN	5	12	6	13	7	1	8	2	9	3	10	4	11
CIMI	6	13	7	1	8	2	9	3	10	4	11	5	12
MANIK	7	1	8	2	9	3	10	4	11	5	12	6	13
LAMAT	8	2	9	3	10	4	11	5	12	6	13	7	1
MULUC	9	3	10	4	11	5	12	6	13	7	1	8	2
OC	10	4	11	5	12	6	13	7	1	8	2	9	3
CHUEN	11	5	12	6	13	7	1	8	2	9	3	10	4
EB	12	6	13	7	1	8	2	9	3	10	4	11	5
BEN	13	7	1	8	2	9	3	10	4	11	5	12	6
IX	1	8	2	9	3	10	4	11	5	12	6	13	7
MEN	2	9	3	10	4	11	5	12	6	13	7	1	8
CIB	3	10	4	11	5	12	6	13	7	1	8	2	9
CABAN	4	11	5	12	6	13	7	1	8	2	9	3	10
ETZNAB	5	12	6	13	7	1	8	2	9	3	10	4	11
CAUAC	6	13	7	1	8	2	9	3	10	4	11	5	12
AHAU	7	1	8	2	9	3	10	4	11	5	12	6	13

FIG. 22.34. *The 260 consecutive days of the Maya liturgical year*

Each day of the religious year had its own specific character. Some were propitious for marriages or military expeditions, others ruled out such events. More generally, an individual's character and prospects were indissolubly linked to the character of the day of his birth, a belief that is still held by many Central American peoples, notably in the Guatemalan uplands.

Why did the pre-Columbian civilisations of Central America choose 260 as the number of days in their liturgical calendars? F. A. Peterson (1961) pointed out that the difference between the religious year (260) and the solar year (365) is 105 days. Moreover, between the tropic of Cancer and the tropic of Capricorn, the sun is at the zenith twice in every year, at intervals of 105 and 260 days precisely. At Copán, an ancient Maya city in Honduras, the relevant dates for the sun passing through its zenith are 13 August and 30 April. The rainy season begins straight after the sun passes through its "spring" zenith; 105 days later, the sun passes through its "autumn" zenith. So the year could be divided into a period of planting and growth that lasted 105 days, and then a period of harvesting and religious feasts that lasted 260.

This astronomical observation, even if it has not been accepted as the ultimate source of the Maya calendar, is certainly very interesting. Unfortunately the correlation of the sun's zenith with the rainy season only fits at Copán, which is on the fringes of the Maya area.

Other scholars have pointed out that 260 must be thought of as the product of 13 and 20, the divine (since there are 13 divinities in the "upper world" of the Maya) and the human (since Maya numbering is vigesimal, and the name of the number 20, *uinic*, means " a man").

Alongside the ritual calendar, the Maya used a solar-year calendar called the *Haab*, and referred to as the "secular" or "civil" or "approximate" calendar. It had a year of 365 days divided into eighteen *uinal* (twenty-day periods), plus a short "extra" period of five days added at the end of the eighteenth *uinal*. The names of the Maya twenty-day "months" were:

Pop	*Yaxkin*	*Mac*
Uo	*Mol*	*Kankin*
Zip	*Chen*	*Muan*
Zotz	*Yax*	*Pax*
Tzec	*Zac*	*Kayab*
Xul	*Ceh*	*Cumku*

These names referred to various agricultural or religious events, and they were represented by the hieroglyphs of the tutelary god or animal-spirit associated with the event.

POP   XUL   ZAC   PAX

UO   YAXKIN   CEH   KAYAB

ZIP   MOL   MAC   CUMKU

ZOTZ   CHEN   KANKIN

TZEC   YAX   MUAN

or

UAYEB

Literally: "That which has no name"

Glyph and name of the five-day period regularly added to the
eighteenth twenty-day "month" to make
up the *Haab* of 365 days.

FIG. 22.35. *Glyphs and names of the eighteen 20-day "months" of the Maya solar calendar.*
[See Gallenkamp (1979), p. 80; Peterson (1961), p. 225]

The "extra" five-day period was called *Uayeb*, meaning "The one that has
no name", and it was represented by a glyph associated with the idea of
chaos, disaster and corruption. They were thought of as "ghost" days, and
considered empty, sad and hostile to human life. Anyone born during
Uayeb was destined to have bad luck and to remain poor and miserable
all his life long. Peterson quotes Diego de Landa, who reported that during
Uayeb the Maya never washed, combed their hair, or picked their nits;
they did no regular or demanding work, for fear that something untoward
would happen to them.

The first day of each "month", including *Uayeb*, was represented by
the glyph for the "month", that is to say the sign for its tutelary divinity,
together with a special sign:

This sign, which is usually translated by specialists as "0", signified that the god who had carried the burden of time up to that point was passing it on to the following month-god. So since *Zip* and *Zotz* are the names of two consecutive "months" in the "approximate" Maya calendar, the hieroglyph:

FIG. 22.37.   0   ZOTZ

meant that *Zip* was handing over the weight of time to *Zotz*.

As a result the remaining days of each "month", including *Uayeb*, were numbered from 1 to 19, with the second day having the number 1, the third day the number 2, and so on (see Fig. 22.40 below). As a result, the following "date" in the secular or civil calendar:

FIG. 22.38.   4   XUL

signified not the fourth, but the fifth day in the twenty-day "month" of *Xul*!

Each of the twenty days of the basic series (laid out in Fig. 22.32 above) kept exactly the same rank-number in each of the eighteen "months" of the civil or secular year. If the "zero day" of the first month of the year was *Eb*, for example, then the "zero day" of the following seventeen months was also *Eb*. But because of the extra five days added on in each annual cycle, the day-names stepped back by five positions each year. So, for example, if *Ahau* was day 8 in year N, it became day 3 in year N + 1, day 18 in year N + 2, day 13 in year N + 3, and day 8 again in year N + 4. The full cycle thus took four years to complete, and only in the fifth year did the correspondence between the names and the numbers of the days of the "months" return to its starting position.

Within the system, there were only four day-names from the basic series that could correspond to the calendrical expression:

FIG. 22.39.   0   POP

These "new year days" were *Eb, Caban, Ik* and *Manik,* and they were referred to as the "year-bearer" days.

POP	UO	ZIP	ZOTZ	TZEC	XUL	YAXKIN	MOL	CHEN	YAX	ZAC	CEH	MAC	KANKIN	MUAN	PAX	KAYAB	CUMKU	UAYEB
0	0	0	0	0	0	0	0	0	0	0	0	0	0	0	0	0	0	0
1	1	1	1	1	1	1	1	1	1	1	1	1	1	1	1	1	1	1
2	2	2	2	2	2	2	2	2	2	2	2	2	2	2	2	2	2	2
3	3	3	3	3	3	3	3	3	3	3	3	3	3	3	3	3	3	3
4	4	4	4	4	4	4	4	4	4	4	4	4	4	4	4	4	4	4
5	5	5	5	5	5	5	5	5	5	5	5	5	5	5	5	5	5	
6	6	6	6	6	6	6	6	6	6	6	6	6	6	6	6	6	6	
7	7	7	7	7	7	7	7	7	7	7	7	7	7	7	7	7	7	
8	8	8	8	8	8	8	8	8	8	8	8	8	8	8	8	8	8	
9	9	9	9	9	9	9	9	9	9	9	9	9	9	9	9	9	9	
10	10	10	10	10	10	10	10	10	10	10	10	10	10	10	10	10	10	
11	11	11	11	11	11	11	11	11	11	11	11	11	11	11	11	11	11	
12	12	12	12	12	12	12	12	12	12	12	12	12	12	12	12	12	12	
13	13	13	13	13	13	13	13	13	13	13	13	13	13	13	13	13	13	
14	14	14	14	14	14	14	14	14	14	14	14	14	14	14	14	14	14	
15	15	15	15	15	15	15	15	15	15	15	15	15	15	15	15	15	15	
16	16	16	16	16	16	16	16	16	16	16	16	16	16	16	16	16	16	
17	17	17	17	17	17	17	17	17	17	17	17	17	17	17	17	17	17	
18	18	18	18	18	18	18	18	18	18	18	18	18	18	18	18	18	18	
19	19	19	19	19	19	19	19	19	19	19	19	19	19	19	19	19	19	

FIG. 22.40. *The 365 consecutive days of the Maya "civil" year*

List of the 20 basic days	1st year 𝒰	1st year UAYEB	2nd year 𝒰	2nd year UAYEB	3rd year 𝒰	4th year 𝒰	5th year 𝒰	5th year UAYEB
Eb	0	0	. 15		. 10	. 5	0	0
Ben	1	1	. 16		. 11	. 6	1	1
Ix	2	2	. 17		. 12	. 7	2	2
Men	3	3	. 18		. 13	. 8	3	3
Cib	4	4	. 19	UAYEB	. 14	. 9	4	4
Caban	5		0 .	0	. 15	. 10	5	
Etznab	6		1 .	1	. 16	. 11	6	
Cauac	7		2 .	2	. 17	. 12	7	
Ahau	8		3 .	3	. 18	. 13	8	
Imix	9		4 .	4	. 19	. 14	9	
Ik	10		5 .		0 . 0	. 15	10	
Akbal	11		6 .		1 . 1	. 16	11	
Kan	12		7 .		2 . 2	. 17	12	
Chicchan	13		8 .		3 . 3	. 18	13	
Cimi	14		9 .		4 . 4	. 19	14	
Manik	15		10 .		5 .	0 . 0	15	
Lamat	16		11 .		6 .	1 . 1	16	
Muluc	17		12 .		7 .	2 . 2	17	
Oc	18		13 .		8 .	3 . 3	18	
Chuen	19		14 .		9 .	4 . 4	19	

𝒰 : any given month of the 18-month year

FIG. 22.41. *Successive positions of the twenty basic days in the Maya "civil" calendar*

THE SACRED CYCLE OF
MESO-AMERICAN CULTURES

The Maya, as we have seen, used two different calendars simultaneously, the *Tzolkin*, or religious calendar, of 260 days, and the *Haab*, or civil calendar, of 365 days. So to express a date in full, they combined the signs of its place in the religious calendar with the signs of its place in the civil year, thus:

Position of the day in the "ritual" year

Position of the day in the "civil" year

FIG. 22.42.　　13 AHAU　　　　18 CUMKU

Since both these cycles permuted the days in regular recurrent order, the correspondence between the two calendars returned to its starting positions after a fixed period of time, which elementary arithmetic shows must be 18,980 days, or 52 "approximate" or civil years. In other words, the amount of time required for a given date in the civil calendar to match a given date in the religious calendar a second time round was equal to 52 years of 365 days or 72 years of 260 days.

You can imagine how this worked by thinking of a huge bicycle, with a chain wheel of 365 numbered teeth pulling round a sprocket with 260 numbered teeth. For the same chain link at the front sitting on tooth 1 to match the same chain link at the back also sitting on tooth 1, the pedals will have to turn 52 times, or (which is necessarily the same thing) the back wheel will have to go round 73 times.

The number of days in this cycle is equal to the lowest common multiple of 260 and 365. Since both these numbers are divisible by 5 and since 5 is moreover the highest common factor of 260 and 365, the number sought is

$$\frac{260 \times 365}{5} = 18,980 = 52 \text{ civil years} = 73 \text{ religious years}$$

That is the origin of the celebrated *sacred cycle of fifty-two years*, otherwise known as the Calendar Round, which played such an important role in Maya and Aztec religious life. (The Aztecs, for example, believed that the end of each Round would be greeted by innumerable cataclysms and catastrophes; so at the approach of the fateful date, they sought to appease the gods by making huge human sacrifices to them, in the hope of being allowed to live on through another cycle.)

We must mention, finally, that Maya astronomers also took the Venusian calendar into consideration. They had observed that after each period of 65

Venusian years, the start of the solar year, of the religious year and of the Venusian year all coincided precisely with the start of a new sacred cycle of 52 "civil" years. Such a remarkable occurrence was celebrated with enormous festivities.

TIME AND NUMBERS ON MAYA *STELAE*

Alongside their two calendars, the Maya also used a third and rather amazing way of calculating the passage of time on their *stelae* or ceremonial columns. This "Long Count", as it is called by Americanists, began at zero at the date of 13 *baktun*, 4 *ahau*, 8 *cumku*, corresponding quite precisely, according to the concordance established by J. E. Thompson (1935), to 12 August 3113 BCE in the Gregorian calendar. It is generally assumed that this date corresponded to the Mayas' calculation of the creation of the world or of the birth of their gods [S. G. Morley (1915)]. However, this kind of reckoning did not use solar years, nor lunar years, nor even the revolutions of Venus, but multiples of recurrent cycles.

Its basic unit was the "day" and an approximate "year" of 360 days. Time elapsed since the start of the Mayan era was reckoned in *kin* ("day"), *uinal* (20-day "month"), *tun* (360-day "year"), *katun* (20-"year" period), *baktun* (400-"year" period), *pictun* (8,000-"year" cycle), and so on as laid out in Fig. 22.43.

The *katun* (= 20 *tun*) obviously did not correspond exactly to twenty years as we reckon them, but to 20 years less 104.842 days; similarly, the *baktun* (= 20 *katun* = 400 *tun*) was not exactly 400 years, but 400 years less 2,096.84 days. However, Mayan astronomers were perfectly aware of the discrepancies and of the corrections needed to the "Long Count" to make it correspond properly to actual solar years.

Order of magnitude	Names and definitions		Equivalences	Number of days
First	*kin*	DAY		1
Second	*uinal*	"MONTH" OF 20 DAYS	20 *kin*	20
Third	*tun*	"YEAR" OF 18 "MONTHS"	18 *uinal*	360
Fourth	*katun*	CYCLE OF 20 "YEARS"	20 *tun*	7,200
Fifth	*baktun*	CYCLE OF 400 "YEARS"	20 *katun*	144,000
Sixth	*pictun*	CYCLE OF 8,000 "YEARS"	20 *baktun*	2,880,000
Seventh	*calabtun*	CYCLE OF 160,000 "YEARS"	20 *pictun*	57,600,000
Eighth	*kinchiltun*	CYCLE OF 3,200,000 "YEARS"	20 *calabtun*	1,152,000,000
Ninth	*alautun*	CYCLE OF 64,000,000 "YEARS"	20 *kinchiltun*	23,040,000,000

FIG. 22.43. *The units of computation of time used in Maya calendrical inscriptions (the "Long Count" system)*

As we have seen, when counting people, animals or objects, the Maya used a strictly vigesimal system (see Fig. 22.10 above); but their time-counting method had an irregularity at the level of the third order of magnitude, which made the whole system cease to be vigesimal:

1 *kin*		1 =	1 day
1 *uinal*	= 20 *kin*	20 =	20 days
1 *tun*	= 18 *uinal*	18 × 20 =	360 days
1 *katun*	= 20 *tun*	20 × 18 × 20 =	7,200 days
1 *baktun*	= 20 *katun*	20 × 20 × 18 × 20 =	144,000 days
1 *pictun*	= 20 *baktun*	20 × 20 × 20 × 18 × 20 =	2,880,000 days

If they had used a *tun* of 20 instead of 18 *uinal*, that is to say, using a truly vigesimal system, then their "year" would have had 400 days, and would have thus been even further "out" from the true solar year than was the 360-day *tun* of their calendrical computations.

FIG. 22.44. *Detail of lintel 48 from Yaxchilán showing a bizarre representation of the expression "16* kin*" ("16 days"): a squatting monkey (a zoomorphic glyph sometimes associated with the word* kin*) holding the head of the god 6 in his hands and, in his legs, the death's-head which represents the number 10*

Each of these units of time had a special sign, which, like most Mayan hieroglyphs, had at least two different realisations, depending on whether it was being written with some kind of ink or paint on a codex, or carved in stone on a monument or ceremonial column. In other words, each of these units of time could be figured :

- by a relatively simple graphical sign, which could be more or less motivated by what it represented, or else an abstract geometrical shape;
- by the head of a god, a man, or an animal – otherwise called cephalomorphic glyphs, which were used for carved inscriptions;
- exceptionally, at Quiriguà and Palenque, by anthropo-morphic glyphs, that is to say, by a god, man, or animal drawn in full.

SYMBOLIC	CEPHALOMORPHIC	ANTHROPOMORPHIC

FIG. 22.45. *Various hieroglyphs for* kin, *"day"*

To represent the numerical coefficients of the units of time in the "Long Count", Maya scribes and sculptors used numerals which, like the unit-signs themselves, had more than one visual realisation.

Kin   Uinal   Tun   Katun   Baktun

FIG. 22.46. *Hieroglyphs for the units of time (from the Quiriguà stelae).*

Method One for showing the numbers was to use the cephalomorphic signs for the thirteen gods of the upper world (the set of gods and signs known as the *Oxlahuntiku*) for numbers 1 to 13. The maize-god, for instance, was associated with and therefore represented the number 5, and the god of death represented number 10.

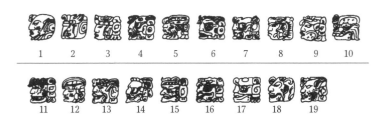

FIG. 22.47. *Maya cephalomorphic numerals 1 to 19 (found on pieces of pottery and sculpture, on* stelae *J and F at Quiriguà, and on the "hieroglyphic staircase" at Palenque). [See Peterson (1961), p. 220, Fig. 52; Thompson (1960), p. 173, Fig. 13]*

For the numbers 14 to 19, however, the system used the numbers 4 to 9 with a modification that can be seen in the following figure:

VARIANTS OF THE GLYPH FOR "9"

VARIANTS OF THE GLYPH FOR "19"

FIG. 22.48.

If you look closely you can see that the jawbone of the "nine-god" has been removed to make the glyph represent the number 19. Arithmetically, this is elementary, because, as the jawbone symbolised the god of death, it enabled an "extra ten" to be shown in the sign:

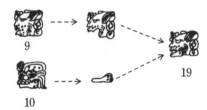

FIG. 22.49.

Method One was not used very often; more frequently, even in calendrical inscriptions, the dot-and-line system (see above, Fig. 22.21) is found.

In any case, dates and lengths of time could be expressed fairly simply within the systems explained so far. Americanists call these expressions "initial series". Our first example of an initial series comes from the "hieroglyphic staircase" of Palenque, where the numbers are represented by heads of the divinities, as shown in Fig. 22.50.

FIG. 22.50. *Initial series on the "hieroglyphic staircase" at Palenque. The date is given in cephalomorphic figures [From Peterson (1961), p. 232, Fig. 58]*

The inscription begins with the glyph called the "initial series start sign", or POP:

POP

FIG. 22.51.

This sign corresponds to the name of the divinity "responsible" for the "month" of the "civil" calendar on the day that the inscription was carved (or, to be more precise, the name of the month in the "secular" calendar in which the last day of the inscribed date falls).

Then, at the foot of the inscription, we can read the position of the date with respect to the "civil" and to the "religious" year, thus:

8 AHAU   13 POP

FIG. 22.52.

As for the number of days elapsed since the initial date of the Mayan era, it is expressed in the "Long Count" as follows:

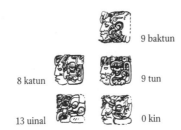

9 baktun

8 katun   9 tun

13 uinal   0 kin

FIG. 22.53.

The date is read from top to bottom, and in descending order of magnitude of the counting units of the Maya calendar. It can be transcribed as follows:

9 *baktun*	=	9 × 144,000 days	....	1,296,000
8 *katun*	=	8 × 7,200 days	....	57,600
9 *tun*	=	9 × 360 days	....	3,240
13 *uinal*	=	13 × 20 days	....	260
0 *kin*	=	0 × 1 day	....	0
Total				1,357,100 days

A fairly simple calculation reveals this to be the year 603 CE.

The Leyden Plate provides another example:

SIDE 1   SIDE 2

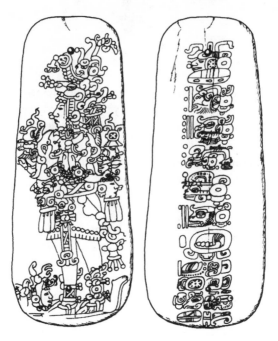

FIG. 22.54. *The Leyden Plate. This thin jade pendant, 21.5 cm high, was found in Guatemala, near Puerto Barrios, and is thought to have been carved at Tikal. On side 1 it shows a richly-clad Maya (probably a god) trampling a prisoner, and, on side 2, a date corresponding to 320 CE. Rijksmuseum voor Volkenkunde, Leyden, Holland*

As in the illustration from Palenque, this expression begins with an "introductory glyph", in this case the name of the god whose "burden" it was to carry the "month" of YAXKIN, during which the building on which this inscription was carved was completed:

FIG. 22.55.   YAXKIN

The date of completion is also expressed in terms of its position in the civil and religious calendars, thus:

1 EB   0 YAXKIN

FIG. 22.56.

As for the corresponding date in the "Long Count" system, it is given in this form:

8 baktun

14 katun

3 tun

1 uinal

12 kin

Fig. 22.57.

This date is also to be read from top to bottom in descending order of magnitudes, and from left to right within each glyph, and produces the following numbers:

8 *baktun*	= 8 × 144,000 days	....	1,152,000
14 *katun*	= 14 × 7,200 days	....	100,800
3 *tun*	= 3 × 360 days	....	1,080
1 *uinal*	= 1 × 20 days	....	20
12 *kin*	= 12 × 1 day	....	12
Total			1,253,912 days

Once again, a simple calculation reveals that in view of the number of days since the beginning of the Mayan era this inscription was carved in the year 320 CE.

It was long thought that the Leyden Plate was the oldest dated artefact from Maya civilisation. However, in 1959, archaeological excavations in the ruins of the city of Tikal, in Guatemala, turned up an even older dated inscription. *Stela* no. 29 carries an inscription which can be translated as:

8 *baktun*	= 8 × 144,000 days	....	1,152,000
12 *katun*	= 12 × 7,200 days	....	86,400
14 *tun*	= 14 × 360 days	....	5,040
8 *uinal*	= 8 × 20 days	....	160
0 *kin*	= 0 × 1 day	....	0
Total			1,243,600 days

which works out at the year 292 CE.

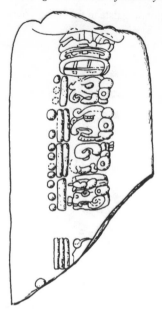

FIG. 22.58. *Side 2 of* stela *29 from Tikal (Guatemala), the oldest dated Mayan inscription found so far. The date written on it – usually transcribed as 8.12.14.8.0 – matches the year 292 CE. [See Shook (1960), p. 33]*

There are many other examples of calendrical inscriptions on the numerous *stelae* of the Maya, each one teeming with fantastical and elaborate signs. To conclude this section, let us look at one date found on *stela E* from Quiriguà.

The date of the *stela*'s erection begins on the top line with two glyphs: the first, on the left, is composed of the figure 9 with the head of the god representing *baktun*, and the other of the figure 17 with the head of the god representing *katun*. It then goes on, on the next line, with two compound glyphs signifying "0 *tun*" and "zero *uinal*" respectively; and, on the bottom line, the date ends with a sign meaning "0 *kin*".

 9 BAKTUN
$9 \times 144,000$
(= 1,296,000 days)

 17 KATUN
$17 \times 7,200$
(= 122,400 days)

 0 TUN
0×360
(= 0 days)

 0 UINAL
0×20
(= 0 days)

 0 KIN
0×1
(= 0 days)

FIG. 22.59.

So the people who put up this column expressed the number of days elapsed since the start of the Mayan era up to the date on which they made this inscription, which is tantamount to expressing the latter date as:

9 *baktun*	$= 9 \times 144{,}000$ days	....	1,296,000
17 *katun*	$= 17 \times 7{,}200$ days	....	122,400
0 *tun*	$= 0 \times 360$ days	....	0
0 *uinal*	$= 0 \times 20$ days	....	0
0 *kin*	$= 0 \times 1$ day	....	0
Total			1,418,400 days

So one million four hundred and eighteen thousand and four hundred days had passed since the "beginning of time" and, given that we know what the start-date was, we can calculate fairly easily that *stela E* at Quiriguà was completed on 24 January 771 CE.

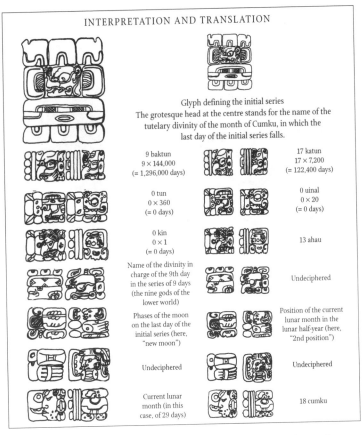

INTERPRETATION AND TRANSLATION

Glyph defining the initial series
The grotesque head at the centre stands for the name of the tutelary divinity of the month of Cumku, in which the last day of the initial series falls.

9 baktun
9 × 144,000
(= 1,296,000 days)

17 katun
17 × 7,200
(= 122,400 days)

0 tun
0 × 360
(= 0 days)

0 uinal
0 × 20
(= 0 days)

0 kin
0 × 1
(= 0 days)

13 ahau

Name of the divinity in charge of the 9th day in the series of 9 days (the nine gods of the lower world)

Undeciphered

Phases of the moon on the last day of the initial series (here, "new moon")

Position of the current lunar month in the lunar half-year (here, "2nd position")

Undeciphered

Undeciphered

Current lunar month (in this case, of 29 days)

18 cumku

FIG. 22.60. *Detail from* stela E *at Quiriguà, giving an initial series together with a complementary series that provides other details on the date of the stela's erection. The date is 9.17.0.0.0 and 13 ahau, 18* cumku, *which matches 24 January 771 CE, in the Gregorian calendar. [See Morley (1915), Fig. 25]*

We should note that these *stelae* contain some of the most interesting Mayan inscriptions that have been found. If we compare the oldest and newest dates found in particular places, we can get an idea of the duration of the great Maya cities. For example, at Tikal, the oldest date found is 292 CE and the latest is 869 CE; at Uaxactún, the limit-dates are 328 CE and 889 CE; at Copán, the relevant inscriptions are of 469 CE and 800 CE;* and so on. The important point in this long digression is to note that in their calendrical inscriptions the Maya represented the "zero", that is to say the absence of units in any one order, by glyphs and signs of the most diverse kinds.

FIG. 22.61. *Hieroglyphs for "zero" found on various Maya stelae and sculptures. Left to right: the first six, the commonest, are symbolic notations; the seventh and eighth are cephalomorphic, and the last is anthropomorphic. [See Peterson (1961), Fig. 51; Thompson (1960). Fig. 13]*

FIG. 22.62. *Detail of a plaque found at the Palenque Palace: an unusual anthropomorphic representation of the expression "0 kin" ("no days"). From Peterson (1961), fig. 14, p. 72.*

* See M. D. Coe, *op.cit.*, p. 68

MAYA MATHEMATICS:
A SCIENCE IN THE SERVICE OF
ASTRONOMY AND MYSTICISM

The Maya system for counting time and for expressing the date did not really require a zero: the date expressed in Fig. 22.60 above, for example, could have been represented just as easily and just as unambiguously by:

9 *baktun*, 17 *katun*

as by the glyphs we actually have, which say

9 *baktun*, 17 *katun*, 0 *tun*, 0 *uinal*, 0 *kin*

So why did Maya calendrical computation bother to invent a zero?

The answers have to do with the religious, aesthetic and graphical ideas and customs of the Maya.

In religious terms, each of the time-units was imagined as a burden carried by one of the gods, the "tutelary god" of that cycle of time. At the end of the relevant cycle, the god passed on the burden of time to the god designated by the calendar as his successor.

On the date of "9 *baktun*, 11 *katun*, 7 *tun*, 5 *uinal*, and 2 *kin*", for instance, the god of the "days" carried the number 2, the god of the "months" carried the number 5, the god of the "years" carried the number 7, and so on, in this manner:

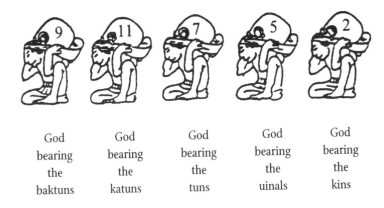

| God bearing the baktuns | God bearing the katuns | God bearing the tuns | God bearing the uinals | God bearing the kins |

FIG. 22.63.

FIG. 22.64. Stela A *at Quiriguà. Erected in 775 CE, this column has gods carved on its front and back, and calendrical, astronomical and other glyphs carved on its other two sides. From Thompson (1960), Fig. 11, p. 163*

If we were to transpose this system to our own Gregorian calendar, we would need six gods to carry the "burden" of the date "31 December 1899". One god – the "day-god" – would "carry" the number 31; the second would bear the number 12, for the months; the third would carry the number 9, for the years; the fourth would "carry" the decades; and we would need two more, for the centuries and the millennia. At the end of the day of the 31 December 1899, these gods would have rested for a moment before setting off on a new cycle. The day-god would resume his burden, but with the number 1, and similarly for the month-god. But as the decade and the century would change (to 1900), the year-god and the decade-god would be released from their burdens for a period of time, the century-god would now bear the burden of the number 8, and the millennium-god would carry on with his 1 as he had been doing for the previous 900 years.

In Maya mystical thought, the fact that the gods occasionally had a rest from their burdens did not justify simply eradicating them from the representation of the task of carrying the burdens of time. Failing to put them in their right places in the inscription might have angered them! It would also have destroyed the absolute regularity of the system, in which calendrical expressions always ran from top to bottom in descending orders of magnitude. The aesthetically pleasing sequence of symbols in an unchanging order would have been altered if there had been no sign for zero. So we can say, in conclusion, that the demands of the writing system itself, the aesthetic appearance of inscriptions intended to be ceremonial, and a set of religious beliefs made the invention of a "zero-count" an absolute necessity (see Fig. 22.65).

Nonetheless, the calendrical system of the Maya is also part of a long and slow evolution leading towards the discovery of a place-value system. The Mayan units of time were always placed in precisely the same position in an inscription, with the same regularity as the tokens in an abacus or "counting table". And Mayan astronomer-priests did not fail to notice the arithmetical potential of their system.

When writing manuscripts, as opposed to inscriptions carved in stone, they eventually came to omit the glyphs representing the units of time (or the gods that were responsible for them), and wrote down only the corresponding numerical coefficients, since the order of the magnitudes was firm and fixed. So dates in the manuscripts are expressed just by numbers. For example, instead of writing the date "8 *baktun*, 11 *katun*, 0 *tun*, 14 *uinal*, 0 *kin*" as follows:

8 BAKTUN   11 KATUN   0 TUN   14 UINAL   0 KIN

Fig. 22.65A.

they wrote (top to bottom, and with the numerical expressions rotated through 90°) simply:

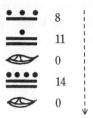

•••__	8
•___	11
⬳	0
••••__	14
⬳	0

FIG. 22.65B.

Omitting the glyphs of the tutelary gods must have had less religious consequence in manuscripts than in the ceremonial and sacred *stelae*.

Taken outside of the context of mysticism and theology, the Maya system constitutes a remarkable written numeral system, incorporating both a true zero and the place-value principle. However, since it had been developed exclusively to express dates and to serve astronomical and calendrical computation, the system retained an irregular value in its third position, which, as we may recall, was $20 \times 18 = 360$, and not $20 \times 20 = 400$. This flaw made the system unsuitable for arithmetical operations and blocked any further mathematical development.

It is true that Maya scholars were concerned above all with matters religious and prophetic; but have not astrology and religion opened the path to philosophical and scientific developments in many places in the world? So we must pay homage to the generations of brilliant Mayan astronomer-priests who, without any Western influence at all, developed concepts as sophisticated as zero and positionality, and, despite having only the most rudimentary equipment, made astronomical calculations of quite astounding precision.

CHAPTER 23

THE FINAL STAGE OF
NUMERICAL NOTATION

THE LEGEND OF SESSA

In Arabic and Persian literature it is often written that the Indian world may glory in three achievements:

- the positional decimal notation and methods of calculation;
- the tales of the *Panchatantra* (from which probably came the well-known fable of *Kalila wa Dimna*);
- the *Shaturanja*, the ancestor of chess, about which a famous legend (adapted into modern terms) will give us an apt introduction to this very important chapter.

In order to prove to his contemporaries that a monarch, no matter how great his power, was as nothing without his subjects, an Indian Brahmin of the name of Sessa one day invented the game of *Shaturanja*.

This game is played between four players on an eight by eight chessboard, with eight pieces (King, Elephant, Horse, Chariot, and four Soldiers), which are moved according to points scored by rolling dice.

When the game was shown to the King of India, he was so amazed by the ingenuity of the game and by the myriad variety of its possible plays that he summoned the Brahmin, that he might reward him in person.

"For your extraordinary invention," said the King, "I wish to make you a gift. Choose your reward yourself and you shall receive it forthwith. I am so rich and so powerful that I can fulfil your wildest desire."

The Brahmin reflected on his reply, and then astonished everyone by the modesty of his request.

"My good Lord," he replied, "I wish that you would grant me as many grains of wheat as will fill the squares on the board: one grain for the first square, two for the second, four for the third, eight for the fourth, sixteen for the fifth, and so on, putting into each square double the number of grains that were put in the square before."

"Are you mad to suggest so modest a demand?" exclaimed the astonished King. "You could offend me with a request so unworthy of my generosity, and so trivial compared with all that I could offer you. But let it be! Since that is your wish, my servant will bring you your bag of wheat before nightfall."

The Brahmin made the merest smile, and withdrew from the Palace. That evening, the King remembered his promise and asked his Minister if the madman Sessa had received his meagre reward. "Lord," replied the Minister, "your orders are being carried out. The mathematicians of your august Court are at this moment working out the number of grains to give to the Brahmin."

The King's brow darkened. He was not used to such delay in obeying his orders. Before retiring to bed, the King asked once more whether the Brahmin had received his bag of grain.

"O King," replied the Minister, hesitating, "Your mathematicians have still not completed their calculations. They are working at it unceasingly, and they hope to finish before dawn."

The calculations proved to take far longer than had been expected. But the King, who did not wish to hear about the details, ordered that the problem should be solved before he awoke.

The next morning, however, his order of the night before remained unfulfilled, and the monarch, incensed, dismissed the calculators who had been working at the task.

"O good Lord," said one of his Counsellors, "you were right to dismiss these incompetents. They were using ancient methods! They are still counting on their fingers and moving counters on an abacus. I permit myself to suggest that the calculators of the central province of your Kingdom have for generations already been using a method far better and more rapid than theirs. They say it is the most expeditious, and the easiest to remember. Calculations which your mathematicians would need days of hard work to complete would trouble those of whom I speak for no more than a brief moment of time."

On this advice, one of these ingenious arithmeticians was brought to the Palace. He solved the problem in record time, and came to present his result to the King.

"The quantity of wheat which has been asked of you is enormous," he said in a grave voice. But the King replied that, no matter how huge the amount, it would not empty his granaries.

He therefore listened with amazement to the words of the sage.

"O Lord, despite all your great power and riches, it is not within your means to provide so great a quantity of grain. This is far beyond what we know of numbers. Know that, even if every granary in your Kingdom were emptied, you would still only have a negligible part of this huge quantity. Indeed, so great a quantity cannot be found in all the granaries of all the kingdoms of the Earth. If you desire absolutely to give this reward, you should begin by emptying all the rivers, all the lakes, all the seas and the oceans, melting the snows which cover the mountains and all the regions of the world, and turning all this into fields of corn. And then, when you have sown and reaped 73 times over this whole area, you will finally be quit of this huge debt. In fact, so huge a quantity of grain would have to be stored in a volume of twelve billion and three thousand million cubic metres, and require a granary 5 metres wide, 10 metres long and 300 million kilometres high (twice the distance from the Earth to the Sun)!"

The calculator revealed to the King the characteristics of the revolutionary method of numeration of his native region.

"The method of representing numbers traditionally used in your Kingdom is very complicated, since it is encumbered with a panoply of different signs for the units from 10 upwards. It is limited, since its largest number is no greater than 100,000. It is also totally unworkable, since no arithmetical operation can be carried out in this representation. On the other hand, the system which we use in our province is of the utmost simplicity and of unequalled efficacity. We use the nine figures 1, 2, 3, 4, 5, 6, 7, 8, and 9, which stand for the nine simple units, but which have different values according to the position in which they are written in the representation of a number, and we use also a tenth figure, 0, which means "null" and stands for units which are not present. With this system we can easily represent any number whatever, however large it may be. And this same simplicity is what makes it so superior, along with the ease which it brings to every arithmetical operation."

With these words, he then taught the King the principal methods of the calculation of the reward, and explained his operations as follows.

According to the demand of the Brahmin, we must place

> 1 grain of corn on the first square;
> 2 grains on the second square;
> 4 (2×2) on the third;
> 8 ($2 \times 2 \times 2$) on the fourth;
> 16 ($2 \times 2 \times 2 \times 2$) on the fifth;

and so on, doubling each time from one square to the next. On the sixty-fourth square, therefore, must be placed as many grains as there are units in the result of 63 multiplications by 2 (namely 2^{63} grains). So the quantity the Brahmin demanded is equal to the sum of these 64 numbers, namely

$$1 + 2 + 2^2 + 2^3 + \ldots + 2^{63}.$$

"If you add one grain to the first square," explained the calculator, "you would have two grains there, therefore 2×2 in the first two squares. By the third square you would then find a total of $2 \times 2 + 2 \times 2$ grains, or $2 \times 2 \times 2$ in all. By the fourth the total would be $2 \times 2 \times 2 + 2 \times 2 \times 2$, or $2 \times 2 \times 2 \times 2$ in all. Proceeding in this way, you can see that by the time you reach the last square of the board the total would be equal to the result of 64 multiplications by 2, or 2^{64}. Now, this number is equal to the six-fold product of 10 successive multiplications by 2, further multiplied by the number 16:

$$2^{64} = 2^{10} \times 2^{10} \times 2^{10} \times 2^{10} \times 2^{10} \times 2^{10} \times 2^4$$
$$= 1,024 \times 1,024 \times 1,024 \times 1,024 \times 1,024 \times 1,024 \times 16$$

"And so," he concluded, "since this number has been obtained by adding one to the quantity sought, the total number of grains is equal to this number diminished by one grain. By completing these calculations in the way I have shown you, you may satisfy yourself, O Lord, that the number of grains demanded is exactly eighteen quadrillion, four hundred and forty-six trillion, seven hundred and forty-four billion, seventy-three thousand seven hundred and nine million, five hundred and fifty-one thousand, six hundred and fifteen (18,446,744,073,709,551,615)!"

"Upon my word!" replied the King, very impressed, "the game this Brahmin has invented is as ingenious as his demand is subtle. As for his methods of calculation, their simplicity is equalled only by their efficiency! Tell me now, my wise man, what must I do to be quit of this huge debt?" The Minister reflected a moment, and said:

"Catch this clever Brahmin in his own trap! Tell him to come here and count for himself, grain by grain, the total quantity of wheat which he has been so bold to demand. Even if he works without a break, day and night, one grain every second, he will gather up just one cubic metre in six months, some 20 cubic metres in ten years, and, indeed, a totally insignificant part of the whole during the remainder of his life!"

THE MODERN NUMBER-SYSTEM:
AN IMPORTANT DISCOVERY

The legend of Sessa thus attributes to Indian civilisation the honour of making this fundamental realisation which we may call the modern number-system. We shall see in due course that, despite the mythical character of the story, this fact is completely true.

But first we must weigh the importance of this written number-system, which nowadays is so commonplace and familiar that we have come to forget its depth and qualities.

Anyone who reflects on the universal history of written number-systems cannot but be struck by the ingeniousness of this system, since the concept of zero, and the positional value attached to each figure in the representation of a number, give it a huge advantage over all other systems thought up by people through the ages.

To understand this, we shall go back to the beginning of this history. But instead of following its different stages purely chronologically, and according to the various civilisations involved, we shall for the moment let ourselves be guided by a kind of logic of time, the regulator of historic data, which has made of human culture a profound unity.

THE EARLIEST NUMERICAL RULE: ADDITION

This story begins about five thousand years ago in Mesopotamia and in Egypt, in advanced societies in full expansion, where it was required to determine economic operations far too varied and numerous to be entrusted to the limited capabilities of human memory. Making use of archaic concrete methods, and feeling the need to preserve permanently the results of their accounts and inventories, the leaders of these societies understood that some completely new approach was required.

To overcome the difficulty, they had the idea of representing numbers by graphic signs, traced on the ground or on tablets of clay, on stone, on sheets of papyrus, or on fragments of pottery. And so were born the earliest number-systems of history.

Independently or not, several other peoples embarked on this road during the millennia which followed. And it all worked out as though, over the ages and across civilisations, the human race had experimented with the different possible solutions of the problem of representing and manipulating numbers, until finally they settled on the one which finally appeared the most abstract, the most perfected and the most effective of all.

To begin with, written number-systems rested on the *additive principle*, the rule according to which the value of a numerical representation is obtained by adding up the values of all the figures it contains. They were therefore very primitive. Their basic figures were totally independent of each other (each one having only one absolute value), and had to be duplicated as many times as required.

The Egyptian hieroglyphic number-system, for example, assigned a special sign to unity and to each power of 10: a vertical stroke for 1, a sign like an upside-down U for 10, a spiral for 100, a lotus flower for 1,000, a raised finger for 10,000, a tadpole for 100,000, and a kneeling man with arms outstretched to the sky for 1,000,000. To write the number 7,659 required 7 lotus flowers, 6 spirals, 5 signs for 10 and 9 vertical strokes of unity, all of which required a total of 27 distinct figures (Fig. 23.1).

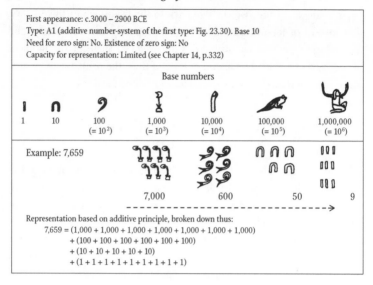

First appearance: c.3000 – 2900 BCE
Type: A1 (additive number-system of the first type: Fig. 23.30). Base 10
Need for zero sign: No. Existence of zero sign: No
Capacity for representation: Limited (see Chapter 14, p.332)

Base numbers

1   10   100 $(= 10^2)$ 1,000 $(= 10^3)$ 10,000 $(= 10^4)$ 100,000 $(= 10^5)$ 1,000,000 $(= 10^6)$

Example: 7,659   7,000   600   50   9

Representation based on additive principle, broken down thus:
7,659 = (1,000 + 1,000 + 1,000 + 1,000 + 1,000 + 1,000 + 1,000)
+ (100 + 100 + 100 + 100 + 100 + 100)
+ (10 + 10 + 10 + 10 + 10)
+ (1 + 1 + 1 + 1 + 1 + 1 + 1 + 1 + 1)

FIG. 23.1. *Egyptian hieroglyphic number-system*

The Sumerian number-system (which used base 60, with 10 as auxiliary base) gave a separate sign to each of the following numbers, in the order of their successive unit orders of magnitude:

$$1 \quad 10 \quad 60 \quad 600 \quad 3,600 \quad 36,000 \quad 216,000$$
$$= 10 \times 60 \quad = 60^2 \quad = 10 \times 60^2 \quad = 60^3$$

But it too was limited to repeating the figures as many times as required to make up the number. The number 7,659 was therefore represented according to the following arithmetical decomposition, which twice repeats the sign for 3,600, seven times the sign for 60, three times that for 10 and nine times the sign for unity, so that 21 distinct signs are required to represent this number (Fig. 23.2):

$$7,659 = (3,600 + 3,600)$$
$$+ (60 + 60 + 60 + 60 + 60 + 60 + 60)$$
$$+ (10 + 10 + 10)$$
$$+ (1 + 1 + 1 + 1 + 1 + 1 + 1 + 1 + 1)$$

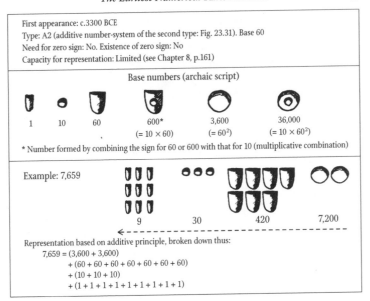

FIG. 23.2. *Sumerian number-system*

Other similar notations include the Proto-Elamite, the Cretan systems (Hieroglyphic and Linear A and Linear B), the Hittite hieroglyphic system, and even the Aztec number-system (which differed from the others only in that it used a base of 20) (Fig. 23.3 to 23.6).

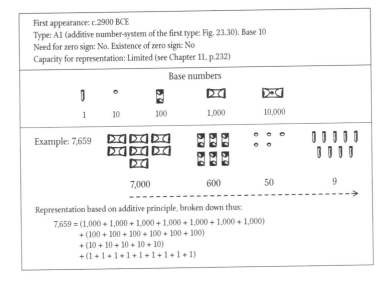

FIG. 23.3. *Proto-Elamite number-system*

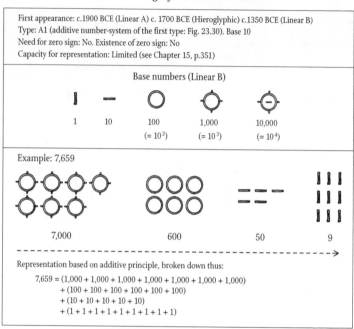

FIG. 23.4. *Cretan number-system (hieroglyphic and Linear A and B)*

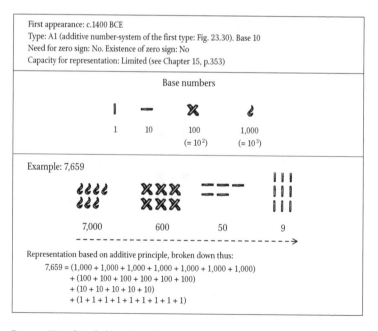

FIG. 23.5. *Hittite hieroglyphic number-system*

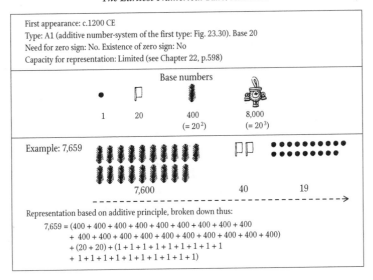

First appearance: c.1200 CE
Type: A1 (additive number-system of the first type: Fig. 23.30). Base 20
Need for zero sign: No. Existence of zero sign: No
Capacity for representation: Limited (see Chapter 22, p.598)

Base numbers

| 1 | 20 | 400 ($= 20^2$) | 8,000 ($= 20^3$) |

Example: 7,659

7,600   40   19

Representation based on additive principle, broken down thus:

$7,659 = (400 + 400 + 400 + 400 + 400 + 400 + 400 + 400 + 400$
$+ 400 + 400 + 400 + 400 + 400 + 400 + 400 + 400 + 400 + 400)$
$+ (20 + 20) + (1 + 1 + 1 + 1 + 1 + 1 + 1 + 1 + 1$
$+ 1 + 1 + 1 + 1 + 1 + 1 + 1 + 1 + 1 + 1)$

FIG. 23.6. *Aztec number-system*

To avoid the encumbrance of such a multitude of symbols, certain peoples introduced supplementary signs which corresponded to inter-mediate units. Such was the case for the Greeks, the Shebans, the Etruscans, and the Romans, who assigned a separate symbol to each of the numbers 5, 50, 500, 5,000, and so on, in addition to those they already had for the different powers of 10 (Fig. 23.7 and 23.8).

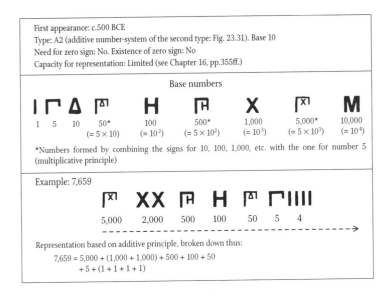

First appearance: c.500 BCE
Type: A2 (additive number-system of the second type: Fig. 23.31). Base 10
Need for zero sign: No. Existence of zero sign: No
Capacity for representation: Limited (see Chapter 16, pp.355ff.)

Base numbers

| 1 | 5 | 10 | 50* ($= 5 \times 10$) | 100 ($= 10^2$) | 500* ($= 5 \times 10^2$) | 1,000 ($= 10^3$) | 5,000* ($= 5 \times 10^3$) | 10,000 ($= 10^4$) |

*Numbers formed by combining the signs for 10, 100, 1,000, etc. with the one for number 5 (multiplicative principle)

Example: 7,659

5,000   2,000   500   100   50   5   4

Representation based on additive principle, broken down thus:

$7,659 = 5,000 + (1,000 + 1,000) + 500 + 100 + 50$
$+ 5 + (1 + 1 + 1 + 1)$

FIG. 23.7. *Greek acrophonic number-system*

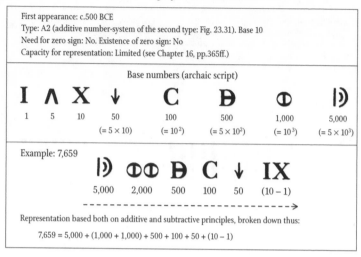

First appearance: c.500 BCE
Type: A2 (additive number-system of the second type: Fig. 23.31). Base 10
Need for zero sign: No. Existence of zero sign: No
Capacity for representation: Limited (see Chapter 16, pp.365ff.)

Base numbers (archaic script)

I	Λ	X	↓	C	Ƅ	⊕	Ɩ)
1	5	10	50	100	500	1,000	5,000
			$(= 5 \times 10)$	$(= 10^2)$	$(= 5 \times 10^2)$	$(= 10^3)$	$(= 5 \times 10^3)$

Example: 7,659

Ɩ) ⊕⊕ Ƅ C ↓ IX
5,000 2,000 500 100 50 (10 − 1)

- →

Representation based both on additive and subtractive principles, broken down thus:

7,659 = 5,000 + (1,000 + 1,000) + 500 + 100 + 50 + (10 − 1)

FIG. 23.8. *Roman number-system*

LARGE ROMAN NUMBERS

To note down large numbers the Romans and the Latin peoples of the Middle Ages developed various conventions. Here are the principal ones (see Chapter 16, pp. 385ff.):

1. *Overline rule*
This consisted in multiplying by 1,000 every number surmounted by a horizontal bar:

$\overline{X} = 10,000$ $\overline{C} = 100,000$ $\overline{CXXVII} = 127 \times 1,000 = 127,000$

2. *Framing rule*
This consisted in multiplying by 100,000 every number enclosed in a sort of open rectangle:

$\lceil X \rceil = 1,000,000$ $\lceil CCLXIV \rceil = 264 \times 100,000 = 26,400,000$

3. *Rule for multiplicative combinations*
The rule is occasionally found in Latin manuscripts in the early centuries CE, but most often in European mediaeval accounting documents. To indicate multiples of 100 and 1,000, first the number of hundreds and thousands to be entered are noted down, then the appropriate letter (C or M) is placed as a coefficient or superscript indication:

100	C		1,000	M
200	II.C or II[c]		2,000	II.M or II[m]
300	III.C or III[c]		3,000	III.M or III[m]
. . . .			. . . .	
900	VIIII.C or VIIII[c]		9,000	VIIII.M or VIIII[m]

Examples taken from Pliny the Elder's *Natural History*, first century CE (VI, 26; XXXIII, 3).

LXXXIII.M   for 83,000
CX.M   for 110,000

FIG. 23.9A. *Latin notation of large numbers (late period)*

The same system is to be found in the Middle Ages, notably in King Philip le Bel's Treasury Rolls, one of the oldest surviving Treasury registers. In this book, dated 1299, we find what is reproduced here below, drawn up in Latin (from Registre du Trésor de Philippe le Bel, BN, Paris, Ms. lat. 9783, fo. 3v, col.1, line 22):

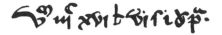

V^m. IIIe.XVI.l(ibras). VI.s(olidos)   "5,316 livres, 6 sols &
I. d(enarios). p(arisiensium)   1 denier parisis"

FIG. 23.9B.

But this was out of the frying pan into the fire, for such systems required even more tedious repetitions of identical signs. In the Roman system, the conventions for writing numbers proliferated so much that the system finally lost coherence (Fig. 23.8 and 23.9). Furthermore, since it made use at the same time of two logically incompatible principles (the additive and the subtractive), this system finally represented a regression with respect to the other historic systems of number representation.

The first notable advance in this respect is in fact due to the scribes of Egypt who, seeking means for rapid writing, early sought to simplify both the graphics and the structure of their basic system. Starting from excessively complicated hieroglyphic signs, they strove to devise extremely schematic signs which could be written in a continuous trace, without interruption, such as are obtained by small rapid movements and often by a single stroke of the brush. Great changes thus occurred in the forms of the hieroglyphic numbers, so that the later forms had only a vague resemblance to their prototypes. This finally resulted in a very abbreviated numerical notation, as in the Egyptian hieratic number-system, giving a separate sign to each of the following numbers (Fig. 23.10):

1	2	3	4	5	6	7	8	9
10	20	30	40	50	60	70	80	90
100	200	300	400	500	600	700	800	900
1,000	2,000	3,000	4,000	5,000	6,000	7,000	8,000	9,000

It was a cursive notation, and was succeeded by an even more abbreviated one, known as the demotic number-system.

First appearance: c.2500 BCE
Type: A3 (additive number-system of the third type: Fig. 23.32). Base 10
Need for zero sign: No. Existence of zero sign: No
Capacity for representation: Limited (see Chapter 14, pp.334ff.)

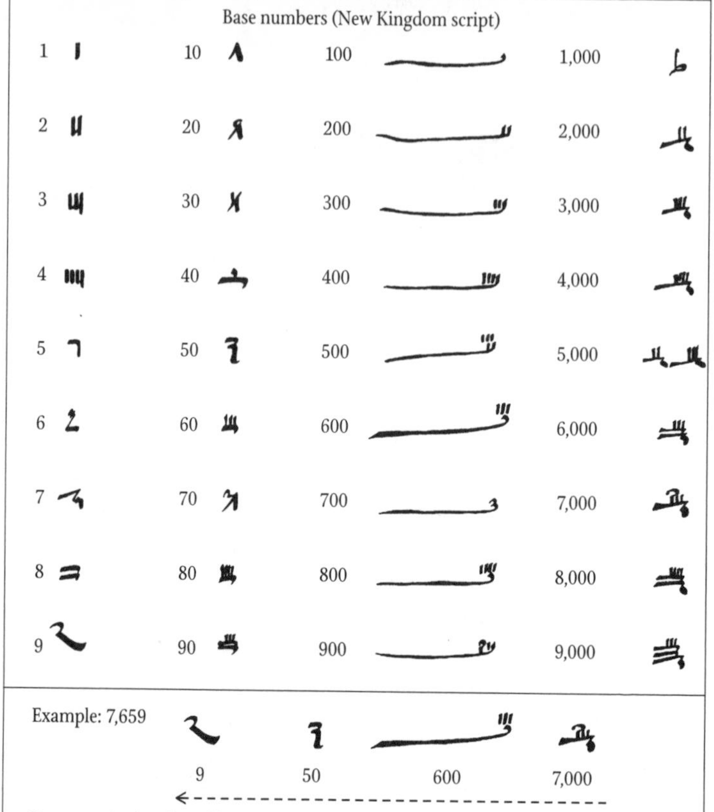

Base numbers (New Kingdom script)

1	10	100	1,000
2	20	200	2,000
3	30	300	3,000
4	40	400	4,000
5	50	500	5,000
6	60	600	6,000
7	70	700	7,000
8	80	800	8,000
9	90	900	9,000

Example: 7,659

9   50   600   7,000

Representation based on additive principle, broken down thus:
7,659 = 7,000 + 600 + 50 + 9

FIG. 23.10. *Egyptian hieratic number-system*

In both cases, there were nine special signs for the units, nine more for the tens, nine more for the hundreds, and so on. Such systems allowed numbers to be represented with much greater economy of symbols. The number 7,659 now only needed four signs (as opposed to the 27 required by the hieroglyphic system), since it only requires writing down the symbols for 7,000, 600, 50, and 9 according to the decomposition

$$7,659 = 7,000 + 600 + 50 + 9.$$

The inconvenience of such a notation is, of course, the burden on the memory of retaining all the different symbols of the system.

The Greeks and the Jews, and later the Syriacs, the Armenians and the Arabs, used notations which are mathematically equivalent to this system (Fig. 23.11 to 23.13, and Fig. 19.4 above). But, instead of proceeding as the Egyptians had done to the progressive refinement of the forms of their figures, they constructed their systems on the basis of the letters of their alphabets. Taking these letters in their usual order (the Phoenician "ABC") associates the first nine letters with the nine units, the next nine with the nine tens, and so on.

First appearance: c. fourth century BCE
Type: A3 (additive number-system of the third type: Fig. 23.32). Base 10
Need for zero sign: No. Existence of zero sign: No
Capacity for representation: Limited (see Chapter 17, p.431)

Base numbers

A	B	Γ	Δ	E	Ϛ	Z	H	Θ
1	2	3	4	5	6	7	8	9
I	K	Λ	M	N	Ξ	O	Π	Ϟ
10	20	30	40	50	60	70	80	90
P	Σ	T	Y	Φ	X	Ψ	Ω	ϡ
100	200	300	400	500	600	700	800	900

Example: 7,659

'Z   X   N   Θ
- - - - - - - - - - - - - - - - - ->
7,000   600   50   9

Representation based both on additive principle, broken down thus:

$$7,659 = 7,000 + 600 + 50 + 9$$

(The notation for the number 7,000 has been derived from that for 7, applying to this a small distinctive sign upper left.)

FIG. 23.11. *Greek alphabetic number-system*

First appearance: c. second century BCE
Type: A3 (additive number-system of the third type: Fig. 23.32). Base 10
Need for zero sign: No. Existence of zero sign: No
Capacity for representation: Limited (see Chapter 17, p.420)

Base numbers

א	ב	ג	ד	ה	ו	ז	ח	ט
1	2	3	4	5	6	7	8	9
י	כ	ל	מ	נ	ס	ע	פ	צ
10	20	30	40	50	60	70	80	90
ק	ר	ש	ת					
100	200	300	400					

Example: 7,659

ט   נ   ר   ת   ז̈
<- - - - - - - - - - - - - - - - - - -
9   50   200   400   7,000

Representation based both on additive principle, broken down thus:

$$7,659 = 7,000 + 400 + 200 + 50 + 9$$

(The notation for the number 7,000 has been derived from that for 7, placing two dots above this.)

FIG. 23.12. *Hebraic alphabetic number-system*

First appearance: c.400 CE
Type A3 (additive number-system of the third type: Fig. 23.32). Base 10
Need for zero sign: No. Existence of zero sign: No
Capacity for representation: Limited (see Chapter 17, pp.439ff.)

Base numbers
(Line 1, lower case; line 2, upper case)

| 1 | 2 | 3 | 4 | 5 | 6 | 7 | 8 | 9 |

| 10 | 20 | 30 | 40 | 50 | 60 | 70 | 80 | 90 |

| 100 | 200 | 300 | 400 | 500 | 600 | 700 | 800 | 900 |

| 1,000 | 2,000 | 3,000 | 4,000 | 5,000 | 6,000 | 7,000 | 8,000 | 9,000 |

Example: 7,659

Lower case:

Upper case:

7,000   600   50   9

Representation based on additive principle, broken down thus:
7,659 = 7,000 + 600 + 50 + 9

Fig. 23.13. *Armenian alphabetic number-system*

Such procedures allow the words of the language to be converted into numbers, which provides ample raw material for every kind of speculation, occultist fantasy or magical imagining, and for superstitious beliefs and practices. But, leaving aside this inconvenient by-product, the procedure gives a more or less acceptable solution to the problem according to the needs of the time. As with the Egyptian hieratic and demotic systems, the number 7,659 requires only four signs to be written down.

THE DISCOVERY OF THE
MULTIPLICATIVE PRINCIPLE

There was still a long road ahead before people could arrive at a system so well perfected as our own. Means for a numeric notation were still limited. Various peoples, it must be said, remained deeply attached to the old additive principle and were therefore in a blind alley. One major reason for this blockage concerned the problem of representing large numbers, which lie beyond the capability of the imagination when one is restricted solely to the additive principle. For this reason, some peoples made a radical change in their number-systems by adopting a hybrid principle which involved both multiplication and addition.

This change took place in two stages. The introduction of the new principle at first served only to extend the capabilities of number-systems which had been very primitive (Fig. 23.14 and 23.15).

The Sumerians

From c.3300 BCE the Sumerians tended to represent the units of different orders in their number-system by means of objects of conventional size and shape.

They had begun by using *calculi* to symbolise 1, 10, 60, and 60^2 (see Chapter 10, p.193):

| 1 | 10 | 60 | 3,600 |

But, not wishing to duplicate the original symbols, they invoked the multiplicative principle to represent the order of 600 and of 36,000:

600 ($= 10 \times 60$)   36,000 ($= 10 \times 60^2$)

They had thus come up with the idea (very abstract for the time) of symbolising × 10 by making in the soft clay a small circular impression ("written" symbol for the pebble representing 10) within the large cone representing the value 60 or within the sphere representing the value 3,600.

And they used the same idea in representing these same numbers when they embarked on a written number-system in archaic script as well as in cuneiform (see Chapter 8, p.161):

Curviform number-symbols

Cuneiform number-symbols

600 ($= 10 \times 60$)   36,000 ($= 10 \times 60^2$)

The Cretans (second millennium BCE):

The Cretans introduced the number for 10,000 by combining the horizontal stroke of 10 with the sign for 1,000 (see Chapter 15, p.351):

10,000 ($= 1,000 \times 10$)

The Greeks (from the fifth century BCE):

The Greeks invoked the same principle, completing their acrophonic number-system by introducing a notation with its own traits for each of the numbers 5, 50, 500, and 5,000 (see Chapter 16, pp.355ff.):

| 5 | 50 | 500 | 5,000 |
| | ($= 5 \times 10$) | ($= 5 \times 10^2$) | ($= 5 \times 10^3$) |

FIG. 23.14. *First emergence of the multiplicative principle*

Thus from the beginning of history, people have sometimes introduced the multiplication rule into systems essentially based on the additive principle. But during this first stage, the habit was confined to certain particular cases and the rule served only to form a few new symbols.

But in the subsequent stage, it gradually became clear that the rule could be applied to avoid not only the awkward repetition of identical signs, but also the unbridled introduction of new symbols (which always ends up requiring considerable efforts of memory).

And that is how certain notations that were rudimentary to begin with were often found to be extensible to large numbers.

The Greeks

This idea was exploited by ancient Greek mathematicians whose "instrument" was their alphabetic number-system: in order to set down numbers superior to 10,000, they invoked the multiplicative rule, placing a sign over the letter M (initial of the Greek word for 10,000, μυριοι) to indicate the number of 10,000s (see Chapter 17, p.431):

$\overset{\alpha}{M}$	$\overset{\beta}{M}$	$\overset{\gamma}{M}$	$\overset{\iota\beta}{M}$
10,000	20,000	30,000	120,000
$(= 1 \times 10,000)$	$(= 2 \times 10,000)$	$(= 3 \times 10,000)$	$(= 12 \times 10,000)$

The Arabs

Using the twenty-eight letters of their number-alphabet the Arabs proceeded likewise, but on a smaller scale: to note down the numbers beyond 1,000, all they had to do was to place beside the letter *ghayin* (worth 1,000 and corresponding to the largest base number in their system) the one representing the corresponding number of units, tens or hundreds (see Chapter 19, p.481):

بغ	جغ	يغ	نغ
2,000	3,000	10,000	50,000
$(= 2 \times 1,000)$	$(= 3 \times 1,000)$	$(= 10 \times 1,000)$	$(= 50 \times 1,000)$

The ancient Indians

The same idea was invoked by the Indians from the time of Emperor Asoka until the beginning of the Common Era in the numerical notation that related to Brâhmî script (see Chapter 24, pp.743ff.). To write down multiples of 100, they used the multiplicative principle, placing to the right of the sign for 100 the sign for the corresponding units. For numbers beyond 1,000 they wrote to the right of the sign for 1,000 the sign for the corresponding units or tens:

400	4,000	6,000	10,000
$(= 100 \times 4)$	$(= 1,000 \times 4)$	$(= 1,000 \times 6)$	$(= 1,000 \times 10)$

FIG. 23.15A. *First extension of the multiplicative principle*

The Egyptian hieroglyphic system (late period)

In Egyptian monumental inscriptions we find (at least from the beginning of the New Kingdom) a remarkable diversion from the "classical" system: when a tadpole (hieroglyphic sign for 100,000) was placed over a lower number-sign, it behaved as a multiplicator. In other words, by placing a tadpole over the sign for 18, for instance, the number 100,018 (= 100,000 + 18) was no longer being expressed, but rather the number 100,000 × 18 = 1,800,000 (a number which in the classical system would have been expressed by setting eight tadpoles adjacent to the hieroglyphic for 1,000,000).

Example: 27,000,000
Expressed in the form:
100,000 × 270

100,000

270

Taken from a Ptolemaic hieroglyphic inscription (third – first century BCE)

The Egyptian hieratic system

But the preceding irregularity was actually the result of the way the hieroglyphic system was influenced by hieratic notation: this used a more systematic method to note down numbers above 10,000 according to the rule in question. (See Chapter 14, pp. 334ff.)

	Early Kingdom	Middle Kingdom	New Kingdom
70,000			
90,000			
200,000			
700,000			
1,000,000			
2,000,000			
10,000,000			

Example: The number 494,800

(From the Great Harris Papyrus; 73, line 3. New Kingdom)

$$800 + 4{,}000 + 10{,}000 \times 9 + 100{,}000 \times 4$$

Fig. 23.15b.

The Assyro-Babylonians and the Aramaeans provide a case in point. They had a separate symbol for each of the numbers 1, 10, 100 and 1,000, but instead of representing the hundreds or thousands by separate signs or by repeating the 100 or 1,000 symbol as often as required, they had the idea of placing the signs for 100 or 1,000 side by side with the symbols for the units, thereby arriving at a multiplicative principle representing arithmetical combinations such as

$$1 \times 100 \qquad\qquad 1 \times 1{,}000$$
$$2 \times 100 \qquad\qquad 2 \times 1{,}000$$
$$3 \times 100 \qquad\qquad 3 \times 1{,}000$$
$$4 \times 100 \qquad\qquad 4 \times 1{,}000$$
$$5 \times 100 \qquad\qquad 5 \times 1{,}000$$
$$\ldots \qquad\qquad\qquad \ldots$$
$$9 \times 100 \qquad\qquad 9 \times 1{,}000$$

However, they continued to write numbers below 100 according to the old additive method, repeating the sign for 1 or for 10 as often as required. The number 7,659, for example, was written according to the following decomposition (Fig. 23.16 and 23.17):

$$7{,}659 = (1 + 1 + 1 + 1 + 1 + 1 + 1) \times 1{,}000$$
$$+ (1 + 1 + 1 + 1 + 1 + 1) \times 100$$
$$+ (10 + 10 + 10 + 10 + 10)$$
$$+ (1 + 1 + 1 + 1 + 1 + 1 + 1 + 1 + 1)$$

First appearance: c.2350 BCE
Type: B1 (hybrid number-system of the first type: Fig. 23.33). Base 10
Need for zero sign: No. Existence of zero sign: No
Capacity for representation: Limited (see Chapter 13, pp.266ff: Chapter 18, p.449)

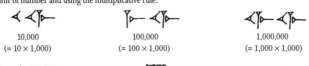

Base numbers				
1	10	100 $(= 10^2)$	1,000* $(= 10^3)$	*Symbol made up of that for 100 and that for 10

Example: 7,659

7   1,000   6   100   50   9

Representation based (in part) on hybrid principle, broken down thus:

$$7{,}659 = (1 + 1 + 1 + 1 + 1 + 1 + 1) \times 1{,}000$$
$$+ (1 + 1 + 1 + 1 + 1 + 1) \times 100$$
$$+ (10 + 10 + 10 + 10 + 10)$$
$$+ (1 + 1 + 1 + 1 + 1 + 1 + 1 + 1 + 1)$$

NOTATION FOR LARGE NUMBERS

This notation has succeeded in extending to the thousands by virtue of considering 1,000 as a fresh unit of number and using the multiplicative rule:

10,000 $(= 10 \times 1{,}000)$ 100,000 $(= 100 \times 1{,}000)$ 1,000,000 $(= 1{,}000 \times 1{,}000)$

Example: 305,412

$$= (3 \times 100 + 5) \times 1{,}000 + 4 \times 100 + 10 + 2$$

(From Assyrian tablets dating from King Sargon II)

† No doubt influenced by the structure of their oral number-system, the Mesopotamian Semites were the first to consider extending the multiplicative rule to the notion of other orders of units, thus creating the first hybrid number-system in history.

FIG. 23.16. *Common Assyro-Babylonian number-system*†

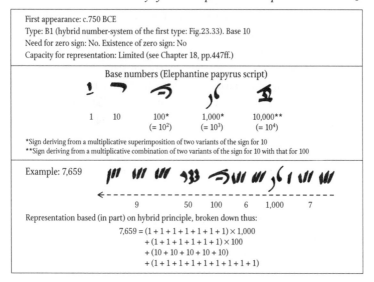

FIG. 23.17. *Aramaean number-system*

By such partial use of the multiplicative principle, the Assyro-Babylonian number-system was therefore of the "partial hybrid" type.

At a later period, the inhabitants of Ceylon went through the same change, but starting from a much better system than those above. They assigned a separate sign not only to every power of 10, but also to each of the nine units and to each of the nine tens, and then applied the same principle as above. In this way, the number 7,659 can be broken down (Fig. 23.18) as

$$7 \times 1,000 + 6 \times 100 + 50 + 9.$$

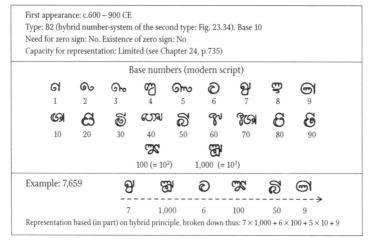

FIG. 23.18. *Singhalese number-system*

However, it was the Chinese, and the Tamils and Malayalams of southern India, who made the best use of this approach. They too had special signs for the numbers 1, 2, 3, 4, 5, 6, 7, 8, 9, 10, 100, 1,000, 10,000 but, instead of representing the tens by special signs, they had the idea of extending the multiplicative principle to all the orders of magnitude of their system, from the unit upwards. For intermediate numbers, they placed the sign for 10 between the sign for the number of units and the sign for the number of hundreds, the sign for 100 between the sign for the number of hundreds and the sign for the number of thousands, and so on. For the number 7,659 this gave rise to a decomposition of the type

$$7,659 = 7 \times 1,000 + 6 \times 100 + 5 \times 10 + 9.$$

Such systems are of "complete hybrid" type, in which the representation of a number resembles a polynomial whose variable is the base of the number-system (Fig. 23.19 to 23.21).

First appearance: c.1450 BCE

Type: B5 (hybrid number-system of the fifth type: Fig. 23.37). Base 10

Need for zero sign: No, when the hybrid principle is rigorously applied. Yes, when the simplified rule below is applied. Existence of zero sign: Yes, at a later date

Capacity for representation: Limited in the case of the unsimplified system (see Chapter 21, pp.514ff.)

Base numbers (modern script)

一	二	三	四	五	六	七	八	九	十	百	千	萬
1	2	3	4	5	6	7	8	9	10	100 $(= 10^2)$	1,000 $(= 10^3)$	10,000 $(= 10^4)$

Example: 7,659
Normal script

七 千 六 百 五 十 九

- ->

7　1,000　6　100　5　10　9

Representation based entirely on hybrid principle, broken down thus: $7 \times 1,000 + 6 \times 100 + 5 \times 10 + 9$

Abridged script in use since modern times

The above representation was sometimes produced in the simplified form below, thus tending towards an application of the positional principle with base 10:

七 六 五 九

- - - - - - - - - - - ->

7　6　5　9

NOTATION FOR LARGE NUMBERS

With the thirteen basic characters of this number-system, considering 10,000 as a fresh unit of number, the Chinese were able to give a rational expression to all the powers of 10 right up to 100,000,000,000 (and hence of all numbers from 1 to 999,999,999,999).

10,000 =	1 wàn =	$1 \times 10,000$
100,000 =	10 wàn =	$10 \times 10,000$, etc

Example: 487,390,629

四 萬 八 千 七 百 三 十 九 萬 六 百 二 十 九

$(4 \times 10^4 + 8 \times 10^3 + 7 \times 10^2 + 3 \times 10 + 9) \times 10^4 + (6 \times 10^2 + 2 \times 10 + 9)$

FIG. 23.19. *Common Chinese number-system*

The discovery of such hybrid principles was a great step forward, in the context of the needs of the time, since it not only avoided tedious repetitions of identical signs but also lightened the burden on the memory, no longer required to retain a large number of different signs.

By the same token, the written representation of numbers could be brought into line with their verbal expression (the linguistic structure of the majority of spoken numbers had conformed, since the earliest times, to this kind of mixed rule).

The principal benefit, however, of this procedure was greatly to extend the range of numbers that could be represented (Fig. 23.15,16 and 19).

THE DIFFICULTIES OF THE PRECEDING SYSTEMS

Despite the considerable advance which these changes represent, the capabilities of numerical notation remained very limited.

By making use of certain conventions of writing, the Greek mathematicians managed to extend their alphabetic notation to cope with larger numbers. Archimedes provides an important example. In his short arithmetical treatise *The Psammites*, he conceived a rule which would allow him to express very large numbers by means of the numeric letters of the Greek alphabet, such as the number of grains of sand which would be contained in the Sphere of the World (whose diameter is the distance from the earth to the nearest fixed stars). In our modern notation, this number would be expressed as a 1 followed by 64 zeros.

Chinese mathematicians also succeeded in extending their number-system to accommodate numbers which could exceed 10^{4096}, a number which is far beyond any quantity that could be physically realised.

None of these systems, however, succeeded in achieving a rational notation for all numbers, since they did not have the unlimited capacity for representation which our own system has. The greater the order of magnitude required, the more special symbols must be invented, or further conventions of writing imposed.

We can therefore appreciate the undoubted superiority of our modern system of numerical notation, which is one of the foundations of the intellectual equipment of modern humankind. With the aid of a very small number of basic symbols, any number whatever, no matter how large, may be represented in a simple, unified and rational manner without the need for any further artifice.

Yet another reason for the superiority of our system is that it is directly adapted to the written performance of arithmetic.

It is precisely this fact which underlies the difficulty, or even impossibility, of doing arithmetic with the ancient number-systems, which remained blocked in this respect for as long as they were in use.

For example, let us try to perform an addition using Roman numerals:

$$
\begin{array}{r}
\text{CCLXVI} \\
+ \quad \text{DCL} \\
+ \quad \text{MLXXX} \\
+ \text{MDCCCVII} \\
\hline
= ?\,?\,?\,?\,?\,? \\
\end{array}
$$

Clearly, unless we translate this into our modern notation, this would be very hard:

$$
\begin{array}{r}
266 \\
+ \quad 650 \\
+ 1{,}080 \\
+ 1{,}807 \\
\hline
= 3{,}803 \\
\end{array}
$$

But this is a mere addition! What about multiplication or division?

In systems of this kind, we are barely able to do arithmetic. This is due to the static nature of the number-signs, which have no operational significance but are more like abbreviations which can be used to write down the results of calculations performed by other means.

In order to do arithmetical calculations, the ancients generally made use of auxiliary aids such as the abacus or a table with counters. This requires long and difficult training and practice, and remains beyond the reach of ordinary mortals. It therefore remained the preserve of a privileged caste of specialist professional calculators. This is not to say, however, that such systems did not allow any written calculation.

The above addition can be carried out in the Roman system. This involves proceeding by stages, by counting and then reducing the results from each order of magnitude (five "I" replaced by one "V", two "V" by one "X", five "X" by one "L", two "L" by one "C", and so on):

			CC	L	X	V	I
+	M	D	CCC			V	II
+		D	C	L			
+	M			L	XXX		
	MMM	D	CCC				III

The Romans probably did use such a method. But since it is at bottom a reduction to written form of operations performed on an abacus, they probably preferred to continue to use that instrument whose counters, for all their inconvenience, were nonetheless easier to manipulate than the symbols in their primitive representation of numbers.

We know also that, despite its very primitive character, the Egyptian number-system allowed arithmetical calculations. The methods certainly had the advantage of not obliging calculators to rely on memory. To multiply or to divide, it was in fact enough to know how to multiply or to divide by 2. Their methods, however, were slow and complicated compared with our modern ones. Worse, though, they lacked flexibility, unification and coherence.

On the other hand, the Graeco-Byzantine mathematicians certainly succeeded in devising various rules for multiplication and division in terms of the number-letters of their alphabet. There again, however, their procedures were much more complicated, and above all far more artificial and less coherent than ours.

These are all, therefore, mere attempts to invent rules of calculation during the ancient times. But, "The fact is that the difficulties encountered in former times were inherent in the very nature of the number-systems themselves, which did not lend themselves to simple straightforward rules" [T. Dantzig (1967)].

Therefore it was the discovery of our modern number-system, and above all its popularisation, which allowed the human race to overcome the obstacles and to dispense with all auxiliary aids to calculation such as we have been considering.

DECISIVE FIRST STEP:
THE PRINCIPLE OF POSITION

In order to achieve a system as ingenious as our own, it is first necessary to discover the *principle of position*. According to this, the value of a figure varies according to the position in which it occurs, in the representation of a number. In our modern decimal notation, a "3" has value 3 units, 3 tens or 3 hundreds depending on whether it is in the first, second or third position. To write seven thousand, six hundred and fifty-nine, all we have to do is to write down the figures 7, 6, 5, and 9 in that order, since according to the rule the representation 7,659 denotes the value

$$7 \times 1,000 + 6 \times 100 + 5 \times 10 + 9.$$

Because of this fundamental convention, only the coefficients of the powers of the base, into which the number has been decomposed, need appear.

This, therefore, is the principle of position. Apparently as simple as Columbus's egg; but it had to be thought of in the first place!

Nowadays, this principle seems to us to have such an obvious simplicity that we forget how the human race has stammered, hesitated and groped through thousands of years before discovering it, and that civilisations as advanced as the Greek and the Egyptian completely failed to notice it.

SYSTEMS WHICH COULD HAVE BEEN POSITIONAL

For all that, even in the earliest times a goodly number of different number-systems could have led on to the discovery of the principle of position.

Consider for example the Tamil and Malayalam systems from south India. According to the hybrid principle, the figure representing the number of tens was placed to the left of the symbol for 10, the one representing the number of hundreds to the left of the symbols for 100, and so on (Fig. 23.20 and 23.21).

First appearance: c.600 – 900 CE

Type: B5 (hybrid number-system of the fifth type: Fig.23.37). Base 10

Need for zero sign: No, when the hybrid principle is rigorously applied. Yes, when the simplified rule below is applied.

Existence of zero sign: Not before the modern era

Capacity for representation: Limited in the case of the unsimplified system (see Chapter 24, p.731)

System used among the Tamils (southern India)

Base numbers (modern script)

1	2	3	4	5	6	7	8	9

10	100 ($= 10^2$)	1,000 ($= 10^3$)

Example: 7,659
Normal script

7	1,000	6	100	5	10	9

Representation based entirely on hybrid principle, broken down thus:

$$7 \times 1,000 + 6 \times 100 + 5 \times 10 + 9$$

Abridged script in use since modern times

The above representation was sometimes produced in the simplified form below, thus tending towards an application of the positional principle with base 10:

7	6	5	9

FIG. 23.20. *Tamil number-system*

First appearance: c.600 – 900 CE

Type: B5 (hybrid number-system of the fifth type: Fig. 23.37). Base 10

Need for zero sign: No, when the hybrid principle is rigorously applied. Yes, when the simplified rule below is applied.

Existence of zero sign: Not before the modern era

Capacity for representation: Limited in the case of the unsimplified system (see Chapter 24, p.733)

System used among the Malayalam (southern India, Malabar coast)

Base numbers (modern script)

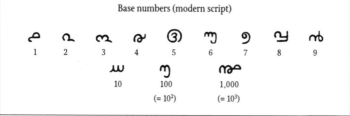

Example: 7,659
Normal script

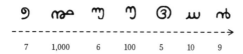

Representation based entirely on hybrid principle, broken down thus:

$$7 \times 1,000 + 6 \times 100 + 5 \times 10 + 9$$

Abridged script in use since modern times

The above representation was sometimes produced in the simplified form below, thus tending towards an application of the positional principle with base 10:

FIG. 23.21. *Malayalam number-system*

In this way, the number 6,657, for example, would usually be written as follows:

Tamil   Malayalam

which corresponded to the decomposition

$$6 \times 1,000 + 6 \times 100 + 5 \times 10 + 7.$$

Now, when we look at certain Tamil or Malayalam writings, we find that the symbols for 10, 100, and 1,000 have in many cases been suppressed [L. Renou and J. Filliozat (1953)]. The number 6,657 would then appear in the abbreviated notation

𑀓𑁆	𑀓𑁆	𑀕	எ		ന	ന	③	൭
6	6	5	7		6	6	5	7
- - - - - - - - - - →					- - - - - - - - - - →			
Tamil					Malayalam			

The result of this simplification is that the figures 6, 6, 5, and 7 have been assigned values as follows:

- seven units to the figure 7 in the first place;
- five tens to the figure 5 in the second place;
- six hundreds to the figure 6 in the third place;
- six thousands to the figure 6 in the fourth place.

Thus the Tamil and Malayalam figures could be assigned values which depended on where they occurred in the representation of a number.

This remarkable potential for evolution towards a positional number-system is characteristic of hybrid numbering systems.

In such systems, in fact, the signs which indicate the powers of the base (10, 100, 1,000) are always written in the same order, either increasing or decreasing. Therefore it is natural that the people who used these systems would be led, for the sake of abbreviation, to suppress these signs leaving only the figures representing their coefficients.

This is what led certain Aramaic stone-cutters of the beginning of our era to sometimes leave out the sign for 100 in their numeric inscriptions.

The inscription of Sa'ddiyat is a remarkable piece of evidence for this. We know that in this region a hybrid system was used, whose basic signs had the following forms and values:

1	5	10	20	100

But we see in this inscription (which dates from the 436th year of the Seleucid era, or 124–125 CE) that the number 436 is written in the form [B. Aggoula (1972), plate II]

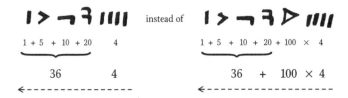

For the same reason, the scribes of Mari often left out the cuneiform figure for 100. This is all the more remarkable since the Mari system, uniquely among Mesopotamian systems, was in use around the nineteenth century BCE, therefore earlier than the period in which the Babylonian positional system appeared (J.-M. Durand).

First appearance: c.2000 BCE

Type: B3 (hybrid number-system of the third type: Fig.23.35). Base 100

Need for zero sign: No, when the hybrid principle is rigorously applied. Yes, when the simplified rule below is applied. Existence of zero sign: No

Capacity for representation: Limited (see Chapter 13, p.277)

Base numbers

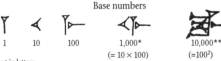

1	10	100	1,000*	10,000**
			(= 10 × 100)	(=100²)

*Number spelt out in letters

**Symbol derived by allocating a multiplicative function to the combination of the 1,000 symbol with that for 10

Example: 7,659
Normal script

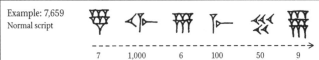

    7 1,000 6 100 50 9

Representation based entirely on hybrid principle, broken down thus:

$$7,659 = (1 + 1 + 1 + 1 + 1 + 1 + 1) \times 1,000$$
$$+ (1 + 1 + 1 + 1 + 1 + 1) \times 100$$
$$+ (10 + 10 + 10 + 10 + 10)$$
$$+ (1 + 1 + 1 + 1 + 1 + 1 + 1 + 1 + 1)$$

Abridged script

The above representation was sometimes produced in the simplified form below, with the number 100 omitted.

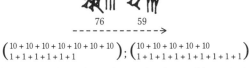

     76   59

$$\begin{pmatrix} 10 + 10 + 10 + 10 + 10 + 10 + 10 \\ 1 + 1 + 1 + 1 + 1 + 1 \end{pmatrix} ; \begin{pmatrix} 10 + 10 + 10 + 10 + 10 \\ 1 + 1 + 1 + 1 + 1 + 1 + 1 + 1 + 1 \end{pmatrix}$$

Put differently, the notation thus tends towards a partial application of the positional principle with base 100:

$$7,659 = [76 ; 59] = 76 \times 100 + 59$$

FIG. 23.22. *Mari number-system*

At Mari they used a hybrid system whose basic signs had the following forms and values (Fig. 23.22):

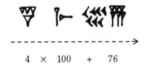

| 1 | 10 | 100 | 1,000 | 10,000 |

The number 476 would therefore be represented as:

4 × 100 + 76

At any rate, that is the normal representation of this number. But, as we have only recently discovered, the Mari gave an abbreviated representation to this number [D. Soubeyran (1984), tablet ARM, XXII 26]:

4 ; 76

This simplification was, nevertheless, only made for the hundreds figure, not for all the powers of the base. For this reason, the Mari system never became positional in the full sense. This system of notation remained strongly bound to the methods of the old additive principle, and was therefore held back from taking the one vital further step forward from this significant advance.

A similar simplification can be found in certain Chinese writers, who also simplified their writing of numbers by suppressing the signs indicating the tens, hundreds, thousands, etc. (see Fig. 23.19 above). For the number 67,859 we therefore find [E. Biot (1839); K. Menninger (1969)]:

| 6 | 7 | 8 | 5 | 9 | instead of | $6 \times 10,000 + 7 \times 1,000 + 8 \times 100 + 5 \times 10 + 9$ |

Finally, consider the Maya priests and astronomers. In order to simplify the "Long Count" of their representation of dates, they too were led to suppress all indications of the glyphs for their units of time, leaving only the series of corresponding coefficients.

Let us take, for example, the Maya period of time expressed, in days, as $5 \times 144{,}000 + 17 \times 7{,}200 + 6 \times 360 + 11 \times 20 + 19$. This would usually be shows on the *stelae* as:

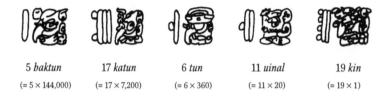

5 *baktun*	17 *katun*	6 *tun*	11 *uinal*	19 *kin*
$(= 5 \times 144{,}000)$	$(= 17 \times 7{,}200)$	$(= 6 \times 360)$	$(= 11 \times 20)$	$(= 19 \times 1)$

But, in their manuscripts, these astronomer-priests often preferred the following form in which appear only the numerical coefficients associated with the different time periods *kin* (days), *uinal* (periods of 20 days), *tun* (periods of 360 days), *katun* (periods of 7,200 days), etc. This gives a strictly positional representation:

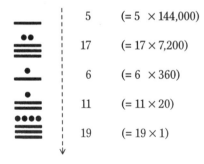

5	$(= 5 \ \times 144{,}000)$
17	$(= 17 \times 7{,}200)$
6	$(= 6 \ \times 360)$
11	$(= 11 \times 20)$
19	$(= 19 \times 1)$

This proves clearly that hybrid numbering systems had the potential to lead to the discovery of the principle of position. However, a simplification of a partial hybrid system could only lead to an incomplete implementation of the rule of position, whereas the simplification of a fully hybrid system was capable of leading to its complete implementation.

The simplification of the Maya notation for "Long Count" dates did give rise to a positional system, as also did changes in certain other systems (but these were belated and therefore of no consequence for the universal history); apart from these marginal exceptions it must be said that none of these earlier systems arrived at the level of a truly positional numbering system.

We therefore see yet again how people who have been widely separated in time or space have, by their tentative researches, been led to very similar if not identical results.

In some cases, the explanation for this may be found in contacts and influences between different groups of people. But it would not be correct to suppose that the Maya were in a position to copy the ideas of the people of the Ancient World. The true explanation lies in what we have previously referred to as the profound unity of human culture: the intelligence of *homo sapiens* is universal, and its potential is remarkably uniform in all parts of the world. The Maya simply found themselves in favourable conditions, strictly identical to those of others who obtained the same results.

THE EARLIEST POSITIONAL NUMBER-SYSTEMS OF HISTORY

The civilisation which developed the basis of our modern number system was therefore neither the first nor the only one to discover the principle of position.

In fact, three peoples came to its full discovery earlier, and independently. The numerical rule which is the basis of the positional system was created:

- for the first time, some 2,000 years BCE, by the Babylonians;
- for the second time, slightly before the Common Era, by Chinese mathematicians;
- for the third time, between the fourth and the ninth century CE, by the Mayan astronomer-priests.

The Babylonian sexagesimal system represented a number such as 392 by writing the number 6 in the second (sixties) place, and the number 32 in the first place, corresponding to a notation which might be transcribed (Fig. 23.23) as [6; 32] (= $6 \times 60 + 32$).

First appearance: c. 1800 BCE
Type: C1 (positional number-system of the first type: Fig. 23.38). Base 60
Need for zero sign: Yes. Existence of zero sign: Yes, but only later on (from the fourth century BCE)
Capacity for representation: Unlimited (see Chapter 13, pp.284ff.)

Significant numbers

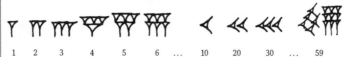

1   2   3   4   5   6 ... 10 20 30 ... 59

(Symbols formed according to the additive principle, starting from two basic symbols, one representing the number 1, the other the number 10)

Positional values

1st rank:	1	
2nd rank:	60	
3rd rank:	$60^2 =$	3,600
4th rank:	$60^3 =$	216,000
5th rank:	$60^4 =$	12,960,000, etc.

Example: 7,659

2 ; 7 ; 39

$(7,659 = 2 \times 60^2 + 7 \times 60 + 39)$

Representation based on positional principle, broken down thus:

$$[1 + 1 ; 1 + 1 + 1 + 1 + 1 + 1 + 1 ; 10 + 10 + 10 + 1 + 1 + 1 + 1 + 1 + 1 + 1 + 1 + 1]$$

FIG. 23.23. *Learned Babylonian number-system (the first positional number-system in history)*

First appearance: c. 200 BCE

Type: C1 (positional number-system of the first type: Fig. 23.38). Base 10

Need for zero sign: Yes. Existence of zero sign: Yes, but only later on (from the eighth century, under Indian influence)

Capacity for representation: Unlimited (see Chapter 21, pp. 545ff.)

Significant numbers

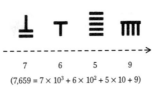

or	or	or	or	or	or	or	or	or
1	2	3	4	5	6	7	8	9

(Symbols formed according to the additive principle, starting from two basic symbols, one representing the number 1, the other the number 5)

Positional values

1st rank:	1	
2nd rank:	10	
3rd rank:	$10^2 =$	100
4th rank:	$10^3 =$	1,000
5th rank:	$10^4 =$	10,000, etc.

Example: 7,659

$$\underline{\underline{\perp}} \quad T \quad \equiv \quad \text{ш}$$

- →

7   6   5   9

$(7,659 = 7 \times 10^3 + 6 \times 10^2 + 5 \times 10 + 9)$

Representation based on positional principle, broken down thus:

$[5 + 1 + 1 ; 5 + 1 ; 1 + 1 + 1 + 1 + 1 ; 5 + 1 + 1 + 1 + 1]$

FIG. 23.24. *Learned Chinese number-system*

This is very much as we might today write 392' = 6 × 60' + 32' in the form 6° 32' (6 degrees, 32 minutes).

The Chinese system was based on the same principle, with the difference that the base of the number-system was decimal instead of being equal to 60. To write 392 in this system, we therefore place the figures 3, 9 and 2 in this order in a notation which we may (Fig. 23.24) transcribe as [3; 9; 2] (= 3 × 100 + 9 × 10 + 2).

In the Maya system with base 20, we may write (Fig. 23.25) [19; 12] (= 19 × 20 + 12). These Babylonian, Chinese and Maya systems were, therefore, the earliest positional number-systems of history.

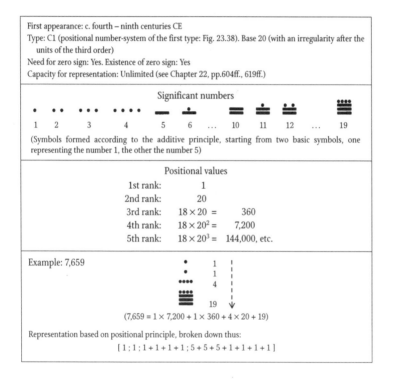

First appearance: c. fourth – ninth centuries CE

Type: C1 (positional number-system of the first type: Fig. 23.38). Base 20 (with an irregularity after the units of the third order)

Need for zero sign: Yes. Existence of zero sign: Yes

Capacity for representation: Unlimited (see Chapter 22, pp.604ff., 619ff.)

Significant numbers

| 1 | 2 | 3 | 4 | 5 | 6 | ... | 10 | 11 | 12 | ... | 19 |

(Symbols formed according to the additive principle, starting from two basic symbols, one representing the number 1, the other the number 5)

Positional values

1st rank:	1
2nd rank:	20
3rd rank:	18 × 20 =	360
4th rank:	18 × 20² =	7,200
5th rank:	18 × 20³ =	144,000, etc.

Example: 7,659

$$\begin{array}{c} 1 \\ 1 \\ 4 \\ 19 \end{array}$$

(7,659 = 1 × 7,200 + 1 × 360 + 4 × 20 + 19)

Representation based on positional principle, broken down thus:

[1 ; 1 ; 1 + 1 + 1 + 1 ; 5 + 5 + 5 + 1 + 1 + 1 + 1]

FIG. 23.25. *Learned Maya number-system*

SYSTEMS WHICH DID NOT SUCCEED

Having made this fundamental and essential discovery, the way was in fact open to each of these three peoples to represent any number whatever, no matter how large, by means of a small set of basic signs. But none of these three succeeded in taking advantage of their discovery.

The Babylonians indeed discovered the principle of position and applied it to base 60. But it never occurred to them, for more than two thousand years, to attach a particular symbol to each unit in their sexagesimal system. Instead of fifty-nine different figures, they in reality had only two: one for unity, and one for 10. All the rest had to be composed by duplicating these as many times as necessary up to 59 (Fig. 23.23).

The Chinese also discovered the principle of position and applied it to base 10. But they did no better, for, instead of assigning a different sign to each of the nine units, they preserved their ideographic system, in which the number 8 was represented by the symbol for 5 with three copies of the symbol for unity (Fig. 23.24).

Likewise the Maya system used the principle of position applied to base 20. But they again had only two distinct figures, one for unity and the other for 5, instead of the nineteen which are required for full dynamic notation in base 20 (Fig. 23.25).

For each of these three, it is somewhat as if the Romans had applied the rule of position to their first few number-signs, for example writing 324 in the form III II IIII, which would surely have led to confusion with:

I IIII IIII	(144)
II III IIII	(234)
II IIII III	(243)
III III III	(333)
III IIII II	(342)
IIII I IIII	(414)
IIII II III	(423) etc.

The Maya system had another source of difficulty inherent in its very structure. The rule of position was not applied to the powers of the base, but to values which were in fact adapted to the requirements of the calendar and of astronomy.

Each number greater than 20 was written in a vertical column with as many levels as there were orders of magnitude: the units were at the bottom level, the twenties on the second level, and so on.

This system therefore became irregular from the third level onwards, and was not rigorously founded on base 20. Instead of giving the multiples of $20^2 = 400$, $20^3 = 8,000$, and so on, the different levels from the third upwards in fact indicated multiples of $360 = 18 \times 20$, $7,200 = 18 \times 20^2$, and so on.

But there was no such problem with the Babylonian and Chinese systems, whose positional values corresponded exactly to the progression of the values of their base:

Units of	Learned Babylonian system (base 60)	Learned Chinese system (base 10)	Regular positional system (base 20)	Learned Maya system (irregular use of base 20)
1st order	1	1	1	1
2nd order	60	10	20	20
3rd order	60^2	10^2	20^2	18×20
4th order	60^3	10^3	20^3	18×20^2
5th order	60^4	10^4	20^4	18×20^3
6th order	60^5	10^5	20^5	18×20^4

If the Maya positional system had been constructed regularly on base 20, the expression [7; 9; 3] would surely have signified

$$7 \times 20^2 + 9 \times 20 + 3 = 7 \times 400 + 9 \times 20 + 3.$$

But for the Maya priests this corresponded to $7 \times 360 + 9 \times 20 + 3$.

This is one of the reasons why their system remained unsuited to practical written calculation.

A MAJOR SECOND STEP: DEVELOPMENT OF A DYNAMIC NOTATION FOR THE UNITS

From what we have seen so far, it is clear that for a numerical notation to be well adapted to written calculation, it must not only be based on the principle of position but must also have distinct symbols corresponding to graphic characters which have no intuitive visual meaning.

Otherwise put, the graphical structure of the number-signs must be like that of our modern written numbers, in that "9", for example, is not composed of nine points nor of nine bars, but is a purely conventional symbol with no ideographic significance (Fig. 23.26):

1 2 3 4 5 6 7 8 9

First appearance: c. fourth century CE
Type: C2 (positional number-system of the second type: Fig. 23.28). Base 10
Need for zero sign: Yes. Existence of zero sign: Yes
Capacity for representation: Unlimited

Base numbers (present-day script)

1   2   3   4   5   6   7   8   9

(Symbols devoid of all direct visual intuition)

Positional values

| 1st rank: 1 | 3rd rank: $10^2 = 100$ |
| 2nd rank: 10 | 4th rank: $10^3 = 1,000$, etc. |

Example: 7,659

7   6   5   9

- - - - - - - - - - - - - - - →

$(7,659 = 7 \times 10^3 + 6 \times 10^2 + 5 \times 10 + 9)$

FIG. 23.26. *Modern number-system*

THE FINAL FUNDAMENTAL DISCOVERY: ZERO

A no less fundamental condition for any number-system to be as well developed and as effective as our own is that it must possess a *zero*.

For so long as people used non-positional notations, the necessity of this concept did not make itself felt. The fact that there were signs for values greater than the base of the system meant that these systems could avoid the stumbling block which occurs whenever units of a certain order of magnitude are absent. To write, for instance, 2,004 in Egyptian hieroglyphics, it was sufficient to put two lotus flowers (for the thousands) and four vertical bars (for the units), the total of the values thus being

$$1,000 + 1,000 + 1 + 1 + 1 + 1 = 2,004.$$

In the Roman numerals, this number would be written MMIIII, and there was no need to have a special symbol to show that there were no hundreds and no tens. In the Chinese system, they would represent this number in the hybrid system, as a "2" followed by the symbol for 1,000 followed by a "4", corresponding to the decomposition $2,004 = 2 \times 1,000 + 4$.

On the other hand, once one has begun to apply place values on a regular basis, it is not long before one faces the requirement to indicate that tens, or hundreds, etc., may be missing. The discovery of zero was therefore a necessity for the strict and regular use of the rule of position, and it was therefore a decisive stage in an evolution without which the progress of modern mathematics, science and technology would be unimaginable.

Consider our decimal system. To write thirty, we have to place "3" in the second position, to have the value of three tens. But how do we show that it is in the second position if there is nothing at all in the first position? Therefore it is essential to have a special sign whose purpose is to indicate the absence of anything in a particular position. This thing which signifies nothing, or rather empty space, is in fact the zero. To arrive at the realisation that empty space may and must be replaced by a sign whose purpose is precisely to indicate that it is empty space: this is the ultimate abstraction, which required much time, much imagination, and beyond doubt great maturity of mind.

In the beginning, this concept was simply synonymous with empty space thus filled. But it was gradually perceived that "empty" and "nothing", originally thought of as distinct, are in reality two aspects of one and the same thing. Thus it is that the zero sign originally introduced to mark empty space finally symbolises in our eyes the value of the null number, a concept at the heart of algebra and modern mathematics.

Nowadays this is so familiar that we are no longer aware of the difficulties which its lack caused to the early users of positional number-systems.

Its discovery was far from a foregone conclusion, for apart from India, Mesopotamia and the Maya civilisation, no other culture throughout history came to it by itself. We can gain some idea of its importance when we recall that it escaped the eyes of the Chinese mathematicians, who nonetheless succeeded in discovering the principle of position. Only since the eighth century of our era, under the influence of our modern number-system, did this concept finally appear in Chinese scientific writings.

The Babylonians themselves were unaware of it for more than a thousand years, leading as one can imagine to numerous errors and confusions.

They certainly tried to get round the difficulty by leaving empty space where the missing units of a certain order would normally be found. Therefore they wrote much as if we wrote the number one hundred and six as 1. .6. But this was not enough to solve the problem in practice, since scribes could easily overlook it in copying, through fatigue or carelessness. Moreover it was difficult to indicate precisely the absence of two or more consecutive orders of magnitude, since one empty space beside another empty space is not easily distinguished from a single empty space.

It was therefore necessary to await the fourth century BCE to see the introduction of a special sign dedicated to this purpose. This was a cuneiform sign, which looked like a double oblique chevron, which was used not only in the medial and final positions but also in the initial position to indicate sexagesimal fractions of unity.

Medial: $[3; 0; 9; 2] = 3 \times 60^3 + 0 \times 60^2 + 9 \times 60 + 2$

$[3; 0; 0; 2] = 3 \times 60^3 + 0 \times 60^2 + 0 \times 60 + 2$

Final: $[3; 1; 5; 0] = 3 \times 60^3 + 1 \times 60^2 + 5 \times 60 + 0$

$[3; 1; 0; 0] = 3 \times 60^3 + 1 \times 60^2 + 0 \times 60 + 0$

Initial: $[0; 3; 4; 2] = 0 + 3 \times \dfrac{1}{60} + 4 \times \dfrac{1}{60^2} + 2 \times \dfrac{1}{60^3}$

This epoch, late in the history of Mesopotamia, saw the emergence of an eminently abstract concept, the Babylonian zero, the first zero of all time, to be followed some centuries later by the Maya zero.

IMPERFECT ZEROS

Even so, neither the Babylonians nor the Maya managed to get the most from their prime discovery (Fig. 23.27).

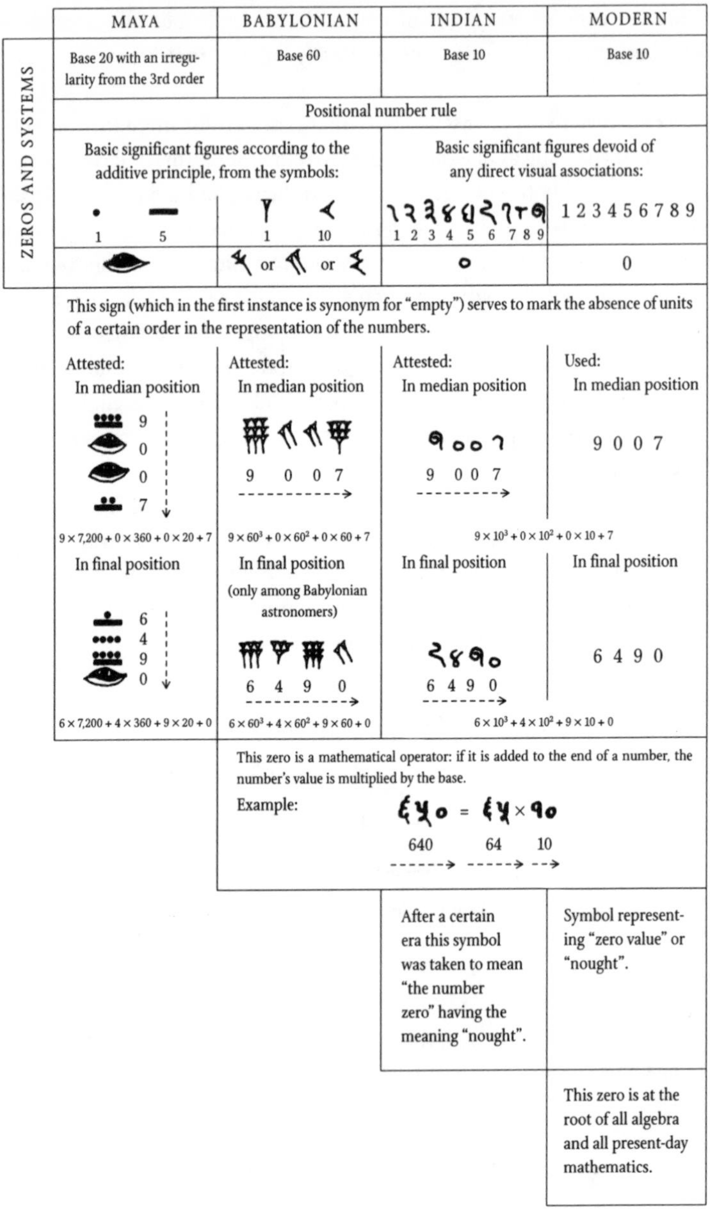

	MAYA	BABYLONIAN	INDIAN	MODERN
ZEROS AND SYSTEMS	Base 20 with an irregularity from the 3rd order	Base 60	Base 10	Base 10
	Positional number rule			
	Basic significant figures according to the additive principle, from the symbols:		Basic significant figures devoid of any direct visual associations:	

This sign (which in the first instance is synonym for "empty") serves to mark the absence of units of a certain order in the representation of the numbers.

Attested: In median position	Attested: In median position	Attested: In median position	Used: In median position
9 0 0 7	9 0 0 7	9 0 0 7	9 0 0 7
$9 \times 7{,}200 + 0 \times 360 + 0 \times 20 + 7$	$9 \times 60^3 + 0 \times 60^2 + 0 \times 60 + 7$	$9 \times 10^3 + 0 \times 10^2 + 0 \times 10 + 7$	
In final position	In final position (only among Babylonian astronomers)	In final position	In final position
6 4 9 0	6 4 9 0	6 4 9 0	6 4 9 0
$6 \times 7{,}200 + 4 \times 360 + 9 \times 20 + 0$	$6 \times 60^3 + 4 \times 60^2 + 9 \times 60 + 0$	$6 \times 10^3 + 4 \times 10^2 + 9 \times 10 + 0$	

This zero is a mathematical operator: if it is added to the end of a number, the number's value is multiplied by the base.

Example:

640   64   10

After a certain era this symbol was taken to mean "the number zero" having the meaning "nought".

Symbol representing "zero value" or "nought".

This zero is at the root of all algebra and all present-day mathematics.

FIG. 23.27. *Classification of zeros*

The Maya of course understood that it was a genuine zero sign, since they used it in medial as well as in final position. But, because of the anomalous progression they introduced at the third position of their positional system, this concept lost all operational usability.

The Babylonian zero not only had this possibility, it even filled the role of an arithmetical operator, at least in the hands of the astronomers (adjoining the zero sign at the end of a representation multiplied the number represented by sixty, i.e. by the value of the base). But it was never understood as a number synonymous with "empty", and never corresponded to the meaning of "null quantity" (Fig.23.27).

Despite these fundamental discoveries, therefore, none of these peoples was able to take the decisive step which would result in the ultimate perfection of numerical notation. Because of these imperfections, neither the Babylonian nor the Chinese nor the Maya positional system ever became adapted to arithmetical calculation, nor could ever give rise to mathematical developments such as our own.

NUMBER-SYSTEMS WHICH COULD HAVE BECOME DYNAMIC

We saw above how the complete adaptation of modern numerical notation to practical arithmetic comes not only from the principle of position and from the zero, but also from the fact that its figures correspond to graphic signs which have no direct intuitive visual meaning.

Once again, the inventors of this system have neither the privilege nor the honour of priority, since certain other systems had already enjoyed this property since the earliest times.

With the Egyptians, as we have seen, the transition from hieroglyphic to hieratic, and then to demotic script, radically changed the notation for the first whole numbers. Starting with groupings of identical strokes representing the nine units, in the end we find cursive signs, independent of each other, with no apparent intuitive meaning [G. Möller (1911–12); R. W. Erichsen]:

	1	2	3	4	5	6	7	8	9
Hieroglyphic notation	I	II	III	II II	III II	III III	IIII III	IIII IIII	III III III
Hieratic notation									
Demotic notation									

The Egyptian cursive notations could therefore have risen to the status of a number-system mathematically equivalent to our modern one if they had only eliminated all the signs for numbers greater than or equal to 10, replacing their additive principle by a principle of position which would then have been applied to the signs for the first nine units. However, this did not take place, since the Egyptian scribes remained profoundly attached to their old and traditional method.

The same characteristic was present in yet another number-system, the Singhalese, whose first nine number-signs certainly correspond to independent graphics stripped of any capacity to directly and visually evoke the corresponding units (Fig. 23.18):

	1	2	3	4	5	6	7	8	9
Singhalese notation	ග	ගා	ගං	ඟ	ගො	ෙ	ඎ	ඏ	ගෟ

But this system too preserved its initial hybrid principle, and therefore remained stuck throughout its existence.

Why therefore did not well-conceived systems like the Tamil or the Malayalam take this decisive step, and why did they not become positional number-systems worthy of the name?

This is all the more surprising since both underwent simplification conducive to such an end, since they had distinct signs for the nine units which had no immediate visual associations as we have seen (Fig. 23.20 and 23.21):

	1	2	3	4	5	6	7	8	9
Tamil notation	க	உ	௩	௪	ரு	௬	எ	அ	கூ
Malayalam notation	൧	൨	൩	൪	൫	൬	൭	൮	൯

The reason is that this simplification was not extended to all the numbers. The largest order of magnitude represented in these systems was 1,000. Numbers greater than or equal to 10,000 were either spelled out in full, or else they used the hybrid principle with the signs for 10, 100 and 1,000. These systems therefore remained firmly attached to their original principle, and for this reason they too remained blocked.

Furthermore, because there was no zero, the rule of simplification would only work on condition that every missing power of the base was followed by the sign for the order of magnitude immediately below.

In order to avoid confusion between the abbreviated Tamil notation for

3,605, and the number 365, it was necessary to keep the indicator for the hundreds in the representation of the former:

ſħ	*&n*	*Ⓖ*		*ſħ*	*&n*	*m*	*Ⓖ*
3	6	5		3	6 × 100		5

```
------------>              -------------------->
     365                           3,605
```

These systems were surely capable of rising to the level of our own if only they had eliminated the signs for the numbers greater than or equal to 10, and if the principle of position had been rigorously applied to the remaining figures. For a while there would have been difficulties due to the absence of zero, but, as necessity is the mother of invention, these would have been overcome by the invention of zero.

The common Chinese system of numeration (which, as we have seen, is in the same category as the two above) indeed went through this change.

In a table of logarithms, which is part of a collection of mathematical works put together on the orders of the Emperor Kangshi (1662–1722 CE) and published in 1713, we see the number 9,420,279,060 written in the form [K. Menninger (1957), II, pp. 278–279]:

九四二〇二七九〇六〇
9 4 2 0 2 7 9 0 6 0

By fully suppressing the classical signs for 10, 100, 1,000, and 10,000, by systematising the rule of position for all numbers, and by introducing a sign in the form of a circle to signify absence of an order of magnitude, the ordinary Chinese notation has been transformed into a number system equipped with a structure which is strictly identical to our own (Fig. 23.19). These number representations are perfectly adapted to arithmetical calculation.

The following example is taken from a work entitled *Ding zhu suan fa* ("Ding zhu's Method of Calculation"), published in 1335. It gives a table showing the multiplication of 3,069 by 45 laid out in a way which no one will have any difficulty in recognising [K. Menninger (1957), II, p. 300]:

TRANSLATION

三 〇 六 九   3069

四 五   45

一 五 三 四 五   15345

一 二 二 七 六   12276

一 三 八 一 〇 五   138105

This change only took place very late in the history of number-systems, however; the "push in the right direction" to the traditional Chinese system in fact came from the influence of the modern number-system.

THE "INVENTION" OF THE MODERN SYSTEM: AN IMPROBABLE CONJUNCTION OF THREE GREAT IDEAS

This fundamental "discovery" did not, therefore, appear all at once like the fully formed act of a god or a hero, or single act of an imaginative genius. These pages show clearly that it had an origin and a very long history. Fruit of a veritable cascade of inventions and innovations, it emerged little by little, following thousands of years during which an extraordinary profusion of trials and errors, of sudden breakthroughs and of standstills, regressions and revolutions occurred.

The discovery is the "fruit of slow maturation of primitive systems, initially well conceived, and patiently perfected through long ages. With the passage of time, some scholars succeeded when the circumstances were right in perfecting the primitive instrument they had inherited from their ancestors. Their motive for this effort was the passion they had to be able to express large numbers. Other scholars, coming after them, realistic and persistent, managed to get this revolutionary novelty accepted by the calculators of their time. We inherit from both" (G. Guitel).

Finally it all came to pass as though, across the ages and the civilisations, the human mind had tried all the possible solutions to the problem of writing numbers, before universally adopting the one which seemed the most abstract, the most perfected, and the most effective of all (Fig. 23.26, 23.27, 23.28, and 23.29).

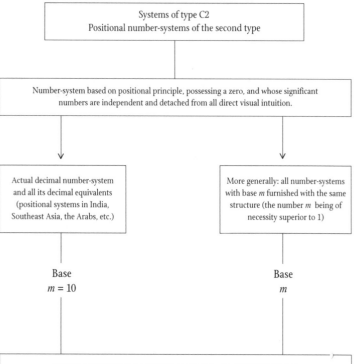

Systems of type C2
Positional number-systems of the second type

Number-system based on positional principle, possessing a zero, and whose significant numbers are independent and detached from all direct visual intuition.

Actual decimal number-system and all its decimal equivalents (positional systems in India, Southeast Asia, the Arabs, etc.)

More generally: all number-systems with base *m* furnished with the same structure (the number *m* being of necessity superior to 1)

Base
m = 10

Base
m

FUNDAMENTAL MATHEMATICAL PROPERTIES OF POSITIONAL NUMBER-SYSTEMS WITH BASE *m*

1. The number of digits (including zero) is equal to *m*.

2. Every integer *x* may be represented uniquely as a polynomial of degree $k - 1$, with the base *m* as variable, and with coefficients all less than *m*. In other words, any number *x* may be written uniquely in the form:

$$x = u_k\, m^{k-1} + u_{k-1}\, m^{k-2} + \ldots + u_4\, m^3 + u_3\, m^2 + u_2\, m + u_1$$

where the integers $u_k, u_{k-1}, \ldots u_2, u_1$, all less than *m*, are symbolised by digits of in the system under consideration. One may adopt the convention of writing the number *x* in the following manner (where the horizontal bar avoids confusion with the product $u_k\, u_{k-1} \ldots u_4\, u_3\, u_2\, u_1$):

$$x = \overline{u_k\, u_{k-1} \ldots u_4\, u_3\, u_2\, u_1}$$

3. The four fundamental arithmetical operations (addition, subtraction, multiplication and division) are easily carried out in such a system, according to simple rules entirely independent of the base *m* of the system.

4. This positional notation may be extended easily to fractions which have a power of the base as denominator, and thus to a simple and coherent notation for all numbers, rational and irrational, by introducing a "decimal point" according to an expansion in positive and negative powers of *m*, by analogy with decimal numbers.

Fig. 23.28. *Classification of positional number-system (Type C2)*

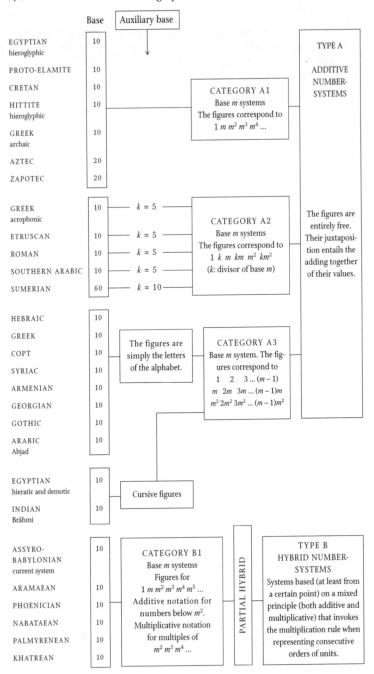

FIG. 23.29A. *Classification of written number-systems*

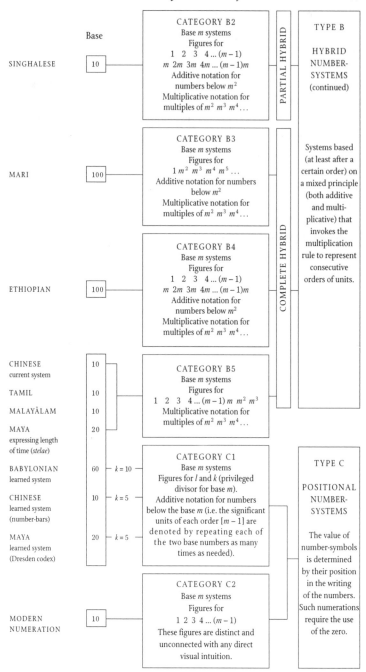

FIG. 23.29B. *Classification of written number-systems*

The story begins with primitive systems whose structure was based on the realities encountered in the course of accounting operations in ancient times. A certain amount of progress in the right direction was made, resulting in the creation of number-systems distinctly superior to the incoherent Roman numerals. But the paths which were taken led to dead ends, because these procedures incorporated only addition.

The awkwardnesses of these representations, together with the need for rapid writing, then brought about the development of hybrid systems, very conveniently mirroring spoken language, of which they can be seen as a more or less faithful transcription, sometimes showing a polynomial structure identical to that of the counting table, and at the same time extending considerably the power to express large numbers. Here too, however, the road was blocked. The principle they incorporated was inappropriate for arithmetical calculation, allowing addition and subtraction at best though at the cost of complicated manoeuvres, but useless for multiplication or division. In short, these systems were really only adequate for noting and recording numbers.

The decisive step in the adoption of systems of numerical notation with unlimited capacity, simple, rational, and immediately useable for calculation, could only be taken by inventing a well-conceived positional notation. This step was finally taken by simplifying hybrid notation, or by abbreviating systems for transferring numbers to the abacus, by the suppression of the signs indicating the powers of the base or by eliminating the columns of the abacus itself.

On the other hand this progress demanded a much higher level of abstraction, and the most delicate concept of the whole story: the zero. This was the supreme and belated discovery of the mathematicians who soon would come to extend it, from its first role of representing empty space, to embrace the truly numeric meaning of a null quantity (Fig. 23.27).

THE KEYSTONE OF OUR MODERN NUMBER-SYSTEM

Number and culture are one, for "to know how a people counts is to know what kind of people it is" (to adapt Charles Morazé). At least from this point of view, the degree of civilisation of a people becomes something measurable.

Thus it now appears to us indisputable that the Babylonians, the Chinese and the Maya were superior to the Egyptians, the Hebrews and the Greeks. For, while the former took the lead with their fundamental discoveries of the principle of position and the zero, the others remained locked up for centuries with number-systems which were primitive, incoherent, and unuseable for practically any purpose save writing numbers down.

The measure of the genius of Indian civilisation, to which we owe our modern system, is all the greater in that it was the only one in all history to have achieved this triumph.

Some cultures succeeded, earlier than the Indian, in discovering one or at best two of the characteristics of this intellectual feat. But none of them managed to bring together into a complete and coherent system the necessary and sufficient conditions for a number-system with the same potential as our own.

We shall see in Chapter 24 that this system began in India more than fifteen centuries ago, with the improbable conjunction of three great ideas (Fig. 23.26), namely:

- the idea of attaching to each basic figure graphical signs which were removed from all intuitive associations, and did not visually evoke the units they represented;
- the idea of adopting the principle according to which the basic figures have a value which depends on the position they occupy in the representation of a number;
- finally, the idea of a fully operational zero, filling the empty spaces of missing units and at the same time having the meaning of a null number.

This fundamental realisation therefore profoundly changed human existence, by bringing a simple and perfectly coherent notation for all numbers and allowing anyone, even those most resistant to elementary arithmetic, the means to easily perform all sorts of calculations; also by henceforth making it possible to carry out operations which previously, since the dawn of time, had been inconceivable; and opening up thereby the path which led to the development of mathematics, science and technology.

It is also the ultimate perfection of numerical notation, as we shall see in the classification of the numerical notations of history to follow. In other words, no further improvement of numerical notation is necessary, or even possible, once this perfect number-system has been invented. Once this discovery had been made, the only possible changes remaining could only affect

- the choice of base (which could be 2, 8, 12, or any other number greater than 2);
- the graphical form of the figures.

But no further change is possible in the essential structure of the system, now once and for all unchangeable by virtue of its mathematical perfection.

Apart from the base (which is only a matter of how things are to be grouped, and therefore of the number of different basic figures for the units), a number-system structurally identical to ours is completely independent of its symbolism. It does not matter if the symbols are conventional graphic signs, letters of the alphabet, or even spoken words, provided it rests strictly and rigorously on the principle of position and it incorporates the full concept of the symbol for zero.

Here is an instructive example. It concerns the great Jewish scholar Rabbi Abraham Ben Meir ibn Ezra of Spain, better known as Rabbi Ben Ezra. He was born at Toledo around 1092, and in 1139 undertook a long journey to the East, which he completed after passing some years in Italy. Then he lived in the South of France, before emigrating to England where he died in 1167. No doubt influenced by his encounters while travelling, he instructed himself in the methods of calculation which had come out of India (precursors of our own). He then set out the principal rules of these in a work in Hebrew entitled *Sefer ha mispar* ("The Book of Number") [M. Silberberg (1895); M. Steinschneider (1893)].

Instead of conforming strictly to the graphics of the original Indian figures, he preferred to represent the first nine whole numbers by the first nine letters of the Hebrew alphabet (which, of course, he knew well since childhood). And, instead of adopting the old additive principle, on which the alphabetic Hebrew number-system had always been based (Fig. 23.12), he eliminated from his own system every letter which had a value greater than or equal to 10. He kept only the following nine, to which he applied the principle of position, and he augmented the series with a supplementary sign in the shape of a circle, which he called either *sifra* (from the Arab word for "empty") or *galgal* (the Hebrew word for "wheel"):

א	ב	ג	ד	ה	ו	ז	ח	ט
1	2	3	4	5	6	7	8	9
aleph	bet	gimmel	dalet	he	vov	zayin	het	tet

Thus, instead of representing the number 200,733 in the traditional Hebrew form (below, on the right), he wrote it as follows (below, left):

ג	ג	ז	0	0	ב	instead of	ג	ל	ש	ת	רֹ
3	3	7	0	0	2		3	30	300	400	200,000

Thus it was that the Hebrew number-system, in his hands, changed from a very primitive static decimal notation, by becoming adapted to the principle of position and the concept of zero, into a system with a structure rigorously identical to our own and, therefore, infinitely more dynamic.

However, this remarkable transformation seems not to have been followed by anyone other than Rabbi Ben Ezra himself, a unique case, it would seem, in the history of this system.

This isolated case, nonetheless, provides us with a model for a situation which must have come about many times following the invention and propagation of the positional system originating in India, mother of the modern system and of all those influenced by it. This is the situation in which scholars and calculators making contact, individually or in groups, with Indian civilisation and then, becoming aware of the ingenuity and many merits of their positional number-system, decide either to adopt it (individually or collectively) in its entirety or else to borrow its structure in order to perfect their own traditional systems.

Now that we can stand back from the story, the birth of our modern number-system seems a colossal event in the history of humanity, as momentous as the mastery of fire, the development of agriculture, or the invention of writing, of the wheel, or of the steam engine.

THE CLASSIFICATION OF THE WRITTEN
NUMBER-SYSTEMS OF HISTORY

With this survey we shall close our chapter. Its aim is to systematise the various comparisons we have made up to this point in a more formal and mathematical manner.

Before I enter into the heart of the matter, I wish at this point to render special homage to Geneviève Guitel, whose remarkable *Classification hiérar-chisée des numérations écrites* has, for the first time, permitted me to bring together, intellectually speaking, systems which distance and time have separated almost totally.

This classification was published in her monumental *Histoire comparée des numérations écrites*, which has been an essential contribution to my understanding of this field.

Prior to her, as Charles Morazé has emphasised, there were certainly other histories of the number-systems, but none has attributed such importance to the comparisons which she has established on the basis of a principle of classification "which has the double merit of being both mathematically rigorous and remarkably relevant to the historical data which were to be put in order".

This classification, which I take up in my turn (while presenting it under a new light and amending certain details, resulting especially from the most recent archaeological discoveries), reveals that the numerical notations devised over five thousand years of history and evolution were not of unlim-ited variety. They may in fact be divided into three main types, of which each may be subdivided into various categories (Fig. 23.29):

- *additive systems*, which fundamentally are simply transcrip-tions of even more ancient concrete methods of counting (Fig. 23.30 to 23.32);
- *hybrid systems*, which were merely written transcriptions of more or less organised verbal expressions of number (Fig. 23.33 to 23.37);
- *positional systems*, which exhibit the ultimate degree of abstraction and therefore represent the ultimate perfection of numerical notation (Fig. 23.28 and 23.38).

NUMBER-SYSTEMS OF THE ADDITIVE TYPE

These are the ones based on the *principle of addition*, where each figure has a characteristic value independent of its position in a representation. Number-systems of this type in turn fall into three categories.

Additive number-systems of the first kind

Our model for this is the Egyptian hieroglyphic system, which assigns a separate symbol to unity and to each power of 10, and uses repetitions of these signs to denote other numbers (Fig. 23.1).

CLASSIFICATION OF ADDITIVE NUMBER-SYSTEMS

> Number-systems of this type fall into three kinds whose mathematical characteristics are summed up in Fig. 23.30 to 23.32: they require the adoption of a new writing convention based on a certain order of magnitude in order to note down high numbers.

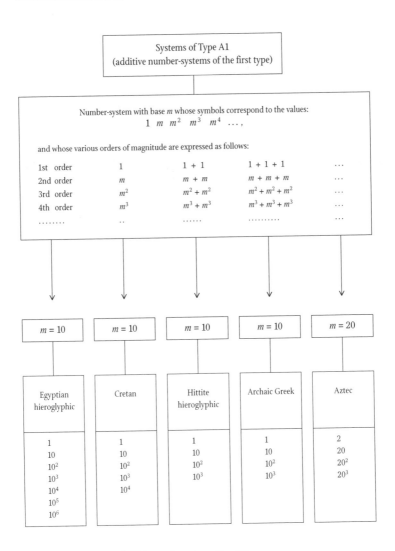

FIG. 23.30. *Classification of additive number-systems (Type A1)*

This is exactly what also happens in the Cretan number-system and in the Hittite hieroglyphic and archaic Greek systems. All of these systems are therefore strictly identical, and they differ only in the written forms of their respective figures (Fig. 23.3 to 23.5).

When they are in base 10, the additive systems of the first kind are therefore characterised by a notation which is based on arithmetical decompositions of the type:

Table 1

1st decimal order (units)	2nd decimal order (tens)	3rd decimal order (hundreds)	4th decimal order (thousands)
1	10	10^2	10^3
1 + 1	10 + 10	$10^2 + 10^2$	$10^3 + 10^3$
1 + 1 + 1	10 + 10 + 10	$10^2 + 10^2 + 10^2$	$10^3 + 10^3 + 10^3$

Special notation for 1, 10, 10^2, 10^3, etc.
Additive notation for all other numbers.

Now if we consider the Aztec number-system, we find that even though it uses a different base (base 20), still like the others it assigns a special symbol only to unity and to the powers of the base (Fig. 23.6):

Base   *Aztec system*

20	1	20	20^2	20^3	20^4	...
m	1	*m*	m^2	m^3	m^4	...
10	1	10	10^2	10^3	10^4	...

Egyptian hieroglyphic system

Since this is an additive system and proceeds by repetition of identical signs, it is characterised by a notation which depends on arithmetical decompositions of the type:

Table 2

1st vigesimal order (units)	2nd vigesimal order (twenties)	3rd vigesimal order (four hundreds)	4th vigesimal order (eight thousands)
1	20	20^2	20^3
1 + 1	20 + 20	$20^2 + 20^2$	$20^3 + 20^3$
1 + 1 + 1	20 + 20 + 20	$20^2 + 20^2 + 20^2$	$20^3 + 20^3 + 20^3$

Special notation for 1, 20, 20^2, 20^3, etc.
Additive notation for all other numbers.

The Aztec system, therefore, is intellectually related to the preceding ones, and differs only in having base 20 instead of base 10.

All of these notations therefore belong to the same type (Fig. 23.30).

CLASSIFICATION OF ADDITIVE NUMBER-SYSTEMS (continued)

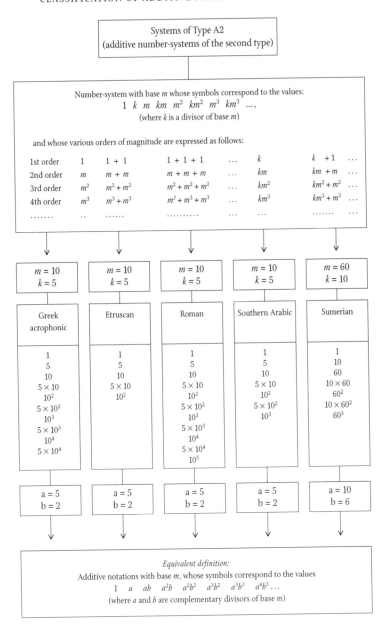

FIG. 23.31. *Classification of additive number-systems (type A2)*

Additive systems of the second kind

A characteristic example is the Greek acrophonic system. It is in base 10, and adopts the principle of addition assigning a special symbol to each of the numbers 1, 10, 100, 1,000, etc., as well as to each of the following: 5, 50, 500, 5,000, and so on (Fig. 23.7). Intellectually, therefore, it is of the same kind as the Southern Arabic system, the Etruscan, and the Roman, characterised by arithmetical decompositions of the type (Fig. 16.18, 16.35 and 23.8):

Table 3

1st decimal order (units)	2nd decimal order (tens)	3rd decimal order (hundreds)	4th decimal order (thousands)
1	10	10^2	10^3
1 + 1	10 + 10	$10^2 + 10^2$	$10^3 + 10^3$
1 + 1 + 1	10 + 10 + 10	$10^2 + 10^2 + 10^2$	$10^3 + 10^3 + 10^3$
...	...	...	...
5	5×10	5×10^2	5×10^3
5 + 1	$5 \times 10 + 10$	$5 \times 10^2 + 10^2$	$5 \times 10^3 + 10^3$
5 + 1 + 1	$5 \times 10 + 10 + 10$	$5 \times 10^2 + 10^2 + 10^2$	$5 \times 10^3 + 10^3 + 10^3$

Special notation for 1, 5, 10, 5×10, 10^2, 5×10^2, etc.
Additive notation for all other numbers.

Denoting by k the divisor of the base m which thus acts as auxiliary base (here, $m = 10$ and $k = 5$), we see that these systems assign a special symbol not only to each power of the base (1, m, m^2, m^3, ...) but also to the product of each of these with k (k, km, km^2, km^3, ...). As the following table shows, this is exactly the structure which can be discerned in the regular progression of the Sumerian number-system (Fig. 23.2):

Sumerian system (where m=60 and k=10)

1	10	60	10×60	60^2	10×60^2	60^3	10×60^3
1	k	m	km	m^2	km^2	m^3	km^3
1	5	10	5×10	10^2	5×10^2	10^3	5×10^3

Greek acrophonic numerals (where m = 10 and k = 5)

Looking at it from another point of view, the succession of numbers receiving a particular sign in the Sumerian system may be expressed as:

1st order	1	< · >	1
	10	< · >	10
2nd order	60	< · >	10.6
	10×60	< · >	10.6.10
3rd order	60^2	< · >	10.6.10.6
	10×60^2	< · >	10.6.10.6.10
4th order	60^3	< · >	10.6.10.6.10.6
	10×60^3	< · >	10.6.10.6.10.6.10

and so on, alternating the numbers 10 and 6.

Let a and b denote the divisors of m which act as alternating auxiliary bases (where, in the Sumerian case, we have $m = 60$, $a = 10$ and $b = 6$). This succession therefore exactly corresponds to that of the Greek acrophonic system (where $m = 10$, $a = 5$ and $b = 2$):

Table 4

Sumerian		Mathematical Characterisation			Greek
1	< · · · >	1		< · · · >	1
10	< · · · >	a		< · · · >	5
10.6	< · · · >	a.b		< · · · >	5.2
10.6.10	< · · · >	a^2b	= a.b.a	< · · · >	5.2.5
10.6.10.6	< · · · >	a^2b^2	= a.b.a.b	< · · · >	5.2.5.2
10.6.10.6.10	< · · · >	a^3b^2	= a.b.a.b.a	< · · · >	5.2.5.2.5
10.6.10.6.10.6	< · · · >	a^3b^3	= a.b.a.b.a.b	< · · · >	5.2.5.2.5.2
⋮		$a = 10$	$a = 5$		⋮
· · · · · · · · · · · · · · · · >		$b = 6$	$b = 2$	< · · · · · · · · · · · · · · ·	

The Greek structure is thus mathematically identical to that of the Sumerian, corresponding to arithmetical decompositions of the type:

1st sexagesimal order (units)	2nd sexagesimal order (sixties)	3rd sexagesimal order (multiples of 60)	4th sexagesimal order (multiples of 60)
1	60	60^2	60^3
1 + 1	60 + 60	$60^2 + 60^2$	$60^3 + 60^3$
1 + 1 + 1	60 + 60 + 60	$60^2 + 60^2 + 60^2$	$60^3 + 60^3 + 60^3$
...	...	...	...
10	10×60	10×60^2	10×60^3
10 + 1	$10 \times 60 + 60$	$10 \times 60^2 + 60^2$	$10 \times 60^3 + 60^3$
10 + 10	$10 \times 60 + 10 \times 60$	$10 \times 60^2 + 10 \times 60^2$	$10 \times 60^3 + 10 \times 60^3$

Special notation for 1, 10, 60, 10×60, 60^2, 10×60^2, etc.
Additive notation for all other numbers.

All these systems therefore belong to the same category (Fig. 23.31).

CLASSIFICATION OF ADDITIVE NUMBER-SYSTEMS (concluded)

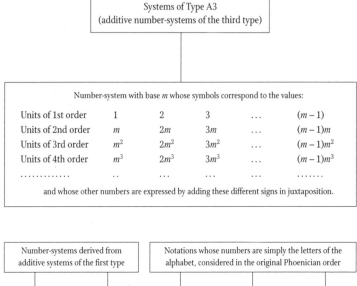

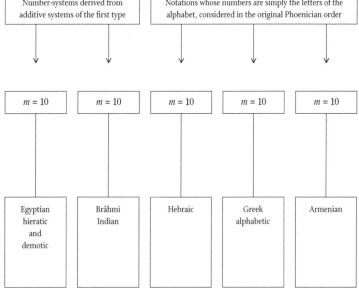

Fıg. 23.32. *Classification of additive number-systems (Type A3)*

Additive systems of the third kind

The Egyptian hieratic system and the Greek alphabetic system are typical examples of this type. Intellectually, they correspond to the following characterisation (Fig. 23.10 to 23.13, and 23.32):

1st decimal order (units)	2nd decimal order (tens)	3rd decimal order (hundreds)	4th decimal order (thousands)
1	10	100	1,000
2	20	200	2,000
3	30	300	3,000 etc.

Special notation for each unit of each number.

1	2	3	4	5	6	7	8	9
10	2.10	3.10	4.10	5.10	6.10	7.10	8.10	9.10
10^2	2.10^2	3.10^2	4.10^2	5.10^2	6.10^2	7.10^2	8.10^2	9.10^2
10^3	2.10^3	3.10^3	4.10^3	5.10^3	6.10^3	7.10^3	8.10^3	9.10^3
10^4	2.10^4	3.10^4	4.10^4	5.10^4	6.10^4	7.10^4	8.10^4	9.10^4

· × · · · · · · · · · · ·

Additive notation for all other numbers.

SYSTEMS OF HYBRID TYPE

These are founded on a mixed system in which both addition and multiplication are involved. On this basis, the multiples of the powers of the base are, from a certain order of magnitude onwards, expressed multiplicatively. This type of system can be divided into five categories.

CLASSIFICATION OF HYBRID NUMBER-SYSTEMS

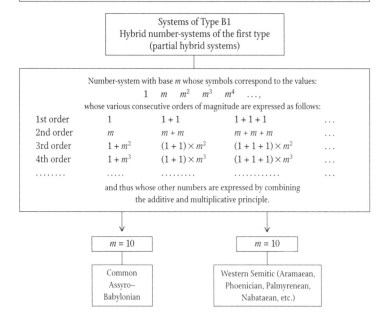

Number-systems of this type fall into five kinds whose mathematical characteristics are summed up in Fig. 23.33 to 23.37

Systems of Type B1
Hybrid number-systems of the first type
(partial hybrid systems)

Number-system with base m whose symbols correspond to the values:
$$1 \quad m \quad m^2 \quad m^3 \quad m^4 \quad \ldots,$$
whose various consecutive orders of magnitude are expressed as follows:

1st order	1	$1 + 1$	$1 + 1 + 1$	...
2nd order	m	$m + m$	$m + m + m$	...
3rd order	$1 + m^2$	$(1 + 1) \times m^2$	$(1 + 1 + 1) \times m^2$	...
4th order	$1 + m^3$	$(1 + 1) \times m^3$	$(1 + 1 + 1) \times m^3$	...
....	.....	....	....	...

and thus whose other numbers are expressed by combining the additive and multiplicative principle.

$m = 10$

Common Assyro--Babylonian

$m = 10$

Western Semitic (Aramaean, Phoenician, Palmyrenean, Nabataean, etc.)

FIG. 23.33. *Classification of hybrid number-systems (Type B1, partial hybrid)*

Hybrid systems of the first kind

The common Assyro-Babylonian system and that of the western Semitic peoples (Aramaeans, Phoenicians, etc.) are typical examples of this type. They have base 10, and assign a special symbol to each of the numbers 1, 10, 100, 1,000, etc., and use multiplicative notation for consecutive multiples of each of these powers of 10. At the same time, the units and the tens are still represented according to the old principle of additive juxtaposition.

When in base 10, hybrid systems of the first kind are characterised by arithmetical decompositions of the type (Fig. 23.16, 23.17, and 23.33):

Table 5

1st order (units)	2nd order (tens)	3rd order (hundreds)	4th order (thousands)
1	10	1×10^2	1×10^3
$1 + 1$	$10 + 10$	$(1 + 1) \times 10^2$	$(1 + 1) \times 10^3$
$1 + 1 + 1$	$10 + 10 + 10$	$(1 + 1 + 1) \times 10^2$	$(1 + 1 + 1) \times 10^3$

Special notation for 1, 10, 10^2, 10^3, etc.   Additive notation for the numbers 1 to 99.
Multiplicative notation for the multiples of the powers of 10, starting with 100.
A notation involving both addition and multiplication for other numbers

CLASSIFICATION OF HYBRID NUMBER-SYSTEMS (continued)

Systems of Type B2
Hybrid number-systems of the second type
(partial hybrid systems)

Number-system with base m whose symbols correspond to the values:

$$\begin{array}{cccc} 1 & 2 & 3 & \dots \ (m-1) \\ m & 2m & 3m & \dots \ (m-1)m \\ & m^2 & m^3 & m^4 \quad \text{etc.}, \end{array}$$

whose various consecutive orders of magnitude are expressed as follows:

1st order	1	2	3	...	$(m-1)$
2nd order	m	$2m$	$3m$	...	$(m-1)m$
3rd order	$1 \times m$	$2 \times m$	$3 \times m$	...	$(m-1) \times m$
4th order	$1 \times m$	$2 \times m$	$3 \times m$	...	$(m-1) \times m$
....	.....	.....	.....	...	....

and thus whose other numbers are expressed by combining
the additive and multiplicative principle.

$m = 10$

Singhalese

FIG. 23.34. *Classification of hybrid number-systems (Type B2, partial hybrid)*

Hybrid systems of the second kind

The model for this type is the Singhalese system. It has base 10, and assigns a special symbol to each unit, to each of the tens, and to each of the powers of 10. The notation for the hundreds, thousands, etc. follows the multiplicative rule (Fig. 23.18).

When in base 10, hybrid systems of this kind are characterised by a notation which is based on arithmetical decompositions of the type (Fig. 23.34):

1st order (units)	2nd order (tens)	3rd order (hundreds)	4th order (thousands)
1	10	1×10^2	1×10^3
2	20	2×10^2	2×10^3
3	30	3×10^2	3×10^3

Special notation for units, tens, 10^2, 10^3, etc.   Additive notation for the numbers 1 to 99.
Multiplicative notation for the multiples of the powers of 10, starting with 100.
A notation involving both addition and multiplication for other numbers

CLASSIFICATION OF HYBRID NUMBER-SYSTEMS (continued)

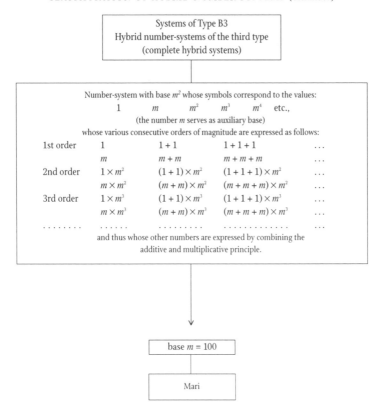

Systems of Type B3
Hybrid number-systems of the third type
(complete hybrid systems)

Number-system with base m^2 whose symbols correspond to the values:

$$1 \qquad m \qquad m^2 \qquad m^3 \qquad m^4 \quad \text{etc.,}$$

(the number m serves as auxiliary base)
whose various consecutive orders of magnitude are expressed as follows:

1st order	1	$1+1$	$1+1+1$	$\ldots$
	m	$m+m$	$m+m+m$	$\ldots$
2nd order	$1 \times m^2$	$(1+1) \times m^2$	$(1+1+1) \times m^2$	$\ldots$
	$m \times m^2$	$(m+m) \times m^2$	$(m+m+m) \times m^2$	$\ldots$
3rd order	$1 \times m^3$	$(1+1) \times m^3$	$(1+1+1) \times m^3$	$\ldots$
	$m \times m^3$	$(m+m) \times m^3$	$(m+m+m) \times m^3$	$\ldots$

and thus whose other numbers are expressed by combining the
additive and multiplicative principle.

base $m = 100$

Mari

F<small>IG</small>. 23.35.*Classification of hybrid number-systems (Type B3, complete hybrid)*

Hybrid systems of the third kind

The model for this type is the Mari system. It uses base 100, and gives a special symbol for each unit, for 10, and for each power of 100. The notation for the hundreds, the ten thousands, etc. uses the multiplicative rule. The system is characterised by a notation based on arithmetical decompositions of the type (Fig. 23.22 and 23.35):

1st centennial order		2nd centennial order	
units	tens	hundreds	thousands
1	10	1×10^2	1×10^3
$1 + 1$	$10 + 10$	$(1 + 1) \times 10^2$	$(1 + 1) \times 10^3$
$1 + 1 + 1$	$10 + 10 + 10$	$(1 + 1 + 1) \times 10^2$	$(1 + 1 + 1) \times 10^3$
...	...	...	...

Special notation for 1, 10, 10^2, 10^3, etc.
Additive notation for the numbers from 1 to 99.
Additive notation for the numbers 1 to 99.
Multiplicative notation for multiples of the powers of 10^2, starting with the first (100).
A notation involving both addition and multiplication for other numbers.

CLASSIFICATION OF HYBRID NUMBER-SYSTEMS (continued)

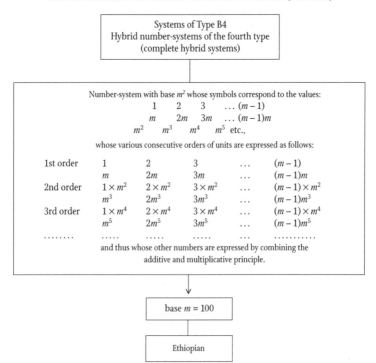

Systems of Type B4
Hybrid number-systems of the fourth type
(complete hybrid systems)

Number-system with base m^2 whose symbols correspond to the values:

	1	2	3	...	$(m-1)$
	m	$2m$	$3m$	...	$(m-1)m$
	m^2	m^3	m^4	m^5 etc.,	

whose various consecutive orders of units are expressed as follows:

1st order	1	2	3	...	$(m-1)$
	m	$2m$	$3m$	...	$(m-1)m$
2nd order	$1 \times m^2$	$2 \times m^2$	$3 \times m^2$	...	$(m-1) \times m^2$
	m^3	$2m^3$	$3m^3$	...	$(m-1)m^3$
3rd order	$1 \times m^4$	$2 \times m^4$	$3 \times m^4$	...	$(m-1) \times m^4$
	m^5	$2m^5$	$3m^5$	...	$(m-1)m^5$
....	.....	.....	.....	...	....

and thus whose other numbers are expressed by combining the
additive and multiplicative principle.

base $m = 100$

Ethiopian

FIG. 23.36. *Classification of hybrid number-systems (Type B4, complete hybrid)*

Hybrid systems of the fourth kind

The model for this type is the Ethiopian system. It has base 100, and assigns a special sign to each unit and to each of the tens, and also to each power of 100. The notation for the hundreds, the ten thousands, etc. uses a multiplicative rule applied to these figures. The system is characterised by a notation based on arithmetical decompositions of the type (Fig. 23.36):

1st centennial order		2nd centennial order	
units	tens	hundreds	thousands
1	10	1×10^2	1×10^3
2	20	2×10^2	2×10^3
3	30	3×10^2	3×10^3
4	40	4×10^2	4×10^3
5	50	5×10^2	5×10^3
6	60	6×10^2	6×10^3
7	70	7×10^2	7×10^3
8	80	8×10^2	8×10^3
9	90	9×10^2	9×10^3

Special notation for each unit, each ten and for each of 10^2, 10^3, etc. Additive notation for the numbers from 1 to 99. Multiplicative notation for multiples of the powers of 10^2, starting with the first (100). A notation involving both addition and multiplication for other numbers.

CLASSIFICATION OF HYBRID NUMBER-SYSTEMS (concluded)

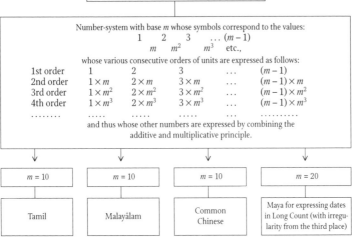

Systems of Type B5
Hybrid number-systems of the fifth type
(complete hybrid systems)

Number-system with base m whose symbols correspond to the values:
1   2   3   ... $(m-1)$
m m^2 m^3 etc.,

whose various consecutive orders of units are expressed as follows:

1st order	1	2	3	...	$(m-1)$
2nd order	$1 \times m$	$2 \times m$	$3 \times m$	...	$(m-1) \times m$
3rd order	$1 \times m^2$	$2 \times m^2$	$3 \times m^2$	...	$(m-1) \times m^2$
4th order	$1 \times m^3$	$2 \times m^3$	$3 \times m^3$	...	$(m-1) \times m^3$

and thus whose other numbers are expressed by combining the additive and multiplicative principle.

$m = 10$	$m = 10$	$m = 10$	$m = 20$
Tamil	Malayâlam	Common Chinese	Maya for expressing dates in Long Count (with irregularity from the third place)

Unlike the hybrid numbering of the first two types, those of the third, fourth and fifth types invoke the multiplicative principle in the notation of all units superior or equal to the base. Hence they are termed "complete hybrid". Thus the representation of the numbers is made here by expressing the various numerical values of a polynomial with the corresponding base m as a variable (apart, that is, from the Maya system which embraces an irregularity in the values of its consecutive orders of units).

FIG. 23.37. *Classification of hybrid number-systems (Type B5, complete hybrid)*

Hybrid systems of the fifth kind

The model for this type is the common Chinese system, as well as the Tamil and Malayalam systems. These systems have base 10, and assign a special symbol to each unit and to each power of 10. The notation for the tens, the hundreds, the thousands, etc. uses the multiplicative principle.

When in base 10, hybrid systems of the fifth kind are characterised by a notation based on arithmetical decompositions of the following type (Fig. 23.37):

1st order (units)	2nd order (tens)	3rd order (hundreds)	4th order (thousands)
1	1×10	1×10^2	1×10^3
2	2×10	2×10^2	2×10^3
3	3×10	3×10^2	3×10^3
4	4×10	4×10^2	4×10^3
5	5×10	5×10^2	5×10^3
6	6×10	6×10^2	6×10^3
7	7×10	7×10^2	7×10^3
8	8×10	8×10^2	8×10^3
9	9×10	9×10^2	9×10^3

Special notation for each unit of the first order and for each of the numbers 10, 10^2, 10^3.
Multiplicative notation for multiples of powers of the base, starting with 10. Notation involving both addition and multiplication for other numbers.

Unlike hybrid systems of the first kind, which only partially use the multiplicative principle, those of types 3, 4 and 5 bring the principle into play in the notation for all the orders of magnitude greater than or equal to the base. Additionally, the representation of other numbers is based on the coefficients of a polynomial whose variable is the base. For these reasons systems of this type are also called complete hybrid systems.

POSITIONAL SYSTEMS

The systems are based on the principle that the value of the figures is determined by their position in the representation of a number.

Historically, there have been only four originally created positional systems:

- the system of the Babylonian scholars;
- the system of the Chinese scholars;
- the system of the Mayan astronomer-priests;
- and finally our modern system which, as we shall see in the next chapter, originated in India.

These systems (which require the use of a zero) may be divided into two categories.

CLASSIFICATION OF POSITIONAL NUMBER-SYSTEMS

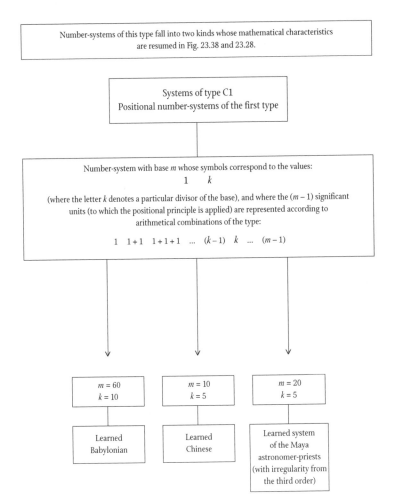

FIG. 23.38. *Classification of positional number-systems (Type C1)*

Positional systems of the first kind

This type includes:

1. – The *system of the Babylonian scholars*

This has base 60. The notation for the units of the first order (from 1 to 59) corresponds to arithmetical decompositions of the following type for the two basic figures, of which one represents unity and the other 10 (Fig. 23.23):

| 1 | 1 + 1 | 1 + 1 + 1 | 1 + 1 + 1 + 1 ... |

| 10 | 10 + 10 | 10 + 10 + 10 | 10 + 10 + 10 + 10 ... |

10 + 10 + 10 + 10 + 10 + 1 + 1 + 1 + 1 + 1 + 1 + 1 + 1 + 1

2. – The *system of the Chinese scholars*

This has base 10. The notation for the units of the first order (from 1 to 9) corresponds to arithmetical decompositions of the following type for the two basic figures, of which one represents unity and the other 5 (Fig. 23.24):

| 1 | 1 + 1 | 1 + 1 + 1 | 1 + 1 + 1 + 1 | 5 |

| 5 + 1 | 5 + 1 + 1 | 5 + 1 + 1 + 1 | 5 + 1 + 1 + 1 + 1 |

3. – The *system of the Maya scholars*

This has base 20. The notation for the units of the first order (from 1 to 19) corresponds to arithmetical decompositions of the following type using two base figures, one representing unity and the other the number 5. In addition there is an irregularity starting with the third order in the succession of positional values (Fig. 23.25):

| 1 | 1 + 1 | 1 + 1 + 1 | 1 + 1 + 1 + 1 | 5 |

| 5 + 1 | 5 + 1 + 1 | 5 + 1 + 1 + 1 | 5 + 1 + 1 + 1 + 1 | 5 + 5 |

| 5 + 5 + 1 | 5 + 5 + 1 + 1 | 5 + 5 + 1 + 1 + 1 | 5 + 5 + 1 + 1 + 1 + 1 | 5 + 5 + 5 |

| 5 + 5 + 5 + 1 | 5 + 5 + 5 + 1 + 1 | 5 + 5 + 5 + 1 + 1 + 1 | 5 + 5 + 5 + 1 + 1 + 1 + 1 |

Systems of this type with base m use the principle of position, but they only possess two digits in the strict sense: one for unity, and the other for a particular divisor of the base, here denoted by k. The $m-1$ units are represented according to the additive principle (Fig. 23.38).

All of these systems clearly require a zero, and in the end have come to possess one, independently or not of outside influence.

Positional systems of the second kind

This category includes our own modern decimal notation, whose nine units are represented by figures (Fig. 23.26):

1 2 3 4 5 6 7 8 9

augmented by a tenth sign, written 0. Known as *zero*, this is used to mark the absence of units of a given rank, and at the same time enjoys a true numerical meaning, that of null number.

The fundamental characteristic of this system is that its conventions can be extended into a notation both simple and completely consistent for all numbers: integers, fractions, and irrationals (whether these be transcendental or not). In other words, the discovery of this system enables us to write down, in a simple and rational way, and using a completely natural extension of the principle of position and of the zero, not only fractions but entities such as $\sqrt{2}$, $\sqrt{3}$ or Π.

A decimal fraction is a fraction of which the denominator is equal to 10 or to a power of 10. 3/10, 1/100, 251/10,000 are therefore decimal fractions.

Now, the sequence of decimal fractions of unity (those which have numerator 1 and denominator a power of 10) has its terms called successively one tenth (or decimal unit of the first order), one hundredth (or decimal unit of the second order), one thousandth (or decimal unit of the third order), and so on:

$$\frac{1}{10} \quad \frac{1}{10^2} \quad \frac{1}{10^3} \quad \frac{1}{10^4} \quad \frac{1}{10^5} \quad \frac{1}{10^6} \quad \text{etc.}$$

Thus we have a sequence where each term is the product of its predecessor by 1/10, which means that the convention of our decimal notation applies here also, ten units of any order being equal to one unity of the order immediately above. These decimal units may therefore be unambiguously represented by a convention which extends the convention which applies to the integers, so that we may represent them in the form:

$$0.1 \quad\quad 0.01 \quad\quad 0.001 \quad\quad 0.0001, \text{ etc}$$
$$(= 10^{-1}) \quad (= 10^{-2}) \quad\quad (= 10^{-3}) \quad\quad (= 10^{-4})$$

If we now consider any decimal fraction, for example 39,654/1,000, we find its arithmetical decomposition according to the positive and negative powers of 10:

$$\frac{39,654}{1,000} = \frac{39,000}{1,000} + \frac{600}{1,000} + \frac{50}{1,000} + \frac{4}{1,000}$$

We observe therefore that this may be written in the form:

$$\frac{39{,}654}{1{,}000} = 39 + \frac{600}{1{,}000} + \frac{50}{1{,}000} + \frac{4}{1{,}000}$$

or, in accordance with the preceding convention:

$$\frac{39{,}654}{1{,}000} = 39 + 0.6 + 0.05 + 0.004$$

$$= 39 + 6 \times 10^{-1} + 5 \times 10^{-2} + 4 \times 10^{-3}$$

This number is therefore composed of 39 units, 6 tenths, 5 hundredths and 4 thousandths. Adopting the convention for the representation of the integers, one may make the convention of separating the integer units from the decimal units by a point, so that the fraction in question may now be put in the form:

$$\frac{39{,}654}{1{,}000} = 39.654$$

It is therefore expressed as a *decimal number* which can be read as 39 units and 654 thousandths.

Thus we see how the principle of position allows us to extend its application to decimal numbers.

One can also show that any number whatever can be expressed as a decimal number whose development may be finite or infinite (i.e. having a finite or an infinite number of figures following the decimal point).

One can therefore see the many mathematical advantages which flow from the discovery of our number-system.

But, clearly, this system is only a special case of the systems in this category. These are nowadays known as *systems* with base m, the number m being at least equal to 2 ($m > 1$). Historically speaking, these are simply positional systems with base m furnished with a fully operational zero, whose ($m-1$) figures are independent of each other and without any direct visual significance (Fig. 23.28).

The written positional systems of the second kind are therefore the most advanced of all history. They allow the simple and completely rational representation of any number, no matter how large. Above all, they bring within the reach of everyone a simple method for arithmetical operations. And all this is independent of the choice of base (Fig. 23.29). It is precisely in these respects that our modern written number-system (or any one of its equivalents) is one of the foundations of the intellectual equipment of the modern human being.